Lecture Notes in Mathematics 2116

Editors-in-Chief:
J.-M. Morel, Cachan
B. Teissier, Paris

Advisory Board:
Camillo De Lellis (Zürich)
Mario di Bernardo (Bristol)
Alessio Figalli (Austin)
Davar Khoshnevisan (Salt Lake City)
Ioannis Kontoyiannis (Athens)
Gabor Lugosi (Barcelona)
Mark Podolskij (Heidelberg)
Sylvia Serfaty (Paris and NY)
Catharina Stroppel (Bonn)
Anna Wienhard (Heidelberg)

T0236332

More information about this series at
http://www.springer.com/series/304

Bo'az Klartag • Emanuel Milman
Editors

Geometric Aspects
of Functional Analysis

Israel Seminar (GAFA) 2011-2013

 Springer

2011–2013

Editors
Bo'az Klartag
School of Mathematical Sciences
Tel Aviv University
Tel Aviv, Israel

Emanuel Milman
Technion - Israel Institute of Technology
Haifa, Israel

ISBN 978-3-319-09476-2   ISBN 978-3-319-09477-9 (eBook)
DOI 10.1007/978-3-319-09477-9
Springer Cham Heidelberg New York Dordrecht London

Lecture Notes in Mathematics ISSN print edition: 0075-8434
                                        ISSN electronic edition: 1617-9692

Library of Congress Control Number: 2014952086

Mathematics Subject Classification (2010): 80M35, 26A51, 32-XX, 46-XX, 60-XX

Printed on acid-free paper

Springer is part of Springer Science+Business Media (www.springer.com)

Preface

Since the mid-1980s, the following volumes containing collections of papers reflecting the activity of the Israel Seminar in Geometric Aspects of Functional Analysis have appeared:

1983–1984   Published privately by Tel Aviv University
1985–1986   Springer Lecture Notes in Mathematics, vol. 1267
1986–1987   Springer Lecture Notes in Mathematics, vol. 1317
1987–1988   Springer Lecture Notes in Mathematics, vol. 1376
1989–1990   Springer Lecture Notes in Mathematics, vol. 1469
1992–1994   Operator Theory: Advances and Applications, vol. 77, Birkhäuser
1994–1996   MSRI Publications, vol. 34, Cambridge University Press
1996–2000   Springer Lecture Notes in Mathematics, vol. 1745
2001–2002   Springer Lecture Notes in Mathematics, vol. 1807
2002–2003   Springer Lecture Notes in Mathematics, vol. 1850
2004–2005   Springer Lecture Notes in Mathematics, vol. 1910
2006–2010   Springer Lecture Notes in Mathematics, vol. 2050

The first six were edited by Lindenstrauss and Milman, the seventh by Ball and Milman, the subsequent four by Milman and Schechtman, and the last one by Klartag, Mendelson and Milman.

As in the previous Seminar Notes, the current volume reflects general trends in the study of Geometric Aspects of Functional Analysis. Most of the papers deal with different aspects of Asymptotic Geometric Analysis, understood in a broad sense; many continue the study of geometric and volumetric properties of convex bodies and log-concave measures in high-dimensions, and in particular the mean-norm, mean-width, metric entropy, spectral-gap, thin-shell and slicing parameters, with applications to Dvoretzky and Central-Limit-type results. The study of spectral properties of various systems, matrices, operators and potentials is another central theme in this volume. As expected, probabilistic tools play a significant role, and probabilistic questions regarding Gaussian noise stability, the Gaussian Free Field and First Passage Percolation are also addressed. The connection to the field of Classical Convexity is also well represented with new properties and applications of mixed-volumes. The interplay between the real convex and complex pluri-subharmonic settings continues to manifest itself in several additional articles. All contributions are original research papers and were subject to the usual refereeing standards.

We are grateful to Vitali Milman for his help and guidance in preparing and editing this volume.

Tel Aviv, Israel   Bo'az Klartag
Haifa, Israel   Emanuel Milman

Contents

Dyson Processes Associated with Associative Algebras: The Clifford Case

Dominique Bakry and Marguerite Zani

Abstract We consider Brownian motions and other processes (Ornstein-Uhlenbeck processes, spherical Brownian motions) on various sets of symmetric matrices constructed from algebra structures, and look at their associated spectral measure processes. This leads to the identification of the multiplicity of the eigenvalues, together with the identification of the spectral measures. For Clifford algebras, we thus recover Bott's periodicity.

1 Introduction

Many works on random matrix theory deal with the law of the spectrum on some specific sets: real symmetric, Hermitian, orthogonal or unitary, etc. There exists a large literature on this topic. We shall mention the early works of Wigner [22], Dyson [7], Mehta [20], Marčenko and Pastur [19], Soshnikov [21] and more recently Anderson et al. [1], Erdös et al. [8–10], Forrester [11] and references therein. For related topics, one can also refer to the work of Ledoux [15, 16] studying empirical measures via Markov generators, and for a classification of random matrices based on a quantum mechanical setting, see [23, 24]. One may also consider stochastic processes on these sets of matrices (Euclidean Brownian motions, Ornstein-Uhlenbeck operators, spherical Brownian motions, Brownian motions on groups, see e.g. Doumerc [6], or Chapon and Defosseux [5] on minors of Hermitian Brownian Motion): they are diffusion processes which are reversible under their invariant measures. One may then consider the stochastic processes which is the empirical measure of the spectrum of the matrix. In many situations, it turns out that these empirical measures are again stochastic diffusion processes, called Dyson processes

---

D. Bakry (✉)
Institut de Mathématiques de Toulouse, Université Paul Sabatier, 118 route de Narbonne, 31062 Toulouse, France
e-mail: bakry@math.univ-toulouse.fr

M. Zani
Laboratoire d'Analyse et Mathématiques appliquées, UMR CNRS 8050, Université Paris-Est -Créteil, 61 av du Général de Gaulle, 94010 Créteil Cedex, France
e-mail: zani@u-pec.fr

© Springer International Publishing Switzerland 2014   1
B. Klartag, E. Milman (eds.), *Geometric Aspects of Functional Analysis*,
Lecture Notes in Mathematics 2116, DOI 10.1007/978-3-319-09477-9__1

(introduced in [7]) and their reversible measures are the image of the invariant measure on the set of matrices. It may then be an easy way of computing those spectral measures, but the study of the spectral processes is by itself a topic of interest.

For, say Gaussian, real symmetric, Hermitian or quaternionic symmetric matrices, when one considers the law of their eigenvalues $(\lambda_1, \cdots, \lambda_n)$, ordered for example as $\lambda_1 < \cdots < \lambda_n$, it has a density with respect to the Lebesgue measure $d\lambda_1 \cdots d\lambda_n$ which is $C(\prod_{i<j} |\lambda_i - \lambda_j|)^\beta e^{-1/2 \sum_i \lambda_i^2}$, where C is a normalizing constant and $\beta = 1, 2, 4$ according to the fact that we are in the real, complex or quaternionic case. This factor $(\prod_{i<j} |\lambda_i - \lambda_j|)^\beta$ also appears in many other situations $(SO(n), SU(n), Sp(n)$ matrices, for example). There are even some results for octonionic 2×2 matrices where $\beta = 8$. On the other hand, if one considers the real symmetric, Hermitian and quaternionic $n \times n$ matrices as real matrices (with size $n \times n, (2n) \times (2n)$ and $(4n) \times (4n)$ respectively), their eigenspaces have dimension 1, 2 and 4 respectively (and 8 in the special case of octonionic matrices). Therefore, one may think that this factor β is due to the multiplicity of the eigenvalues for the real form of the matrices. We shall see that this is not the case.

One may construct symmetric matrices with the help of some other algebra structures endowed with an Euclidean metric, and it turns out that in many cases one may again construct the associated Dyson processes. There is then a rich interplay between the algebra structure, the Euclidean structure, the Brownian motions and the law of the associated Dyson processes. In this paper for example, we go beyond the cases of real symmetric, Hermitian and quaternionic cases, and consider some associative dimension 2^p algebras of the Clifford type, which extend the cases of real, complex and quaternions. One may then define real symmetric matrices with size $(n \times 2^p) \times (n \times 2^p)$, which have eigenspaces with dimension 2^k, where k is related to p in some specific way described below, and may be chosen as large as we wish. We then consider Brownian motion on these set of matrices, compute the Dyson processes, which appear to be symmetric diffusion processes, or union of pairs of independent processes. This depends indeed on the algebra structure and leads to the computation of the law of the spectrum. It turns out that there is still a factor $(\prod_{i<j} |\lambda_i - \lambda_j|)^\beta$ with $\beta = 1, 2, 4$, and we do not produce in this way other values for β: these values reflect in fact a phenomenon which is known as Bott's periodicity, and has nothing to do with the dimension of the eigenspaces (for references on Bott's Periodicity Theorem, see [2, 3, 14]).

In order to deal with spectral measures, we consider processes on the characteristic polynomials $P(X) = \det(M - X\mathrm{Id})$. Indeed, functions of the spectral measures are nothing else than symmetric functions of the roots of P. Usually, in order to characterize the laws of the spectral measure, one works with its moments, that is the functions $\sum_i^n \lambda_i^k$, where $(\lambda_1, \cdots, \lambda_n)$ denote the eigenvalues of the matrix. We find more convenient to deal with the elementary symmetric functions of the roots, that is the coefficients of P. Curiously, this approach, very close to the study of Stieltjes transform of the measure, is not that popular. Although many of these computations are well known, and for the sake of completeness, we chose to present the basic objects allowing to manipulate the various quantities appearing in the formulae. For this, we first consider the diagonal cases (that is when we start directly on some processes on the roots of $P(X)$), that we analyze in the case of flat Brownian motion, Ornstein-Uhlenbeck operators and spherical Brownian motions. Then, we pass to the analysis of the laws of the characteristic polynomials in various sets of symmetric matrices, starting with the classical settings (real symmetric, Hermitian and quaternionic), before looking at the case of general Clifford algebras.

The paper is organized as follows. Section 2 is a short introduction to the methods and language of symmetric diffusion processes. In Sects. 3, 4 and 5 we consider the simpler cases of diagonal matrices in the flat Euclidean, Gaussian and spherical cases respectively. We describe how to handle and identify various quantities (discriminants, metric structures, etc) which appear in the computations through the characteristic polynomials. Sections 6 and 7 are devoted to the analysis of the real symmetric, Hermitian and quaternionic cases (although the quaternionic case is just sketched since it is not simpler to handle than the general Clifford case). The real symmetric case allows to develop the techniques used for dealing with characteristic polynomials, while the Hermitian case shows how to deal with multiple eigenvalues. In Sect. 8 we introduce the general Clifford algebras, through a presentation which is quite handy for our purpose and does not seem to be classical. Finally, in Sect. 9, we give the complete description of the laws of the spectra for the standard Clifford algebras, where we recover Bott's periodicity through the various laws appearing in the spectra of the matrices, for different values of the dimension of the algebra. For references on Clifford algebras see [18] and for some generalized Clifford algebras see [13]. The general conclusion is that the study of the law of the Dyson processes is helpful to decipher the algebra structure itself.

2 Symmetric Diffusion Generators and Their Images

The general setting for symmetric diffusion generators is inspired from [4] and is the following. The ambient space E is some measure space, endowed with a σ-finite measure μ. If μ is finite, then it is always assumed to be a probability measure. In what follows, E will be a compact connected smooth (that is $\mathscr{C}^\infty$), with or without boundary, or some open connected set in $\mathbb{R}^n$, or $\mathbb{R}^n$ itself. When a boundary appears, it will be at least piecewise $\mathscr{C}^1$. On E is given an algebra $\mathscr{A}_0$ of function $E \mapsto \mathbb{R}$; typically when E is an open set in $\mathbb{R}^n$, $\mathscr{A}_0$ is the set of smooth ($\mathscr{C}^\infty$) compactly supported functions, but it may be also the set of polynomial functions when E is bounded and μ is a probability measure, or more generally when polynomials are dense in $\mathscr{L}^2(\mu)$.

A fundamental assumption on $\mathscr{A}_0$ is that $\mathscr{A}_0 \subset \cap_{1 \le p \le \infty} \mathscr{L}^p(\mu)$, and that $\mathscr{A}_0$ is dense in very $\mathscr{L}^p(\mu)$, $1 \le p < \infty$. Moreover, $\mathscr{A}_0$ is stable under any transformation $(f_1, \cdots, f_n) \mapsto \Phi(f_1, \cdots, f_n)$, where Φ is a smooth function $\mathbb{R}^n \mapsto \mathbb{R}$ such that $\Phi(0) = 0$ (in the case of polynomials, one restricts to polynomial functions Φ). For any linear operator $\mathrm{L} : \mathscr{A}_0 \mapsto \mathscr{A}_0$, one defines its carré du champ operator

$$\Gamma(f, g) = \frac{1}{2}\Big(\mathrm{L}(fg) - f\mathrm{L}(g) - g\mathrm{L}(f)\Big).$$

We have the following

Definition 1. A symmetric diffusion operator is a linear operator L: $\mathscr{A}_0 \oplus 1 \mapsto \mathscr{A}_0$, such that

1. $L(1) = 0$,
2. $\forall f, g \in \mathscr{A}_0 \oplus 1, \ \int f L(g) \, d\mu = \int g L(f) \, d\mu$,
3. $\forall f \in \mathscr{A}_0, \Gamma(f, f) \geq 0$,
4. $\forall f = (f_1, \cdots, f_n)$, where $f_i \in \mathscr{A}_0$, $\forall \Phi : \mathbb{R}^n \mapsto \mathbb{R}$,

$$L(\Phi(f)) = \sum_i \partial_i \Phi(f) L(f_i) + \sum_{i,j} \partial^2_{ij} \Phi(f) \Gamma(f_i, f_j). \tag{1}$$

Such an operator describes the law of a stochastic Markov process (ξ_t) with values in E, with generator L and reversible measure μ. It is often important to be able to extend L on a larger class of functions (typically $\mathscr{C}^\infty$ functions, without any restriction on the size of the function or condition on its support). There is in general a unique way of doing this (in an obvious way, see Eq. (2) below). We shall use this extension without further comments. For a consistent reference on these operators, see [4].

In the case of open sets of $\mathbb{R}^n$ or manifolds, L is in general given locally on some set $\Omega \subset E$ in a local system of coordinates (x^i), as

$$L(f) = \sum_{ij} g^{ij}(x) \partial^2_{ij} f + b^i(x) \sum_i \partial_i f, \tag{2}$$

where

$$g^{ij}(x) = \Gamma(x^i, x^j), \ b^i(x) = L(x^i).$$

This is in particular the case for open subsets $\Omega \subset \mathbb{R}^n$ or in a local chart Ω on a manifold, and is a direct consequence of the change of variable formula 4 in Definition 1.

The positivity condition (point 3 in Definition 1) imposes that for any $x \in \Omega$, the symmetric matrix $(g^{ij}(x))$ is non negative. Moreover, provided μ has a smooth positive density ρ with respect to the Lebesgue measure $dx^1 \cdots dx^n$, the operator L is entirely described, up to a normalizing factor, by

$$L(f) = \frac{1}{\rho} \sum_{ij} \partial_j (g^{ij} \rho \partial_j f). \tag{3}$$

Observe that, in an open subset of $\mathbb{R}^n$, this equation determines entirely the density ρ up to some constant whenever it exists. Indeed, identifying Eqs. (2) and (3) leads to

$$\sum_j g^{ij} \partial_j \log \rho = b_i - \sum_j \partial_j g^{ij}, \tag{4}$$

and, when g^{ij} is invertible, this leads to the identification of $\partial_i \log(\rho)$.

This description (2) is not restricted to local system of coordinates, that is in the manifold case to local diffeomorphisms. Suppose indeed that one may find functions $(a^1, \cdots, a^k)$ $(a^i \in \mathscr{A})$ such that for some smooth functions B^i and G^{ij}

$$L(a^i) = B^i(a^1, \cdots, a^k), \quad \Gamma(a^i, a^j) = G^{ij}(a^1, \cdots, a^k),$$

then, writing $a = (a^1, \cdots, a^k)$, we get readily from Eq. (1)

$$L(f(a)) = \sum_{ij} G^{ij}(a) \partial^2_{ij} f(a) + \sum_i B^i(a) \partial_i f(a),$$

and the operator

$$\hat{L} = \sum_{ij} G^{ij} \partial^2_{ij} + \sum_i B^i \partial_i$$

is nothing else than the operator L acting on functions depending only on $a = (a^1, \cdots, a^k)$. We shall say that $\hat{L}$ is the image of the operator L through $a = (a^1, \cdots, a^k)$, seen as a function $E \mapsto \mathbb{R}^k$. As a consequence, the operator $\hat{L}$ is symmetric with respect to the image measure of μ through $a = (a^1, \cdots, a^k)$. With the help of Eq. (4), it will be a good way to identify the image measure, provided it has a density ρ with respect to the Lebesgue measure.

A special case concerns Laplace operators, where (g^{ij}) is everywhere non degenerate and ρ is by definition $\det(g)^{-1/2}$. When $x \mapsto a = (a^1, \cdots, a^n)$ is a local diffeomorphism (a simple change of coordinates), then the image of L is again a Laplace operator, (the Laplace operator written in the new system of coordinates), and the density measure is again $\det(G)^{-1/2}$ is the new system of coordinates. It is however important to notice that we shall use this procedure even when $x \mapsto a$ is not a diffeomorphism, for example with the map $M \mapsto P(X)$, where M is a matrix and P its characteristic polynomial.

Those operators L are related with the associated Markov diffusion processes (ξ_t) by the requirement that, for any function $f \in \mathscr{A}$,

$$f(\xi_t) - \int_0^t L(f)(\xi_s) \, ds \tag{5}$$

is a local martingale (a true martingale if for example f and $L(f)$ are bounded). In good situations (see below), this is enough to describe the law of (ξ_t), or the joint laws of $(\xi_{t_1}, \cdots, \xi_{t_n})$ from the starting point $\xi_0 = x$ and the knowledge of the operator L, see [4]. In this case, the law of ξ_t when $\xi_0 = x$ is described through the formula $\mathbb{E}(f(\xi_t)/\xi_0 = x) = P_t(f)(x)$, where $P_t = \exp(tL)$ is the semigroup with generator L, which is uniquely determined by the knowledge of L acting on $\mathscr{A}_0$. This linear operator P_t is a bounded operator on any $\mathscr{L}^p(\mu)$, for $1 \leq p \leq \infty$. Good situation here refers to the fact that the operator L is essentially self-adjoint on $\mathscr{L}^2(\mu)$ with respect to $\mathscr{A}_0$. When we deal with $E = \mathbb{R}^n$ with some elliptic operator with smooth coefficients, this is always the case when $\mathscr{A}_0$ is the set of compactly supported smooth functions.

In presence of boundaries however, there may be several self-adjoint extensions of this operator L defined on $\mathscr{A}_0$, and therefore may possible semigroups: for example, one may deal with Dirichlet or Neuman boundary conditions. It shall be the case for example when looking at the spectral measures, which live on the Weyl chambers $\{\lambda_1 < \cdots < \lambda_n\}$. Fortunately, in this case the operator $\hat{L}$ is defined as the image of L through the map a. Moreover, the (unique) semigroup P_t itself preserves functions of a, which will be easily checked in all our situations. That is, one may define a new semigroup $\hat{P}_t$ as $(\hat{P}_t f)(a) = P_t(f(a))$, and by construction, this semigroup $\hat{P}_t$ has generator $\hat{L}$. We shall therefore use this semigroup $\hat{P}_t$ as the semigroup generated by L. In the case of spectral measures, it corresponds indeed to the Neuman boundary conditions for the operator $\hat{L}$ defined in the interior of the Weyl chamber.

Anyhow, we shall not really use this interpretation in terms of stochastic processes in what follows, since we shall mainly concentrate on the properties of the local operator L and of it's various images $\hat{L}$, that we want to describe in some local systems of coordinates. One may observe that thanks to formula (3), this is enough to describe the image measure of the measure μ through the map a. As mentioned above, we shall deal with operators acting on the space of polynomials i.e. on $\mathbb{R}^n$, when identifying $\mathbb{R}^n$ with the set of the coefficients of the polynomial. The polynomials shall be monic in general, i.e. $P(X) = X^n + \sum_0^{n-1} a_i X^i$, where $(a_0(\xi_t), \cdots, a_{n-1}(\xi_t))$ is the stochastic process, and X may be considered as a parameter. The coefficients (a_i) can be viewed as coordinates in this set of polynomials, writing for example $L(P(X)) = \sum_i X^i L(a_i)$, and acting similarly for the operator $\Gamma(P(X), P(Y))$. One may also consider a fixed X and see $P(X)$ as an application $\mathbb{R}^n \mapsto \mathbb{R}$, for which one can take $\log(P(X))$ or $P(X)^r$. Last, those function can be described as series in the variable X (even formal series, regardless of their domain of convergence), with coefficients being some polynomial functions of $(a_0, \cdots, a_{n-1})$, and all computations on these expressions boiling down to algebraic computations involving L, Γ and polynomial expressions in the coefficients a_i (see for example Lemma 2 below).

3   The Image of the Euclidean Laplacian Under Elementary Symmetric Functions

As a warm up, let us start with the case of diagonal matrices. The computations that we shall perform in this section are well known, and we refer to [1] or [11] for further details. See also Gamburd [12] on applications of symmetric functions to random matrices. Let $x = (x_1, \cdots, x_n) \in \mathbb{R}^n$ and

$$P(X) = \prod_{i=1}^{n}(X - x_i) = \sum_{i=0}^{n} a_i X^i,$$

such that $(-1)^i a_i(x_1, \cdots, x_n)$ are the elementary symmetric functions. If we want to describe the image of the Laplace operator $\Delta_{\mathbb{E}}$ on $\mathbb{R}^n$ under symmetric functions of $(x_1, \cdots, x_n)$, we may look at smooth functions $F(a_0, \cdots, a_{n-1})$. At least in the Weyl chamber $\{x_1 < x_2 \cdots < x_n\}$, the application $(x_1, \cdots, x_n) \mapsto \Phi(x_1, \cdots, x_n) = (a_0, \cdots, a_{n-1})$ is a local diffeomorphism. We first have to look at the image of the Lebesgue measure $dx = dx_1 \cdots dx_n$ under Φ. For that purpose, we recall the definition of the discriminant of a polynomial P, considering only monic polynomials (when the coefficient of the leading term is 1). For two monic polynomials $P(X) = \sum_{i=0}^{n} a_i X^i$ and $Q(X) = \sum_{i=0}^{p} b_i X^i$, the resultant $R(P, Q)$ is a polynomial in the coefficients $(a_0, \cdots, a_{n-1}, b_0, \cdots, b_{p-1})$ which vanishes exactly when P and Q have a common root (in the complex plane). Indeed, $R(P, Q)$ is the determinant of the $(n + p) \times (n + p)$ Sylvester matrix

$$\begin{pmatrix}
1 & a_{n-1} & a_{n-2} & \cdots & a_0 & 0 & \cdots & 0 \\
0 & 1 & a_{n-1} & \cdots & a_1 & a_0 & \cdots & 0 \\
0 & 0 & 1 & \cdots & a_2 & a_1 & \cdots & 0 \\
\cdots & \cdots & & \cdots & \cdots & \cdots & & \cdots \\
\cdots & \cdots & & \cdots & a_{p-2} & \cdots & a_1 & a_0 \\
1 & b_{p-1} & b_{p-2} & \cdots & b_0 & 0 & \cdots & 0 \\
0 & 1 & b_{p-1} & \cdots & b_1 & b_0 & \cdots & 0 \\
0 & 0 & 1 & \cdots & b_2 & b_1 & \cdots & 0 \\
\cdots & \cdots & & \cdots & \cdots & \cdots & & \cdots \\
\cdots & \cdots & & \cdots & \cdots & \cdots & b_1 & b_0
\end{pmatrix}$$

It can be viewed as the determinant of the following system of linear equations in the unknown variables $\{1, X, \cdots, X^{n+p-1}\}$:

$$\{P(X) = 0, XP(X) = 0, \cdots, X^{p-1}P(X) = 0, Q(X) = 0, XQ(X) = 0, \cdots, X^{n-1}Q(X) = 0\}.$$

It turns out that, when $P(X) = \prod_i(X - x_i)$ and $Q(X) = \prod_j(X - y_j)$, then $R(P, Q) = \prod_{i,j}(x_i - y_i)$.

The discriminant $\text{discr}(P) = (-1)^{n(n-1)/2}R(P, P')$ and expresses a necessary and sufficient condition for P to have a double root. Then, when $P(X) = \prod(X - x_i)$, one has $\text{discr}(P) = \prod_{i<j}(x_i - x_j)^2$.

The discriminant is not an easy expression of the coefficients $(a_0, \cdots, a_{n-1})$, and the following computations are here to make them easier.

The first central result concerns the image of the Lebesgue measure.

Proposition 1. *The image measure of dx under Φ is*

$$d\mu_0 = n!|\mathrm{discr}(P)|^{-1/2}\mathbb{1}_{\mathcal{D}>0}da_0\cdots da_{n-1},$$

where D is the connected component of the set $\{\mathrm{discr}(P) > 0\}$ where all the roots of the polynomial P are real.

Proof. This is a classical result known as Weyl integration formula (see [1, 11, 12]). We give a proof here for the sake of completeness. Let us first observe that for a polynomial P having only distinct real roots, $\mathrm{discr}(P) > 0$. The condition $\mathrm{discr}(P) > 0$ is not sufficient to assert that P has only real roots: it is also positive for example when P has an even number of pairs of complex conjugate roots. The set where all the roots are real and distinct is obviously connected, and therefore there is only one connected component of this set where all the roots are real. This is our set $\mathcal{D}$, the set which contains for example the point $\prod_0^{n-1}(X - i)$.

To compute the image measure of Φ, it is enough to identify the image of Φ when restricted to the Weyl chamber $\{x_1 < \cdots < x_n\}$, since the same computation will hold true in any other Weyl chamber (that is the image of the first one under a permutation of the coordinates), and there are $n!$ of such chambers.

First, it is clear that the support of the image measure is included in the closure of $\mathcal{D} \subset \{\mathrm{discr}(P) > 0\}$, and, from the explicit expression of the discriminant in terms of the roots, is strictly positive in any Weyl chamber. The boundary of this set is a subset of the algebraic surface (in the space of the (a_i) coordinates) $\{\mathrm{discr}(P) = 0\}$.

Indeed, it is quite easy to see the result of Proposition 1 by induction on the degree n. It is clear that it is true for $n = 1$, since $a_1 = -x_1$

Let us assume that the result is true for n and set $P(X) = (X - x_{n+1})Q(x) = \sum_0^{n+1} a_i X^i$, where $Q(X) = \sum_0^n b_i X^i = \prod_1^n(X - x_i)$. Then

$$a_n = b_{n-1} - x_{n+1}, a_{n-1} = b_{n-2} - x_{n+1}b_{n-1}, \cdots, a_1 = b_0 - x_{n+1}b_1, a_0 = -x_{n+1}b_0.$$

The Jacobian of the transformation $(x_{n+1}, b_0, b_1, \cdots, b_{n-1}) \mapsto (a_0, \cdots, a_n)$ is easily seen to be

$$|x_{n+1}^n + b_n x_{n+1}^{n-1} + \cdots + b_0| = |Q(x_{n+1})| = |\prod_1^n(x_{n+1} - x_i)|.$$

Therefore, if $db_0 \cdots db_{n-1} = |\prod_{1\leq i<j\leq n}(x_i - x_j)|dx_1 \cdots dx_n$, then

$$da_0 \cdots da_n = \prod_{1\leq i<j\leq n+1} |x_i - x_j|dx_1 \cdots dx_{n+1}.$$

From what precedes, it is clear that the Jacobian of the transformation $(x_1, \cdots, x_n) \mapsto (a_0, \cdots, a_{n-1})$ is non zero on any Weyl chamber, and then the boundary of the image is included in the algebraic set $\{\text{discr}(P) = 0\}$. This is enough to identify the support of the image measure as the closure of $\mathscr{D}$.

Let us now compute the image of the Laplace operator $\Delta_{\mathbb{E}}$ on $\mathbb{R}^n$ under Φ. In what follows, and throughout the paper, $\Gamma_{\mathbb{E}}$ denotes the Euclidean carré du champ, that is, in the standard system of coordinates,

$$\Gamma_{\mathbb{E}}(f, g) = \sum_i \partial_i f \partial_i g.$$

Fix $X \in \mathbb{R}$, and consider the function

$$\mathbb{R}^n \setminus \{\exists i, x_i = X\} \mapsto \mathbb{R}$$

$$(x_1, \cdots, x_n) \mapsto \log P(X) = \sum_i \log(X - x_i)$$

We get

Proposition 2.

1. For any $X \in \mathbb{R}$, $\Delta_{\mathbb{E}}(P(X)) = 0$
2. For any $(X, Y) \in \mathbb{R}^2$,

$$\Gamma_{\mathbb{E}}(\log(P(X), \log P(Y)) = \frac{1}{Y - X}\left(\frac{P'(X)}{P(X)} - \frac{P'(Y)}{P(Y)}\right). \tag{6}$$

Proof. The first assertion is immediate, since every function a_i is an harmonic function on $\mathbb{R}^n$ (as a polynomial of degree 1 in any coordinate x_i).

For the second, one has

$$\Gamma_{\mathbb{E}}(\log P(X), \log P(Y)) = \sum_i \partial_{x_i} \log P(X) \partial_{x_i} \log P(Y) = \sum_i \frac{1}{(X - x_i)(Y - x_i)}.$$

But

$$\frac{1}{(X - x_i)(Y - x_i)} = \frac{1}{Y - X}\left(\frac{1}{X - x_i} - \frac{1}{Y - x_i}\right)$$

and

$$\sum_i \frac{1}{(X - x_i)(Y - x_i)} = \frac{1}{Y - X}\left(\frac{P'(X)}{P(X)} - \frac{P'(Y)}{P(Y)}\right).$$

Corollary 1. *Setting* $\alpha_{i,j} = (i + 1)a_{i+1}a_j$, *where* $a_i = 0$ *if* $i > n$ *and* $a_n = 1$, *one has*

$$\sum_{i,j} \Gamma_{\mathbb{E}}(a_i, a_j)X^i Y^j = \sum_{i \neq j} \alpha_{i,j}\frac{X^i Y^j - X^j Y^i}{Y - X},$$

from which

$$\Gamma_{\mathbb{E}}(a_k, a_p) = \sum_{(p-k)_+ \leq l \leq p} \alpha_{p-l,k+l+1} - \sum_{(k-p)_+ \leq l \leq k} \alpha_{p+l+1,k-l}. \tag{7}$$

Moreover

$$\partial_{a_k} \Gamma_{\mathbb{E}}(a_k, a_p) = \mathbb{1}_{p \geq k}(k - p - 2)a_{p+2} + \mathbb{1}_{p=k-1}k a_{p+2}. \tag{8}$$

and

$$\sum_k \partial_{a_k} \Gamma_{\mathbb{E}}(a_k, a_p) = -\frac{1}{2}(p + 1)(p + 2)a_{p+2}. \tag{9}$$

Proof. This is straightforward from Eq. (6), which gives

$$\Gamma_{\mathbb{E}}(P(X), P(Y)) = \frac{1}{Y - X}(P'(X)P(Y) - P'(Y)P(X)).$$

On the other hand, by bilinearity, one has

$$\Gamma_{\mathbb{E}}(P(X), P(Y)) = \sum_{ij} \Gamma_{\mathbb{E}}(a_i, a_j)X^i Y^j.$$

We then obtain

$$\sum_{i,j} \Gamma_{\mathbb{E}}(a_i, a_j)X^i Y^j = \sum_{i \neq j} \alpha_{i,j} \frac{X^i Y^j - X^j Y^i}{Y - X},$$

from which formula (7) follows easily.

Formulae (8) and (9) are easy consequences of the first one.

The application $(x_1, \cdots, x_n) \mapsto (a_0, \cdots, a_{n-1})$ is a local diffeomorphism. Therefore, the image of the Laplace operator in coordinates $(x_1, \cdots, x_n)$ is the Laplace operator in coordinates $(a_0, \cdots, a_{n-1})$. This leads to some formulae which are less immediate.

Corollary 2. *One has*

1. $\mathrm{discr}(P) = \det(\Gamma_{\mathbb{E}}(a_i, a_j))$.
2. For any $i \in \{0, \cdots, n-1\}$, $\sum_j \Gamma_{\mathbb{E}}(a_i, a_j)\partial_{a_j} \log \mathrm{discr}(P) = 2\sum_j \partial_{a_j} \Gamma_{\mathbb{E}}(a_i, a_j)$.
3. For any $i \in \{0, \cdots, n-1\}$, $\sum_{i,j} X^i \Gamma_{\mathbb{E}}(a_i, a_j)\partial_{a_j} \log \mathrm{discr}(P) = -P''(X)$.

Proof. One the one hand, we know that in the coordinates $(a_0, \cdots, a_{n-1})$, the Laplace operator has carré du champ $\Gamma_{\mathbb{E}}(a_i, a_j)$ and reversible measure (here the Riemann measure) density $\mathrm{discr}(P)^{-1/2}da_0 \cdots da_{n-1}$.

One the other hand, we know that this density measure is always, (for any Laplace operator) $\det(\Gamma_{\mathbb{E}}(a_i, a_j)^{-1/2})$. This is enough to get point 1.

Point 2 comes from the observation that in these coordinates (a_i), $\Delta(a_i) = 0$. Setting $G^{ij} = \Gamma_{\mathbb{E}}(a_i, a_j)$, and $\rho = \det(G^{ij})^{-1/2}$, the Laplace operator writes

$$\Delta(f) = \sum_{ij} G^{ij} \partial_{ij}^2 f + \sum_{ij} \partial_j (G^{ij}) \partial_i f + \sum_{ij} G^{ij} \partial_i f \partial_j \log \rho.$$

From this, we know that

$$\Delta(a_i) = \sum_j \partial_j G^{ij} - \frac{1}{2} \sum_j G^{ij} \partial_j \log \operatorname{discr}(P).$$

Applying $\Delta(a_i) = 0$, point 2 is the direct translation of the previous.

For point 3, it suffices to combine point 2 together with formula (7) to obtain, for $i = 0, \cdots, n-1$

$$\sum_j \Gamma_{\mathbb{E}}(a_i, a_j) \partial_{a_j} \log \operatorname{discr}(P) = -(i+2)(i+1)a_{i+2}.$$

As a consequence, we obtain the central point of many of the computations below.

Proposition 3. $\Gamma_{\mathbb{E}}(P, \log \operatorname{discr}(P)) = -P''$.

Proof. This is just a rephrasing of point 3 in Corollary 2.

4 The Image of the Ornstein-Uhlenbeck Operator Under Elementary Symmetric Functions

In Euclidean spaces, the Ornstein-Uhlenbeck operator is defined as

$$L_{OU}(f) = \Delta_{\mathbb{E}}(f) - \frac{1}{2}\Gamma_{\mathbb{E}}(\|x\|^2, f) = \Delta_{\mathbb{E}}(f) - \sum_{i=1}^n x_i \partial_i f.$$

It shares the same carré du champ operator than the Laplace operator, but has the standard Gaussian measure as reversible measure. It also admits a complete orthonormal system of eigenvectors, namely the Hermite polynomials, which polynomials in the variables $(x_1, \cdots, x_n)$ with total degree k, with associated eigenvalue $L_{OU}(H_k) = -kH_k$.

Proposition 4. *For $P(X) = \prod_{i=1}^n (X - x_i) = \sum_{i=0}^n a_i X^i$, one has*

1. $L_{OU}(P) = -\sum_i x_i \partial_i P = \sum_i (n-i)a_i X^i = -nP(X) + XP'(X)$.
2. $\frac{1}{2}\Gamma_{\mathbb{E}}(a_{n-1}^2 - 2a_{n-2}, P(X)) = nP(X) - XP'(X)$.

3. $\forall i = 0, \cdots, n-1$, $a_{n-1} \Gamma_{\mathbb{E}}(a_i, a_{n-1}) - \Gamma_{\mathbb{E}}(a_i, a_{n-2}) = (n-i)a_i$.
4. *The image of the Gaussian measure is*

$$d\gamma_s := \frac{n!}{(2\pi)^{n/2}} \exp(a_{n-2} - \frac{a_{n-1}^2}{2}) \mathrm{discr}(P)^{-1/2} \mathbb{1}_{\mathscr{D}} \prod_0^{n-1} da_i. \tag{10}$$

Proof. The first item 1 comes from the fact that since $\Delta(a_i) = 0$, one has

$$L_{\mathbb{OU}}(P) = \sum_i X^i L_{\mathbb{OU}}(a_i) = -\sum_{ij} X^i x_j \partial_j(a_j).$$

But the functions a_j are homogeneous polynomial of degree $n-i$ in the variables x_i, and therefore $\sum_j x_j \partial_j(a_j) = (n-i)a_i$.

If we observe that $\sum_{ij} x_j \partial_i f = \frac{1}{2} \Gamma(\|x\|^2, f)$, and that $\|x\|^2 = a_{n-1}^2 - 2a_{n-2}$, one sees that

$$\frac{1}{2} \Gamma_{\mathbb{E}}(a_i, a_{n-1}^2 - 2a_{n-2}) = (n-i)a_i,$$

from which

$$\frac{1}{2} \Gamma_{\mathbb{E}}(a_{n-1}^2 - 2a_{n-2}, P(X)) = nP(X) - XP'(X).$$

Points 3 and 4 are immediate consequences.

Remark 1. Observe also that if $U = a_{n-1}^2 - 2a_{n-2}$, then, for the Euclidean quantities, $\Gamma_{\mathbb{E}}(U, U) = 4U$ and $\Delta_{\mathbb{E}}(U) = 2n$, as expected.

It is worth to observe that, contrary to the Gaussian measure, the image measure $d\gamma_s$ given in Eq. (10) does not have in general exponential moments. Nevertheless, polynomials are dense in $\mathscr{L}^2(d\gamma_s)$. Indeed, any function f in $\mathscr{L}^2(d\gamma_s)$ which is orthogonal for $d\gamma_s$ to any polynomial $Q(a_0, \cdots, a_{n-1})$ may be lifted into a function $\hat{f} : \mathbb{R}^n \mapsto \mathbb{R}$ which is invariant under permutation of the variables $(x_1, \cdots, x_n)$ and orthogonal to any symmetric polynomial. But such a function would then be orthogonal to any polynomial, (even non symmetric), and therefore zero since polynomials are dense for the Gaussian measure.

Observe that $\Gamma_{\mathbb{E}}(a_i, a_j)$ are bilinear functions of the (a_i), and $L_{\mathbb{OU}}(a_i)$ are linear. From the change of variable formula, if $Q(a_i)$ is a polynomial of total degree less than k in the variables a_i, then so is $L_{\mathbb{OU}}(Q)$. Since the set of polynomials in the variables a_i are dense in $\mathscr{L}^2(\mu)$, the operator $L_{\mathbb{OU}}$ may be then diagonalized in a basis formed of orthogonal polynomials in the variables $(a_0, \cdots, a_{n-1})$.

Remark 2. It is worth to observe that if Q is the mean value of P, i.e. $Q := \langle P \rangle = \int P(X) d\mu(P)$, where μ is the image of the Gaussian measure, then since $\int L_{\mathbb{OU}}(P) d\mu(P) = 0$

$$XQ' = nQ.$$

Then, $Q(X) = X^n$, which was obvious from the explicit expression of the a_i in terms of x_i, which are independent centered Gaussian. But we shall see later (Remark 5) that the same computation performed on symmetric matrices leads to a more interesting result.

5 The Image of the Spherical Laplacian Under Elementary Symmetric Functions

In an Euclidean n-dimensional space, the spherical Laplace operator $\Delta_\mathbb{S}$ may be written as the restriction to the unit sphere $\|x\|^2 = 1$ of some combinations of $\Delta_\mathbb{E}$ and the Euler operator $D(f) = \sum_i x_i \partial_i f$, namely

$$\Delta_\mathbb{S}(f) = \Delta_\mathbb{E}(f) - D^2(f) - (n-2)Df.$$

Indeed, when considering the restriction to the sphere of coordinate x_i as a function $\mathbb{S}^{n-1} \mapsto \mathbb{R}$, one may describe the spherical Laplace operator $\Delta_\mathbb{S}$ from

$$\Gamma_\mathbb{S}(x_i, x_j) = \delta_{ij} - x_i x_j, \quad \Delta_\mathbb{S}(x_i) = -(n-1)x_i.$$

From this, when considering a smooth function $F(x_1, \cdots, x_n)$, the change of variable formula gives immediately $\Delta_\mathbb{S}(F) = \Delta_\mathbb{E}(f) - D^2 f - (n-2)Df$.

In the study of Ornstein-Uhlenbeck processes, we already computed the action of D on $P(X)$. Then, we get

Proposition 5. *For the Laplace-Beltrami operator $\Delta_\mathbb{S}$ acting on the unit sphere $\mathbb{S}^{n-1} \subset \mathbb{R}^n$, and for the polynomial $P(X) = \prod(X - x_i)$, one has*

1. $\Gamma_\mathbb{S}(\log P(X), \log P(Y)) = \dfrac{1}{Y-X}\left(\dfrac{P'(X)}{P(X)} - \dfrac{P'(Y)}{P(Y)}\right) - \left(n - X\dfrac{P'(X)}{P(X)}\right)$
$\left(n - Y\dfrac{P'(Y)}{P(Y)}\right)$

2. $\Delta_\mathbb{S}(P(X)) = -2n(n-1)P(X) + 3(n-1)XP'(X) - X^2 P''(X)$.

Proof. From

$$\Gamma_\mathbb{S}(P(X), P(Y)) = \Gamma_\mathbb{E}(P(X), P(Y)) - DP(X)DP(Y),$$

we get point 1. Moreover,

$$\Delta_\mathbb{S}(P(X) = -D(nP(X) - XP'(X)) = -(n(nP(X) - XP'(X)) + D(\sum_i ia_i X^i).$$

And

$$D(\sum_i i a_i X^i) = \sum_i i(n-i)a_i X^i = nXP'(X) - X^2 P''(X) - XP'(X),$$

which gives point 2.

The support of the image measure is included in the set $\{a_{n-1}^2 - 2a_{n-2} = 1\}$. It does not have a density with respect to the Lebesgue measure $da_0 \cdots da_{n-1}$. To deal with it, it is more convenient to look at a more general operator, defined on the unit ball $\mathbb{B} \subset \mathbb{R}^n$, and defined, for any $p > 1$ from

$$\Gamma_{\mathbb{B}}(x_i, x_j) = \delta_{ij} - x_i x_j, \quad \Delta(x_i) = -p x_i.$$

Indeed, for any integer $p > n - 1$, this operator is nothing else than the projection on the unit disk of the previous spherical operator on the sphere $\mathbb{S}^p$. If one observes that

$$\Gamma_{\mathbb{B}}(1 - \|x\|^2, x_i) = -2(1 - \|x\|^2)x_i,$$

comparing with Eq. (3), which gives for the density ρ of the associated invariant measure $\Gamma(\log \rho, x_i) = (n + 1 - p)x_i$, one sees that this operator has invariant measure

$$C_{p,n}(1 - \|x\|^2)^{(p-n-1)/2} dx_1 \cdots dx_n$$

on the unit ball $\mathbb{B}_n$.

Then, letting p converging to $n - 1$, the measure concentrates on the uniform measure on the unit sphere $\mathbb{S}^{n-1}$.

One may then consider the image measure through $(a_0, \cdots, a_{n-1})$ of this last operator, which is written as

$$\begin{cases} \Gamma_{\mathbb{B}}(\log P(X), \log P(Y)) = \frac{1}{Y-X}\left(\frac{P'(X)}{P(X)} - \frac{P'(Y)}{P(Y)}\right) - \left(n - X\frac{P'(X)}{P(X)}\right)\left(n - Y\frac{P'(Y)}{P(Y)}\right), \\ \Delta_{\mathbb{S}}(P(X)) = -n(p+n-1)P(X) + (p+2(n-1))XP'(X) - X^2 P''(X). \end{cases}$$

Its reversible measure is then, up to some scaling factor

$$\mathbb{1}_{\mathscr{D}}D(a_0, \cdots, a_{n-1})^{-1/2}(1 + 2a_{n-2} - a_{n-1}^2)^{(n-p-1)/2} da_0 \cdots da_{n-1},$$

which concentrates on the set $1 + 2a_{n-2} - a_{n-1}^2 = 0$ when $p \to n - 1$.

Remark 3. In random matrix theory, there are indeed many situations where the law of the spectrum does not have a density with respect to the Lebesgue measure. This is in particular the case when dealing with square matrices with random real independent Gaussian entries (the Ginibre ensemble), where the (complex) eigenvalues are conjugate to each other, and where the probability that there are a finite number of real eigenvalue is positive (see [17]). There may be simpler situations where the spectrum is carried by sub manifolds (in general algebraic varieties) and where the law has a density with respect of the surface measure of this manifold. This is for example the case for Wishart matrices, that is matrices of the form $M = mm^*$ where m is an $n \times p$ matrix with independent Gaussian entries. Whenever $n < p$, the spectrum is carried by the space where the discriminant of the characteristic polynomial is 0. When the associated Dyson process is a diffusion process, this situation may be detected through formula (4). Indeed, whenever the invariant measure of the starting matrix process is a probability measure, then the image measure is a probability measure. Then, when solving Eq. (4) which provides the density ρ of the measure, it may happen that the unique (up to a normalizing constant) solution is not integrable on the image domain, as it is in the previous case of the spherical Brownian motion process. In complicated situations, this may be a way to detect the support of the image measure.

6 Symmetric Operators on the Spectrum of Real Symmetric Matrices

The space of symmetric real matrices is an Euclidean space with the norm $\|M\|^2 = \text{trace}(M^2)$. The spectrum of such matrices is real and we want to describe the action of the Euclidean and spherical Laplacians and the Ornstein-Uhlenbeck on the spectrum of the matrix. For the moment, we only deal with the Euclidean Laplacian.

For this, we first start with the description of the Euclidean Laplace operator on the entries M_{ij} of the matrix.

One has

$$\Gamma_{\mathbb{E},\mathbb{R}}(M_{ij}, M_{kl}) = \frac{1}{2}(\delta_{ik}\delta_{jl} + \delta_{il}\delta_{jk}), \quad L_{\mathbb{E},\mathbb{R}}(M_{ij}) = 0.$$

This formula just captures the fact that the entries of the matrix are independent Brownian motions, subject to the restriction that the matrix is symmetric. More precisely, it encodes by itself the fact the associated process lives in the space of symmetric matrices. Since we shall use this kind of argument in many places, it is worth to explain why in this simple case. In terms of the stochastic process (X_t) associated with $L_{\mathbb{E},\mathbb{R}}$, which lives a priori in the space of matrices, and for the functions $h_{ij}(M) = M_{ij} - M_{ji}$, it is not hard to see that $L_{\mathbb{E},\mathbb{R}}(h_{ij}) = 0$ and $\Gamma(h_{ij}, h_{ij}) = 0$. Therefore, for the associated Brownian motion (ξ_t), and any smooth function f, $L_{\mathbb{E},\mathbb{R}}(f(h_{ij})) = 0$ and, provided $f(h_{ij})$ and $L_{\mathbb{E},\mathbb{R}}(f(h_{ij}))$ are bounded, $\mathbb{E}(f(h_{ij}(\xi_t)/\xi_0 = x) = f(h_{ij})(x)$. This shows that $h_{ij}(\xi_t)$ remains constant (almost surely). Then, if the process starts from a symmetric matrix, it stays forever in the space of symmetric matrices (we shall not need this fact for what follows).

We start with the following elementary lemma, which will be in full use through the rest of the paper

Lemma 1. *Let $M = (M_{ij})$ be a matrix and M^{-1} be its inverse, defined on the set $\{\det M\} \neq 0$. Then*

1. $\partial_{M_{ij}} \log \det M = M_{ji}^{-1}$,
2. $\partial_{M_{ij}} \partial_{M_{kl}} \log \det M = -M_{jk}^{-1} M_{li}^{-1}$.

Proof. We first start with

$$\partial_{M_{ij}} M_{kl}^{-1} = -M_{ik}^{-1} M_{lj}^{-1}. \tag{11}$$

To see this, consider the formula $\sum_k M_{ik}^{-1} M_{kj} = \delta_{ip}$, that we derive with respect to M_{pq}, leading to

$$\sum_k (\partial_{M_{pq}} M_{ik}^{-1}) M_{kj} + M_{ik}^{-1} \delta_{kp} \delta_{jq} = 0.$$

Fixing p and q, if DM denotes the matrix $\partial_{M_{pq}} M^{-1}$, one gets for every p, q ($DM \times M)_{ij} + M_{ip}^{-1} \delta_{jq} = 0$, which we may now multiply from the right by M^{-1} to get (11).

Now, we observe that $P_{ij} = \det(M) M^{-1}$ is the comatrix, and therefore the M_{ji} entry of this matrix is a polynomial in the entries of the matrix which does not depend on M_{ij}.

One gets

$$(\partial_{M_{ij}} \det M) M_{ji}^{-1} - \det(M) M_{ji}^{-1} M_{ji}^{-1} = 0,$$

which is an identity between rational functions in the entries M_{ij}. Therefore, we may as well divide both terms by M_{ij}^{-1} to obtain item 1. Item 2 follows immediately.

We are now in position to consider the action of the Laplace operator on the spectral measure of M, that is to characterize the Dyson real process.

If we consider the characteristic polynomial $P(X) = \det(M - X\text{Id})$, one has

Proposition 6.

1. $\Gamma_{\mathbb{E},\mathbb{R}}(\log P(X), \log P(Y)) = \frac{1}{Y-X}\left(\frac{P'(X)}{P(X)} - \frac{P'(Y)}{P(Y)}\right)$
2. $L_{\mathbb{E},\mathbb{R}} P(X) = -\frac{1}{2} P''$.

In other words, the diffusion associated to the spectrum of M has the same operator carré du champ, and the invariant measure has a constant density with respect to the Lebesgue measure $\mathbb{1}_D da_0 \cdots da_{n-1}$.

Proof. We start with the first formula. Fix X and Y and write $U = (M - X\text{Id})^{-1}$, $V = (M - Y\text{Id})^{-1}$. One has

$$\Gamma_{\mathbb{E},\mathbb{R}}(\log P(X), \log P(Y)) = \sum_{ijkl} \partial_{M_{ij}} \log P(X) \partial_{M_{kl}} \log P(Y) \Gamma_{\mathbb{E},\mathbb{R}}(M_{ij}, M_{kl}).$$

Since $P(X) = \det(M - X\mathrm{Id})$, and from the value of $\Gamma_{\mathbb{E},\mathbb{R}}(M_{ij}, M_{kl})$, this writes

$$\frac{1}{2}\sum_{ijkl} U_{ji}V_{lk}(\delta_{ik}\delta_{jl} + \delta_{il}\delta_{jk}) = \frac{1}{2}\sum_{ijkl}(U_{ji}V_{ji} + U_{ji}V_{ij}) = \mathrm{trace}(UV).$$

But, if x_i are the eigenvalues of M, then $P(X) = \prod_i(x_i - X)$, $P(Y) = \prod_i(x_i - Y)$, and

$$\mathrm{trace}(UV) = \sum_i \frac{1}{(x_i - X)(x_i - Y)} = \frac{1}{X - Y}\sum_i \frac{1}{x_i - X} - \frac{1}{x_i - Y}$$
$$= \frac{1}{Y - X}\left(\frac{P'(X)}{P(X)} - \frac{P'(Y)}{P(Y)}\right).$$

It remains to compute

$$L_{\mathbb{E},\mathbb{R}}(\log P(X)) = \sum_{ij} \partial_{M_{ij}} \log P(X) L_{\mathbb{E},\mathbb{R}}(M_{ij}) + \sum_{ijkl} \partial_{M_{ij}} \partial_{M_{kl}} \Gamma_{\mathbb{E},\mathbb{R}}(M_{ij}, M_{kl}),$$

which writes

$$-\frac{1}{2}\sum_{ijkl} U_{jk}U_{li}(\delta_{ik}\delta_{jl} + \delta_{il}\delta_{jk}) = -\frac{1}{2}\left(\mathrm{trace}(U^2) + (\mathrm{trace}U)^2\right).$$

Writing

$$\frac{L_{\mathbb{E},\mathbb{R}}P(X)}{P(X)} = L_{\mathbb{E},\mathbb{R}}\log P(X) + \Gamma_{\mathbb{E},\mathbb{R}}(\log P(X), \log P(X)),$$

and noticing that

$$\mathrm{trace}(U^2) = \frac{P'(X)^2}{P(X)^2} - \frac{P''(X)}{P(X)} = \Gamma_{\mathbb{E},\mathbb{R}}(\log P(X), \log P(X)), \quad \mathrm{trace}U = -\frac{P'(X)}{P(X)}$$

one gets the formula for $L_{\mathbb{E},\mathbb{R}}P(X)$.

Comparing with Corollary 2, one sees that $L_{\mathbb{E},\mathbb{R}}(P) = \frac{1}{2}\Gamma_{\mathbb{E},\mathbb{R}}(P, \log \mathrm{discr}(P))$. The last point then is just the observation that if ρ is the reversible measure for the image of $L_{\mathbb{E},\mathbb{R}}$, then ρ has, up to a constant, density $\mathrm{discr}(P)^{1/2}$ with respect to the Riemann measure, which is just $C\,\mathrm{discr}(P)^{-1/2}$.

Remark 4. Moving back the measure to the Weyl chamber $\{\lambda_1 < \cdots < \lambda_n\}$, one sees that the density of the spectral measure with respect to $d\lambda_1 \cdots d\lambda_n$ is $C\,\mathrm{discr}(P)^{1/2} = C\prod |\lambda_i - \lambda_j|$.

If one wants to extend the previous computation to the Gaussian or spherical case, one has to consider also the image of $D = \sum_{ij} M_{ij}\partial_{M_{ij}} = \frac{1}{2}\Gamma_{\mathrm{E,R}}(\|M\|^2, \cdot)$ on the spectral function $P(X) = \det(M - X\,\mathrm{Id})$. One gets, with $U(X) = (M - X\,\mathrm{Id})^{-1}$,

$$D(\log P(X)) = \sum_{ij} M_{ij}\partial_{M_{ij}}P(X) = \sum_{ij} M_{ij}U(X)_{ji} = \mathrm{trace} M U(X)$$

$$= \mathrm{trace}(\mathrm{Id} + XU(X)) = n - X\frac{P'}{P},$$

from which $D(P) = nP - XP'$. Therefore, if $P(X) = \sum_i a_i X^i$, $D(a_i X^i) = (n - i)a_i$. If one wants to consider also the action of the spherical Laplace operator, one needs also to consider $D^2(P)$. But $D^2(a_i X^i) = (n - i)^2 a_i$, from which

$$D^2(P) = n^2 P - (2n - 1)XP' + X^2 P'',$$

that is with $DP = (n\,\mathrm{Id} - X\partial_X)P$, $D^2(P) = (n\,\mathrm{Id} - X\partial_X)^2 P$, although there is no reason a priori for this last identity, since DP is no longer the characteristic polynomial of the symmetric matrix. Observe that the action of D on P is similar that the one in the diagonal case. This is not surprising since, setting $U = a_{n-1}^2 - 2a_{n-2} = \|M\|^2$, $D(F) = \frac{1}{2}\Gamma_{\mathrm{E,R}}(U, F)$, for any function F.

Then, one has for the spherical operator on symmetric matrices, $L_{\mathrm{S,R}} = L_{\mathrm{E,R}} - D^2 - (N - 2)D$, where $N = n(n + 1)/2$, from which

$$L_{\mathrm{S,R}}(P) = -(\frac{1}{2} + X^2)P'' + \frac{(n + 6)(n - 1)}{2}XP' - \frac{n(n + 4)(n - 1)}{2}P.$$

and

$$\Gamma_{\mathrm{S,R}}(P(X), P(Y))$$

$$= \frac{1}{Y - X}(P'(X)P(Y) - P'(Y)P(X)) - (nP(X) - XP'(X))(nP(Y) - XP'(Y)).$$

Remark 5. If we perform the same computation from the Gaussian measure γ instead of the Lebesgue one (that is if we start from an Ornstein-Uhlenbeck process instead of the Brownian motion on matrices), we end up for this Ornstein-Uhlenbeck Dyson process with $L_{\mathrm{OU,R}}(P) = -\frac{1}{2}P'' + XP' - nP$.

Now, if we consider now the mean value polynomial, that is $Q = \langle P \rangle = \int Q\,d\gamma$, one gets $Q'' - 2XQ' = -2nQ$. From which we see that, up to some constant, $Q(X/\sqrt{2})$ is an Hermite polynomial.

On the other hand, the same computation for the spherical case leads to

$$(\frac{1}{2} + X^2)Q'' - \frac{(n+6)(n-1)}{2}XQ' = -\frac{n(n+4)(n-1)}{2}Q.$$

In the same way that Hermite polynomials are the orthogonal polynomial family associated with Gaussian measure, one would expect some connection between those polynomials Q and the one-dimensional projection of the uniform measure on a sphere (in some dimension), i.e. Jacobi polynomials, but it does not seem to be the case.

Remark 6. One may wonder about what is hidden behind the fact that the spectrum of the random matrix is again a diffusion process. Indeed, it is certainly easier to see this by considerations on the associated semigroup P_t. From the form of the generator, it is clear that the law of the process (ξ_t) on the symmetric matrices in those three cases is invariant under $\phi_P : \xi \mapsto P^{-1}\xi P$, where P is any given invertible matrix. Then, if a function $M \mapsto f(M)$ on the space of symmetric matrices is invariant under those operations ϕ_P, such is $P_t(f)$. But those functions are nothing else than functions of the spectral measure. The semigroup acting on those functions describes a Markov process, which is the Dyson process, on the co-set space, which is the spectral measure of the matrix. This is a particular case of a general situation, when we have an action on some space E which commutes with the action of the generator L, see [4].

In the more complicated situations of the Clifford algebras described below, it would be more difficult to describe exactly for which such actions we do have an invariance for the generator, but the computation of the operator on the spectrum provides directly the image process.

7 Symmetric Operators on the Spectrum of Hermitian and Quaternionic Matrices

7.1 Hermitian Matrices

In this section, we extend the previous computations to Hermitian matrices. We mainly consider an Hermitian matrix on $\mathbb{C}^n$ as a real symmetric matrix on $\mathbb{R}^{2n}$. Indeed, considering a vector $Z = S + iT$ in $\mathbb{C}^n$, where S and T are the real and imaginary part of Z, an Hermitian matrix H may be seen as $M + iA$, where M is an $n \times n$ real symmetric and A is $n \times n$ real antisymmetric matrix. Then, writing H as a bloc matrix, we have $H = \begin{pmatrix} M & A \\ -A & M \end{pmatrix}$, which is a real symmetric matrix with special structure. Indeed, any real eigenspace for this matrix is at least two dimensional (and is exactly two-dimensional in the generic case), since if $Z = (S, T)$ is an eigenvector, so is $(-T, S)$, which corresponds to the eigenvector iZ.

Moreover, the determinant $P(X)$ of $H - X\mathrm{Id}$ may be written as $Q(X)^2$, where $Q(X)$ is a polynomial whose coefficients are polynomials in the entries of M and A. Actually, $Q(X)$ is the Pfaffian of the anti-symmetric $2n \times 2n$ matrix $\begin{pmatrix} A & M - X\mathrm{Id} \\ -M + X\mathrm{Id} & A \end{pmatrix}$. Therefore, if we consider the entries of M and A uniformly distributed under the Lebesgue measure, as we did for real symmetric ones, the spectrum of the matrix H is certainly not absolutely continuous with respect to the Lebesgue measure, and we are indeed more interested in the law of the roots of Q than in the law of the roots of P.

As before, we look at the Euclidean Laplace operator $L_{E,\mathbb{C}}$ acting on $H = M + iA$, with $M = (M_{ij})$ and $A = (A_{ij})$ and we encode the symmetries via the following formulae

$$\begin{cases} L_{E,\mathbb{C}}(M_{ij}) = L_{E,\mathbb{C}}(A_{ij}) = 0 \\ \Gamma_{E,\mathbb{C}}(M_{ij}, M_{kl}) = \frac{1}{2}(\delta_{ij}\delta_{kl} + \delta_{ik}\delta_{jl}) \\ \Gamma_{E,\mathbb{C}}(A_{ij}, A_{kl}) = \frac{1}{2}(\delta_{ij}\delta_{kl} - \delta_{ik}\delta_{jl}) \\ \Gamma_{E,\mathbb{C}}(M_{ij}, A_{kl}) = 0 \end{cases}$$

It is worth to observe that any power (and therefore the inverse when it exists) of an Hermitian matrix is again an Hermitian matrix, and one may perform the same computation as before on $P(x) = \det(H - X\mathrm{Id})$ (still considered as a $2n \times 2n$ matrix).

We obtain, with again $U(X) = (H - X\mathrm{Id})^{-1}$,

Proposition 7.

$$\Gamma_{E,\mathbb{C}}(\log P(X), \log P(Y)) = 2\mathrm{trace}(U(X)U(Y)) = \frac{2}{Y - X}\left(\frac{P'(X)}{P(X)} - \frac{P'(Y)}{P(Y)}\right).$$

$$L_{E,\mathbb{C}}(P(X)) = \frac{3}{2}\frac{P'(X)^2}{P(X)} - 2P''(X).$$

We do not give the details of the proof here since we shall give a more general result in the setting of Clifford algebras, of which this is just the simplest example.

But is worth to observe the following. Since $P(X) = \sum_i X^i a_i$, then $L_{E,\mathbb{C}}(P(X)) = \sum_i X^i L_{E,\mathbb{C}}(a_i)$. Therefore, $L_{E,\mathbb{C}}(P(X))$ has to be a polynomial, and then $\frac{P'(X)^2}{P(X)}$ is a polynomial in X. This implies in particular that all the roots of P have multiplicities at least 2, since every root of P is also a root of P'. In particular, the image measure of the Lebesgue measure is not absolutely continuous with respect to $d\mu_0$.

Observe furthermore that if we set $P = P_1^2$, then one gets from the change of variable formula

$$\Gamma_{\mathbb{E},\mathbb{C}}(\log P_1(X), \log P_1(Y)) = \frac{1}{Y-X}\left(\frac{P_1'(X)}{P_1(X)} - \frac{P_1'(Y)}{P_1(Y)}\right), \quad \mathrm{L}_{\mathbb{E},\mathbb{C}}(P_1) = -P_1''.$$

$$(12)$$

In particular, moving back to the Weyl chamber $\{\lambda_1 < \cdots < \lambda_n\}$, the invariant measure is, up to a constant, $\prod_{i<j}(\lambda_i - \lambda_j)^2 d\lambda_1 \cdots d\lambda_n$.

Are we able to deduce directly from the form of the generator that indeed $P(X)$ has almost surely double roots? We shall see that it is indeed the case. It may be seen at this level as a purely formal argument, since we know in advance that in this Hermitian case the roots are double. But later we shall face similar situations, where we do not know in advance the multiplicity of the roots, and we want to be able to deduce them from the generator. More precisely, we shall see that if a generator of the form given in Proposition 7 maps polynomials into polynomials, then those polynomials must have roots with multiplicity 2. This relies on Lemma 2 and Proposition 8.

From the form of the operator, one already sees that there are some algebraic relations between the coefficients a_i of the polynomial P. The following Lemma 2 is quite formal, and allows to devise the multiplicities of the roots of P from the generator. Then, Proposition 8 provides a proof that the multiplicities of the roots are indeed what is expected.

Lemma 2. *Suppose that a diffusion operator* L *on some set of analytic functions* $P(X) = \sum_i a_i X^i$ *in the variable* X *satisfies, for some constants* α, β, γ,

$$L(P) = \alpha P'' + \beta \frac{P'^2}{P}, \quad \Gamma(\log P(X), \log P(Y)) = \frac{\gamma}{Y-X}\left(\frac{P'(X)}{P(X)} - \frac{P'(Y)}{P(Y)}\right).$$

$$(13)$$

Let $a \in \mathbb{R}$, $a \neq 0$, *and set* $P = P_1^a$. *Then*

$$\Gamma(\log P_1(X), \log P_1(Y)) = \frac{\gamma/a}{Y-X}\left(\frac{P_1'(X)}{P_1(X)} - \frac{P_1'(Y)}{P_1(Y)}\right)$$

and

$$LP_1 = (\alpha + \gamma \frac{a-1}{a})P_1'' + (a(\alpha + \beta) + \gamma \frac{1-a}{a} - \alpha)\frac{P_1'^2}{P_1}.$$

In particular, if a satisfies

$$a^2(\alpha + \beta) - a(\alpha + \gamma) + \gamma = 0, \quad (14)$$

then,

$$L(P_1) = a(\alpha + \beta)P_1''.$$

Therefore, one may expect that, whenever L maps polynomials into polynomials, and precisely for those values of a solutions of Eq. (14), the roots of P have multiplicity a.

Proof. The formula for $\Gamma(\log P_1(X), \log P_1(Y))$ is immediate. The formula for $L(P)$ follows easily from the remark that

$$\frac{LP}{P} = \alpha \partial_X^2 \log P + (\alpha + \beta)(\partial_X \log P)^2, \ \Gamma(\log P, \log P) = -\gamma \partial_X^2 \log P.$$

Then,

$$a \frac{LP_1}{P_1} = \frac{LP}{P} + \frac{1-a}{a} \Gamma(\log P, \log P),$$

and this gives

$$\frac{LP_1}{P_1} = (\alpha + \gamma \frac{a-1}{a}) \frac{P_1''}{P_1} + (a(\alpha + \beta) + \gamma \frac{1-a}{a} - \alpha) \frac{P_1'^2}{P_1^2}.$$

In the next Proposition 8, we consider polynomials $P(X)$ with coefficients a_i which are polynomials in some variables (x_i) (in our case the entries of a matrix). Then, when writing $P(X) = \prod(X - \lambda_i)^{\alpha_i}$, with $\lambda_1 < \lambda_2 < \cdots < \lambda_k$, the multiplicities α_i may only change on some algebraic surface in the set of coefficients (x_i). Those algebraic surfaces having Lebesgue measure 0, and our operators L being local, we may as well (up to some localization procedure and outside a set of Lebesgue measure 0) consider them as constants.

Proposition 8. *Let L be a diffusion operator acting on a set of degree d monic polynomials, with values in the set of degree d polynomials, and satisfying Eq. (13). Then, every root of P has multiplicity α_1 or α_2, where α_i, $i = 1, 2$ are the roots of Eq. (14).*

In particular, Eq. (13) may only hold for polynomials whenever Eq. (14) has at least one integer solution.

In practise, Eq. (14) will have only one integer positive root, which will allow us to identify the multiplicity without any ambiguity. Moreover, in this situation, we may set $P = P_1^a$, where P_1 is a polynomial, and Lemma 2 applies within the set of polynomials.

Proof. Let us consider $\lambda_1 < \lambda_1 \cdots < \lambda_k$ the different roots of $P(X)$, and set

$$P(X) = \prod_1^k (X - \lambda_i)^{\alpha_i},$$

where $\alpha_i \geq 1$. Then

$$\Gamma(\log P(X), \log P(Y)) = \sum_{ij} \frac{\alpha_i \alpha_j}{(X - \lambda_i)(Y - \lambda_j)} \Gamma(\lambda_i, \lambda_j).$$

On the other hand, we know from Lemma 2 that

$$\Gamma(\log P(X), \log P(Y)) = \frac{\gamma}{Y - X} \left(\frac{P'(X)}{P(X)} - \frac{P'(Y)}{P(Y)} \right),$$

which translates into

$$\Gamma(\log P(X), \log P(Y)) = \frac{\gamma}{Y - X} \sum_i \frac{\alpha_i}{X - \lambda_i} - \frac{\alpha_i}{Y - \lambda_i}.$$

Identifying both expressions leads to

$$\Gamma(\lambda_i, \lambda_j) = \gamma \frac{\delta_{ij}}{\alpha_i}.$$

Also, on the one hand,

$$L(\log P) = \sum_i \frac{\alpha_i}{X - \lambda_i} L\lambda_i - \sum_i \frac{\alpha_i}{(X - \lambda_i)^2} \Gamma(\lambda_i, \lambda_i),$$

and on the other, from (13)

$$\begin{aligned} L(\log P) &= \frac{LP}{P} - \Gamma(\log P, \log P) \\ &= (\alpha + \gamma)\partial_X^2 \log P + (\alpha + \beta)(\partial_X \log P)^2 \\ &= -(\alpha + \gamma) \sum_i \frac{\alpha_i}{(X - \lambda_i)^2} + (\beta + \alpha)\left(\sum_i \frac{\alpha_i}{X - \lambda_i} \right)^2. \end{aligned}$$

Identifying the terms in $(X - \lambda_i)^{-2}$ leads to

$$\alpha_i^2(\beta + \alpha) - (\alpha + \gamma)\alpha_i + \gamma = 0.$$

In particular, applying Lemma 2 in the case of Hermitian matrices leads, with $\alpha = -2, \beta = 3/2$ and $\gamma = 2$ to $\alpha_i = 2$, and then, setting $P = P_1^2$, to (12) for $L_{\mathbb{E},\mathbb{C}}(P_1)$ and $\Gamma_{\mathbb{E},\mathbb{C}}(\log P_1(X), \log P_1(Y))$. This in turns shows that every root of P has multiplicity 2, and therefore that, up to some sign, P may be written P_1^2, where P_1 is a polynomial for which (12) holds. As a consequence, the image measure for the roots of P_1 has density $\mathrm{discr}(P_1)$ with respect to $d\mu_0$.

Remark 7. It is worth to observe that if P is a monic polynomial whose coefficients are polynomials in some variables $(x_1, \cdots, x_n)$, and if $P = P_1^a$, where P_1 is a polynomial, then P_1 is monic and the coefficients of P_1 are again polynomials in the variables $(x_1, \cdots, x_n)$. In the case of Hermitian matrices, this shows in particular that the determinant of a matrix of the form $\begin{pmatrix} M & A \\ -A & M \end{pmatrix}$, where M is symmetric and A is antisymmetric may be written as Q^2, where Q is a polynomial in the entries of M and A (indeed, it is nothing else up to a sign than the Pfaffian of $\begin{pmatrix} A & M \\ -M & A \end{pmatrix}$, but it is worth deriving it by pure probabilistic arguments).

7.2 Quaternionic Matrices

Here, we are given M, A^1, A^2, A^3 where M is symmetric and A^i are antisymmetric.
The associated real symmetric matrix is then

$$\mathcal{M} = \begin{pmatrix} M & A^1 & A^2 & A^3 \\ -A^1 & M & A^3 & -A^2 \\ -A^2 & -A^3 & M & A^1 \\ -A^3 & A^2 & -A^1 & M \end{pmatrix}$$

The eigenspaces are four-dimensional and the determinant of such a matrix may be written Q^4, where Q is a polynomial in the entries of the various matrices.

A real $4n \times 4n$ matrix $\mathcal{M}$ having this structure will be called a $\mathscr{H}$-symmetric matrix. It is quite immediate that if $\mathcal{M}$ is $\mathscr{H}$-symmetric, such is $\mathcal{M}^k$ for any $k \in \mathbb{N}$, and also such is $\mathcal{M}^{-1}$ on the set where $\mathcal{M}$ is invertible.

On the entries of M_{ij} and A^k_{ij}, we shall impose the metric coming from the euclidean metric on $\mathcal{M}$. This gives

$$\begin{cases} \Gamma_{\mathbb{E},\mathbb{H}}(M_{ij}, M_{kl}) = \frac{1}{2}(\delta_{ij}\delta_{kl} + \delta_{ik}\delta_{jl}) \\ \Gamma_{\mathbb{E},\mathbb{H}}(A^p_{ij}, A^q_{kl}) = \frac{\delta_{pq}}{2}((\delta_{ij}\delta_{kl} - \delta_{ik}\delta_{jl}) \\ \Gamma_{\mathbb{E},\mathbb{H}}(M_{ij}, A^p_{kl}) = 0 \end{cases}.$$

We also impose

$$\mathrm{L}_{\mathbb{E},\mathbb{H}} M_{ij} = \mathrm{L}_{\mathbb{E},\mathbb{H}} A^p_{ij} = 0.$$

Setting $U(X) = (\mathcal{M} - X\mathrm{Id})^{-1}$, $P(X) = \det(\mathcal{M} - X\mathrm{Id})$, one has

$$\Gamma_{\mathbb{E},\mathbb{H}}(P(X), P(Y)) = \frac{4}{Y - X}(P'(X)P(Y) - P'(Y)P(X)),$$

$$\frac{\mathrm{L}_{\mathbb{E},\mathbb{H}}(P)}{P} = \frac{\Gamma_{\mathbb{E},\mathbb{H}}(P, P)}{P^2} + \mathrm{trace}(U(X)^2) - \frac{1}{2}(\mathrm{trace}U(X))^2.$$

And in the end

$$L_{\mathbb{E},\mathbb{H}} P = \frac{9}{2} \frac{P'^2}{P} - 5P''.$$

Looking for which a, one has $P = P_1^a$, Eq. (14) on a leads to $a^2 - 2a - 8 = 0$, for which the unique positive solution is $a = 4$, leading to

$$L_{\mathbb{E},\mathbb{H}} P_1 = -2P_1'', \quad \Gamma_{\mathbb{E},\mathbb{H}}(\log P_1(X), \log P_1(Y)) = \frac{1}{Y - X}\left(\frac{P_1'(X)}{P_1(X)} - \frac{P_1'(Y)}{P_1(Y)}\right),$$

and Proposition 8 shows that all the roots of P have multiplicity 4, and that P_1 is indeed a polynomial.

In the end, one sees that the reversible measure for the image operator has density is discr$(P)^2$ with respect to the Riemann measure, or in other terms discr$(P)^{3/2}$ with respect to the measure $da_0 \cdots da_{n-1}$. Back to the Weyl chamber $\{\lambda_1 < \cdots < \lambda_n\}$, the invariant measure has density $\prod_{i<j}(\lambda_i - \lambda_i)^4$ with respect to the Lebesgue measure.

8 Symmetric Matrices on General Clifford Algebras

In this section, we extend the previous computations made for real complex and quaternionic matrices to more general sets of symmetric matrices. Indeed, the special structure of the real forms of the symmetric matrices described in the complex and quaternionic cases do only depend on the algebra structure of the complex and quaternionic fields, and not really on the field property. However, the associativity of the algebra plays a fundamental role in the computations, and things would be completely different for non associative algebras (such as the octonion algebra).

There are many natural algebras with dimension 2^p. Among them, let us mention exterior algebras, Cayley-Dickson algebras and Clifford algebras. Both Cayley-Dikson and Clifford are extensions of the real, complex and quaternionic cases, but in different ways. Cayley-Dikson algebras are non associative when $p \geq 3$, and do not seem to play a fundamental role beyond the important case of octonions ($p = 3$), whereas Clifford algebras and exterior algebras are central in many places in geometry and topology.

Since 2^p is the cardinal of $\mathscr{P}(\{1, \cdots, p\})$, it is natural to look for a basis ω_A for such algebras, where $A \subset E$, and $|E| = p$. If we denote by $A \triangle B$ the symmetric difference $A \cup B \setminus (A \cap B)$, in those three cases one has $\omega_A \omega_B = (A|B)\omega_{A \triangle B}$, where $(A|B)$ takes values in $\{-1, 0, 1\}$.

We define general Clifford algebras are the ones where the algebra is associative and $(A|B) \in \{-1, 1\}$. We shall impose $\omega_\emptyset$ to be the unitary element of the algebra.

The associativity imposes that, for any triple (A, B, C) of elements of $\mathscr{P}(E)$, one has

$$(A|\emptyset) = (\emptyset|A) = 1, \ (A|B\Delta C)(B|C) = (A|B)(A\Delta B|C).$$

It is worth to reduce to the case $E = \{1, \cdots, p\}$ (that is to decide that E is an ordered set), such that up to a change of sign in ω_A, one may always suppose that $\omega_A = \omega_{i_1} \cdots \omega_{i_k}$ when $A = (i_1, \cdots, i_k)$. Therefore, one sees that all the multiplication rules are just given by $e_i e_j = \pm e_j e_i$ and $e_i^2 = \pm e_i$. In which case, we are reduced to $(i|j) = 1$ if $i < j$ and $(A|B) = \prod_{i \in A, j \in B} (i|j)$, from which we get

$$(A\Delta B|C) = (A|C)(B|C), \ (A|B\Delta C) = (A|B)(A|C).$$

The general Clifford algebra is then just determined by the choice of the various signs in $(i|j)(j|i)$ for $i < j$ and $(i|i)$. But many such different choices may give rise to isomorphic algebra: for example, given any Clifford algebra and any choice $(A_1, \cdots, A_p)$ which generates $\mathscr{P}(E)$ by symmetric difference would produce a Clifford algebra isomorphic to the starting one with signs $(A_i|A_j)$ instead of $(i|j)$ (think for example of $A_i = \{1, \cdots, i\}$).

The Clifford algebra $Cl(E)$ is then $\{\sum_{A \subset E} x_A \omega_A, x_A \in \mathbb{R}\}$ that we endow with the standard Euclidean metric in $\mathbb{R}^{2^p}$ (that is $(\omega_A, A \in \mathscr{P}(E))$ form an orthonormal basis).

We now consider on $\mathbb{R}^n \otimes Cl(E)$ matrices $\sum_A M^A \omega_A$, where M^A are $n \times n$ matrices, acting on $\mathbb{R}^n \otimes Cl(E)$ by

$$\left(\sum_A M^A \omega_A\right)\left(\sum_B X^B \omega_B\right) = \sum_{A,B} M^A X^B (A|B) \omega_{A\Delta B} = \sum_{A,B} (A\Delta B|B) M^{A\Delta B} X^B \omega_A,$$

and we end up with bloc matrices $(M_{ij}^{A,B})$, where $M^{A,B} = (A\Delta B|B) M^{A\Delta B}$. Indeed, what we did is to associate to a matrix M with coefficient in the algebra $Cl(E)$ a matrix $\phi(M)$ with real coefficients, in a linear injective way. It turns out that, thanks to the associativity of the algebra $Cl(E)$, this is an algebra homomorphism, that is $\phi(MN) = \phi(M)\phi(N)$.

Endowing $\mathbb{R}^n \otimes Cl(E)$ with the associated Euclidean metric, we may therefore look at those matrices $\phi(M)$ which are symmetric. One sees that the requirement is that $(M^A)^t = (A|A)M^A$, and then the associated bloc matrix is $\mathscr{M} = ((A\Delta B|B)M^{A\Delta B})$. We shall call those symmetric matrices $Cl(E)$-symmetric matrices.

We now chose the Euclidean metric on those $Cl(E)$-symmetric matrices, and look at the associated Laplace operator. One then sets

$$\Gamma_{\mathrm{E},Cl}(M_{ij}^A, M_{kl}^B) = \frac{1}{2}\delta_{A,B}(\delta_{ik}\delta_{jl} + (A|A)\delta_{il}\delta_{jk}), \mathrm{L}(M_{ij}^A) = 0. \tag{15}$$

In terms of the associated stochastic processes, these formulae just say that the various entries of the matrices are independent Brownian motions, subject to the restrictions that the matrices must satisfy the symmetry relations which are imposed by the algebra structure of $Cl(E)$.

The aim is now to compute when possible the image of this Laplace operator on the spectrum of $\mathcal{M}$. We shall see that it strongly depends on the sign structure of the algebra $Cl(E)$. In the next section, we shall reduce our analysis to standard Clifford algebras, that is when $(i|i) = (i|j)(j|i) = -1$, for any $(i, j) \in E^2$. But is worth to describe first the computations in the general case. Indeed, as mentioned before, since many different sign structures lead to isomorphic algebras, the various quantities which will appear in the computations will to be invariant under those isomorphisms, and it is worth to identify them.

The first task is to observe that, if $\mathcal{M}$ is $Cl(E)$-symmetric, so is $\mathcal{M}^k$ for any k. This is a direct consequence of the algebra homomorphism, since if $\mathcal{M} = \phi(M)$, then $\mathcal{M}^k = \phi(M^k)$. Therefore, on the set where $\det(\mathcal{M}) \neq 0$, the inverse $\mathcal{M}^{-1}$ is also $Cl(E)$-symmetric, and the same is true for $U(X) = (\mathcal{M} - X\mathrm{Id})^{-1}$, when X is not in the spectrum of $\mathcal{M}$. Indeed, for $\|\mathcal{M}\|$ close to 0, $(\mathrm{Id} - \mathcal{M})^{-1} = \sum_k \mathcal{M}^k$, and consequently, for $X \neq 0$ and $\mathcal{M}$ small enough, then $U(X)$ is $Cl(E)$-symmetric. Since the property of being $Cl(E)$-symmetric is linear in the coefficients of $\mathcal{M}$, and since the coefficients of $U(X)$ are rational functions of the coefficients of $\mathcal{M}$, the $Cl(E)$-symmetry of $U(X)$ may be extended from small values of $\mathcal{M}$ to any $\mathcal{M}$ which is $Cl(E)$-symmetric.

Once this is observed, and still denoting $U(X) = (\mathcal{M} - X\mathrm{Id})^{-1}$, it may be written as a block matrix $((A \triangle B|B)U(X)_{ij}^{A \triangle B})$, where $U(X)^A = U(X)^{A,\emptyset}$ is such that $(U(X)^A)^t = (A|A)U(X)$. Then, the method used for real symmetric matrices may be extended to $Cl(E)$-symmetric matrices and we get

Proposition 9. *Let* $P(X) = \det(\mathcal{M} - X\mathrm{Id})$ *and* $U(X) = (\mathcal{M} - X\mathrm{Id})^{-1}$. *Then*

$$\Gamma_{\mathbb{E},Cl}(P(X), P(Y)) = \frac{2^p}{Y - X}(P'(X)P(Y) - P'(Y)P(X)),$$

and

$$\frac{\mathrm{L}_{\mathbb{E},Cl}(P)}{P} = \Gamma_{\mathbb{E},Cl}(\log P)$$

$$-\frac{1}{2}\Big(\sum_{A \subset E}(A|A)\Big)\mathrm{trace}\big(U(X)^2\big) - 2^{p-1}\sum_{C \subset E}(C|C)H(C)\big(\mathrm{trace}U(X)^C\big)^2,$$

where

$$H(C) = \sum_{A \subset E}(A|C)(C|A).$$

Moreover,

$$\text{trace}(U(X)^2) = \frac{P'^2}{P^2} - \frac{P''}{P}.$$

Proof. Let us start with the formula for Γ. If $U(X) = (U(X)^{A,B}$, where $U(X)^{A,B} = (A\Delta B|B)U(X)^{A\Delta B}$, using the change of variable formula and Eq. (1), one has

$$\Gamma_{\mathbb{E},Cl}(\log P(X), \log P(Y)) = \sum_{A,B,C,D,i,j,k,l} U(X)_{ji}^{B,A} U(Y)_{lk}^{D,C} \Gamma_{\mathbb{E},Cl}(\mathscr{M}_{ij}^{A,B}, \mathscr{M}_{kl}^{C,D}).$$

Now, since $\mathscr{M}^{A,B} = (A\Delta B|B)M^{A\Delta B}$ and $\mathscr{M}^{C,D} = (C\Delta D|D)M^{C\Delta D}$, and from (15), one gets

$$\Gamma_{\mathbb{E},Cl}(\mathscr{M}_{ij}^{A,B}, \mathscr{M}_{kl}^{C,D}) = I_{\{A\Delta B\Delta C\Delta D=\emptyset\}}(A\Delta B|B\Delta D)\frac{1}{2}(\delta_{ik}\delta_{jl} + (A\Delta B|A\Delta B)\delta_{il}\delta_{jk}).$$

On the other hand

$$\text{trace}U(X)U(Y) = \sum_{A,B,i,j} U(X)_{i,j}^{A,B} U(Y)_{ji}^{B,A} = \sum_{A,B}(A\Delta B|A\Delta B)\text{trace}U(X)^{A\Delta B} U(Y)^{A\Delta B}.$$

From this, we get

$$\Gamma_{\mathbb{E}}(\log P(X), \log P(Y)) = \sum_{A\Delta B\Delta C\Delta D=\emptyset} (A\Delta B|A\Delta B)\text{trace}\left(U(X)^{A\Delta B} U(Y)^{A\Delta B}\right)$$

$$= 2^p \sum_{A,B}(A\Delta B|A\Delta B)\text{trace}\left(U(X)^{A\Delta B} U(Y)^{A\Delta B}\right)$$

$$= 2^p \text{trace}U(X)U(Y).$$

If we denote by λ_i the eigenvalues of $\mathscr{M}$, then

$$\text{trace}U(X)U(Y) = \sum_i \frac{1}{(\lambda_i - X)(\lambda_i - Y)} = \frac{1}{Y - X} \sum_i \frac{1}{\lambda_i - Y} - \frac{1}{\lambda_i - X}$$

$$= \frac{1}{Y - X}\left(\frac{P'(X)}{P(X)} - \frac{P'(Y)}{P(Y)}\right).$$

For the formula for $L_{\mathbb{E},Cl}(P)$, we start with

$$L_{\mathbb{E},Cl}(\log P) = \sum_{A,B,i,j} U(X)_{ji}^{B,A} LM_{ij}^{A,B} - \sum_{A,B,C,D,i,j,k,l} U(X)_{jk}^{B,C} U(X)_{l,i}^{D,A} \Gamma(\mathscr{M}_{ij}^{A,B}, \mathscr{M}_{kl}^{C,D})$$

$$= -\frac{1}{2} \sum_{A,B,C,D} E(A,B,C,D)U(X)_{jk}^{B\Delta C} U(X)_{li}^{A\Delta D}(\delta_{ik}\delta_{jl} + (A\Delta B|A\Delta B)\delta_{il}\delta_{jk}),$$

where

$$E(A, B, C, D) = \mathbb{1}_{A\Delta B\Delta C\Delta D=\emptyset}(B\Delta C|C)(A\Delta D|A)(A\Delta B|B)(C\Delta D|D)$$

$$= \mathbb{1}_{A\Delta B\Delta C\Delta D=\emptyset}(A\Delta C|A\Delta C)$$

We obtain in the end

$$L_{\mathbb{E},Cl}(\log P) = -\frac{1}{2}\left[\sum_{A,B,C} (A\Delta C|A\Delta C)(B\Delta C|B\Delta C)\mathrm{trace}\left(U(X)^{B\Delta C}\right)^2 \right.$$

$$\left. +(B\Delta C|A\Delta C)(A\Delta B|B\Delta C)\left(\mathrm{trace}U(X)^{B\Delta C}\right)^2 \right]$$

$$= -\frac{1}{2}\left(\sum_A (A|A)\right)\mathrm{trace}\left(U(X)^2\right) - 2^{p-1}\sum_C (C|C)H(C)\left(\mathrm{trace}U(X)^C\right)^2.$$

If we are interested in images of the Gaussian measure, we consider the Ornstein-Uhlenbeck operator

$$L_{\mathbb{OU},Cl}(P) = L_{\mathbb{E},Cl}(P) - D(P)$$

where

$$D = \frac{1}{2}\Gamma_{\mathbb{E},Cl}(\|M\|^2, \cdot)$$

and

$$\|M\|^2 = \sum_{i,j,A,B} (M_{i,j}^{A,B})^2.$$

If one is interested in images of the uniform measure on the unit sphere, we consider instead the spherical operator

$$L_{\mathbb{S},Cl}\mathbb{S}(P) = L_{\mathbb{E},Cl}(P) - D^2(P) - (N-2)D(P), \tag{16}$$

where N is the dimension on the Euclidean space in which the sphere is embedded, that is

$$N = \frac{n2^p(n2^p + 1)}{2}.$$

Observing their action on the characteristic polynomial, we have

$$D(\log P) = \frac{1}{2} \sum_{i,j,k,l,A,B,C,D} 2M_{i,j}^{A,B} \partial_{M_{k,l}^{C,D}}(\log P)\Gamma_{\mathbb{E}}(\mathcal{M}_{ij}^{A,B}, \mathcal{M}_{kl}^{C,D})$$

$$= \frac{1}{2} \sum_{i,j,A\Delta B\Delta C\Delta D=\emptyset} M_{i,j}^{A,B}(A\Delta B|B\Delta D)(U(X)_{j,i}^{D,C} + (A\Delta B|A\Delta B)U(X)_{i,j}^{D,C})$$

$$= \sum_{A\Delta B\Delta C\Delta D=\emptyset} (A\Delta B|A\Delta B)\mathrm{trace}\left(M^{A\Delta B}U(X)^{A\Delta B}\right) = 2^p\mathrm{trace}\left(MU(X)\right)$$

$$= 2^p\left(n2^p - X\frac{P'}{P}\right)$$

Hence

$$D(P) = 2^p(n2^p P - XP') \tag{17}$$

The carré du champ operator is the same for the Ornstein-Uhlenbeck operator than for the Laplace operator, whereas for the sphere, the carré du champ operator acting on the characteristic polynomial P becomes

$$\Gamma_{\mathbb{S},Cl}(P(X), P(Y)) = \frac{2^p}{Y-X}\left(P'(X)P(Y) - P'(Y)P(X)\right)$$
$$-2^{2p}\left(n2^p P(X) - XP'(X)\right)\left(n2^p P(Y) - XP'(Y)\right)$$

From Proposition 9, one sees that the final expression depends on some specific factors for $Cl(E)$: the value of $\sum_A(A|A)$, and, for various $C \subset E$, the value of $H(C) = \sum_A(A|C)(C|A)$. We shall therefore restrict our attention to standard Clifford algebras, for which those computations may be explicitly done through some basic combinatorial arguments.

9   Symmetric Matrices on Standard Clifford Algebras

Recall that for standard Clifford algebras, and with $E = \{1, \cdots, n\}$, one has, for any pair $(i, j) \in E^2$, $(i|j) = \mathbb{1}_{i<j} - \mathbb{1}_{j\leq i}$.

From $(A|B) = \prod_{i\in A, j\in B}(i|j)$, this immediately leads to

$$(A|A) = (-1)^{|A|(|A|+1)/2}, \quad (A|B)(B|A) = (-1)^{|A||B|+|A\cap B|}. \tag{18}$$

Notice also that for any $i \in E = \{1, \cdots, p\}$, $(i|E) = (-1)^i$.

Proposition 10. *In a standard Clifford algebra with $|E| = p$, one has*

$$\sum_A(A|A) = \begin{cases} 2^{2m}(-1)^m & \text{if } p = 4m \\ 0 & \text{if } p = 4m+1 \\ 2^{2m+1}(-1)^{m+1} & \text{if } p = 4m+2 \\ 2^{2m+2}(-1)^{m+1} & \text{if } p = 4m+3 \end{cases}$$

Proof. From (18), one has

$$\sum_A (A|A) = \sum_k \binom{p}{k}(-1)^{k(k+1)/2} = \sum_k \binom{p}{2k}(-1)^k - \sum_k \binom{p}{2k+1}(-1)^k.$$

Comparing with

$$(1+i)^p = \sum_k \binom{2k}{p}(-1)^k + i\sum_k \binom{2k+1}{p}(-1)^k,$$

we see that $\sum_k \binom{2k}{p}(-1)^k$ is the real part of $(1+i)^p$, while $\sum_k \binom{2k+1}{p}(-1)^k$ is its imaginary part. But $1+i = \sqrt{2}e^{i\pi/4}$, and therefore, for $p = 4m$, $(1+i)^p = 2^{2m}(-1)^m$, for $p = 4m+1$, $(1+i)^p = (1+i)2^{2m}(-1)^m$, for $p = 4m+2$, $(1+i)^p = i2^{2m+1}(-1)^m$ and for $p = 4m+3$, $(1+i)^p = (1+i)2^{2m}(-1)^m$. It remains to collect the various cases.

The following will also be useful

Proposition 11. *For a standard Clifford algebra $Cl(E)$, and for $B, C \subset E$, let*

$$\begin{cases} S^e(B,C) = \sum_{A\subset C, |A|=2k}(A|B)(B|A), & S^e(B,\emptyset) = 1 \\ S^o(B,C) = \sum_{A\subset C, |A|=2k+1}(A|B)(B|A), & S^o(B,\emptyset) = 0 \end{cases}$$

Then,

$$\begin{cases} S^e(B,C) &= S^e(B\cap C, C), \\ S^o(B,C) &= (-1)^{|B\cap C^c|}S^o(B\cap C, C). \end{cases}$$

and, for $B \subset C$

$$\begin{cases} S^e(B,C) = S^o(B,C) = 0, & B \neq \emptyset, B \neq C \\ S^e(\emptyset,C) = S^o(\emptyset,C) = 2^{|C|-1}, & C \neq \emptyset \\ S^e(\emptyset,\emptyset) = 1, S^o(\emptyset,\emptyset) = 0, & \\ S^e(B,B) = -S^o(B,B) = 2^{|B|-1} & |B| = 2k, B \neq \emptyset \\ S^e(B,B) = S^o(B,B) = 2^{|B|-1} & |B| = 2k+1 \end{cases}$$

Proof. For the first point, we decompose $B = (B\cap C)\cup(B\cap C^c) = B_1 \cup B_2$. Then, if $B_1 \cap B_2 = \emptyset$

$$(A|B_1 \cup B_2)(B_1 \cup B_2|A) = (A|B_1)(B_1|A).(A|B_2)(B_2|A),$$

and, if $A \cap B_2 = \emptyset$, then $(A|B_2)(B_2|A) = (-1)^{|A||B_2|}$.

It remains to study $S^e(B,C)$ and $S^o(B,C)$ for $B \subset C$. Replacing E by C, we are therefore bound to study the same quantity for a standard Clifford algebra $\mathscr{L}(C)$.

Let us then fix C et $B \subset C$. For $C \neq \emptyset$ $S^e(\emptyset, C) = S^o(\emptyset, C) = 2^{|C|-1}$. When $B \neq \emptyset$ chose some point $i \in B$, and cut $\mathscr{P}(C)$ into $\{A \subset C, x \in A\}$ and $\{A \subset C \setminus \{x\}\}$.

Summing on $\mathscr{P}(A)$, we get

$$\begin{cases} S^e(B,C) & = S^e(B \setminus x, C \setminus x) + (-1)^{|B|} S^o(B \setminus x, C \setminus x), \\ S^o(B) & = (-1)^{|B|-1} S^e(B \setminus x, C \setminus x) - S^o(B \setminus x, C \setminus x). \end{cases}$$

In another way, setting $U(B,C) = \begin{pmatrix} S^e(B,C) \\ S^o(B,C) \end{pmatrix}$, U_k if $|B| = k$, we get

$$U(B,C) = M_{\epsilon_k} U\big(B \setminus \{x\}, C \setminus \{x\}\big),$$

where $\epsilon_k = (-1)^{|B|}$ and

$$M_1 = \begin{pmatrix} 1 & 1 \\ -1 & -1 \end{pmatrix} M_{-1} = \begin{pmatrix} 1 & -1 \\ 1 & -1 \end{pmatrix}. \tag{19}$$

Setting $S = \begin{pmatrix} 0 & 1 \\ 1 & 0 \end{pmatrix}$, which satisfies $S^2 = 1$, one has

$$M_1 M_{-1} = 2(1 - S), M_{-1} M_1 = 2(1 + S), M_1 M_{-1} M_1 = 2M_1,$$
$$M_{-1} M_1 M_{-1} = 2M_{-1}, (1 + S)M_{-1} = 2M_{-1}$$

from which

$$(M_1 M_{-1})^k = 2^{2k-1}(1 - S), (M_{-1} M_1)^k = 2^{2k-1}(1 + S),$$

and also

$$M_{-1}^2 = 0, M_1^2 = 0, M_1 M_{-1} M_1 = 2M_1, M_{-1} M_1 M_{-1} = 2M_{-1}.$$

In the end, we get

$$\begin{cases} |B| = 2k, & U(B,C) = 2^{2k-1}(1 + S)U(\emptyset, C \setminus B) \\ |B| = 2k + 1, & U(B,C) = 2^{2k+1}(1 - S)M_{-1}U(\emptyset, C \setminus B) \end{cases}.$$

Il remains to collect all the possible cases.

From Proposition 11, in a standard Clifford algebra $Cl(E)$ with $|E| = n$, one sees that for any $B \subset E$, with $B \neq \emptyset, E$, one has

$$\sum_A (A|B)(B|A) = S^e(B, E) + S^o(B, E) = 0.$$

Moreover, $S^e(E, E) + S^o(E, E) = 0$ when $|E| = 2k$. Therefore, $H(C) = 0$ unless $C = \emptyset$ or $C = E$, and $H(E) = 0$ when $|E| = 2k$. This leads to

Proposition 12. *When $p = |E|$ is even, if $\mathscr{M}$ is $Cl(E)$-symmetric, with $P(X) = \det(M - XI)$ and $U(X) = (M - X\mathrm{Id})^{-1}$,*

$$\Gamma_{\mathbb{E},Cl}\big(P(X), P(Y)\big) = \frac{2^p}{Y - X}\big(P'(X)P(Y) - P'(Y)P(X)\big),$$

and

$$\begin{cases} \frac{L_{\mathbb{E},Cl}(P)}{P} = \Gamma_{\mathbb{E},Cl}(\log P) - 2^{2m}(-1)^{m+1}(\mathrm{trace}U(X)^2) - \frac{1}{2}(\mathrm{trace}U(X))^2 & when \ |E| = 4m + 2 \\ \frac{L_{\mathbb{E},Cl}(P)}{P} = \Gamma_{\mathbb{E},Cl}(\log P) - 2^{2m-1}(-1)^{m}(\mathrm{trace}U(X)^2) - \frac{1}{2}(\mathrm{trace}U(X))^2 & when \ |E| = 4m \end{cases}$$

In particular,

$$\frac{L_{\mathbb{E},Cl}(P)}{P} = \begin{cases} (2^p + 2^{2m}(-1)^m)\big(\frac{P'^2}{P^2} - \frac{P''}{P}\big) - \frac{1}{2}\frac{P'^2}{P^2} & when \ p = |E| = 4m + 2 \\ (2^p + 2^{2m-1}(-1)^{m+1})\big(\frac{P'^2}{P^2} - \frac{P''}{P}\big) - \frac{1}{2}\frac{P'^2}{P^2} & when \ p = |E| = 4m \end{cases}$$

As a consequence, one has $P(X) = Q(X)^a$, where Q is a polynomial, where

$$\begin{cases} a = 2^{4q}, & when \ p = 8q \\ a = 2^{4q+2}, & when \ p = 8q + 2 \\ a = 2^{4q+3}, & when \ p = 8q + 4 \\ a = 2^{4q+3}, & when \ p = 8q + 6 \end{cases}$$

Moreover, in those case, for $\hat{L} = \frac{a}{2^p}L_{\mathbb{E},Cl}$, Q satisfies

$$\hat{\Gamma}(Q(X), Q(Y)) = \frac{1}{Y - X}\big(Q'(X)Q(Y) - Q'(Y)Q(X)\big)$$

and

$$\begin{cases} \hat{L}(Q) = -\frac{1}{2}Q'', \ p = 8q, \ p = 8q + 6 & (real \ case) \\ \hat{L}(Q) = -2Q'', \ p = 8q + 2, \ p = 8q + 4 & (quaternionic \ case) \end{cases}$$

Proof. Using Proposition 9, the only term in the formula for $L_{E,Cl}(P)$ which is not immediate to identify is $\text{trace}U(X)^{\emptyset}$. But $\text{trace}U(X) = 2^p\text{trace}U(X)^{\emptyset}$, since only $U(X)^{\emptyset}$ appear in the diagonal blocs of $U(X)$. Then, everything boils down to the computation of $\sum_A (A|A)$ given in Proposition 10. Then, the identification of a comes from Eq. (14) in Lemma 2. It is worth to observe that those equations have indeed integer roots in every case. Moreover, Proposition 8 allows to assert that effectively, $P = Q^a$, and the rest is given again in Lemma 2.

It remains to deal with the case where $p = |E|$ is odd. Then, the term $(\text{trace}U(X)^E)^2$ appears in the formula for $L_{E,Cl}(P)$. But, when $(E|E) = -1$, then $U(X)^E$ is antisymmetric and $\text{trace}U(X)^E = 0$. Since $(E|E) = (-1)^{|E|(|E|+1)}$, this happens as soon as $|E| = 4m + 1$.

This leads to

Proposition 13. *Suppose that $p = |E| = 4m + 1$. Then,*

$$\Gamma_{E,Cl}(P(X), P(Y)) = \frac{2^p}{Y - X}(P'(X)P(Y) - P'(Y)P(X)),$$

$$\frac{L_{E,Cl}(P)}{P} = 2^p\left(\frac{P'^2}{P^2} - \frac{P''}{P}\right) - \frac{1}{2}\frac{P'^2}{P^2}.$$

Setting $a = 2^{2m+1}$ and $P = Q^a$, and for $\hat{L} = a2^{-p}L$, one has $\hat{L}(Q) = -Q''$. Then, Q is a polynomial and the model corresponds to the complex case.

It remains to deal with the case $p = 4m + 3$, which turns out to be more delicate. Indeed, in those cases, $Cl(E)$ is no longer simple, and splits into the direct sum of two ideals. From Propositions 9 and 11, we see that the set E plays a special role in the analysis of $L_{E,Cl}(P(X))$.

We already saw that in this situation, $(E|E) = 1$, and for any $A \subset E$, $(A|E)(E|A) = (-1)^{|A|(|E|+1)} = 1$, so that ω_E commutes to every element in the algebra and satisfies $\omega_E^2 = 1$. Then, one may decompose the algebra $Cl(E)$ into the sum of the two ideals $Cl(E)_+ = \{x \in Cl(E), \omega_E x = x\}$ and $Cl(E)_- = \{x \in Cl(E), \omega_E x = -x\}$. Symmetric matrices will also split into the direct sum of two symmetric matrices, and therefore the characteristic polynomial will be the product of characteristic polynomials.

We are therefore bound to consider separately the action on $Cl(E)_+$ and $Cl(E)_-$. We concentrate on the first one. First observe that in this situation, for any $A \subset E$, $(A|E) = (E|A) = (A\Delta E|E) = (E|A\Delta E)$. From this, it is easy to see that

$$Cl(E)_+ = \{X = \sum_A \lambda_A(\omega_A + (A|E)\omega_{A\Delta E})\}.$$

The action of the matrix $\sum_A M^A\omega_A$ on $Cl(E)_+$ is the same as $\sum_A (A|E)M^{A\Delta E}\omega_{A\Delta E}$, and therefore one may concentrate on matrices $\sum_A M^A\omega_A$ such that $M^{A\Delta E} = (A|E)M^A$. This condition is clearly compatible with $(M^A)^t = (A|A)M^A$. We therefore chose

$$\Gamma_{\mathbb{E},Cl} M_{ij}^A, M_{kl}^B) = \frac{1}{2}(\mathbb{1}_{A\triangle B=\emptyset} + (A|E)\mathbb{1}_{A\triangle B=E})(\delta_{ik}\delta_{jl} + (A|A)\delta_{il}\delta_{jk}).$$

We may start the computation again, but it is simpler to observe that, setting $\sigma_A = \frac{1}{2}(\omega_A + (A|E)\omega_{A\triangle E})$, one gets $\sigma_A\sigma_B = (A|B)\sigma_{A\triangle B}$, and therefore the family (σ_A) generates a standard Clifford algebra with size $|E| - 1$. Then, we boil down to a standard Clifford algebra with size $4m + 2$, and we see that we obtain the quaternionic case when $p = 8q + 3$ and the real one when $p = 8q + 7$.

If one wants to describe the law of $P(X)$ in the case $p = 4m + 3$, then we write $P = P_1 P_2$, where P_1 and P_2 behave independently as the previous ones.

We have then described all the laws of the spectra for symmetric matrices on standard Clifford algebras.

We thus recover Bott periodicity: in the following table, we give the algebra structure of $Cl(p)$, together with the dimension d of the irreducible spaces in the third column, the multiplicity α of the roots of the characteristic polynomial in the fourth, computed from the generator. In the last column, we indicate the parameter β for which the law of the simple roots $(\lambda_1 < \cdots < \lambda_d)$ have density $\prod |\lambda_i - \lambda_j|^\beta$ with respect to the Lebesgue measure $d\lambda_1 \cdots d\lambda_d$

$|E|$	Structure	d	α	β
$Cl(1)$	$\mathbb{C}$	2	2	2
$Cl(2)$	$\mathbb{H}$	4	4	4
$Cl(3)$	$\mathbb{H} \oplus \mathbb{H}$	4	4	4
$Cl(4)$	$\mathbb{H}[2]$	8	8	4
$Cl(5)$	$\mathbb{C}[4]$	8	8	2
$Cl(6)$	$\mathbb{R}[8]$	8	8	1
$Cl(7)$	$\mathbb{R}[8] \oplus \mathbb{R}[8]$	8	8	1
$Cl(8)$	$\mathbb{R}[16]$	16	16	1

then we tensorize by $\mathbb{R}[16]$ through Bott's periodicity: $Cl(p + 8) = \mathbb{R}[16] \otimes Cl(p)$. (Here, $K[n]$ denotes the irreducible algebra of square $n \times n$ matrices with coefficients in the field K.) We may then observe that the multiplicity of the roots corresponds as expected to the dimension of the irreducible spaces, and that the parameter β corresponds to the structure algebra: when the irreducible components are $K[n]$, then $\beta = 1, 2, 4$ corresponding to the case where $K = \mathbb{R}, \mathbb{C}$ or $\mathbb{H}$.

Remark 8. Considering the O-U operator $L_{\text{OU},Cl}$ described in the previous section, one gets here

$$L_{\text{OU}}(P) = -CP'' + (C - \frac{1}{2})\frac{P'^2}{P} - n2^{2p}P + 2^p XP' \qquad (20)$$

and analogously to Lemma 2, we get

$$a\frac{L_{\mathrm{OU},Cl}(P_1)}{P_1} = \frac{L_{\mathrm{OU},Cl}(P)}{P} + \frac{1-a}{a}\Gamma_{\mathrm{E},Cl}(\log P, \log P)$$

which leads to

$$L_{\mathrm{OU},Cl}(P_1) = -(C + 2^p\frac{(1-a)}{a})P_1'' + ((C + 2^p\frac{(1-a)}{a} - \frac{1}{2}a)\frac{P_1'^2}{P_1} - n\frac{2^{2p}}{a}P_1 + 2^p XP_1' \tag{21}$$

where $P = P_1^a$ and the constant C differs according to n (see Propositions 12, 13). Then, choosing a as before, we can boil down to the following relation:

$$L_{\mathrm{OU},Cl}(P_1) = -\frac{a}{2}P_1'' - n\frac{2^{2p}}{a}P_1 + 2^p XP_1' \tag{22}$$

References

1. G.W. Anderson, A. Guionnet, O. Zeitouni, An introduction to random matrices, in *Cambridge Studies in Advanced Mathematics*, vol. 118 (Cambridge University Press, Cambridge, 2010)
2. M.F. Atiyah, R. Bott, A. Shapiro, Clifford modules. Topology **3**, 3–38 (1964)
3. M.F. Atiyah, I.M. Singer, Index theory for skew-adjoint fredholm operators. Inst. Hautes Études Sci. Publ. Math. **37**, 5–26 (1969)
4. D. Bakry, I. Gentil, M. Ledoux, *Analysis and Geometry of Markov Diffusion Operators*. Grundlehren der mathematischen Wissenschaften, vol. 348 (Springer, Berlin, 2013)
5. F. Chapon, M. Defosseux, Quantum random walks and minors of hermitian brownian motion. Can. J. Math. **64**(4), 805–821 (2012)
6. Y. Doumerc, Matrices aléatoires, processus stochastiques et groupes de réflexions. Ph.D. thesis, Université Toulouse 3, 2005
7. F.J. Dyson, A brownian-motion model for the eigenvalues of a random matrix. J. Math. Phys. **3**, 1191–1198 (1962)
8. L. Erdős, S. Péché, J.A. Ramírez, B. Schlein, H.-T. Yau, Bulk universality for wigner matrices. Commun. Pure Appl. Math. **63**, 895–925 (2010)
9. L. Erdős, J. Ramírez, B. Schlein, T. Tao, V. Vu, H.-T. Yau, Bulk universality for wigner hermitian matrices with subexponential decay. Math. Res. Lett. **17**, 667–674 (2010)
10. L. Erdős, A. Knowles, H.-T. Yau, J. Yin, The local semicircle law for a general class of random matrices. Electron. J. Probab. **18**, 1–58 (2013)
11. P.J. Forrester, *Log-gases and random matrices*. London Mathematical Society Monographs Series, vol. 34 (Princeton University Press, Princeton, 2010)
12. A. Gamburd, Some applications of symmetric functions theory in random matrix theory, in *Ranks of Elliptic Curves and Random Matrix Theory*, ed. by J.B. Conrey, D.W. Farmer, F. Mezzadri, N.C. Snaith. London Mathematical Society, Lecture Notes Series, vol. 341 (Cambridge University Press, Cambridge, 2007), pp. 143–169
13. R. Jagannathan, On generalized clifford algebras and their physical applications, in *The legacy of Alladi Ramakrishnan in the Mathematical Sciences* (Springer, New York, 2010), pp. 465–489
14. M. Karoubi, Algèbres de clifford et K-théorie. Ann. Sci. École Norm. Sup. (4) **1**, 161–270 (1968)
15. M. Ledoux, Differential operators and spectral distributions of invariant ensembles from the classical orthogonal polynomials. The continuous case. Electron. J. Probab. **9**, 177–208 (2004)

16. M. Ledoux, Differential operators and spectral distributions of invariant ensembles from the classical orthogonal polynomials. the discrete case. Electron. J. Probab. **10**, 1116–1146 (2005)

17. N. Lehmann, H.-J. Sommers, Eigenvalue statistics of random real matrices. Phys. Rev. Lett. **67**(8), 941–944 (1991)

18. P. Lounesto, *Clifford Algebras and Spinors*. London Mathematical Society Lecture Note Series, 2nd edn., vol. 286 (Cambridge University Press, Cambridge, 2001)

19. L.A. Marčenko, V.A. Pastur, Distribution of eigenvalues in certain sets of random matrices. Matematicheskiĭ Sbornik (N.S.) **72**(114), 507–536 (1967)

20. M.L. Mehta, *Random Matrices*. Pure and Applied Mathematics (Amsterdam), vol. 142 (Elsevier/Academic, Amsterdam, 2004)

21. A. Soshnikov, Universality at the edge of the spectrum in wigner random matrices. Commun. Math. Phys. **207**, 697–733 (1999)

22. E.P. Wigner, On the distribution of the roots of certain symmetric matrices. Ann. Math. **67**, 325–327 (1958)

23. M.R. Zirnbauer, Symmetry classes (2010). arXiv:1001.0722

24. M.R. Zirnbauer, Symmetry classes in random matrix theory (2004). arXiv:0404058

Gaussian Free Field on Hyperbolic Lattices

Itai Benjamini

Abstract It is shown that the maximum of the height of the Gaussian free field in a ball of a two dimensional hyperbolic lattice, grows linearly with the radius, while only as a square root of the radius on higher dimensional hyperbolic lattices.

1 Introduction

Localization and delocalization of random interfaces is a well studied subject, with very deep results and still many fundamental open problems. These are usually modeled by random functions, via Gibbs measures with respect to a given Hamiltonian, over Euclidean lattices (see e.g. [9]). The Gaussian free field (GFF) over Euclidean lattices is a canonical random surface and is most amenable to analysis. The typical height difference of the GFF on a n^d lattice tori is order $\log^{1/2} n$, in two dimensions, and is tight in higher dimensions.

In the this note we consider the GFF on lattices in real hyperbolic spaces, a transition for the order of the maximal height difference occurs between two and higher dimensions, as in Euclidean lattices, but the magnitude is different due both to hyperbolicity and to the exponential volume growth. For background, on hyperbolic geometry see [3] and on the Gaussian free field see [6]. We will assume knowledge of basic electric networks theory see [4, 7].

The *Gaussian free field* on a graph can be **defined** as follows.

Let $G = (V, E)$ be an oriented connected graph. For each directed edge $e = (u, v)$ let $\xi_e = \xi_{uv}$ be a standard normal random variable. We assume that $\xi_{uv} = -\xi_{vu}$ and otherwise ξ_e are independent. Condition on the event

$$\{\sum_{e \in \gamma} \xi_e = 0; \text{ for all closed paths } \gamma\}. \tag{1}$$

I. Benjamini (✉)
Department of Mathematics and Computer Science, Weizmann Institute of Science,
Rehovot 76100, Israel
e-mail: itai.benjamini@weizmann.ac.il

© Springer International Publishing Switzerland 2014   39
B. Klartag, E. Milman (eds.), *Geometric Aspects of Functional Analysis*,
Lecture Notes in Mathematics 2116, DOI 10.1007/978-3-319-09477-9_2

Pick a base vertex v_0 and define the *height* function $h : V \to R$ by the following equation

$$h(v_0) = 0; \text{for } v \in V \; h(v) = \sum_{\gamma_v} \xi_e,$$

where γ_v is any path from v_0 to v in the graph. Note that the height is well defined.

Definition. Two metric spaces (X, d_X) and (Y, d_Y) are said to be *quasi isometric* if there exists some constant $K > 0$ and a map $\phi : X \to Y$ such that for any two points x, x' in X

$$K^{-1} d_X(x, x') - K \leq d_Y(\phi(x), \phi(x')) \leq K d_X(x, x') + K$$

and

$$\forall y \in Y, \exists x \in X, d_Y(y, \phi(x)) \leq K.$$

We are interested in the behavior of the maximum of the height function on graphs which are quasi isometric to real hyperbolic spaces, denoted $\mathbb{H}^d$, e.g. co-compact hyperbolic lattices. We may assume that there are no double edges in all the graphs considered.

Theorem 1. *Let G be a graph quasi isometric to $\mathbb{H}^d$, $d > 1$, let $M_n = \max_{d(x, v_0) \leq n} \{|h(x)|\}$ then*

(i) $EM_n^2 = \Theta(n^2)$ *for $d = 2$.*
(ii) $EM_n^2 = \Theta(n)$ *for $d > 2$.*

The theorem is related to the *free uniform spanning tree* transition from a tree on lattices in $\mathbb{H}^2$ to a forest on lattices in $\mathbb{H}^d$, $d > 2$, see [2].

Following the same lines of proof the upper bound attained for dimension 3 and above can also be proven for the following graphs. Lattices in $\mathbb{H}^2 \times \mathbb{R}$ and also any uniformly transient exponential growing graph, admitting no non constant harmonic Dirichlet functions.

It is of interest to understand random graph homomorphisms or random Lipschitz functions (see [8]) on hyperbolic lattices. For lattices in dimensions at least three, the GFF maxima is achieved due to a large difference along a single edge. *Maybe random graph homomorphisms has only polylogarithmic maximum, since the nearest neighbourhood difference is 1?*

2 Proof

First note the following two useful theorems.

Theorem 2 ([6], p. 137). *For the GFF model described above, the height difference $h(u) - h(v) = \xi_{uv}$ between two vertices in G has a distribution $N(0, \sigma_{uv}^2)$, where the* *variance σ_{uv}^2 equals the effective resistance between u and v if G is regarded as an electrical network with unit resistance along the edges.*

Moreover, $\tilde{\xi}_{uv}$ can be expressed as a linear combination $\phi_e \xi_e$, where ϕ_e is a component along e of a unit electric current flowing from u to v through the network.

We also need the following theorem due to Fernique [5]

Theorem 3. *Let X, Y be two n-dimensional Gaussian random vectors such that for every i, j*

$$E|Y_i - Y_j|^2 \leq E|X_i - X_j|^2,$$

then for every non-negative convex increasing function f on $\mathbb{R}_+$,

$$Ef(\max_{i,j} |Y_i - Y_j|) \leq Ef(\max_{i,j} |X_i - X_j|).$$

View the graph as an electrical network by letting each edge be a 1 Ohm conductors. Denote by $\mathcal{R}(u, v)$ the electric resistance between v and u.

Lemma 1. *Let G be a graph quasi isometric to hyperbolic space of dimension d. Then there exist $c, R > 0$ such that for all $u, v \in V$*

(i) $\mathcal{R}(u, v) \geq c\, d(u, v)$ for $d = 2$.
(ii) $\mathcal{R}(u, v) \leq R$ for $d > 2$.

Proof of Lemma. (i) Model the hyperbolic plane with a metric on the half plane,

$$ds^2 = \frac{dx^2 + dy^2}{y^2}, \quad y > 0.$$

A quasi isometric approximation of this model is shown in Fig. 1. Consider the graph which is dual to the combinatorial approximation and denote it by G. This

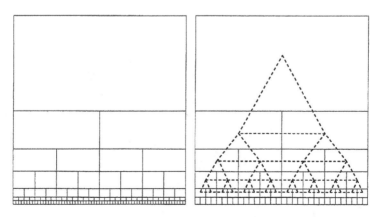

Fig. 1 Combinatorial approximation of the half-space model of the hyperbolic plane and its dual

graph is quasi isometric to $\mathbb{H}^2$. Any other graph quasi isometric to $\mathbb{H}^2$ will be also quasi isometric to G and so if resistance grows proportionally to distance in G then it will be true for any other such graph. Here we use the *Dirichlet principle* (see e.g. [4]) for the effective conductance

$$\mathcal{C}(u,v) = \frac{1}{\mathcal{R}(u,v)} = \min(\sum_{e \in E} C(e)(dF(e))^2),$$

where the minimum is over functions, F, such that $F(u) = 0, F(v) = 1$. Note that in our case $C(e) \equiv 1$ (no double edges) and for an edge $e = \{u,v\}$, $dF(e) = F(u) - F(v)$.

Choosing F, Dirichlet principle gives an upper bound for the effective conductance and, consequently, a lower bound for resistance

$$\mathcal{R}(u,v) \geq (\sum_{e \in E}(dF(e))^2)^{-1}.$$

Since G is quasi isometric to the hyperbolic plane which is 2-transitive (for any two pairs of point with same distance, there is an isometry mapping one pair onto the other). Effective resistance is a function of the distance, up to a multiplicative constant. Note that quasi isometry changes effective resistance by a multiplicative constant, see e.g. [4]. Thus, it is enough to bound the resistance between the top vertex in the graph in Fig. 2 (we omit horizontal edges from the schematic figure). Denoted u and another vertex distance n away on the left boundary of the triangle, denoted v. To do that, pick F as follows.

Set $F(u) = 0, F(v) = 1$, interpolate linearly along the left boundary; on the right boundary of the triangle let F equal 0. Next, let us denote by $G(k)$ the components shown in Fig. 2. Each of $G(k)$ is a binary tree with added edges. For every $k = 0, \ldots, n$ let F equals k/n on the right boundary of $G(k)$. Now interpolate linearly on each horizontal line.

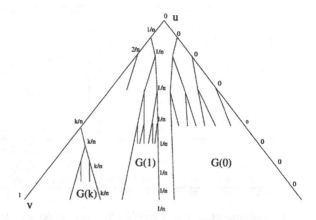

Fig. 2 The function F

Let us compute $\sum (dF(e))^2$ for horizontal edges in $G(k)$, this quantity does not depend on k because only the differences matter. On the line with the distance i from the root of $G(k)$ we have 2^i edges, including the edge connecting $G(k)$ and $G(k+1)$. The difference of F on the ends of this line is $1/n$, therefore, for each edge on the line the difference between adjacent vertices is $1/(n2^i)$. The summation on all horizontal lines gives for any $G(k)$:

$$\sum_{e \in G(k), e \text{ is a horizontal edge}} (df(e))^2 = \sum_{i=1}^{\infty} \frac{2^i}{(n2^i)^2} = \frac{1}{n^2},$$

It is not hard to check that the sum restricted to the vertical edges is of the same order (actually the half of the first sum).

Summing all the $G(k)$'s, the whole sum will be of order n^{-1}.

Finally we get that $\mathcal{C}(u,v) < c/n$, hence $\mathcal{R}(u,v) > n/c, c > 0$. So (i) follows.

(ii) In [2] it is shown that for hyperbolic lattices of dimension 3 and higher, wired limit current and free limit current (see [2] for definitions) coincide. Wired resistance to infinity in transient transitive graphs is finite and uniformly bounded (since every hyperbolic lattice is transient) and this implies that the free or effective resistance is uniformly bounded. □

Proof of Theorem 1. (i) Assume G is quasi isometric to $\mathbb{H}^2$. G contains as a subgraph a stretched binary tree. That is, a binary tree, denoted by T where each edge is replaced by a path with K edges, for some $K < \infty$. For some $\alpha < \infty$ we have,

$$d_G(u,v) \leq d_T(u,v) \leq \alpha d_G(u,v) \quad \text{for all } u, v \in T, \tag{2}$$

This can be shown using the combinatorial model above, but also follows from a general result in [1].

By Theorem 2 we know that the variance of $h(u) - h(v)$ equals the effective resistance between these two points, $(\mathcal{R}(u,v))$. On the other hand by Lemma 1, the resistances on the graph G are proportional to the distance.

So from Eq. (2) we can conclude that

$$E|h(u) - h(v)|^2 \geq cd_G(u,v) \geq c\alpha d_T(u,v).$$

Obviously

$$E(\max_{u,v \in G} |h(u) - h(v)|)^2 \geq E(\max_{u,v \in T} |h(u) - h(v)|)^2.$$

Thus, it is enough to look at the field on the tree T. Place on the edges of a binary tree T_2 independent Gaussian random variables with zero expectation and with the variance $cK\alpha$.

Denote by h the height function on T as well. By Fernique (Theorem 3),

$$E(\max_{u,v\in T} |h(u) - h(v)|)^2 \geq E(\max_{u,v\in T_2} |h(u) - h(v)|)^2.$$

The following is well known. Recursively, it is easy to see that the maximum of the Gaussian free field on the binary tree of depth n is of order n,

Let $M'_n = \max_{d(u,0)\leq n} \{h(u); u \in T_2\}$.

Then $M'_n \geq M'_{n-1} + \max(\xi', \xi'')$ where ξ', ξ'' are the variables assigned to the edges emanating from the vertex at which M_{n-1} is attained. So $M'_n \geq \sum_1^n \eta_i$, $\eta_i = \max(\xi'_i, \xi''_i)$.

Recall that

$$\max_x |f(x)| \leq \max_{x,y} |f(x) - f(y)| \leq \max_x 2|f(x)|,$$

provided that there exists x_0 such that $f(x_0) = 0$. The upper bound follows from the fact that removing edges can only increase the effective resistance.

This ends the proof of the first part.

(ii) From the lemma, the resistance between any two points is uniformly bounded in this case.

The upper bound follows by comparing to an exponential number of independent standard normals, via Fernique's theorem.

For the lower bound, define a set $W \subset V$ such that for every $w_1, w_2 \in W$ we have $\mathcal{R}(w_1, w_2) > 2c$ for some $c > 0$, in the case of a transitive graph we take $W = V$. By Fernique, the distribution of the maximum of the height function dominates the distribution of the maximum in the process $\{\eta_w, w \in W\}$, where all random variables are independent Gaussians with all with variance c. Since the graph is quasi isometric to a transitive graph, the number of vertices in W grows exponentially; the estimate follows.   □

In relation to the fact that the electric resistance in planar hyperbolic lattices is proportional to the distance, *we ask*: Assume G is a vertex transitive graph, admitting non constant harmonic Dirichlet function. Is the resistance proportional to distance along a sequence of pairs of vertices with distance growing to infinity? See [7] for background.

Acknowledgements Thanks to Omer Angel, Alessandro Carderi and Yuval Peres. Sergey Khristo helped with the writing and figures.

References

1. I. Benjamini, O. Schramm, Every graph with a positive Cheeger constant contains a tree with a positive Cheeger constant. Geom. Funct. Anal. **7**, 403–419 (1997)

2. I. Benjamini, R. Lyons, Y. Peres, O. Schramm, Uniform spanning forests. Ann. Probab. **29**, 1–65 (2001)
3. J. Canon, W. Floyed, R. Kenyon, W. Parry, Hyperbolic geometry. Math. Sci. Res. Inst. Publ. **31**, 59–115 (1997)
4. P. Doyle, S. Snell, *Random Walks and Electric Networks* (Mathematical Association of America, America, 1984)
5. X. Fernique, *Regularite des trajectoires des fonctions aleatoires gaussiennes*. Lecture Notes in Mathematics, vol. 480 (Springer, Berlin, 1975)
6. S. Janson, *Gaussian Hilbert Spaces* (Cambridge University Press, Cambridge, 1998)
7. R. Lyons, Y. Peres, *Probability on Trees and Networks* (Cambridge University Press, Cambridge, 2013). Current version available at http://mypage.iu.edu/~rdlyons/
8. R. Peled, High-dimensional Lipschitz functions are typically flat. Ann. Probab. (to appear). http://arxiv.org/abs/1005.4636
9. Y. Velenik, Localization and delocalization of random interfaces. Probab. Surv. **3**, 112–169 (2006)

Point-to-Point Distance in First Passage Percolation on (Tree)×Z

Itai Benjamini and Pascal Maillard

Abstract We consider first passage percolation (FPP) on $\mathbb{T}_d \times G$, where $\mathbb{T}_d$ is the d-regular tree ($d \geq 3$) and G is a graph containing an infinite ray $0, 1, 2, \ldots$. It is shown that for a fixed vertex v in the tree, the fluctuation of the distance in the FPP metric between the points $(v, 0)$ and (v, n) is of the order of at most $\log n$. We conjecture that the real fluctuations are of order 1 and explain why.

Denote by $\mathbb{T}_d$ the d-regular tree ($d \geq 3$), rooted at a vertex ρ. We consider FPP on $\mathbb{T}_d \times G$, where G is a graph containing an infinite ray $0, 1, 2, \ldots$ (for example, $G = \mathbf{N}, \mathbf{Z}$ or an infinite tree). That is, attach to each edge e a random variable X_e, all X_e being independent copies of a random variable $X \geq 0$ with $\mathbf{E}[X] < \infty$. For a path γ, denote by $|\gamma|$ the number of edges on the path and define $|\gamma|_X = \sum_{e \in \gamma} X_e$. Then define the random (pseudo-)metric

$$d_X(v, w) = \min \left\{ |\gamma|_X \,\middle|\, \gamma \text{ is a path from } v \text{ to } w \right\}.$$

Write $D(n) = d_X((\rho, 0), (\rho, n))$, i.e. the minimal distance between two points which are n steps apart in the direction of the infinite ray.

We say hypothesis (H) is verified if

1. $\mathbf{E}[X^{1+\varepsilon}] < \infty$ for some $\varepsilon > 0$, and
2. there exist constants $C, K < \infty$, such that $\mathbf{E}[|\gamma_m|] < Cn^K$ for all n, where γ_m is the path that minimizes $D(n)$ in $\mathbb{T}_d \times G$.

Note that 2 is verified for example if $X \geq c$ for some $c > 0$, with $K = 1$, because then $\mathbf{E}[|\gamma_m|] \leq c^{-1}\mathbf{E}[|\gamma_m|_X] \leq c^{-1}\mathbf{E}[X]n$.

Theorem. *Suppose hypothesis (H) is verified. Then, $(D(n) - \mathbf{E}[D(n)])/\log n$ is tight in n.*

I. Benjamini (✉) • P. Maillard
Department of Mathematics and Computer Science, Weizmann Institute of Science, POB 26, Rehovot 76100, Israel
e-mail: itai.benjamini@weizmann.ac.il; pascal.maillard@weizmann.ac.il

© Springer International Publishing Switzerland 2014   47
B. Klartag, E. Milman (eds.), *Geometric Aspects of Functional Analysis*,
Lecture Notes in Mathematics 2116, DOI 10.1007/978-3-319-09477-9_3

We *conjecture* that $D(n) - \mathbf{E}[D(n)]$ is tight (without rescaling). The rationale for this conjecture is that this is indeed the case for the graph $\mathbb{T}_{d-1,d} \times G$, where $\mathbb{T}_{d-1,d}$ is the rooted d-ary tree, i.e. the tree where the root has degree $d - 1$ and every other vertex has degree d. This is the statement of the following proposition, which even does not need hypothesis (H):

Proposition. *In $\mathbb{T}_{d-1,d} \times G$, the sequence $D(n) - \mathbf{E}[D(n)]$ is tight in n.*

First passage percolation is a model of random perturbation of a given geometry. It has mostly been studied in Euclidean space and lattices (see e.g. Howard [8] for a review, although a bit outdated), and on trees, where it is also called the branching random walk [5, 10, 11]. Other setups considered include the complete graph [3, 9], the Erdös-Rényi graph [4] and a class of graphs admitting a certain recursive structure [1] (see more on that below). However, to our knowledge, the present note is the first example where the fluctuation of the point-to-point distance in FPP on the Cayley graph of a finitely generated group (i.e. $\mathbb{T}_4 \times \mathbb{Z}$) is shown to be small.

The graph $\mathbb{T}_d \times G$ can also be seen as an example of a "large" graph. The results of this note therefore add support to the common belief that the point-to-point distance of FPP in high-dimensional Euclidean space has small fluctuations.

We remark that the method leading to the proof of the above theorem in general is not applicable for the study of the distance in the FPP metric between the points $(\rho, 0)$ and $(v_n, 0)$, for v_n a vertex at distance n from the root in $\mathbb{T}_d$. For example, in the case $G = \mathbb{Z}$, the minimizing path looks like a path in $\mathbb{Z}^2$ with additional "handles", and it is not clear for us whether the fluctuations are actually small (recall that in FPP on $\mathbb{Z}^2$, the fluctuations are believed to be of order $n^{1/3}$ (see e.g. [6]) and up to now have only been proven to be of order at most $\sqrt{n/\log n}$ [2]). We therefore ask the following question:

Question. *In $\mathbb{T}_d \times \mathbb{Z}$ or $\mathbb{T}_{d-1,d} \times \mathbb{Z}$, how big are the fluctuations of $d_X((\rho, 0)$, $(v_n, 0))$, where v_n is a vertex at distance n from the root in the tree?*

We finally remark that even in $\mathbb{T}_d \times \mathbb{T}_d$, the current proof does not extend to the study to the FPP distance between two arbitrary vertices at distance n apart in $\mathbb{T}_d \times \mathbb{T}_d$, although here we also conjecture that the fluctuations are of order 1.

Proofs

The proof of the proposition uses a variant of an argument by Dekking and Host [7] on point-to-sphere distance in FPP on a tree, which was generalized by Benjamini and Zeitouni [1] to a large class of graphs, including $\mathbb{T}_{d-1,d} \times G$. For the point-to-sphere distance, the argument applies to every rooted graph G containing two vertex-disjoint rooted subgraphs G_1 and G_2 which are isomorphic to G.[1] This argument can be adapted for the point-to-point distance in $\mathbb{T}_{d-1,d} \times G$ in the direction of the infinite ray in G. It fails for $\mathbb{T}_d \times G$, but not completely: It can be applied to an auxiliary graph, which "almost" looks like $\mathbb{T}_d \times G$. This however induces an error, which is the reason of the $\log n$ term appearing in the statement of the theorem.

Before turning to the details, we introduce some more notation: Let $\mathbb{T}$ be a rooted tree and v, w two vertices in $\mathbb{T}$. We say that w is a *descendant* of v, if v is contained in the direct path from the root to w. We then denote by $\mathbb{T}|_v$ the subtree of $\mathbb{T}$ rooted at v, i.e. the subgraph of $\mathbb{T}$ spanned by the descendants of v, rooted at v.

Proof of the Proposition. Write for short $\mathbb{T} = \mathbb{T}_{d-1,d}$. Let 1,2 denote two distinct children of the root in $\mathbb{T}$. For $i = 1, 2$, let $D_i(n)$ be the distance between $(i, 0)$ and (i, n) in the FPP metric restricted to the subgraph $\mathbb{T}|_i \times G$. Then $D_i(n)$ has the same law as $D(n)$. Furthermore, $D_1(n)$ and $D_2(n)$ are independent. Now, for $i = 1, 2$ and $j \in \mathbf{N}$, let $e_{i,j}$ be the edge between (ρ, j) and (i, j). Then

$$D(n) \leq \min(D_1(n), D_2(n)) + X_{e_{1,0}} + X_{e_{1,n}} + X_{e_{2,0}} + X_{e_{2,n}}.$$

Taking expectations and using the formula $\min(a, b) = (a + b)/2 - |a - b|/2$, we get

$$\mathbf{E}[D(n)] \leq \tfrac{1}{2}(\mathbf{E}[D_1(n)] + \mathbf{E}[D_2(n)]) + 4\mathbf{E}[X] - \tfrac{1}{2}\mathbf{E}|D_1(n) - D_2(n)|.$$

Since $\mathbf{E}[D_1(n)] = \mathbf{E}[D_2(n)] = \mathbf{E}[D(n)]$, this gives

$$\mathbf{E}|D_1(n) - D_2(n)| \leq 8\mathbf{E}[X].$$

Tightness follows from the inequality $\mathbf{E}|Z| \leq \mathbf{E}|Z - Z'|$, which holds for any random variable Z with $\mathbf{E}Z = 0$ and with Z' being an independent copy of Z. $\square$

Proof of the Theorem. In the graph $\mathbb{T}_d \times G$ the situation is trickier: This graph does not contain two vertex-disjoint copies of itself. We will resolve this issue by considering an auxiliary graph first.

Write $\mathbb{T} = \mathbb{T}_d$. Fix an integer k. Let 0, 1 and 2 be three distinct children of the root in $\mathbb{T}$. For $i = 0, 1, 2$, let v_i be a vertex at distance k from the root, which is a descendant of i. The auxiliary graph we consider is $\mathbb{T}' \times G$, where $\mathbb{T}' = \mathbb{T}\backslash(\mathbb{T}|_{v_0})$, and the graph is rooted at $(\rho, 0)$. In contrast to $\mathbb{T} \times G$, this graph does contain two vertex disjoint copies of itself, namely the graphs $\mathbb{T}|_1 \times G$ and $\mathbb{T}|_2 \times G$, rooted at $(v_1, 0)$ and $(v_2, 0)$, respectively.

---

[1]This is Property (1) in [1]. Properties (2) and (3) are actually not needed, since on page 3 of that article, one can bound the right-hand side of the last inequality by $E \min(Z_n, Z'_n) + KC$ and continue from that point on.

Now let $D' = D'(n)$ be the FPP distance between $(\rho, 0)$ and (ρ, n) in $\mathbb{T}' \times G$ and $D'_i = D'_i(n)$ the FPP distance between $(v_i, 0)$ and (v_i, n) in $\mathbb{T}|_i \times G$, for $i = 1, 2$. Then note that D', D'_1 and D'_2 have the same distribution and D'_1 and D'_2 are independent. Let $\gamma_{i,j}$ for $i = 1, 2$ and $j \in \mathbf{N}$ be the (unique) path from (ρ, j) to (v_i, j) in $\mathbb{T} \times \{j\}$ and let $|\gamma_{i,j}|_X$ be its length in the FPP metric. We then have

$$D' \leq \min(D'_1, D'_2) + |\gamma_{1,0}|_X + |\gamma_{1,n}|_X + |\gamma_{2,0}|_X + |\gamma_{2,n}|_X,$$

and taking expectations we get as in the previous proof

$$8\mathbf{E}[X]k \geq \mathbf{E}[|D'_1 - D'_2|] \geq \mathbf{E}[|D' - \mathbf{E}[D']|].$$

We now get back to the graph $\mathbb{T} \times G$. Let C denote some positive constant, whose value may change from line to line. We claim that we can choose $k = k(n) = O(\log n)$, such that

$$\mathbf{E}[|D - \mathbf{E}[D]|] \leq \mathbf{E}[|D' - \mathbf{E}[D']|] + C. \tag{1}$$

Together with the previous inequality, this will imply the statement of the theorem. To prove it, let γ be the path from $(\rho, 0)$ to (ρ, n) of minimal length in the FPP metric on $\mathbb{T} \times G$ and let V_γ be the projection on $\mathbb{T}$ of the set of vertices traversed by γ. Define the event $B = \{v_0 \in V_\gamma\}$. Conditioned on $|V_\gamma|$, we have by symmetry,

$$\mathbf{P}\Big(B \,\Big|\, |V_\gamma|\Big) \leq \frac{|V_\gamma|}{d \times (d-1)^{k-1}},$$

since there are $d \times (d-1)^{k-1}$ vertices at distance k from the root in $\mathbb{T}_r$. Now, since $|V_\gamma| \leq |\gamma|$, we have by hypothesis (H), with $k = \lceil \alpha \log_{d-1} n \rceil$, $\alpha > 0$,

$$\mathbf{P}(B) \leq \mathbf{E}|V_\gamma|/(d \times (d-1)^{k-1}) \leq Cn^K/(d-1)^k \leq Cn^{K-\alpha}. \tag{2}$$

Note that $D \leq D'$ by definition, with $D = D'$ on the complement of B. Together with the triangle inequality, this gives

$$\mathbf{E}|D - \mathbf{E}[D]|] \leq \mathbf{E}[|D' - \mathbf{E}[D']|] + \mathbf{E}[|(D' - D) - \mathbf{E}[D' - D]|] \leq C \log n + 2\mathbf{E}[D'\mathbf{1}_B].$$

If γ_0 is the direct path from $(\rho, 0)$ to (ρ, n) along the ray $0, 1, 2, \ldots$, we have $D' \leq |\gamma_0|_X$. Hypothesis (H) and Minkowski's inequality then give $\mathbf{E}[(D')^{1+\varepsilon}] \leq Cn^{1+\varepsilon}$. Together with Hölder's inequality and (2), this yields the existence of $\alpha > 0$, such that $\mathbf{E}[D'\mathbf{1}_B] < C$ for all n. This proves (1) and therefore finishes the proof of the theorem. $\square$

Acknowledgements We thank an anonymous referee who has spotted some typographical errors.

References

1. I. Benjamini, O. Zeitouni, Tightness of fluctuations of first passage percolation on some large graphs, in *Geometric Aspects of Functional Analysis: Papers from the Israel Seminar (GAFA) Held 2006–2010*, ed. by B. Klartag, S. Mendelson, V.D. Milman. Lecture Notes in Mathematics, vol. 2050 (Springer, Heidelberg, 2012), pp. 127–132
2. I. Benjamini, G. Kalai, O. Schramm, First passage percolation has sublinear distance variance. Ann. Probab. **31**(4), 1970–1978 (2003)
3. S. Bhamidi, R. van der Hofstad, Weak disorder asymptotics in the stochastic mean-field model of distance. Ann. Appl. Probab. **22**(1), 29–69 (2012)
4. S. Bhamidi, R. Van Der Hofstad, G. Hooghiemstra, First passage percolation on the Erdős–Rényi random graph. Comb. Probab. Comput. **20**(05), 683–707 (2011)
5. J.D. Biggins, Branching out, in *Probability and Mathematical Genetics: Papers in Honour of Sir John Kingman*, ed. by N.H. Bingham, C.M. Goldie. London Mathematical Society Lecture Note Series (Cambridge University Press, Cambridge, 2010), pp. 112–133
6. S. Chatterjee, The universal relation between scaling exponents in first-passage percolation. Ann. Math. Second Ser. **177**(2), 663–697 (2013)
7. F.M. Dekking, B. Host, Limit distributions for minimal displacement of branching random walks. Probab. Theory Relat. Fields **90**(3), 403–426 (1991)
8. C.D. Howard, Models of first-passage percolation, in *Probability on Discrete Structures*, ed. by H. Kesten. Encyclopaedia Mathematical Sciences, vol. 110 (Springer, Berlin, 2004), pp. 125–173
9. S. Janson, One, two and three times $\log n / n$ for paths in a complete graph with random weights. Comb. Probab. Comput. **8**(04), 347–361 (1999)
10. Z. Shi, Random walks and trees. ESAIM: Proc. **31**, 1–39 (2011)
11. O. Zeitouni, Branching random walks and Gaussian fields. Lecture Notes. Available at http://www.wisdom.weizmann.ac.il/~zeitouni/pdf/notesBRW.pdf

A Lower Bound for the Bergman Kernel and the Bourgain-Milman Inequality

Zbigniew Błocki

Abstract For pseudoconvex domains in $\mathbb{C}^n$ we prove a sharp lower bound for the Bergman kernel in terms of volume of sublevel sets of the pluricomplex Green function. For $n = 1$ it gives in particular another proof of the Suita conjecture. If Ω is convex then by Lempert's theory the estimate takes the form $K_\Omega(z) \geq 1/\lambda_{2n}(I_\Omega(z))$, where $I_\Omega(z)$ is the Kobayashi indicatrix at z. One can use this to simplify Nazarov's proof of the Bourgain-Milman inequality from convex analysis. Possible further applications of Lempert's theory in this area are also discussed.

1 Introduction

For a domain Ω in $\mathbb{C}^n$ and $w \in \Omega$ we are interested in the Bergman kernel

$$K_\Omega(w) = \sup\{|f(w)|^2 : f \in \mathcal{O}(\Omega), \int_\Omega |f|^2 d\lambda_{2n} \leq 1\}$$

and in the pluricomplex Green function with pole at w

$$G_{\Omega,w} = \sup\{u \in PSH^-(\Omega) : \limsup_{z \to w}(u(z) - \log|z - w|) < \infty\}$$

(Here PSH^- denotes the class of negative plurisubharmonic functions.)

Our main result is the following bound:

Theorem 1. *Assume that Ω is pseudoconvex. Then for $w \in \Omega$ and $a \geq 0$ we have*

$$K_\Omega(w) \geq \frac{1}{e^{2na}\lambda_{2n}(\{G_{\Omega,w} < -a\})}. \tag{1}$$

This estimate seems to be very accurate. It is certainly optimal in the sense that if Ω is a ball centered at w then we get equality in (1) for all a. It is useful and

Z. Błocki (✉)

Uniwersytet Jagielloński, Instytut Matematyki, Łojasiewicza 6, 30-348 Kraków, Poland

e-mail: Zbigniew.Blocki@im.uj.edu.pl; umblocki@cyf-kr.edu.pl

© Springer International Publishing Switzerland 2014

B. Klartag, E. Milman (eds.), *Geometric Aspects of Functional Analysis*,
Lecture Notes in Mathematics 2116, DOI 10.1007/978-3-319-09477-9_4

not trivial already for $n = 1$. Note that in this case if we let a tend to ∞ then we immediately obtain

$$K_\Omega \geq \frac{1}{\pi} c_\Omega^2, \tag{2}$$

where

$$c_\Omega(z) = \exp(\lim_{\zeta \to z}(G_{\Omega,z}(\zeta) - \log|\zeta - z|))$$

is the logarithmic capacity of $\mathbb{C} \setminus \Omega$ with respect to z. This is precisely the inequality conjectured by Suita [19] and recently proved in [5].

A lower bound of the Bergman kernel in terms of the volume of the sublevel sets of the Green function follows from an estimate of Herbort (Proposition 3.6 in [10] with $f \equiv 1$). The main point in (1) is that the constant is optimal. Our proof of (1) uses the L^2-estimate for the $\bar\partial$-equation of Donnelly-Fefferman [8] from which we can first get a weaker version:

$$K_\Omega(w) \geq \frac{c(n,a)}{\lambda_{2n}(\{G_{\Omega,w} < -a\})},$$

where

$$c(n,a) = \left(\frac{\mathrm{Ei}\,(na)}{\mathrm{Ei}\,(na) + 2}\right)^2$$

and

$$\mathrm{Ei}\,(b) = \int_b^\infty \frac{ds}{se^s} \tag{3}$$

(for $b > 0$). Then we employ the tensor power trick and use the fact that

$$\lim_{m \to \infty} c(nm,a)^{1/m} = e^{-2na}.$$

This way we get an optimal constant in (1).

Our new proof of the one-dimensional estimate (2) makes crucial use of many complex variables. The use of the tensor power trick here replaces a special ODE in [5]. It should be noted though that this works only for the Suita conjecture, we do not get the Ohsawa-Takegoshi extension theorem from Theorem 1.

It is probably interesting to investigate the limit of the right-hand side of (1) as a tends to ∞ also in higher dimensions. We suspect that it always exists, at least for sufficiently regular domains. This way we would get a certain counterpart of logarithmic capacity in higher dimensions. Using Lempert's theory [15, 16] one can check what happens with this limit for smooth and strongly convex domains, see Proposition 3 below. This way we get the following bound:

Theorem 2. *Let Ω be a convex domain in $\mathbb{C}^n$. Then for $w \in \Omega$*

$$K_\Omega(w) \geq \frac{1}{\lambda_{2n}(I_\Omega(w))},$$

where

$$I_\Omega(w) = \{\varphi'(0) : \varphi \in \mathcal{O}(\Delta, \Omega),\ \varphi(0) = w\}$$

is the Kobayashi indicatrix (here Δ denotes the unit disc).

One can use Theorem 2 to simplify Nazarov's approach [17] to the Bourgain-Milman inequality [6]. For a convex symmetric body (i.e. open, bounded) L in $\mathbb{R}^n$ its dual is given by

$$L' := \{y \in \mathbb{R}^n : x \cdot y \leq 1 \text{ for all } x \in L\}.$$

The product $\lambda_n(L)\lambda_n(L')$ is called a Mahler volume of L. It is independent of linear transformations and on an inner product in $\mathbb{R}^n$, and thus depends only on the finite dimensional Banach space structure whose unit ball is L. The Blaschke-Santaló inequality says that the Mahler volume is maximized by balls.

On the other hand, the still open Mahler conjecture states that it is minimized by cubes. A partial result in this direction is the Bourgain-Milman inequality [6] which says that there exists $c > 0$ such that

$$\lambda_n(L)\lambda_n(L') \geq c^n \frac{4^n}{n!}. \tag{4}$$

The Mahler conjecture is equivalent to saying that we can take $c = 1$ in (4). Currently, the best known constant in (4) is $\pi/4$ and is due to Kuperberg [14].

Nazarov [17] recently proposed a complex-analytic approach to (4). He considered tube domain $T_L := L + i\mathbb{R}^n$ and proved the following bounds for the Bergman kernel at the origin:

$$K_{T_L}(0) \leq \frac{n!}{\pi^n} \frac{\lambda_n(L')}{\lambda_n(L)} \tag{5}$$

$$K_{T_L}(0) \geq \left(\frac{\pi}{4}\right)^{2n} \frac{1}{(\lambda_n(L))^2}. \tag{6}$$

This gave (4) with $c = (\pi/4)^3$. The upper bound (5) was obtained relatively easily from Rothaus' formula for the Bergman kernel in tube domains (see [18] and [12]):

$$K_{T_L}(0) = \frac{1}{(2\pi)^n} \int_{\mathbb{R}^n} \frac{d\lambda_n}{J_L},$$

where

$$J_L(y) = \int_L e^{-2x \cdot y} d\lambda_n(x).$$

For the lower bound (6) Nazarov used the Hörmander estimate [11] for $\bar{\partial}$.

We will show that (6) follows easily from Theorem 2. It should be noted however that although we are using the Donnelly-Fefferman estimate here, it can be deduced quite easily from the Hörmander estimate (see [1]), so the latter still plays a crucial role.

We conjecture that in fact the following lower bound holds:

$$K_{T_L}(0) \ge \left(\frac{\pi}{4}\right)^n \frac{1}{(\lambda_n(L))^2}. \tag{7}$$

Since we have equality for cubes, this would be optimal. In Sect. 4 we discuss possible applications of Lempert's theory to this problem.

The author learned about the Nazarov paper [17] from professor Vitali Milman during his visit to Tel Aviv in December 2011. He is also grateful to Semyon Alesker for his invitation and hospitality.

2 Proofs of Theorems 1 and 2

Proof of Theorem 1. By approximation we may assume that Ω is bounded and hyperconvex, so that by [7] the Green function $G := G_{\Omega,w}$ is continuous on $\bar{\Omega} \setminus \{w\}$. We may also assume that $a > 0$, as for $a = 0$ it is enough to take $f \equiv 1$ in the definition of the Bergman kernel. Set

$$\varphi := 2nG, \quad \psi := -\log(-G)$$

and

$$\alpha := \bar{\partial}(\chi \circ G) = (\chi' \circ G) \bar{\partial} G,$$

where χ will be determined later. We have

$$i\bar{\alpha} \wedge \alpha \le (\chi' \circ G)^2 i \partial G \wedge \bar{\partial} G \le G^2 (\chi' \circ G)^2 i \partial \bar{\partial} \psi.$$

By the Donnelly-Fefferman estimate [8] (see also [1, 2], and [3] for a formulation with non-smooth weights which is needed here) we can find $u \in L^2_{loc}(\Omega)$ solving $\bar{\partial} u = \alpha$ and such that

$$\int_\Omega |u|^2 d\lambda_{2n} \leq \int_\Omega |u|^2 e^{-\varphi} d\lambda_{2n} \leq C \int_\Omega G^2(\chi' \circ G)^2 e^{-2nG} d\lambda_{2n},$$

where C is an absolute constant (in fact, the optimal one is $C = 4$, see [2,4]). We now set

$$\chi(t) := \begin{cases} 0 & t \geq -a, \\ \int_a^{-t} \frac{e^{-ns}}{s} ds, & t < -a, \end{cases}$$

so that

$$\int_\Omega |u|^2 d\lambda_{2n} \leq C \lambda_{2n}(\{G < -a\}).$$

The function $f := \chi \circ G - u$ is holomorphic and since $\chi \circ G$ is continuous, we see that u must be continuous. We also have $u(w) = 0$ because $e^{-\varphi}$ is not integrable near w (by monotonicity of the Green function we have $G_\Omega(z, w) \leq \log|z - w| - \log r$ if $B(z, r) \subset \Omega$). Therefore

$$f(w) = \chi(-\infty) = \mathrm{Ei}\,(na)$$

with Ei given by (3). We also have (with $\|\cdot\|$ denoting the L^2-norm in Ω)

$$\|f\| \leq \|\chi \circ G\| + \|u\| \leq (\chi(-\infty) + \sqrt{C})\sqrt{\lambda_{2n}(\{G < -a\})}.$$

Therefore

$$K_\Omega(w) \geq \frac{|f(w)|^2}{\|f\|^2} \geq \frac{c(n, a)}{\lambda_{2n}(\{G < -a\})},$$

where

$$c(n, a) = \frac{\mathrm{Ei}\,(na)^2}{(\mathrm{Ei}\,(na) + \sqrt{C})^2}.$$

We are now going to use the tensor power trick. For a big natural number m consider the domain $\widetilde{\Omega} = \Omega^m$ in $\mathbb{C}^{nm}$ and $\widetilde{w} = (w, \dots, w) \in \widetilde{\Omega}$. Then

$$K_{\widetilde{\Omega}}(\widetilde{w}) = (K_\Omega(w))^m$$

and by [13] (see also [9])

$$G_{\widetilde{\Omega}, \widetilde{w}}(z^1, \dots, z^m) = \max_{j=1,\dots,m} G(z^j),$$

therefore

$$\lambda_{2nm}(\{G_{\widetilde{\Omega},\widetilde{w}} < -a\}) = (\lambda_{2n}(\{G < -a\}))^m.$$

It follows from the previous part that

$$(K_\Omega(w))^m \geq \frac{c(nm,a)}{(\lambda_{2n}(\{G < -a\}))^m}$$

and it is enough to use the fact that

$$\lim_{m \to \infty} c(nm,a)^{1/m} = e^{-2na}.$$

$\square$

Theorem 2 follows immediately from Theorem 1 and the following result by approximation.

Proposition 3. *Assume that Ω is a bounded, smooth, strongly convex domain in $\mathbb{C}^n$. Then for any $w \in \Omega$*

$$\lim_{a \to \infty} e^{2na} \lambda_{2n}(\{G_{\Omega,w} < -a\}) = \lambda_{2n}(I_\Omega(w)). \tag{8}$$

Proof. Denote $I := I_\Omega(w)$, $G := G_{\Omega,w}$, we may assume that $w = 0$. By the results of Lempert [15] there exists a diffeomorphism $\Phi : \bar{I} \to \bar{\Omega}$ such that for $v \in \partial I$ the mapping $\Delta \ni \zeta \mapsto \Phi(\zeta v)$ is a geodesic in Ω, that is

$$G(\Phi(\zeta v)) = \log|\zeta|. \tag{9}$$

(Φ can be treated as an exponential map for the Kobayashi distance.) We also have

$$\Phi(\zeta v) = \zeta v + O(|\zeta|^2).$$

By (9)

$$\{G < -a\} = \Phi(e^{-a}\,\text{int}\, I)$$

and therefore

$$\lambda_{2n}(\{G < -a\}) = \int_{e^{-a}I} \text{Jac}\, \Phi \, d\lambda_{2n}.$$

Since $\Phi'(0)$ is the identity, we obtain (8). $\square$

3   Applications to the Bourgain-Milman Inequality

Assume that L is a convex symmetric body in $\mathbb{R}^n$. In view of Theorem 2, in order to prove Nazarov's lower bound (6) it is enough to show the estimate

$$\lambda_{2n}(I_{T_L}(0)) \le \left(\frac{4}{\pi}\right)^{2n} (\lambda_n(L))^2. \tag{10}$$

But this follows immediately from the following:

Proposition 4. $I_{T_L}(0) \subset \dfrac{4}{\pi}(\bar{L} + i\bar{L})$.

Proof. We will use an idea of Nazarov [17] here. Let Φ be a conformal mapping from the strip $\{|\operatorname{Re}\zeta| < 1\}$ to Δ with $\Phi(0) = 0$, so that $|\Phi'(0)| = \pi/4$. For $u \in L'$ we can then define $F \in \mathcal{O}(T_L, \Delta)$ by $F(z) = \Phi(z \cdot u)$. For $\varphi \in \mathcal{O}(\Delta, T_L)$ with $\varphi(0) = 0$ by the Schwarz lemma we have $|(F \circ \varphi)'(0)| \le 1$. Therefore $|\varphi'(0) \cdot u| \le 4/\pi$ and $\frac{\pi}{4} I_{T_L}(0) \subset L_{\mathbb{C}}$, where

$$L_{\mathbb{C}} = \{z \in \mathbb{C}^n : |z \cdot u| \le 1 \text{ for all } u \in L'\} \subset \bar{L} + i\bar{L}.$$

$\square$

It will be convenient to use the notation $J_L := \frac{\pi}{4} I_{T_L}(0)$, so that by the proof of Proposition 4

$$J_L \subset L_{\mathbb{C}} \subset \bar{L} + i\bar{L}. \tag{11}$$

We thus have $\lambda_{2n}(J_L) \le (\lambda_n(L))^2$ but we conjecture that

$$\lambda_{2n}(J_L) \le \left(\frac{\pi}{4}\right)^n (\lambda_n(L))^2. \tag{12}$$

Note that $J_{[-1,1]^n} = \Delta^n$, so that we have equality for cubes. The inequality (12) would give the optimal lower bound for the Bergman kernel in tube domains (7).

We first give an example that (11) cannot give us (12):

Example. Let $L = \{x_1^2 + x_2^2 < 1\}$ be the unit disc in $\mathbb{R}^2$. One can then show that $L_{\mathbb{C}} = \{|z|^2 \le 1 + (x_1 y_2 - x_2 y_1)^2\}$ and

$$\lambda_4(L_{\mathbb{C}}) = \frac{2\pi^2}{3} > \frac{\pi^4}{16} = \left(\frac{\pi}{4}\right)^2 (\lambda_2(L))^2.$$

4   Lempert's Theory in Tube Domains

Our goal is to approach (12) using Lempert's theory. First assume that Ω is bounded, smooth, strongly convex domain in $\mathbb{C}^n$. Then for any $z, w \in \Omega$, $z \neq w$, there exists unique extremal disc $\varphi \in \mathcal{O}(\Delta, \Omega) \cap C^\infty(\bar{\Delta}, \bar{\Omega})$ such that $\varphi(0) = w$, $\varphi(\xi) = z$ for some ξ with $0 < \xi < 1$, and

$$G_{\Omega, w}(\varphi(\zeta)) = \log |\zeta|, \quad \zeta \in \Delta.$$

Lempert [15] showed in particular the following characterization of extremal discs: a disc $\varphi \in \mathcal{O}(\Delta, \Omega) \cap C(\bar{\Delta}, \bar{\Omega})$ is extremal if and only if $\varphi(\partial\Delta) \subset \partial\Omega$ and there exists $h \in \mathcal{O}(\Delta, \mathbb{C}^n) \cap C(\bar{\Delta}, \mathbb{C}^n)$ such that the vector $e^{it}h(e^{it})$ is outer normal to $\partial\Omega$ at $\varphi(e^{it})$ for every $t \in \mathbb{R}$.

Lempert [15] also proved that for every extremal disc φ in Ω there exists a left-inverse $F \in \mathcal{O}(\Omega, \Delta)$ (that is $F(\varphi(\zeta)) = \zeta$ for $\zeta \in \Delta$). It solves the equation

$$(z - \varphi(F(z))) \cdot h(F(z)) = 0, \quad z \in \Omega. \tag{13}$$

Now assume that L is a smooth, strongly convex body in $\mathbb{R}^n$. Although T_L is neither bounded nor strongly convex, we may nevertheless try to apply Lempert's condition for extremal discs (the details have been worked out by Zając [20]). First note that $h \in \mathcal{O}(\Delta, \mathbb{C}^n) \cap C(\bar{\Delta}, \mathbb{C}^n)$ in our case must be such that $e^{it}h(e^{it})$ is an outer normal to T_L and therefore its imaginary part vanishes:

$$\mathrm{Im}\,(e^{-it}h(e^{it})) = 0, \quad t \in \mathbb{R}. \tag{14}$$

It follows that h must be of a very special form:

Lemma 5 ([20]). *If $h \in \mathcal{O}(\Delta) \cap C(\bar{\Delta})$ satisfies (14) then $h(\zeta) = a + b\zeta + \bar{a}\zeta^2$ for some $a \in \mathbb{C}$ and $b \in \mathbb{R}$.*

Proof. Set $a := h(0)$. Then for $\zeta \in \partial\Delta$

$$0 = \mathrm{Im}\left(\frac{h(\zeta)}{\zeta}\right) = \mathrm{Im}\left(\frac{h(\zeta) - a}{\zeta} - \bar{a}\zeta\right)$$

and therefore

$$\frac{h(\zeta) - a}{\zeta} - \bar{a}\zeta = b \in \mathbb{R}, \quad \zeta \in \bar{\Delta}.$$

$\square$

We thus see that in our case h must be of the form

$$h(\zeta) = w + \zeta b + \zeta^2 \bar{w}, \quad \zeta \in \bar{\Delta},$$

for some $w \in \mathbb{C}^n$ and $b \in \mathbb{R}^n$. Take the extremal disc φ for T_L associated with h. Since $e^{it}\overline{h(e^{it})}$ is an outer normal to T_L at $\varphi(e^{it})$ and its imaginary part vanishes, it follows that its real part is an outer normal to L at $\mathrm{Re}\,(\varphi(e^{it}))$. Therefore

$$\mathrm{Re}\,\varphi(e^{it}) = v^{-1}\left(\frac{b + \mathrm{Re}\,(e^{-it}w)}{|b + \mathrm{Re}\,(e^{-it}w)|}\right), \tag{15}$$

where

$$v : \partial L \to \mathbb{S}^{n-1}$$

is the Gauss map.

For $\varphi \in \mathcal{O}(\Delta) \cap C(\bar{\Delta})$ we can recover the values of φ in Δ from the values of $\mathrm{Re}\,\varphi$ on $\partial\Delta$ using the Schwarz formula:

$$\varphi(\zeta) = \frac{1}{2\pi} \int_0^{2\pi} \frac{e^{it} + \zeta}{e^{it} - \zeta} \mathrm{Re}\,\varphi(e^{it})\, dt + i\,\mathrm{Im}\,\varphi(0), \quad \zeta \in \Delta.$$

Therefore extremal discs satisfying (15) are given by

$$\varphi(\zeta) = \frac{1}{2\pi} \int_0^{2\pi} \frac{e^{it} + \zeta}{e^{it} - \zeta} v^{-1}\left(\frac{b + \mathrm{Re}\,(e^{-it}w)}{|b + \mathrm{Re}\,(e^{-it}w)|}\right) dt + i\,\mathrm{Im}\,\varphi(0), \quad \zeta \in \Delta.$$

We now assume that L is in addition symmetric and then consider the case when $b = 0$ and $\mathrm{Im}\,\varphi(0) = 0$:

$$\varphi(\zeta) = \frac{1}{2\pi} \int_0^{2\pi} \frac{e^{it} + \zeta}{e^{it} - \zeta} v^{-1}\left(\frac{\mathrm{Re}\,(e^{-it}w)}{|\mathrm{Re}\,(e^{-it}w)|}\right) dt. \tag{16}$$

Since L is symmetric the function $B(t)$ under the integral in (16) satisfies $B(t + \pi) = -B(t)$. We thus have $\varphi(0) = 0$ and one can show (see [20] for details) that all geodesics of T_L passing through the origin are given by (16). They are bounded and smooth up to the boundary if $\mathrm{Re}\,w$ and $\mathrm{Im}\,w$ are linearly independent in $\mathbb{R}^n$. If $\mathrm{Re}\,w$ and $\mathrm{Im}\,w$ are linearly dependent (and $w \neq 0$) then (16) gives special extremal discs of the form

$$\varphi(\zeta) = \Phi^{-1}(\zeta)\,x, \quad x \in \partial L,$$

where Φ is as in the proof Proposition 4. Left-inverses to these φ are then given by $F(z) = \Phi(z \cdot u)$ for unique $u \in \partial L'$ with $x \cdot u = 1$.

For geodesics (16) we have

$$\varphi'(0) = \frac{1}{\pi} \int_0^{2\pi} e^{it} v^{-1}\left(\frac{\mathrm{Re}\,(e^{it}w)}{|\mathrm{Re}\,(e^{it}w)|}\right) dt. \tag{17}$$

These vectors parametrize the boundary of the Kobayashi indicatrix $I_L(0)$. If $F \in \mathcal{O}(\Omega, \Delta)$ is the left-inverse of φ satisfying (13) we get, since $h'(0) = 0$,

$$F'(0) = \frac{w}{\varphi'(0) \cdot w}.$$

Therefore

$$J_L = \{z \in \mathbb{C}^n : |z \cdot w| \le |\Psi(w)| \text{ for all } w \in (\mathbb{C}^n)_*\}, \tag{18}$$

where

$$\Psi(w) = \frac{1}{4} \int_0^{2\pi} e^{it} w \cdot v^{-1} \left(\frac{\mathrm{Re}\,(e^{it}w)}{|\mathrm{Re}\,(e^{it}w)|} \right) dt.$$

Both (17) and (18) give a description of the set J_L. It would be interesting to try to use it to prove (12). We can at least show this for a ball:

Example. Let $B = \{|x| < 1\}$ be the unit ball in $\mathbb{R}^n$. For $w \in (\mathbb{C}^n)_*$ we have

$$\mathrm{Im}\,\Psi(w) = \frac{1}{4} \int_0^{2\pi} \frac{\mathrm{Im}\,(e^{it}w) \cdot \mathrm{Re}\,(e^{it}w)}{|\mathrm{Re}\,(e^{it}w)|} dt = -\frac{1}{4} \int_0^{2\pi} \frac{d}{dt} |\mathrm{Re}\,(e^{it}w)| dt = 0$$

and thus

$$\Psi(w) = \frac{1}{4} \int_0^{2\pi} |\mathrm{Re}\,(e^{it}w)| dt \le \frac{\pi}{\sqrt{8}} |w|.$$

By (18) J_B is contained in a ball with radius $\pi/\sqrt{8}$ in $\mathbb{C}^n$. Therefore

$$\lambda_{2n}(J_B) \le \frac{\pi^{3n}}{8^n n!}.$$

On the other hand,

$$\lambda_n(B) = \frac{\pi^{n/2}}{\Gamma(\frac{n}{2} + 1)},$$

and we see that (12) holds for B if $n \ge 3$. To show this also for $n = 2$ we have to use in addition Proposition 4: $J_B \subset (\bar{B} + i\bar{B}) \cap (r_0\bar{B})$, where $r_0 = \pi/\sqrt{8}$. With $\rho_0 = \sqrt{r_0^2 - 1}$ we will get

$$\lambda_4(J_B) \le \pi^2 \rho_0^2 + \pi^2 \int_{\rho_0}^1 \rho(r_0^2 - \rho^2) d\rho = \frac{\pi^6}{256} + \frac{\pi^4}{16} - \frac{\pi^2}{2} < \frac{\pi^4}{16} = \left(\frac{\pi}{4}\right)^2 (\lambda_2(B))^2.$$

References

1. B. Berndtsson, Weighted estimates for the $\bar{\partial}$-equation, in *Complex Analysis and Geometry*, Columbus, Ohio, 1999. Ohio State Univ. Math. Res. Inst. Publ., vol. 9 (Walter de Gruyter, Berlin, 2001), pp. 43–57
2. Z. Błocki, A note on the Hörmander, Donnelly-Fefferman, and Berndtsson L^2-estimates for the $\bar{\partial}$-operator. Ann. Pol. Math. **84**, 87–91 (2004)
3. Z. Błocki, The Bergman metric and the pluricomplex Green function. Trans. Am. Math. Soc. **357**, 2613–2625 (2005)
4. Z. Błocki, Estimates for $\bar{\partial}$ and optimal constants, in Complex Geometry, Abel Symposium 2013, Springer (to appear)
5. Z. Błocki, Suita conjecture and the Ohsawa-Takegoshi extension theorem. Invent. Math. **193**, 149–158 (2013)
6. J. Bourgain, V. Milman, New volume ratio properties for convex symmetric bodies in $\mathbb{R}^n$. Invent. Math. **88**, 319–340 (1987)
7. J.-P. Demailly, Mesures de Monge-Ampère et mesures plurisousharmoniques. Math. Z. **194**, 519–564 (1987)
8. H. Donnelly, C. Fefferman, L^2-cohomology and index theorem for the Bergman metric. Ann. Math. **118**, 593–618 (1983)
9. A. Edigarian, On the product property of the pluricomplex Green function. Proc. Am. Math. Soc. **125**, 2855–2858 (1997)
10. G. Herbort, The Bergman metric on hyperconvex domains. Math. Z. **232**, 183–196 (1999)
11. L. Hörmander, L^2 estimates and existence theorems for the $\bar{\partial}$ operator. Acta Math. **113**, 89–152 (1965)
12. C.-I. Hsin, The Bergman kernel on tube domains. Rev. Un. Mat. Argentina **46**, 23–29 (2005)
13. M. Jarnicki, P. Pflug, Invariant pseudodistances and pseudometrics - completeness and product property. Ann. Polon. Math. **55**, 169–189 (1991)
14. G. Kuperberg, From the Mahler conjecture to Gauss linking integrals. Geom. Funct. Anal. **18**, 870–892 (2008)
15. L. Lempert, La métrique de Kobayashi et la représentation des domaines sur la boule. Bull. Soc. Math. France **109**, 427–474 (1981)
16. L. Lempert, Holomorphic invariants, normal forms, and the moduli space of convex domains. Ann. Math. **128**, 43–78 (1988)
17. F. Nazarov, The Hörmander proof of the Bourgain-Milman theorem, in *Geometric Aspects of Functional Analysis, Israel Seminar 2006–2010*, ed. by B. Klartag, S. Mendelson, V.D. Milman. Lecture Notes in Mathematics, vol. 2050 (Springer, Berlin, 2012), pp. 335–343
18. O.S. Rothaus, Some properties of Laplace transforms of measures. Trans. Am. Math. Soc. **131**, 163–169 (1968)
19. N. Suita, Capacities and kernels on Riemann surfaces. Arch. Ration. Mech. Anal. **46**, 212–217 (1972)
20. S. Zając, Complex geodesics in convex tube domains, Ann. Scuola Norm. Sup. Pisa (to appear)

An Improved Estimate in the Restricted Isometry Problem

Jean Bourgain

Abstract It is shown that for the $n \times n$-Hadamard matrix (or, more generally, a bounded orthogonal matrix) the RIP-property for r-space vectors holds, with row restriction to a set S of size

$$|S| < C(\varepsilon)(\log n)^2 (\log r) r.$$

This bound represents a slight improvement over (Rudelson and Vershynin, Commun Pure Appl Math 61:1025–1045, 2008) in that the power of the logarithm is decreased by one unit.

1 Preliminaries

Let H be an $(n \times n)$-matrix such that

$$\|Hx\|_2 = \sqrt{n}\|x\|_2 \text{ and } |H_{ij}| = 1$$

(the argument below can be generalized).

Denote

$$\mathcal{E}_r = \{x \in \mathbb{R}^n; \|x\|_2 = 1 \text{ and } |\operatorname{supp} x| \le r\}.$$

Recall that $S \subset \{1, \ldots, n\}$ satisfies the ε-RIP property with respect to the set $\mathcal{E}_r$, provided

$$(1 - \varepsilon)\sqrt{m}\|x\|_2 \le \left(\sum_{i \in S} |\langle Hx, e_i \rangle|^2\right)^{\frac{1}{2}} \le (1 + \varepsilon)\sqrt{m}\|x\|_2 \text{ for all } x \in \mathcal{E}_r$$

where $m = |S|$.

J. Bourgain (✉)
School of Mathematics, Institute for Advanced Study, Princeton, NJ 08540, USA
e-mail: bourgain@math.ias.edu

© Springer International Publishing Switzerland 2014

B. Klartag, E. Milman (eds.), *Geometric Aspects of Functional Analysis*,
Lecture Notes in Mathematics 2116, DOI 10.1007/978-3-319-09477-9_5

In what follows, we recall a classical entropy result. It is then used to carry out a construction (Lemma 2 below) providing an approximation crucial to our argument (completed in the next section).

Lemma 1 (Entropy Bound).

$$\log N(H\mathcal{E}_r, tB_\infty) \le c\frac{(\log n)^2}{t^2}r \qquad (t \le \sqrt{r}). \tag{1}$$

Proof. Since $\mathcal{E}_r \subset \sqrt{r}B_{1,n}$, it follows

$$N(H\mathcal{E}_r, tB_\infty) \le N\left(HB_{1,n}, \frac{t}{\sqrt{r}}B_\infty\right)$$

and using Lemma 3.9 from [1] (Maurey's lemma)

$$\log N(H\mathcal{E}_r, tB_\infty) \le c(\log n)^2\frac{r}{t^2}$$

$\square$

Lemma 2. *Fix $\varepsilon > 0$, $\log \frac{1}{\varepsilon} \le O(\log r)$.*
To each $x \in H\mathcal{E}_r$, we may associate a vector $\tilde{x}$ of the form

$$\tilde{x} = \sum_j \sum_{i \in I_j}(1 + \varepsilon)^j e_i \tag{2}$$

where the sum is over all $j \in \mathbb{Z}$ satisfying

$$\frac{\varepsilon}{4} < (1 + \varepsilon)^j < \sqrt{r} \tag{3}$$

such that

* *The sets I_j are disjoint*
*

$$|\tilde{x}_i| = \big(1 + O(\varepsilon)\big)|x_i| + O(\varepsilon) \text{ for } 1 \le i \le n \tag{4}$$

* *I_j belongs to a class $\mathcal{F}_j$ of subsets of $\{1, \ldots, n\}$ satisfying*

$$\log |\mathcal{F}_j| \lesssim \varepsilon^{-3}(\log n)^2(1 + \varepsilon)^{-2j}r. \tag{5}$$

Proof. According to Lemma 1, there is a finite set $\mathcal{D}_j \subset H\xi_r$ such that

$$\max_{x \in H\mathcal{E}_r} \min_{x' \in \mathcal{D}_j} \|x - x'\|_\infty < \frac{\varepsilon}{10}(1 + \varepsilon)^j \tag{6}$$

$$\log |\mathcal{D}_j| < c(\log n)^2(1 + \varepsilon)^{-2j}\varepsilon^{-2}r. \tag{7}$$

Let $x \in H\xi_r$. We construct $\tilde{x}$. Take j satisfying (3) and $x(j) \in \mathcal{D}_j$ such that $\|x - x(j)\|_\infty < \frac{\varepsilon}{10}(1+\varepsilon)^j$. Denote

$$I'_j = \{i; (1+\varepsilon)^{j-2} < |x(j)_i| < (1+\varepsilon)^{j+1}\}$$

$$I_j = I'_j \setminus \bigcup_{j'>j} I'_{j'}$$

and let

$$\tilde{x} = \sum_j (1+\varepsilon)^j \sum_{i \in I_j} e_i.$$

Assume $i \in I_j$. Then $i \in I'_j$ and

$$\frac{1}{2}\varepsilon < (1+\varepsilon)^{j-2} < |x(j)_i| < (1+\varepsilon)^{j+1}$$

$$|x(j)_i - x_i| < \varepsilon(1+\varepsilon)^j$$

implying

$$|x_i| = (1+\varepsilon)^j \left(1 + O(\varepsilon)\right) \text{ for } i \in \bigcup I_j.$$

If $i \notin \bigcup I_j, \tilde{x}_i = 0$ and $|x_i| < \varepsilon$. Indeed, if

$$\frac{\varepsilon}{2} < (1+\varepsilon)^{j-1} < |x_i| < (1+\varepsilon)^j$$

we get

$$\frac{\varepsilon}{4} < (1+\varepsilon)^{j-2} < |x(j)_i| < (1+\varepsilon)^{j+1}$$

and therefore $i \in I'_j \subset \bigcup_{j' \geq j} I_{j'}$.

This proves (4).

Next, the number of sets I'_j introduced above is at most $|\mathcal{D}_j|$ and hence the number of sets I_j at most

$$\prod_{j' \geq j} |\mathcal{D}_{j'}|.$$

Therefore (5) follows from (7).   □

We also need the following probabilistic statement

Lemma 3. *Let* $\mathcal{G}_\ell, \ell = 1, 2, \ldots$ *be collections of subsets of* $\{1, \ldots, n\}$ *and* $r_\ell > 0$ *such that* $\sum 2^{-r_\ell} \leq \frac{1}{2}$. *Then there is a subset* $S \subset \{1, \ldots, n\}$, $|S| = m = \delta n$, *satisfying for each* ℓ

$$\max_{I \in \mathcal{G}_\ell} \left| |S \cap I| - \delta |I| \right| \leq C \{ \delta^{\frac{1}{2}} (r_\ell + \log |\mathcal{G}_\ell|)^{\frac{1}{2}} (\max_{I \in \mathcal{G}_\ell} |I|)^{\frac{1}{2}} + r_\ell + \log |\mathcal{G}_\ell| \} \tag{8}$$

Proof. Denote $\mathbb{E}_S$ the expectation over all $S \subset \{1, \ldots, n\}$ of size $|S| = m = \delta n$. Standard probabilistic considerations show that for any $x \in \mathbb{R}^n, q \geq 1$,

$$\left\{ \mathbb{E}_S \left[\left| \sum_{i \in S} x_i - \delta \sum_{i=1}^n x_i \right|^q \right] \right\}^{\frac{1}{q}} \leq C \sqrt{q} \left\{ \mathbb{E}_S \left[\left(\sum_{i \in S} x_i^2 \right)^{q/2} \right] \right\}^{\frac{1}{q}}. \tag{9}$$

Take $q = q_\ell = r_\ell + \log |\mathcal{G}_\ell|$ and apply (9) with $x = 1_I, I \in \mathcal{G}_\ell$. It follows that

$$\left\{ \mathbb{E}_S \left[\max_{I \in \mathcal{G}_\ell} \left| |I \cap S| - \delta |I| \right|^q \right] \right\}^{\frac{1}{q}}$$

$$\leq c \sqrt{q} \left\{ \max_{I \in \mathcal{G}_\ell} \mathbb{E}_S \left[|I \cap S|^{\frac{q}{2}} \right] \right\}^{\frac{1}{q}}$$

$$\leq c \sqrt{\delta} \sqrt{q} \left(\max_{I \in \mathcal{G}_\ell} |I| \right)^{\frac{1}{2}} + c \sqrt{q} \left\{ \mathbb{E}_S \left[\max_{I \in \mathcal{G}_\ell} \left| |I \cap S| - \delta |I| \right|^q \right] \right\}^{\frac{1}{2q}}$$

from which one deduces that

$$\left\{ \mathbb{E}_S \left[\max_{I \in \mathcal{G}_\ell} \left| |I \cap S| - \delta |I| \right|^{q_\ell} \right] \right\}^{\frac{1}{q_\ell}} \leq c \sqrt{\delta} \sqrt{q_\ell} \left(\max_{I \in \mathcal{G}_\ell} |I| \right)^{\frac{1}{2}} + c q_\ell. \tag{10}$$

Next, from (10) and Tche'bychev's inequality, we obtain

$$\mathbb{E}_S \left[\max_{I \in \mathcal{G}_\ell} \left| |I \cap S| - \delta |I| \right| > 2 c \sqrt{\delta} \sqrt{q_\ell} \left(\max_{I \in \mathcal{G}_\ell} |I| \right)^{\frac{1}{2}} + 2 c q_\ell \right] \tag{11}$$

$$< 2^{-q_\ell} < 2^{-r_\ell}. \tag{12}$$

Finally, since $\sum_\ell 2^{-r_\ell} < \frac{1}{2}$, (11) implies existence of a set $S \subset \{1, \ldots, n\}$, $|S| = \delta n$ for which (8) holds for each ℓ. $\qquad\square$

2   Proof of the Result

Let $S \subset \{1, \ldots, n\}$, be the subset corresponding to the row restriction $|S| = m = \delta n$. It follows from (4) that

$$\left(1 - O(\varepsilon)\right) \sum_{i \in S} \tilde{x}_i^2 - O(\varepsilon^2 m) < \sum_{i \in S} x_i^2 < \left(1 + O(\varepsilon)\right) \sum_{i \in S} \tilde{x}_i^2 + O(\varepsilon^2 m).$$

Hence, it will suffice to ensure that

$$\sum_{i \in S} \tilde{x}_i^2 = \left(1 + O(\varepsilon)\right) m. \tag{13}$$

Note that

$$\sum_{i=1}^{n} \tilde{x}_i^2 = \left(1 + O(\varepsilon)\right) n = \sum_j (1 + \varepsilon)^{2j} |I_j| \tag{14}$$

and by (2)

$$\sum_{i \in S} \tilde{x}_i^2 = \sum_j (1 + \varepsilon)^{2j} |I_j \cap S|.$$

From (14),

$$|I_j| \le (1 + \varepsilon)^{-2j} n \equiv n_j \tag{15}$$

let

$$\mathcal{F}_{j,s} = \{I \in \mathcal{F}_j; |I| \sim 2^{-s} n_j\} \text{ for } 0 \le s < s_* \tag{16}$$

with s_* taken as to satisfy

$$2^{s_*} \sim \frac{\log r}{\varepsilon^2}. \tag{17}$$

Let then

$$\mathcal{F}_{j,s_*} = \{I \in \mathcal{F}_j, |I| < 2^{-s_*} n_j\}. \tag{18}$$

Apply Lemma 3 to the system $\{\mathcal{F}_{j,s}\}$, taking

$$r_{j,s} = r_j = \varepsilon^{-3} (\log n)^2 (1 + \varepsilon)^{-2j} r \ge \log |\mathcal{F}_j| \ge \log |\mathcal{F}_{j,s}| \tag{19}$$

recalling (5). Hence, by (3)

$$\sum_{j,s \le s_*} 2^{-r_{j,s}} \le s_* \sum_j \frac{1}{r_j} < \log\left(\frac{\log r}{\varepsilon^2}\right) \frac{\varepsilon^2}{(\log n)^2} < \frac{1}{2}. \tag{20}$$

Therefore, by Lemma 3 and (16), (18), we may introduce a set S, $|S| = m$ such that for all j and s and $I \in \mathcal{F}_{j,s}$

$$|I \cap S| = \delta|I| + O\big(r_j^{\frac{1}{2}}(\delta 2^{-s} n_j)^{\frac{1}{2}} + r_j\big). \tag{21}$$

Thus the second term in (21) is bounded by

$$(\log n)\varepsilon^{-3/2}(1+\varepsilon)^{-j}\sqrt{r}\sqrt{\delta.|I_j|} + (\log n)\varepsilon^{-3/2}(1+\varepsilon)^{-2j}\sqrt{r}\sqrt{2^{-s_*}}\sqrt{m}$$
$$+ (\log n)^2\varepsilon^{-3}(1+\varepsilon)^{-2j}r. \tag{22}$$

Recalling (14), we obtain

$$\sum_{i \in S}\tilde{x}_i^2 = \big(1 + O(\varepsilon)\big)m+ \tag{23}$$

where by (2), (3), (17)

$$(23) \lesssim (\log n)\varepsilon^{-3/2}\sqrt{r}\sqrt{\delta}\Big[\sum_j(1+\varepsilon)^j|I_j|^{\frac{1}{2}}\Big] + (\log n)(\log r)^{\frac{1}{2}}\varepsilon^{-\frac{3}{2}}\sqrt{r}\sqrt{m}$$
$$+ (\log n)^2\varepsilon^{-4}r\log r. \tag{24}$$

By Cauchy-Schwarz, (3), (14), the first term in (24) is at most

$$(\log n)\varepsilon^{-2}\sqrt{r}\sqrt{m}\sqrt{\log r}.$$

Hence

$$(23) \lesssim (\log n)\sqrt{\log r}\,\varepsilon^{-2}\sqrt{r}\sqrt{m} + (\log n)^2(\log r)\varepsilon^{-4}.r \tag{25}$$
$$< O(\varepsilon m)$$

provided

$$m > C(\log n)^2(\log r)\varepsilon^{-6}r. \tag{26}$$

Acknowledgements Research supported in part by NSF Grant DMS 1301619.

Reference

1. M. Rudelson, R. Vershynin, On sparse reconstruction from Fourier and Gaussian measurements. Commun. Pure Appl. Math. **61**(8), 1025–1045 (2008)

On Eigenvalue Spacings for the 1-D Anderson Model with Singular Site Distribution

Jean Bourgain

Abstract We study eigenvalue spacings and local eigenvalue statistics for 1D lattice Schrödinger operators with Hölder regular potential, obtaining a version of Minami's inequality and Poisson statistics for the local eigenvalue spacings. The main additional new input are regularity properties of the Furstenberg measures and the density of states obtained in some of the author's earlier work.

1 Introduction

This Note results from a few discussions with A. Klein (UCI, summer 011) on Minami's inequality and the results from [7] on Poisson local spacing behavior for the eigenvalues of certain Anderson type models. Recall that the Hamiltonian H on the lattice $\mathbb{Z}^d$ has the form

$$H = \lambda V + \Delta \tag{1}$$

with Δ the nearest neighbor Laplacian on $\mathbb{Z}^d$ and $V = (v_n)_{n \in \mathbb{Z}^d}$ IID variables with a certain distribution. Given a box $\Omega \subset \mathbb{Z}^d$, H_Ω denotes the restriction of H to Ω with Dirichlet boundary conditions. Minami's inequality, which is a refinement of Wegner's estimate, is a bound on the expectation that H_Ω has two eigenvalues in a given interval $I \subset \mathbb{R}$. This quantity can be expressed as

$$\mathbb{E}\big[Tr\mathcal{X}_I(H_\Omega)\big(Tr\mathcal{X}_I(H_\Omega) - 1\big)\big] \tag{2}$$

where the expectation is taken over the randomness V. An elegant treatment may be found in [6].

Assuming the site distribution has a bounded density, (2) satisfies the expected bound

$$C|\Omega|^2|I|^2. \tag{3}$$

J. Bourgain
Institute for Advanced Study, Princeton, NJ 08540, USA
e-mail: bourgain@math.ias.edu

© Springer International Publishing Switzerland 2014
B. Klartag, E. Milman (eds.), *Geometric Aspects of Functional Analysis*,
Lecture Notes in Mathematics 2116, DOI 10.1007/978-3-319-09477-9_6

More generally, considering a site distribution probability measure μ which is Hölder with exponent $0 < \beta \leq 1$, i.e.

$$\mu(I) \leq C|I|^\beta \text{ for all intervals } I \subset \mathbb{R} \tag{4}$$

it is shown in [6] that

$$(2) \leq C|\Omega|^2|I|^{2\beta}. \tag{5}$$

For the sake of the exposition, we briefly recall the argument. Rewrite (2) as

$$\mathbb{E}_V\left[\sum_{j \in \Omega} \langle \delta_j, \mathcal{X}_I(H_\Omega^{(V)})\delta_j\rangle \left(Tr\mathcal{X}_I(H_\Omega^{(V)}) - 1\right)\right] \tag{6}$$

where (δ_j) denote the unit vectors of $\mathbb{Z}^d$. Introduce a second independent copy $W = (w_n)$ of the potential V. Fixing $j \in \Omega$, denote by $(V_j^\perp, \tau_j)$ the potential with assignments v_n for $n \neq j$ and τ_j for $n = j$. Assuming $\tau_j \geq v_j$, it follows from the interlacing property for rank-one perturbations that

$$Tr\mathcal{X}_I(H_\Omega^{(V)}) \leq Tr\mathcal{X}_I\left(H_\Omega^{(V_j^\perp, \tau_j)}\right) + 1 \tag{7}$$

and hence

$$(6) \leq \mathbb{E}_V\mathbb{E}_W\left[\sum_{j \in \Omega} \langle \delta_j, \mathcal{X}_I(H_\Omega^{(V)})\delta_j\rangle \, Tr\mathcal{X}_I(H_\Omega^{(V_j^\perp, \|v_j\|_\infty + w_j)})\right]. \tag{8}$$

Next, invoking the fundamental spectral averaging estimate (see [6, Appendix A]), we have

$$\mathbb{E}_{v_j}[\langle \delta_j, \mathcal{X}_I(H_\Omega^{(V_j^\perp, v_j)})\delta_j] \leq C|I|^\beta \tag{9}$$

so that

$$(8) \leq C|I|^\beta \sum_{j \in \Omega} \mathbb{E}_{V_j^\perp}\mathbb{E}_{w_j}\left[Tr\mathcal{X}_I\left(H_\Omega^{(V_j^\perp, \|v_j\|_\infty + w_j)}\right)\right]. \tag{10}$$

The terms in (10) may be bounded using a Wegner estimate. Applying again (9), the j-term in (10) is majorized by $C|\Omega||I|^\beta$, leading to the estimate $C|I|^{2\beta}|\Omega|^2$ for (2). It turns out that at least in 1D, one can do better than reapplying the spectral averaging estimate. Indeed, it was shown in [2] that in 1D, SO's with Hölder regular site distribution have a smooth density of states. This suggests in (5) a better $|I|$-dependence, of the form $|I|^{1+\beta}$. Some additional work will be needed in order to turn the result from [2] into the required finite scale estimate. We prove the following (set $\lambda = 1$ in (1)).

Proposition 1. *Let H be a 1D lattice random SO with Hölder site distribution satisfying (4) for some $\beta > 0$. Denote $H_N = H_{[1,N]}$. Then*

$$\mathbb{E}[I \cap Spec\, H_N \neq \phi] \leq Ce^{-cN} + CN|I|. \tag{11}$$

It follows that $\mathbb{E}[Tr\mathcal{X}_i(H_N)] \leq Ce^{-cN} + CN^2|I|$.

The above discussion then implies the following Minami-type estimate.

Corollary 2. *Under the assumption from Proposition 1, we have*

$$\mathbb{E}[Tr\mathcal{X}_I(H_\Omega)(Tr\mathcal{X}_I(H_\Omega) - 1)] \leq C|\Omega|^3|I|^{1+\beta} \tag{12}$$

provided $\Omega \subset \mathbb{Z}$ is an interval of size $|\Omega| > C_1 \log(2 + \frac{1}{|I|})$, where C, C_1 depend on V.

Denote $\mathcal{N}$ the integrated density of states (IDS) of H and $k(E) = \frac{d\mathcal{N}}{dE}$. Recall that k is smooth for Hölder regular site distribution (cf. [2]).

Combined with Anderson localization, Proposition 1 and Corollary 2 permit to derive for H as above.

Proposition 3. *Assuming $\log \frac{1}{\delta} < cN$, we have for $I = [E_0 - \delta, E_0 + \delta]$ that*

$$\mathbb{E}[Tr\mathcal{X}_I(H_N)] = Nk(E_0)|I| + O\left(N\delta^2 + \delta \log\left(N + \frac{1}{\delta}\right)\right) \tag{13}$$

and

Proposition 4.

$$\mathbb{E}[H_\Omega \text{ has at least two eigenvalues in } I] \leq C|\Omega|^2|I|^2 + C|\Omega| \log^2\left(|\Omega| + \frac{1}{|I|}\right).|I|^{1+\beta}. \tag{14}$$

Following a well-known strategy, Anderson localization permits a decoupling for the contribution of pairs of eigenvectors with center of localization that are at least $C \log \frac{1}{|I|}$-apart. Invoking (11), this yields the first term in the r.h.s of (14). For the remaining contribution, use Corollary 2.

With Propositions 3, 4 at hand and again exploiting Anderson localization, the analysis from [7] becomes available and we obtain the following universality statement for 1D random SO's with Hölder regular site distribution.

Proposition 5. *Let $E_0 \in \mathbb{R}$ and $I = [E_0, E_0 + \frac{L}{N}]$ where we let first $N \to \infty$ and then $L \to \infty$. The rescaled eigenvalues*

$$\{N(E - E_0)\mathcal{X}_I(E)\}_{E \in Spec\, H_N}$$

converge to a Poisson point process in the limit $N \to \infty$.

At the end of the paper, we will make some comments on eigenvalue spacings for the Anderson-Bernoulli (A-B) model, where in (1) the v_n are $\{0, 1\}$-valued. Further results in line of the above for A-B models with certain special couplings λ will appear in [4].

2 Proof of Proposition 1

Set $\lambda = 1$ in (1). We denote

$$M_n = M_n(E) = \prod_{j=n}^{1} \begin{pmatrix} E - v_j & -1 \\ 1 & 0 \end{pmatrix} \tag{15}$$

the usual transfer operators. Thus the equation $H\xi = E\xi$ is equivalent to

$$M_n \begin{pmatrix} \xi_1 \\ \xi_0 \end{pmatrix} = \begin{pmatrix} \xi_{n+1} \\ \xi_n \end{pmatrix}. \tag{16}$$

Considering a finite scale $[1, N]$, let $H_{[1,N]}$ be the restriction of H with Dirichlet boundary conditions. Fix $I = [E_0 - \delta, E_0 + \delta]$ and assume $H_{[1,N]}$ has an eigenvalue $E \in I$ with eigenvector $\xi = (\xi_j)_{1 \le j \le N}$. Then

$$M_N(E) \begin{pmatrix} \xi_1 \\ 0 \end{pmatrix} = \begin{pmatrix} 0 \\ \xi_N \end{pmatrix}. \tag{17}$$

Assume $|\xi_1| \ge |\xi_N|$ (otherwise replace M_N by M_N^{-1} which can be treated similarly). It follows from (17) that

$$\|M_N(E)e_1\| \le 1 \tag{18}$$

with (e_1, e_2) the $\mathbb{R}^2$-unit vectors. On the other hand, from the large deviation estimates for random matrix products (cf. [1]), we have that

$$\log \|M_N(E_0)e_1\| > cN \tag{19}$$

with probability at least $1 - e^{-cN}$ (in the sequel, c, C will denote various constants that may depend on the potential).
Write

$$\left| \log \|M_N(E)e_1\| - \log \|M_N(E_0)e_1\| \right| \le \int_{-\delta}^{\delta} \left| \frac{d}{dt} [\log \|M_N(E_0 + t)e_1\|] \right| dt. \tag{20}$$

The integrand in (20) is clearly bounded by

$$\sum_{j=1,2} \sum_{n=1}^{N} \frac{|\langle M_{N-n}^{(v_N,\dots,v_{n+1})}(E_0+t)e_1, e_j\rangle| \cdot |\langle M_{n-1}^{(v_{n-1},\dots,v_1)}(E_0+t)e_1, e_1\rangle|}{\|M_N^{(v_N,\dots,v_1)}(E_0+t)e_1\|} \tag{21}$$

$$\leq 2|E-E_0| \sum_{n=1}^{N} \frac{\|M_{N-n}^{(v_N,\dots,v_{n+1})}(E_0+t)\|}{\|M_{N-n}^{(v_N,\dots,v_{n+1})}(E_0+t)\zeta_n\|} \tag{22}$$

where

$$\zeta_n = \frac{M_{n-1}^{(v_{n-1},\dots,v_1)}(E_0+t)e_1}{\|M_{n-1}^{(v_{n-1},\dots,v_1)}(E_0+t)e_1\|} \tag{23}$$

depends only on the variables $v_1, \dots, v_{n-1}$.

At this point, we invoke some results from [2]. It follows from the discussion in [2, Sect. 5] on SO's with Hölder potential that for $\ell > C = C(V)$, the inequality

$$\mathbb{E}_{v_1,\dots,v_\ell}[\|M_\ell(\zeta)\| < \varepsilon \|M_\ell\|] \lesssim \varepsilon \tag{24}$$

holds for any $\varepsilon > 0$ and unit vector $\zeta \in \mathbb{R}^2$, $M_\ell = M_\ell^{(v_1,\dots,v_\ell)}$.

A word of explanation. It is proved in [2] that if we take n large enough, the map $(v_1,\dots,v_n) \mapsto M_n^{(v_n,\dots,v_n)}$ defines a bounded density on $SL_2(\mathbb{R})$. Fix then some $n = O(1)$ with the above property and write for $\ell > n$,

$$\|M_\ell(\zeta)\| \geq |\langle M_n(\zeta), M_{\ell-n}^* e_j\rangle| \qquad (j=1,2)$$

noting that here M_n and $M_{\ell-n}$ are independent as functions of the potential. Choose j such that $\|M_{\ell-n}^* e_j\| \sim \|M_{\ell-n}^*\| = \|M_{\ell-n}\| \sim \|M_\ell\|$ and fix the vector $M_{\ell-n}^* e_j$. Since then $(v_1,\dots,v_n) \mapsto M_n(\zeta)$ defines a bounded density, inequality (24) holds.

Since always $\|M_\ell\| < C^\ell$ and $\|M_\ell(\zeta)\| > C^{-\ell}$, it clearly follows from (24) that

$$\mathbb{E}_V\left[\frac{\|M_\ell^{(V)}\|}{\|M_\ell^{(V)}(\zeta)\|}\right] \leq C\ell. \tag{25}$$

Therefore

$$\mathbb{E}_V[(22)] < CN^2\delta. \tag{26}$$

Hence, we showed that, assuming (19), $\mathrm{Spec}\, H_N^{(V)} \cap I \neq \phi$ with probability at most $CN\delta$. Therefore $\mathrm{Spec}\, H_N^{(V)} \cap I \neq \phi$ with probability at most $CN\delta + Ce^{-cN}$, proving (11).

3 Proof of Propositions 3 and 4

Assume $\log \frac{1}{|I|} < cN$ and set $M = C \log \left(N + \frac{1}{|I|} \right)$ for appropriate constants c, C. From the theory of Anderson localization in 1D, the eigenvectors ξ_α of $H_N, |\xi_\alpha| = 1$ satisfy

$$|\xi_\alpha(j)| < e^{-c|j-j_\alpha|} \text{ for } |j - j_\alpha| > \frac{M}{10} \qquad (27)$$

with probability at least $1 - e^{-cM}$, with j_α the center of localization of ξ_α.

The above statement is well-known and relies on the large deviation estimates for the transfer matrix. Let us also point out however that the above (optimal) choice of M is not really important in what follows and taking for M some power of the log would do as well.

We may therefore introduce a collection of intervals $(\Lambda_s)_{1 \leq s \lesssim \frac{N}{M}}$ of size M covering $[1, N]$, such that for each α, there is some $1 \leq s \lesssim \frac{N}{M}$ satisfying

$$j_\alpha \in \Lambda_s \text{ and } \|\xi_\alpha|_{[1,N]\setminus\Lambda_s}\| < e^{-cM} \qquad (28)$$

$$\|(H_{\Lambda_s} - E_\alpha)\xi_{\alpha,s}\| < e^{-cM} \qquad (29)$$

with $\xi_{\alpha,s} = \xi_\alpha|\Lambda_s$. Therefore dist $(E_\alpha, \text{Spec } H_{\Lambda_s}) < e^{-cM} < \delta$.

Let us establish Proposition 3. Denoting Λ_1 and Λ_{s_*} the intervals appearing at the boundary of $[1, N]$, one obtains by a well-known argument based on exponential localization

$$\mathbb{E}[Tr\mathcal{X}_I(H_N)] = N.\mathcal{N}(I) + O\left(e^{-cM} + \mathbb{E}[Tr\mathcal{X}_{\tilde{I}}(H_{\Lambda_1})] + \mathbb{E}[Tr\mathcal{X}_{\tilde{I}}(H_{\Lambda_{s_*}})]\right) \qquad (30)$$

with $\tilde{I} = [E_0 - 2\delta, E_0 + 2\delta]$. Invoking then Proposition 1 and Corollary 2, we obtain

$$\mathbb{E}[Tr\mathcal{X}_I(H_{\Lambda_s})] < ce^{-cM} + CM\delta + CM^3\delta^{1+\beta} < CM\delta \qquad (31)$$

by the choice of M and assuming $(\log N)^2\delta^\beta < 1$, as we may.

Substituting (31) in (30) gives then

$$N \int_I k(E)dE + O(M\delta)$$

$$= Nk(E_0)|I| + O\left(N\delta^2 + \delta \log\left(N + \frac{1}{\delta}\right)\right)$$

since k is Lipschitz. This proves (13).

Next, we prove Proposition 4.

Assume $E_\alpha, E_{\alpha'} \in I, \alpha \neq \alpha'$. We distinguish two cases.

Case 1. $|j_\alpha - j_{\alpha'}| > CM$.

Here C is taken large enough as to ensure that the corresponding boxes $\Lambda_s, \Lambda_{s'}$ introduced above are disjoint. Thus

$$\text{Spec } H_{\Lambda_s} \cap I \neq \phi \tag{32}$$

$$\text{Spec } H_{\Lambda_{s'}} \cap I \neq \phi. \tag{33}$$

Since the events (32), (33) are independent, it follows from Proposition 1 that the probability for the joint event is at most

$$Ce^{-cM} + CM^2\delta^2 < CM^2\delta^2 \tag{34}$$

by our choice of M. Summing over the pairs $s, s' \lesssim \frac{N}{M}$ gives therefore the bound $CN^2\delta^2$ for the probability of a Case 1 event.

Case 2. $|j_\alpha - j_{\alpha'}| \leq CM$.

We obtain an interval Λ as union of at most C consecutive Λ_s-intervals such that (28), (29) hold with Λ_s replaced by Λ for both $(\xi_\alpha, E_\alpha), (\xi_{\alpha'}, E_{\alpha'})$. This implies that Spec $H_\Lambda \cap \tilde{I}$ contains at least two elements. By Corollary 2, the probability for this is at most $CM^3\delta^{1+\beta}$. Hence, we obtain the bound $CM^2N\delta^{1+\beta}$ for the Case 2 event.

The final estimate is therefore

$$e^{-cM} + CN^2\delta^2 + CM^2N\delta^{1+\beta}$$

and (14) follows from our choice of M.

4 Sketch of the Proof of Proposition 5

Next we briefly discuss local eigenvalue statistics, following [7].

The Wegner and Minami type estimates obtained in Propositions 3 and 4 above permit to reproduce essentially the analysis from [G-K] proving local Poisson statistics for the eigenvalues of H_N^ω. We sketch the details (recall that we consider a 1D model with Hölder site distribution).

Let $M = K \log N, M_1 = K_1 \log N$ with $K \gg K_1 \gg 1 \ (\to \infty$ with $N)$ and partition

$$\Lambda = [1, N] = \Lambda_1 \cup \Lambda_{1,1} \cup \Lambda_2 \cup \Lambda_{2,1} \cup \ldots = \bigcup_{\alpha \lesssim \frac{N}{M+M_1}} (\Lambda_\alpha \cup \Lambda_{\alpha,1})$$

where Λ_α (resp. $\Lambda_{\alpha,1}$) are M (resp. M_1) intervals

Denote

$$\mathcal{E}_\alpha = \text{eigenvalue of } H_\Lambda \text{ with center of localization in } \Lambda_\alpha$$
$$\mathcal{E}_{\alpha,1} = \underline{\hspace{3cm}} \qquad \Lambda_{\alpha,1}$$

Let Λ'_α (resp. $\Lambda'_{\alpha,1}$) be a neighborhood of Λ_α (resp. $\Lambda_{\alpha,1}$) of size $\sim \log N$ taken such as to ensure that

$$\text{dist}(E, \text{ Spec } H_{\Lambda'_\alpha}) < \frac{1}{N^A} \text{ for } E \in \mathcal{E}_\alpha$$

(A a sufficiently large constant), and

$$\text{dist}(E, \text{ Spec } H_{\Lambda'_{\alpha,1}}) < \frac{1}{N^A} \text{ for } E \in \mathcal{E}_{\alpha,1}. \tag{35}$$

Choosing K_1 large enough, we ensure that the Λ'_α are disjoint and hence $\{\text{Spec } H^\omega_{\Lambda'_\alpha}\}$ are independent.

Consider an energy interval

$$I = \left[E_0, E_0 + \frac{L}{N}\right]$$

Denote

$$P_\Omega(I) = \mathcal{X}_I(H_\Omega)$$

with L a large parameter, eventually $\to \infty$.

We obtain from (11) and our choice of M_1 that

$$\mathbb{P}[\mathcal{E}_{\alpha,1} \cap I \neq \phi] \lesssim M_1|I|$$

and hence

$$\mathbb{P}[\bigcup_\alpha \mathcal{E}_{\alpha,1} \cap I \neq \phi] \lesssim \frac{N}{M}M_1|I| \lesssim \frac{LK_1}{K} = o(1) \tag{36}$$

provided

$$K_1 L = o(K). \tag{37}$$

Also, by (12)

$$\mathbb{P}[|\mathcal{E}_\alpha \cap I| \geq 2] \leq$$

$$\mathbb{P}[H_{\Lambda'_\alpha} \text{ has at least two eigenvalues in } \tilde{I}] \lesssim M^3 |I|^{1+\beta} < M^3 \frac{L^{1+\beta}}{N^{1+\beta}} \tag{38}$$

so that

$$\mathbb{P}[\max_\alpha |\mathcal{E}_\alpha \cap I| \geq 2] \lesssim \frac{N}{M}(38) \lesssim \frac{M^2 L^{1+\beta}}{N^\beta} < N^{-\beta/2}. \tag{39}$$

Next, we introduce the (partially defined) random variables

$$E_\alpha(V) = \sum_{E \in \text{Spec } H_{\Lambda'_\alpha} \cap I} E \quad \text{provided } |\text{Spec } H_{\Lambda'_\alpha} \cap I| \leq 1. \tag{40}$$

Thus the $E_\alpha, \alpha = 1, \ldots, \frac{N}{M+M_1}$ take values in I, are independent since $\{\text{Spec } H_{\Lambda'_\alpha}\}$ are independent and have the same distribution.

Let $J \subset I$ be an interval, $|J|$ of the order of $\frac{1}{N}$. Then by (38) and Proposition 3.

$$\mathbb{E}[1_J(E_\alpha)] = \mathbb{E}[Tr \, P_{\Lambda'_\alpha}(J)] + O\left(\frac{1}{N^{1+\beta/2}}\right) = k(E_0)\left(1 + O\left(\frac{1}{K}\right)\right)|J|M' \tag{41}$$

where $M' = |\Lambda'_\alpha|$.

Therefore $\{N(E_\alpha - E_0)\}_{\alpha \leq \frac{N}{M+M_1}}$ converge in distribution to a Poisson process (in a weak sense), proving Proposition 5.

5   Comments on the Bernoulli Case

Consider the model (1) with $V = (v_n)_{n \in \mathbb{Z}}$ independent $\{0, 1\}$-valued. For large $|\lambda|$, H does not have a bounded density of states. It was shown in [3] that for certain small algebraic values of the coupling constant λ, $k(E) = \frac{d\mathcal{N}}{dE}$ can be made arbitrarily smooth (see [3] for the precise statement). In particular $k \in L^\infty$ and one could ask if Proposition 4 remains valid in this situation. One could actually conjecture that the analogue of Proposition 4 holds for the A-B model in 1D, at small disorder. This problem will be pursued further in [4]. What we prove here is an eigenvalue separation property at finite scale for the A-B model at arbitrary disorder $\lambda \neq 0$. Denote again H_N the restriction of H to $[1, N]$ with Dirichlet boundary conditions. We have

Proposition 6. *With large probability, the eigenvalues of H_N are at least N^{-C} separated, $C = C(\lambda)$.*

A statement of this kind is known for random SO's with Hölder site distribution of regularity $\beta > \frac{1}{2}$, in arbitrary dimension [6]. But note that our proof of

Proposition 6 is specifically 1D, as will be clear below. There are three ingredients, each well-known.

1. Anderson localization

Anderson localization holds also for the 1D A-B model at any disorder. In fact, there is the following quantitative form. Denote $\xi^{(1)}, \ldots, \xi^{(N)}$ the normalized eigenvectors of H_N. Then, with large probability ($> 1 - N^{-A}$), each $\xi^{(j)}$ is essentially localized on some interval of size $C(\lambda) \log N$, in the sense that there is a center of localization $\nu_j \in [1, N]$ such that

$$|\xi_n^{(j)}| < e^{-c(\lambda)|n - \nu_j|} \text{ for } |n - \nu_j| > C(\lambda) \log N. \tag{42}$$

2. Hölder regularity of the IDS

The IDS $\mathcal{N}(E)$ of H is Hölder of exponent $\gamma = \gamma(\lambda) > 0$. There are various proofs of this fact (see in particular [5] and [8]). In fact, it was shown in [2] that $\gamma(\lambda) \to 1$ for $\lambda \to 0$ but we will not need this here. What we use is the following finite scale consequence.

Lemma 7. *Let* $M \in \mathbb{Z}_+$, $E \in \mathbb{R}$, $\delta > 0$. *Then*

$$\mathbb{E}[\text{there is a vector } \xi = (\xi_j)_{1 \leq j \leq M}, \|\xi\| = 1, \text{ such that}$$
$$\|(H_M - E)\xi\| < \delta, |\xi_1| < \delta, |\xi_M| < \delta] \leq CM\delta^\gamma. \tag{43}$$

The derivation is standard and we do briefly recall the argument.

Take $N \to \infty$ and split $[1, N]$ in intervals of size M. Denoting τ the l.h.s. of (43), we see that

$$\mathbb{E}\left[\#(\operatorname{Spec} H_N \cap [E - 5\delta, E + 5\delta])\right] \geq \frac{N}{M}\tau.$$

Dividing both sides by N and letting $N \to \infty$, one obtains that

$$\frac{\tau}{M} \leq \mathcal{N}([E - 5\delta, E + 5\delta])$$

where $\mathcal{N}$ is the IDS of H.

3. A repulsion phenomenon

The next statement shows that eigenvectors with eigenvalues that are close together have their centers far away. The argument is based on the transfer matrix and hence strictly 1D.

Lemma 8. *Let* ξ, ξ' *be distinct normalized eigenvectors of* H_N *with centers* ν, ν',

$$H_N \xi = E \xi$$
$$H_N \xi' = E' \xi. \tag{44}$$

Assuming $|E - E'| < N^{-C(\lambda)}$, it follows that

$$|\nu - \nu'| \gtrsim \log \frac{1}{|E - E'|}. \tag{45}$$

Proof. Let $\delta = |E - E'|$ and assume $1 \leq \nu \leq \nu' \leq N$. Take $M = C(\lambda) \log N$ satisfying (42) and Λ an M-neighborhood of $[\nu, \nu']$ in $[1, N]$.

In particular, we ensure that

$$|\xi_n|, |\xi_n'| < N^{-10} \text{ for } n \notin \Lambda. \tag{46}$$

We can assume that $|\xi_\nu| > \frac{1}{2\sqrt{M}}$. Since $\|\xi_\nu'\xi - \xi_\nu\xi'\| \geq |\xi_\nu| > \frac{1}{2\sqrt{M}}$, it follows from (46) that for some $n_0 \in \Lambda$

$$|\xi_\nu'\xi_{n_0} - \xi_\nu\xi_{n_0}'| \gtrsim \frac{1}{\sqrt{M}\sqrt{|\Lambda|}}. \tag{47}$$

Next, denote for $n \in [1, N]$

$$D_n = \xi_\nu'\xi_n - \xi_\nu\xi_n'$$

and

$$W_n = \xi_n'\xi_{n+1} - \xi_n\xi_{n+1}'.$$

Clearly, using Eq. (44)

$$\|(H_N - E)D\| \leq \delta \tag{48}$$

and

$$\sum_{1 \leq n < N} |W_n - W_{n+1}| < \delta. \tag{49}$$

Let $\nu < N$. Since $D_\nu = 0$, it follows from (48) that

$$|D_n| \leq (2 + |\lambda| + |E|)^{|n-\nu|}(|D_{\nu+1}| + 2\delta). \tag{50}$$

(If $\nu = N$, replace $\nu + 1$ by $\nu - 1$). From (47), (50)

$$\frac{1}{\sqrt{M}\sqrt{|\Lambda|}} \lesssim (2 + |\lambda| + |E|)^{|\Lambda|}(|D_{\nu+1}| + 2\delta)$$

and since $D_{\nu+1} = W_\nu$, it follows that

$$|W_\nu| + 2\delta > 10^{-|\Lambda|}. \tag{51}$$

Invoking (49), we obtain for $n \in [1, N]$

$$|W_n| > 10^{-|\Lambda|} - (|n - \nu| + 1)\delta. \tag{52}$$

On the other hand, by (42)

$$|W_n| \le |\xi_n| + |\xi_{n+1}| < e^{-c\lambda^2|n-\nu|} \text{ for } |n - \nu| > C(\lambda) \log N.$$

Taking $|n - \nu| \sim |\Lambda|$ appropriately, it follows that

$$\delta \gtrsim \frac{1}{|\Lambda|} 10^{-|\Lambda|}$$

and hence

$$|\nu - \nu'| + M \gtrsim \log \frac{1}{\delta}.$$

Lemma 8 follows. $\square$

Proof of Proposition 6. Assume H_N has two eigenvalues E, E' such that

$$|E - E'| < \delta < N^{-C_1}$$

where C_1 is the constant from Lemma 8. It follows that the corresponding eigenvectors ξ, ξ' have resp. centers $\nu, \nu' \in [1, N]$ satisfying

$$|\nu - \nu'| \gtrsim \log \frac{1}{\delta}. \tag{53}$$

Introduce $\delta_0 > \delta$ (to specify), $M = C_2(\lambda) \log \frac{1}{\delta_0}$ and $\Lambda = [\nu - M, \nu + M] \cap [1, N]$, $\Lambda' = [\nu' - M, \nu' + M] \cap [1, N]$. Let $\tilde{\xi} = \frac{\xi|_\Lambda}{\|\xi|_\Lambda\|}, \tilde{\xi}' = \frac{\xi'|_{\Lambda'}}{\|\xi'|_{\Lambda'}\|}$. According to (42), choose M such that

$$\|(H_\Lambda - E)\tilde{\xi}\| < e^{-c\lambda^2 M} < \delta_0 \text{ and } |\xi|_{\partial\Lambda}| < \delta_0 \tag{54}$$

and

$$\|H_{\Lambda'} - E')\tilde{\xi}'\| < \delta_0 \text{ and } |\xi'|_{\partial\Lambda'}| < \delta_0. \tag{55}$$

Requiring

$$\log \frac{1}{\delta} > C_3 M$$

(53) will ensure disjointness of Λ, Λ'. Hence H_Λ, $H_{\Lambda'}$ are independent as functions of V. It follows in particular from (54) that dist $(E, \text{Spec } H_\Lambda) < \delta_0$, hence $|E - E_0| < \delta_0$ for some $E_0 \in \text{Spec } H_\Lambda$. Having fixed E_0, (55) implies that

$$\|(H_{\Lambda'} - E_0)\xi'\| < |E - E'| + 2\delta_0 < 3\delta_0. \tag{56}$$

Apply Lemma 7 to $H_{\Lambda'}$ in order to deduce that the probability for (56) to hold with $E_0 \in \text{Spec } H_\Lambda$ fixed, is at most $CM\delta_0^\gamma$. Summing over all $E_0 \in \text{Spec } H_\Lambda$ and then over all pairs of boxes Λ, Λ' gives the bound

$$O(N^2 M^2 \delta_0^\gamma) = O\left(N^2 \left(\log \frac{1}{\delta_0}\right)^2 \delta_0^\gamma\right) < N^2 \delta_0^{\gamma/2}. \tag{57}$$

It remains to take $\delta_0 = N^{-\frac{5}{\gamma}}$, $\log \frac{1}{\delta} > C \log \frac{1}{\delta_0}$. $\qquad\square$

Acknowledgements The author is grateful to an anonymous referee and A. Klein for comments and to the UC Berkeley mathematics department for their hospitality. This work was partially supported by NSF grant DMS-1301619.

References

1. Ph. Bougerol, J. Lacroix, *Products of Random Matrices with Applications to Schrödinger Operators* (Birkhauser, Boston, 1985)
2. J. Bourgain, On the Furstenberg measure and density of states for the Anderson-Bernoulli model at small disorder. J. Anal. Math. **117**, 273–295 (2012)
3. J. Bourgain, An application of group expansion to the Anderson-Bernoulli model. Preprint 07/13
4. J. Bourgain, On the local eigenvalue spacings for certain Anderson-Bernoulli Hamiltonians. Preprint 08/13
5. R. Carmona, A. Klein, G. Martinelli, Anderson localization for Bernoulli and other singular potentials. Commun. Math. Phys. **108**, 41–66 (1987)
6. J.-M. Combes, F. Germinet, A. Klein, Generalized eigenvalue - counting estimates for the Anderson model. J. Stat. Phys. **135**, 201–216 (2009)
7. F. Germinet, F. Klopp, Spectral statistics for random Schrödinger operators in the localized regime. JEMS (to appear)
8. C. Shubin, T. Vakilian, T. Wolff, Some harmonic analysis questions suggested by Anderson-Bernoulli models. Geom. Funct. Anal. **8**, 932–964 (1988)

On the Local Eigenvalue Spacings for Certain Anderson-Bernoulli Hamiltonians

Jean Bourgain

Abstract The aim of this work is to extend the results from Bourgain (On eigenvalue spacings for the 1D Anderson model with singular site distribution) on local eigenvalue spacings to certain 1D lattice Schrodinger with a Bernoulli potential. We assume the disorder satisfies a certain algebraic condition that enables one to invoke the recent results from Bourgain (An application of group expansion to the Anderson-Bernoulli model. Preprint) on the regularity of the density of states. In particular we establish Poisson local eigenvalue statistics in those models.

1 Introduction

The aim of this Note is to exploit the results from [2] on certain Anderson-Bernoulli (A-B) Hamiltonians, in order to extend some of the eigenvalue spacing properties obtained in [3] for Hamiltonians with Hölder site-distribution to the A-B setting.

As in [3], all models are 1D. Recall that the A-B Hamiltonian with coupling λ is given by

$$H = H_\lambda = \Delta + \lambda V \tag{1}$$

where $V = (v_1)_{n \in \mathbb{Z}}$ are IID-variables ranging in $\{-1, 1\}$, $\mathbb{E}[v_n = -1] = \frac{1}{2} = \text{Prob}[v_n = 1]$. It is believed that for $\lambda \neq 0$ sufficiently small, the integrated density of states (IDS) $\mathcal{N}$ of H is Lipschitz and becomes arbitrary smooth for $\lambda \to 0$. A first result in this direction was obtained in [2], for small λ with certain specific algebraic properties.

Proposition 1 (See [2]). *Let H_λ be the A-B model considered above and restrict $|E| < 2 - \delta_0$ for some fixed $\delta_0 > 0$. Given a constant $C > 0$ and $k \in \mathbb{Z}_+$, there is some $\lambda_0 = \lambda_0(C, k) > 0$ such that $\mathcal{N}(E)$ is C^k-smooth on $[-2 + \delta_0, 2 - \delta_0]$ provided λ satisfies the following conditions.*

J. Bourgain (✉)
Institute for Advanced Study, Princeton, NJ 08540, USA
e-mail: bourgain@math.ias.edu

© Springer International Publishing Switzerland 2014
B. Klartag, E. Milman (eds.), *Geometric Aspects of Functional Analysis*,
Lecture Notes in Mathematics 2116, DOI 10.1007/978-3-319-09477-9_7

$$|\lambda| < \lambda_0 \tag{2}$$

$$\lambda \text{ is an algebraic number of degree } d < C \text{ and minimal polynomial} \tag{3}$$

$$P_d(x) \in \mathbb{Z}[X] \text{ with coefficients bounded by } \left(\frac{1}{\lambda}\right)^C$$

$$\lambda \text{ has a conjugate } \lambda' \text{ of modulus } |\lambda'| \geq 1 \tag{4}$$

In what follows, we assume λ satisfies the conditions of Proposition 1 and the energy E restricted to $[-2 + \delta_0, 2 - \delta_0]$, unless specified differently.

Once we are in the presence of the Hamiltonian with a bounded density of states $k(E) = \frac{d\mathcal{N}}{dE}$, it becomes a natural question to inquire about local eigenvalue statistics for 'truncated' models H_N, denoting H_N the restriction of H to the interval $[1, N]$ with Dirichlet boundary conditions. This problem was explored in [3], assuming the site distribution v_n of V Hölder regular of some exponent $\beta > 0$, and we extended (in 1D) the theorem from [5] on Poisson statistics in this setting. Here we consider the A-B situation.

Proposition 2. *With H as in Proposition 1, the rescaled eigenvalues of H_N*

$$\{N(E - E_0)\mathcal{X}_I(E)\}_{E \in \text{Spec } H_N}$$

where $I = [E_0, E_0 + \frac{L}{N}]$ and we let first $N \to \infty$, then $L \to \infty$, obey Poisson statistics.

This is the analogue of [3, Proposition 5]. Again one could conjecture the above statement to hold under the sole assumption that λ be sufficiently small.

Proposition 2 gives a natural example of a Jacobi Schrödinger operator with bounded density of states where the local eigenvalue spacing distribution differs from that of the potential (cf. S. Jitomirskaya's talk 'Eigenvalue statistics for ergodic localization', Berkeley 11/10/2010).

Even with the smoothness of the IDS at hand, the arguments from [3] do not carry over immediately to the A-B setting. For instance, the 'classical' approach to Minami's inequality (see [4]) rests also on regularity of the single site distribution (in addition to a Wegner estimate) which makes it inapplicable in the A-B case. This will require us to develop an alternative argument in order to deal with near resonant eigenvalues.

Roughly speaking, it turns out that for the analysis below, the following ingredients suffice.

(1) The Furstenberg measures v_E are absolutely continuous with bounded density
(2) The density of states k is C^1

Hence the results from [3] for Hölder site distribution follow from the present treatment. We believe however that the presentation in [3] remains of interest since it is considerably simpler than the method from this paper.

As in [3], the techniques are very much 1D and based on the usual transfer matrix formalism.

Recall that

$$M_n = M_n(E) = \prod_{j=n}^{1} \begin{pmatrix} E - v_j & -1 \\ 1 & 0 \end{pmatrix} \tag{5}$$

and that the equation $H\xi = E\xi$ on the positive side is equivalent to

$$M_n \begin{pmatrix} \xi_1 \\ \xi_0 \end{pmatrix} = \begin{pmatrix} \xi_{n+1} \\ \xi_n \end{pmatrix}. \tag{6}$$

What follows will use extensively ideas and techniques developed in [2,3].

2 Preliminary Estimates

Denote v_E the Furstenberg measure at energy E. This is the unique probability measure on $P_1(\mathbb{R}) \simeq S^1$ which is $\mu = \frac{1}{2}(\delta_{g_+} + \delta_{g_-})$ - stationary, where

$$g_+ = \begin{pmatrix} E + \lambda & -1 \\ 1 & 0 \end{pmatrix} \text{ and } g_- = \begin{pmatrix} E - \lambda & -1 \\ 1 & 0 \end{pmatrix}. \tag{7}$$

Thus

$$v_E = \sum_g \mu(g)\tau_g^*[v_E] \tag{8}$$

where τ_g denotes the projective action of $g \in SL_2(\mathbb{R})$.

It was proven in [2] that in the context of Proposition 1, v_E is absolutely continuous wrt Haar measure on S^1 and moreover $\frac{dv_E}{d\theta}$ becomes arbitrarily smooth for $\lambda \to \infty$.

The results of this section are stated for A-B Hamiltonians without further assumption on the specific nature of the coupling constant however and rely on general random matrix product theory.

Lemma 1 (cf. [1]). *Let ξ, η be unit vectors in $\mathbb{R}^2$. Given E and $\varepsilon > 0$, $N > C(\lambda)\log\frac{1}{\varepsilon}$ we have*

$$\mathbb{E}[|\langle M_N(E)(\xi), \eta\rangle| < \varepsilon\|M_N(E)(\xi)\|] \leq \tau(\varepsilon) \tag{9}$$

where $\tau(\varepsilon) = \max_{|I|=\varepsilon} v_E(I), I \subset S^1$ an interval.

Proof. Let $\xi = e^{i\theta}$, $\eta = e^{i\psi}$. If $M_N = g_N \cdots g_1$, $g_j \in \{g_+, g_-\}$, then

$$\left| \frac{\langle M_N(\xi), \eta \rangle}{\|M_N(\xi)\|} \right| = \left| \cos(\tau_{g_N \cdots g_1}(\theta) - \psi) \right|.$$

Hence the l.h.s. of (9) is bounded by

$$\mathbb{E}_{g_1,\ldots,g_N}\left[|\tau_{g_N \cdots g_1}(\theta) - \psi'| < \varepsilon \right] \qquad (\psi' = \psi^{\perp})$$

$$= \sum_g \mu^{(N)}(g) \, 1_{[\psi'-\varepsilon, \psi'+\varepsilon]}(\tau_g(\theta)). \tag{10}$$

Let $0 \le f \le 1$ be a smooth function on S^1 such that $f = 1$ on $[\psi' - \varepsilon, \psi' + \varepsilon]$, supp $f \subset [\psi' - 2\varepsilon, \psi' + 2\varepsilon]$ and $|\partial^\alpha f| \lesssim_\alpha \left(\frac{1}{\varepsilon}\right)^\alpha$. Then, invoking the large deviation estimate for the μ-random walk, we obtain

$$(10) \le \sum_g \mu^{(N)}(g)(f \circ \tau_g)$$

$$\le \int f \, d\nu_E + e^{-c(\lambda)N}\|f\|_{C^1}$$

$$\le \nu_E([\psi' - 2\varepsilon, \psi' + 2\varepsilon]) + \frac{1}{\varepsilon} e^{-c(\lambda)N}$$

proving the lemma.   □

Lemma 2. *Assume the Lyapounov exponent Hölder regular of exponent $\alpha > 0$. Then*

$$\max_{|E-E_0|<\kappa} \log \|M_N(E)\| < L(E_0)N + c\kappa^\alpha N \tag{11}$$

outside a set Ω of measure at most $e^{-\kappa' N}$.

Proof. Recall the large deviation theorem for the Lyapounov exponent

$$\mathbb{E}\left[\left| \frac{1}{N} \log \|M_N(E)\| - L(E) \right| > \sigma \right] \lesssim e^{-\sigma' N}. \tag{12}$$

Set $\kappa_1 = C\kappa^\alpha$. It follows from (12) that for $|E - E_0| < \kappa$

$$\mathbb{E}[\log \|M_N(E)\| > L(E_0)N + \kappa_1 N] \lesssim e^{-\kappa' N} \tag{13}$$

since $|L(E) - L(E_0)| \lesssim \kappa^\alpha$.

More generally, given indices $N > \ell_1 > \ell_2 > \cdots > \ell_r > 1$ $(r = O(1))$, we have for $|E - E_0| < \kappa$ that

$$\mathbb{E}[\log \| M_{N-\ell_1}^{v_N,\dots,v_{\ell_1+1}}(E)\| + \log \| M_{\ell_1-\ell_2-1}^{(v_{\ell_1-1},\dots,v_{\ell_2+1})}(E)\| + \cdots + \log \| M_{\ell_r-1}^{(v_{\ell_r-1},\dots,v_1)}(E)\|$$

$$> L(E_0)N + \kappa_1 N] \lesssim e^{-\kappa' N}.$$
(14)

Set $\theta = e^{-\frac{1}{2}\kappa' N}$ and let $\mathcal{E}$ be a finite subset of $|E - E_0| < \kappa$, $|\mathcal{E}| < \frac{1}{\theta}$ such that $\max_{|E-E_0|<\kappa} \text{dist}(E, \mathcal{E}) < \theta$. Take $r = r(\kappa)$ and $\Omega \subset \{1, -1\}^N$ such that (14) holds for $V \notin \Omega$, $E \in \mathcal{E}$ and all $N > \ell_1 > \cdots > \ell_r > 1$,

$$|\Omega| \lesssim N^r.|\mathcal{E}| e^{-\kappa' N} < e^{-\frac{1}{3}\kappa' N}.$$
(15)

Take then $E \in [E_0 - \kappa, E_0 + \kappa]$ and $E_1 \in \mathcal{E}$, $|E - E_1| < \theta$. Using a truncated Taylor expansion, we get

$$M_N(E) = M_N(E_1) + \left[\sum_{1 \le \ell \le N} M_{N-\ell}(E_1) \begin{pmatrix} 1 & 0 \\ 0 & 0 \end{pmatrix} M_{\ell-1}(E_1)\right](E - E_1) + \cdots$$

$$+ \frac{1}{r!}\left[\sum_{1 \le \ell_1 < \ell_2 < \cdots < \ell_r \le N} M_{N-\ell_1}(E_1) \begin{pmatrix} 1 & 0 \\ 0 & 0 \end{pmatrix} M_{\ell_1-\ell_2-1}(E_1) \begin{pmatrix} 1 & 0 \\ 0 & 0 \end{pmatrix} \cdots M_{\ell_r-1}(E_1)\right](E - E_1)^r$$

$$+ O(C^N \theta^{r+1}).$$
(16)

Taking $r \sim \frac{1}{\kappa'}$, we ensure the remainder term $< e^{-N}$, while for $V \notin \Omega$

$$\|(16)\| < e^{(L(E_0)+\kappa_1)N} + \theta e^{(L(E_0)+\kappa_1)N} + \cdots + \theta^r e^{(L(E_0)+\kappa_1)N}$$

Lemma 2 follows. □

3 A Wegner Estimate

Proposition 3. *Assume the Furstenberg measures of H have bounded density. Then*

$$\mathbb{E}[\text{Spec } H_N \cap I \ne \phi] < CN.|I| + Ce^{-cN}$$
(17)

if $I \subset R$ is an interval.

Proof. What follows is an adaptation of the argument used in [3]. Let $I = [E_0 - \delta, E_0 + \delta]$ and assume H_N has an eigenvalue $E \in I$ with eigenvector $\xi = (\xi_j)_{1 \le j \le N}$. Then

$$M_N(E) \begin{pmatrix} \xi_1 \\ 0 \end{pmatrix} = \begin{pmatrix} 0 \\ \xi_N \end{pmatrix}.$$

Assume $|\xi_1| \geq |\xi_N|$ (otherwise, we replace M_N by M_N^{-1}). It follows that

$$\|M_N(E)e_1\| \leq 1. \tag{18}$$

On the other hand, from the large deviation theorem

$$\log \|M_N(E_0)e_1\| > L(E_0)N - \kappa N \tag{19}$$

for $V \notin \Omega$, where

$$|\Omega| < e^{-\kappa' N}. \tag{20}$$

Here $\kappa > 0$ is an appropriate constant.

In view of Lemma 2, we may moreover assume that for $|E - E_0| < \delta$

$$\max_n \|M_{N-n}^{(v_N,\ldots,v_{n+1})}(E)\| \, \|M_{n-1}^{(v_{n-1},\ldots,v_n)}(E)\| < e^{L(E_0)N+\kappa N} \tag{21}$$

if $V \notin \Omega$.

Denote

$$B = e^{L(E_0)N - 2\kappa N}.$$

Then, for $V \notin \Omega$,

$$\kappa N \leq \left| \log(\|M_N(E)e_1\| + B) - \log(\|M_N(E_0)e_1\| + B) \right|$$
$$\leq \int_{-\delta}^{\delta} \left| \frac{d}{dt} \log(\|M_N(E_0 + t)e_1\| + B) \right| dt. \tag{22}$$

The integrand in (22) is clearly bounded by

$$\sum_{j=1,2} \sum_{n=1}^{N} \frac{|\langle M_{N-n}^{(v_N,\ldots,v_{n+1})}(E_0 + t)e_1, e_j\rangle| \, |\langle M_{n-1}^{(v_{n-1},\ldots,v_1)}(E_0 + t)e_1, e_1\rangle|}{\|M_N^{(v_N,\ldots,v_1)}(E_0 + t)e_1\| + B} \tag{23}$$

and we estimate the n-term by

$$\frac{\|\left(M_{N-n}^{(v_N,\ldots,v_{n+1})}(E_0 + t)\right)^* e_j\| \cdot \|M_{n-1}^{(v_{n-1},\ldots,v_1)}(E_0 + t)e_1\|}{|\langle \left(M_{N-n}^{v_n,\ldots,v_{n+1}}(E_0 + t)\right)^* e_j, M_{n-1}^{(v_{n-1},\ldots,v_1)}(E_0 + t)e_1\rangle|}. \tag{24}$$

We distinguish two cases. If $n \geq \frac{N}{2}$, set

$$\eta = \frac{\left(M_{N-n}^{(v_N,\ldots,v_{n+1})}(E_0 + t)\right)^* e_j}{\|\left(M_{N-n}^{(v_N,\ldots,v_{n+1})}(E_0 + t)^* e_j\|}$$

which is independent from $v_1, \ldots, v_n$. From Lemma 1, we get the distributional inequality

$$\mathbb{E}_{v_1,\ldots,v_{n-1}}[|\langle M_{n-1}^{v_{n-1},\ldots,v_1)}(E_0+t)e_1, \eta\rangle| < \varepsilon \| M_{n-1}^{(v_{n-1},\ldots,v_1)}(E_0+t)e_1 \|] \leq C\varepsilon \quad (25)$$

since by assumption $\tau(E) \leq C\varepsilon$. If $n < \frac{N}{2}$, set

$$\eta = \frac{M_{n-1}^{v_{n-1},\ldots,v_1)}(E_0+t)e_1}{\| M_{n-1}^{(v_{n-1},\ldots,v_1)}(E_0+t)e_1 \|}$$

and argue similarly, considering $\left(M_{N-n}^{(v_N,\ldots,v_{n-1})}(E_0+t)\right)^*$.

Hence we proved that

$$\mathbb{E}_{v_1,\ldots,v_N}[(24) > u] < Cu^{-1} \text{ for } u < e^{cN}. \quad (26)$$

On the other hand, the n-term in (23) is also bounded by

$$\frac{1}{B} \| M_{N-n}^{(v_N,\ldots,v_{n+1})}(E_0+t) \| . \| M_{n-1}^{(v_{n-1},\ldots,v_1)}(E_0+t) \| < e^{3\kappa N} \quad (27)$$

since $V \notin \Omega$, by (21). Therefore, taking κ in (27) appropriately, (26), (27) imply that

$$\mathbb{E}[(23) 1_{\Omega^c}] \lesssim N^2. \quad (28)$$

Consequently, recalling (22), we obtain from (28) and Tchebychev's inequality

$$\mathbb{E}[\operatorname{Spec} H_N \cap I \neq \phi] \leq |\Omega| + C \frac{\delta}{\kappa N} N^2 < e^{-\kappa' N} + \frac{c}{\kappa} \delta N$$

proving (17). □

Let H be as in Proposition 1 on the sequel.

The energy range is restricted to $[-2 + \delta_0, 2 - \delta_0]$ according to Proposition 1.

Using Proposition 3 and Anderson localization, one deduces then the analogue of Proposition 3 in [3]. We leave the details to the reader (see [3]).

Proposition 4. *Assuming* $\log \frac{1}{\delta} < c(\lambda)N$, *we have for* $I = [E_0 - \delta, E_0 + \delta]$ *that*

$$\mathbb{E}[Tr\mathcal{X}_I(H_N)] = Nk(E_0)|I| + O\left(N\delta^2 + \delta \log^2\left(N + \frac{1}{\delta}\right)\right). \quad (29)$$

4   Near Resonances

In what follows, we develop an alternative to Minami's argument that is applicable in the A-B context (recall that $H = H_\lambda$ with λ satisfying the conditions from Proposition 1).

Lemma 3. *Let $I = [E_0 - \delta, E_0 + \delta]$ be as above. Let $N \in \mathbb{Z}_+$.*

The probability for existence of a pair of orthogonal unit vectors $\xi, \xi' \in \mathbb{R}^n$ satisfying

$$\|(H_N - E_0)\xi\|_2 < \delta, \ \|(H_N - E_0)\xi'\|_2 < \delta \tag{30}$$

$$\max_{j < \sqrt{N}} (|\xi_j|, |\xi_{N-j}|, |\xi'_j|, |\xi'_{N-j}|) < \frac{1}{N^{10}} \tag{31}$$

is at most

$$CN^7\delta^2 + e^{-c\sqrt{N}}. \tag{32}$$

Proof. We take $\sqrt{N} < \nu < N - \sqrt{N}$ such that $|\xi_\nu| \gtrsim \frac{1}{\sqrt{N}}$. Since $\xi \perp \xi'$, $\|\xi_\nu \xi' - \xi'_\nu \xi\|_2 \gtrsim \frac{1}{\sqrt{N}}$ and there is some ν_1 so that

$$|\xi_\nu \xi'_{\nu_1} - \xi'_\nu \xi_{\nu_1}| \frac{1}{N}. \tag{33}$$

Again by (31). $\sqrt{N} < \nu_1 < N - \sqrt{N}$. Set further for $1 \le j \le N$

$$(\xi_j, \xi'_j) = (\xi_j^2 + (\xi'_j)^2)^{\frac{1}{2}} e^{i\theta_j}. \tag{34}$$

Hence (33) certainly implies that

$$|\sin(\theta_\nu - \theta_{\nu_1})| > \frac{1}{N}. \tag{35}$$

Assume $\nu < \nu_1$ (the other alternative is similar). We distinguish two cases.

Case 1.   There is some $\nu < j_1 < \nu_1$ such that

$$|\sin(\theta_\nu - \theta_{j_1})| > \frac{1}{10N} \text{ and } |\sin(\theta_{\nu_1} - \theta_{j_1})| > \frac{1}{10N}. \tag{36}$$

Define the vector

$$\eta = \frac{\xi_{j_1}\xi' - \xi'_{j_1}\xi}{(\xi_{j_1}^2 + (\xi'_{j_1})^2)^{\frac{1}{2}}}.$$

Obviously $\|\eta\|_2 = 1$ and $\eta_{j_1} = 0$. Also, from (30) it easily follows that

$$\|(H_N - E_0)\eta\|_2 < 2\delta. \tag{37}$$

From (36)

$$|\eta_v| = \left(\xi_v^2 + (\xi_v')^2\right)^{\frac{1}{2}} |\sin(\theta_{j_1} - \theta_v)| \gtrsim \frac{1}{N^{3/2}} \tag{38}$$

$$|\eta_{v_1}| \gtrsim \frac{1}{N^2}. \tag{39}$$

Next, we introduce the vectors

$$\eta^{(1)} = \eta|_{[1,j_1[} \text{ and } \eta^{(2)} = \eta|_{[j_1+1,N]}$$

as well as the restrictions

$$H^{(1)} = H_{[1,j_1[} \text{ and } H^{(2)} = H_{[j_1+1,N]} \tag{40}$$

with Dirichlet boundary conditions. By (38), (39), $\|\eta^{(1)}\|_2, \|\eta^{(2)}\|_2 \gtrsim \frac{1}{N^2}$ while by (37) and $\eta_{j_1} = 0$, it follows that $\|(H^{(1)} - E_0)\eta^{(1)}\|_2 < 2\delta$, $\|(H^{(2)} - E_0)\eta^{(2)}\|_2 < 2\delta$.
   Hence

$$\text{dist}(E, \text{Spec } H^{(1)}) < N^2\delta, \text{dist}(E, \text{Spec } H^{(2)}) < N^2\delta. \tag{41}$$

Note that $H^{(1)}$, $H^{(2)}$ are independent as functions of $V = (v_n)_{1 \leq n \leq N}$ and by construction, $\sqrt{N} \leq j_1 \leq N - \sqrt{N}$. Involving Proposition 3, it follows that the probability for the joint event (41) is at most

$$c[j_1 N^2\delta + e^{cj_1}][(N - j_1)N^2\delta + e^{-c(N-j_1)}] < CN^6\delta^2 + e^{-c\sqrt{N}}. \tag{42}$$

Case 2. For all $v \leq j \leq v_1$, either $|\sin(\theta_v - \theta_j)| \leq \frac{1}{10N}$ or $|\sin(\theta_{v_1} - \theta_j)| \leq \frac{1}{10N}$. Take then the smallest $v < j_1 \leq v_1$ for which $|\sin(\theta_{v_1} - \theta_{j_1})| \leq \frac{1}{10N}$. Hence $|\sin(\theta_v - \theta_{j_1-1})| \leq \frac{1}{10N}$. Denote

$$\eta^{(1)} = \left. \frac{\xi_{j_1}\xi' - \xi_{j_1}'\xi}{(\xi_{j_1}^2 + (\xi_{j_1}')^2)^{\frac{1}{2}}} \right|_{1 \leq j \leq j_1-1}$$

$$\eta^{(2)} = \left. \frac{\xi_{j_1-1}\xi' - \xi_{j_1-1}'\xi}{(\xi_{j_1-1}^2 + (\xi_{j_1-1}')^2)^{\frac{1}{2}}} \right|_{j_1 \leq j \leq N}$$

and $H^{(1)} = H_{[1,j_1[}, H^{(2)} = H_{[j_1,N]}$. Since $\eta^{(1)}_{j_1} = 0, \|(H^{(1)} - E_0)\eta^{(1)}\|_2 < 2\delta$.
Also $\|\eta^{(1)}\|_2 \geq |\eta_\nu| \gtrsim \frac{1}{\sqrt{N}}|\sin(\theta_\nu - \theta_{j_1})| \gtrsim \frac{1}{\sqrt{N}}(\frac{1}{N} - \frac{1}{10N}) > \frac{1}{N^2}$, implying
dist(E, Spec $H^{(1)}$) $< N^2\delta$. Similarly dist(E, Spec $H^{(2)}$) $< N^2\delta$ and we conclude as
in Case 1.

Summing (42) over j, Lemma 3 follows.   □

We may now establish an analogue of Proposition 4 in [3] for Anderson-Bernoulli
Hamiltonians as considered above.

Proposition 5. *Let $I = [E_0 - \delta, E_0 + \delta] \subset [-2 + \delta_0], [2 + \delta_0]$ and $\log \frac{1}{\delta} < c\sqrt{N}$.
Then*

$$\mathbb{E}[H_N \text{ has at least two eigenvalues in } I] \leq CN^2\delta^2 + C\delta\log\left(N + \frac{1}{\delta}\right). \quad (43)$$

Proof. Proceeding as in [3], set $M = C\log^2\left(N + \frac{1}{\delta}\right)$ for an appropriate constant
C. From the theory of Anderson localization, the eigenvectors ξ_α of $H_N |\xi_\alpha| = 1$,
satisfy

$$|\xi_\alpha(j)| < e^{-c|j - j_\alpha|} \text{ for } |j - j_\alpha| > \frac{M}{10} \quad (44)$$

with probability at least $1 - e^{-cM}$, with j_α the center of localization of ξ_α. We may
therefore introduce a collection $(\Lambda_s)_{1 \leq s \leq s_1}, s_1 \lesssim \frac{N}{M}$, of size M subinterval of $[1, N]$
such that for each α, there is some $1 \leq s \leq s_1$ satisfying

$$j_\alpha \in \Lambda_s \text{ and } \|\xi_\alpha|_{[1,N]\backslash\Lambda_s}\|_2 < e^{-cM} \quad (45)$$

$$\|(H_{\Lambda_s} - E_\alpha)\xi_{\alpha,s}\|_2 < e^{-cM}. \quad (46)$$

where $\xi_{\alpha,s} = \xi_\alpha|_{\Lambda_s}$. For $1 < s < s_1$, we may moreover ensure that

$$|\xi_\alpha(j)| < e^{-cM} \text{ if dist}(j, \partial\Lambda_s) < \frac{M}{10}. \quad (47)$$

By (45), (46), dist(E_α, Spec H_{Λ_s}) $< e^{-cM}$ and hence *Spec $H_{\Lambda_s} \cap \tilde{I} \neq \phi, \tilde{I} =$*
$[E_0 - 2\delta, E_0 + 2\delta]$, if $E_\alpha \in I$. According to Proposition 3, by our choice of M

$$\mathbb{E}[\text{Spec } H_{\Lambda_s} \cap \tilde{I} \neq \phi] < CM\delta + ce^{-cM} < CM\delta. \quad (48)$$

Note that if $\Lambda_s \cap \Lambda_{s'} = \phi$, then $H_{\Lambda_s}, H_{\Lambda_{s'}}$ are independent.
Hence, by construction and (48),

$$\mathbb{E}[\text{there are } \alpha, \alpha' \text{ s.t. } E_\alpha, E_{\alpha'} \in I \text{ and } |j_\alpha - j_{\alpha'}| > 4M] \leq$$

$$C\sum_{s,s'}(M\delta)^2 < CN^2\delta^2. \quad (49)$$

with the s, s'-sum performed over pairs such that $\Lambda_s \cap \Lambda_{s'} = \phi$.

It remains to consider the case $|j_\alpha - j_{\alpha'}| \leq 4M$. If $\text{dist}(j_\alpha, \{1, N\}) < 2M$ then $Spec\ H_{[1,4M]} \cap \tilde{I} \neq \phi$ or $Spec\ H_{[N-4M,N]} \cap \tilde{I} \neq \phi$.

Again by Proposition 3, the probability for this event is less than

$$CM\delta < C \log\left(N + \frac{1}{\delta}\right)\delta. \tag{50}$$

Next assume moreover $\text{dist}(\{j_\alpha, j_{\alpha'}\}, \{1, N\}) \geq 2M$. Then

$$j_\alpha \in \Lambda_t, j_{\alpha'} \in \Lambda_{t'} \text{ where } 1 < t, t' < s_1 \text{ and } |t - t'| < 10.$$

Introduce an interval Λ obtained as union of at most ten consecutive Λ_s intervals, such that $\Lambda_t, \Lambda_{t'} \subset \Lambda$. By (45), (47), setting $\tilde{\xi}_\alpha = \xi_\alpha|_\Lambda, \tilde{\xi}_{\alpha'} = \xi_{\alpha'}|_\Lambda$, we get

$$\|(H_\Lambda - E_\alpha)\tilde{\xi}_\alpha\|_2 < e^{-cM}, \|(H_\Lambda - E_{\alpha'})\tilde{\xi}_{\alpha'}\|_2 < e^{-cM}$$

so that for $E_\alpha, E_{\alpha'} \in I$

$$\|(H_\Lambda - E_0)\tilde{\xi}_\alpha\|_2 < 2\delta, \|(H_\Lambda - E_0)\tilde{\xi}_{\alpha'}\|_2 < 2\delta. \tag{51}$$

Also, by (47), $\max_{\text{dist}(j,\partial\Lambda) < \frac{M}{10}}(|\xi_\alpha(j)|, |\xi_{\alpha'}(j)|) < e^{-cM} < \frac{1}{|\Lambda|^{10}}$. Hence, Lemma 3 applies to H_Λ. According to (32), the probability that H_Λ satisfies the above property is at most (again by our choice of M)

$$CM^7\delta^2 + e^{-c\sqrt{M}} < CM^7\delta^2. \tag{52}$$

Summing over the different boxes Λ introduced above gives then

$$CN.M^7\delta^2 < CN\left(\log\left(N + \frac{1}{\delta}\right)\right)^7\delta^2. \tag{53}$$

Adding the contributions (49), (50), (53) and noting that the last is majorized by the first two, inequality (43) follows. □

5 Local Eigenvalue Statistics

Following the same argument as in Proposition 5 of [3], Propositions 4 and 5 above permit to establish Poisson statistics for the local eigenvalue spacings. Thus we obtain Proposition 2 stated in Sect. 1.

The proof is completely analogous to that of Proposition 5 in [3], except that instead of choosing $M = K \log N$, $M_1 = K_1 \log N$, we take say $M = (\log N)^4$, $M_1 = (\log N)^3$.

Acknowledgements The author is grateful to the UC Berkeley mathematics department (where the paper was written) for its hospitality. This work was partially supported by NSF grant DMS-1301619.

References

1. P. Bougerol, J. Lacroix, *Products of Random Matrices with Applications to Schrödinger Operators* (Birkhauser, Boston, 1985)
2. J. Bourgain, An application of group expansion to the Anderson-Bernoulli model. Preprint 07/13
3. J. Bourgain, On eigenvalue spacings for the 1D Anderson model with singular site distribution, in *Geometric Aspects of Functional Analysis: Israel Seminar (GAFA) 2011–2013*, ed. by B. Klartag, E. Milman (Springer, Heidelberg, 2014, in this volume)
4. J.-M. Combes, F. Germinet, A. Klein, Generalized eigenvalue - counting estimates for the Anderson model. J. Stat. Phys. **135**, 201–216 (2009)
5. F. Germinet, F. Klopp, Spectral statistics for random Schrödinger operators in the localized regime. JEMS (to appear)

On the Control Problem for Schrödinger Operators on Tori

Jean Bourgain

Abstract We consider the linear Schrödinger equation on the three dimensional torus with a bounded spatially dependent potential and prove controllability. This extends the earlier work due to Burq, Zworski and the author in the two dimensional case.

1 Introduction

The starting point of this Note is the paper [7] on Schrödinger operators (SO for short) on the 2-torus with rough potentials. Let $\mathbb{T}^2 = \mathbb{R}^2/\mathbb{Z}^2$ or, more generally $\mathbb{T}^2 = \mathbb{R}^2/A\mathbb{Z} \times B\mathbb{Z}$, $A, B \in \mathbb{R}\backslash\{0\}$ and $V \in L^\infty(\mathbb{T}^2)$ a time independent potential. Consider the Schrödinger equation on $\mathbb{T}^2$

$$i u_t = \left(- \Delta + V(x) \right)u \text{ with } (x, t) \in \mathbb{T}^2 \times \mathbb{R} \qquad (1)$$

and in initial condition

$$u|_{t=0} = \phi \in L^2(\mathbb{T}^2). \qquad (2)$$

Let $\Omega \subset \mathbb{T}^2$ be a fixed open set and $T > 0$. In [7] (see Theorem 2) the following observability result is proven.

Theorem 1. *There is a constant $C = C(V, \Omega, T)$ such that with above notations*

$$\|\phi\|_2^2 \leq C \int_0^T \|u(x, t)\|_{L^2(\Omega)}^2 dt. \qquad (3)$$

The HUM method implies then a controllability property.

Theorem 2. *For any $\phi \in L^2(\mathbb{T}^2)$, there exists $f \in L^2(\Omega \times [0, T])$ such that the solution of the equation*

J. Bourgain (✉)
School of Mathematics, Institute for Advanced Study, Princeton, NJ 08540, USA
e-mail: bourgain@math.ias.edu

© Springer International Publishing Switzerland 2014   97
B. Klartag, E. Milman (eds.), *Geometric Aspects of Functional Analysis*,
Lecture Notes in Mathematics 2116, DOI 10.1007/978-3-319-09477-9_8

$$\begin{cases} (i\partial_t + \Delta - V(x))u = f\mathbb{1}_{\Omega \times [0,T]} \\ u|_{t=0} = \phi \end{cases} \tag{4}$$

satisfies

$$u|_{t=T} = 0. \tag{5}$$

We note that an earlier paper [8] established the above result for continuous potentials $V = V(x) \in C(\mathbb{T}^2)$, while in [1] the problem is treated in arbitrary dimension, again for continuous potentials $V = V(x) \in C(\mathbb{T}^d)$ (in fact, a larger class but which does not capture $L^\infty(\mathbb{T}^d)$).

There are two obvious remaining problems. The first is the generalization of Theorems 1 and 2 to higher dimension, assuming $V = V(x) \in L^\infty(\mathbb{T}^d)$. The second issue is the controllability question for time dependent potentials $V = V(x,t)$, which will not be discussed here. The main harmonic analysis input in the proof of Theorem 1 is Zygmund's inequality, stating the uniform equivalence of $L^2(\mathbb{T}^2)$ and $L^4(\mathbb{T}^2)$ norms for toral eigenfunctions in dimension $d = 2$.

Thus, denoting $e(u) = e^{2\pi i u}$.

Theorem 3. *There is an absolute constant C such that for all $E \in \mathbb{Z}_+$,*

$$\left\| \sum_{n \in \mathbb{Z}^2, |n|^2 = E} c_n e(nx) \right\|_{L^4(\mathbb{T}^2)} \le C \left(\sum |c_n|^2 \right)^{\frac{1}{2}} \tag{6}$$

for all coefficients $\{c_n\}$.

Proposition. *Does there exist for each $d \ge 3$ some exponent $p = p(d) > 2$ such that for some constant C_d, the inequality*

$$\left\| \sum_{n \in \mathbb{Z}^d, |n|^2 = E} c_n e(n.x) \right\|_{L^p(\mathbb{T}^d)} \le C_d \left(\sum |c_n|^2 \right)^{\frac{1}{2}} \tag{7}$$

holds for all $E \in \mathbb{Z}_+$ and coefficients $\{c_n\}$?

It is conjectured that any $p < \frac{2d}{d-2}$ should satisfy (7) for some constant $c(d, p)$. At this time however, Zygmund's inequality for $d = 2$ seems to be the only known inequality of this type. In [5], the following estimate is obtained.

Theorem 4. *Let $d \ge 3$, $p = \frac{2d}{d-1}$. Then, for any $\varepsilon > 0$, there is a constant C_ε such that*

$$\left\| \sum_{n \in \mathbb{Z}^d, |n|^2 = E} c_n e(n.x) \right\|_{L^p(\mathbb{T}^d)} \le C_\varepsilon E^\varepsilon \left(\sum |c_n|^2 \right)^{\frac{1}{2}}. \tag{8}$$

For $d = 3$, one may in fact take $p = 4$ (see also the recent paper [6]).

Following the approach in [7], an affirmative answer to the above problem in dimension d would allow to generalize Theorems 1 and 2 to that dimension. A statement of the type (8) turns out to be insufficient however, unless one makes the stronger assumption $V \in L^\infty(\mathbb{T}^d) \cap H^s(\mathbb{T}^d)$ for some $s > 0$.

It appears that in fact, in order to prove Theorem 1 in dimension d, it is already sufficient to have an inequality of the form (7), $p > 2$, in dimension $d - 1$. Thus, in view of Theorem 3

Theorem 5. *Theorems 1 and 2 hold in dimension $d = 3$.*

We sketch the argument in the next section. The general case is based on similar considerations (and an induction on the dimension), which we do not elaborate on here since there are no further applications at this point.

2 Proof of Theorem 1 or $d = 3$

Denote u_V^ϕ the solution of the equation on $\mathbb{T}^d \times \mathbb{R}$

$$\begin{cases} i u_t = (-\Delta + V(x))u \\ u|_{t=0} = \phi. \end{cases} \tag{9}$$

It follows from the uniqueness-compactness argument due to Bardos-Lebeau-Rauch [2] (see also [7], Sect. 6 for a quantitative form) that in order to establish (3), it suffices to prove the weaker inequality (we may clearly replace $[0, T]$ by $[-T, T]$)

$$\|\phi\|_2^2 \leq C_1 \int_{-T}^{T} \|u(x, t)\|_{L^2(\Omega)}^2 dt + C_2 \|\phi\|_{H^{-1}}^2. \tag{10}$$

Assuming (10) violated, we obtain a sequence $\{\phi_j\}$ of initial data in $L^2(\mathbb{T}^d)$, $\|\phi_1\|_1 = 1$ such that

$$\phi_j \to 0 \text{ weakly} \tag{11}$$

and

$$\int_{-T}^{T} \|u_V^{\phi_j}(x, t)\|_{L^2(\Omega)}^2 dt \to 0. \tag{12}$$

Denote $W = W_s$ regularizations of V with finitely supported Fourier transform obtained by suitable convolutions of V. Thus

$$\|W_s\|_\infty \leq \|V\|_\infty \text{ and } W_s \to V \text{ in } L^q(\mathbb{T}^d) \text{ for all } q < \infty. \tag{13}$$

By (12), denoting $I = [0, \frac{T}{2}]$

$$\overline{\lim_{j}} \int_{I} \|u_{W}^{\phi_{j}}(x,t)\|_{L^{2}(\Omega)}^{2} dt \leq \overline{\lim_{j}} \|u_{V}^{\phi_{j}} - u_{W}^{\phi_{j}}\|_{L^{2}(\mathbb{T}^{d} \times I)}. \tag{14}$$

Comparing the SO's with respective potentials V and W, write

$$\begin{cases} (i\,\partial_{t} + \Delta - W)u_{V} = (V - W)u_{V} \\ u_{V}|_{t=0} = \phi_{j} = U_{W}|_{t=0} \end{cases} \tag{15}$$

implying by Duhamel's formula and unitarily that

$$\|u_{V}u_{W}\|_{L^{2}(\mathbb{T}^{d} \times I)} \leq \left\| \int_{0}^{t} e^{i(t-\tau)(-\Delta+W)}\big((V-W)u_{V}(\tau)\big)d\tau \right\|_{L^{2}(\mathbb{T}^{d} \times I)}$$

$$\leq T\|(V-W)u_{V}\|_{L^{2}(\mathbb{T}^{d} \times I)}. \tag{16}$$

Hence the r.h.s. of (14) may be bounded by

$$C \overline{\lim_{j}} \|(V - W)u_{V}^{\phi_{j}}\|_{L^{2}(\mathbb{T}^{d} \times I)} \tag{17}$$

The solutions $u_{W}^{\phi_{j}}$ in the l.h.s. of (14) are obtained from the SO $-\Delta + W$ with smooth potential W. Setting $d = 3$ and invoking the analysis from [1] for smooth potential, the quantum limit produced by the sequence $\{u_{W}^{\phi_{j}}\}$ may be represented as a superposition of quantum limits obtained by Fourier restriction to certain sub-lattices in $\mathbb{Z}^{3}$ that are at most two-dimensional (rather than invoking the semi-classical approach from [1], one may alternatively use a 'cluster-decomposition' in lower dimensional clusters as introduced below; note that this part of the analysis is 'soft' because the potential W is smooth). Next, we may apply Theorem 1 to these lower dimensional quantum limits, produced from potentials $\tilde{W}$ obtained by suitable Fourier restriction of $W = W_{s}$ and hence ranging in an L^{2}-pre-compact set. It was indeed noted in [7] that although the constant C in (3) depends on V, it remains bounded for V ranging in a compact subset of $L^{2}(\mathbb{T}^{2})$. The outcome of the above discussion is that the left side of (14) satisfies a positive lower bound, which is independent from the regularization $W = W_{s}$. In order to obtain a contradiction, it remains to show that (17) can be made arbitrarily small for $W = W_{s}, s \to \infty$.

Assume $T < \frac{1}{2}$. The function $u = u_{V}$ solving (9) admits on $\mathbb{T}^{d} \times [-T, T]$ a representation

$$u(x,t) = \sum_{\substack{n \in \mathbb{Z}^{d} \\ m \in \mathbb{Z}}} \hat{u}(n,m)e(n.x + mt) \quad \text{for } (x,t) \in \mathbb{T}^{d} \times [-T, T] \tag{18}$$

where

$$\|u\| = \left(\sum_{\substack{n \in \mathbb{Z}^d \\ m \in \mathbb{Z}}} (1 + ||n|^2 - m|)^2 |\hat{u}(n,m)|^2 \right)^{\frac{1}{2}} \leq C \|\phi\|_2 \tag{19}$$

(cf. [3]). Also, since $\lim \phi_j = 0$ (weakly),

$$\lim_{j \to \infty} u_V^{\phi_j}(t) = 0(\text{weakly}). \tag{20}$$

We will need the following version of the 'cluster structure' of lattice points on quadrics (cf. [4, Lemma 19.10]). Denote $\pi : \mathbb{Z}^d \times \mathbb{Z} \to \mathbb{Z}^d$ the coordinate projection.

Lemma. *Let N be large and consider the set*

$$\mathfrak{S} = \mathfrak{S}_N = \{(n, |n|^2) \in \mathbb{Z}^d \times \mathbb{Z}; |n| > N\}.$$

Let $K < N^c$. There is a partition

$$\mathfrak{S} = \bigcup \mathfrak{S}_\alpha$$

with the following properties.

$\pi(\mathfrak{S}_\alpha)$ *is contained in a lower dimensional affine subspace (depending on α)*
$$\tag{21}$$

$$\text{dist}(\mathfrak{S}_\alpha, \mathfrak{S}_\beta) > K \text{ for } \alpha \neq \beta \tag{22}$$

$$\text{diam } \mathfrak{S}_\alpha < K^C \text{ for each } \alpha \tag{23}$$

where c, C are constants depending on d.

Recalling (12), we take j sufficiently large as to ensure that $u = u_V^{\phi_j}$ satisfies

$$\int_{-T}^{T} \int_\Omega |u(x,t)|^2 dx dt < \varepsilon \tag{24}$$

(fixing some arbitrary $\varepsilon > 0$).

Introduce a smooth function $\omega = \omega(x,t), 0 \leq \omega \leq 1$ such that

$$\text{supp } \hat{\omega} \subset B_R \text{ for some } R = R(\Omega, T, \varepsilon) \tag{25}$$

$$\iint |u|^2 \omega dx dt < \varepsilon \tag{26}$$

$$\int \omega(x,t) dx > \frac{1}{2}|\Omega| \text{ for } t \in I = \left[0, \frac{T}{2}\right]. \tag{27}$$

Set

$$K = 10\left(R + \frac{1}{\varepsilon}\right) \tag{28}$$

and $N > K$ large enough for the lemma to be applicable. By (20), we can take j sufficiently large as to ensure that $\hat{u}(n, m) \approx 0$ for $|n| + |m| < 2N$. In view of (19) and the preceding, the function

$$u'(x, t) = \sum_{\substack{n \in \mathbb{Z}^d, m \in \mathbb{Z} \\ |n| > N, |m - |n|^2| < \frac{1}{\varepsilon}}} \hat{u}(m, n) e(n.x + mt) \tag{29}$$

satisfies

$$\|u - u'\|_{L^2(\mathbb{T}^d \times [-T, T])} < C\varepsilon \tag{30}$$

and therefore, denoting $\| \ \|_2 = \| \ \|_{L^2_{t \in I} L^2_x}$,

$$\|u(V - W)\|_2 \le \|u'(V - W)\|_2 + C\|V\|_\infty \varepsilon \tag{31}$$

$$\iint |u'|^2 \omega < C\varepsilon. \tag{32}$$

Let $\{\mathfrak{S}_\alpha\}$ be the decomposition of $\mathfrak{S}_N$ obtained in the lemma, satisfying (21)–(23). Defining

$$\mathfrak{S}'_\alpha = \left\{(n, m) \in \pi(\mathfrak{S}_\alpha) \times \mathbb{Z}; |m - |n|^2| < \frac{1}{2}\right\} \tag{33}$$

it follows from (29) that

$$u' = \sum_\alpha P_\alpha u' \tag{34}$$

with P_α the Fourier projection operator on $\mathfrak{S}'_\alpha$.
   By (22)

$$\text{dist}(\mathfrak{S}'_\alpha, \mathfrak{S}'_\beta) > \frac{K}{2} > \frac{1}{\varepsilon} \quad \text{for } \alpha \ne \beta$$

and (25) implies

$$C\varepsilon > \iint |u'|^2 \omega = \sum_\alpha \iint |P_\alpha u'|^2 \omega. \tag{35}$$

We distinguish the following two cases
(*) $\pi(\mathfrak{S}_\alpha) = \pi(\mathfrak{S}'_\alpha)$ is a single point $\{n\}$.
   Then clearly

$$\iint |P_\alpha u'|^2 \omega \geq \int_I \|P_\alpha u'\|^2_{L^2_x} \left(\int \omega(x,t)dx \right) dt \overset{(27)}{\geq} c\|P_\alpha u'\|^2_2.$$

Hence, by (35)

$$\sum_{\alpha(*)} \|P_\alpha u'\|^2_2 < C\varepsilon \tag{36}$$

$\binom{*}{*}$ $\pi(\mathfrak{S}_\alpha$ consists of at least two elements.
   Let $(n, |n|^2) \in \mathfrak{S}_\alpha, (n'|n'|^2) \in \mathfrak{S}_\alpha$ with $n \neq n'$. It follows from (23) that

$$0 < |n - n'| = |\xi| < K^C \tag{37}$$

and

$$\left| |n|^2 - |n'|^2 \right| = \left| |n|^2 - |n + \xi|^2 \right| < K^C. \tag{37'}$$

Hence

$$|n.\xi| \leq \frac{1}{2}(K^C + K^{2C}) < K^{2C}. \tag{38}$$

   Note that there are at most $O(K^{Cd})$ vectors $\xi \in \mathbb{Z}^d$ with $0 < |\xi| < K^C$. We showed that for any $n \in \bigcup_{\alpha\binom{*}{*}} \pi(\mathfrak{S}_\alpha)$, there is some $\xi \in \mathbb{Z}^d$ satisfying (37) and (38).
   Denote

$$u'' = \sum_{\alpha\binom{*}{*}} P_\alpha u'. \tag{39}$$

It follows from (36) that

$$\|u' - u''\|_2 = \left\| \sum_{\alpha(*)} P_\alpha u' \right\|_2 = \left(\sum_{\alpha(*)} \|P_\alpha u'\|^2_2 \right)^{\frac{1}{2}} < C\sqrt{\varepsilon}$$

and hence, returning to (31)

$$\|u'(V - W)\|_2 \leq \|u''(V - W)\|_2 + C\sqrt{\varepsilon}. \tag{40}$$

Next, letting $\frac{1}{p} + \frac{1}{q} = 1$, estimate

$$\|u''(V-W)\|_2 \le \|V-W\|_{2q} \Big(\sum_m \Big\| \sum_n \widehat{u''}(n,m)e(n.x) \Big\|_{L_x^{2p}}^2 \Big)^{\frac{1}{2}}. \tag{41}$$

In the inner sum, m is fixed and, by (33), n satisfies

$$\Big| m - |n|^2 \Big| < \frac{1}{\varepsilon}. \tag{42}$$

Also, from the preceding, there is some $\xi \in \mathbb{Z}^d$, $|\xi| < K^C$ such that (38) holds restricting n to at most $K^{C(d+2)}$ affine hyperplanes H_τ. Hence, by (42), n belongs to a union of at most $\frac{2}{\varepsilon} K^{C(d+1)}$ $(d-2)$-spheres $S_\tau \subset H_\tau$. We estimate

$$\Big\| \sum_n \widehat{u''}(n,m)e(n,x) \Big\|_{L_x^{2p}} \le \sum_\tau \Big\| \sum_{n \in S_\tau} \widehat{u''}(n,m)e(n.x) \Big\|_{L_x^{2p}}.$$

At this point, recall that $d = 3$ so that S_τ is a 1-sphere. Taking $p = q = 2$, Zygmund's inequality applies and we get

$$\Big\| \sum_{n \in S_\tau} \widehat{u''}(n,m)e(n.x) \Big\|_{L_x^4} \le C \Big(\sum_{n \in S_\tau} |\widehat{u''}(n,m)|^2 \Big)^{\frac{1}{2}}. \tag{43}$$

This gives the bound

$$(41) \le C\|V-W\|_4 \Big(\frac{2}{\varepsilon} K^{C(d+2)} \Big)^{\frac{1}{2}} \Big(\sum_{n,m} |\widehat{u''}(n,m)|^2 \Big)^{\frac{1}{2}}$$

$$\le C \Big(\frac{1}{\varepsilon} K^{C(d+2)} \Big)^{\frac{1}{2}} \|V-W\|_4 \tag{44}$$

which can be made arbitrarily small by appropriate choice of W.

From (31), (40), (44), it follows that for a suitable smoothing W of V,

$$\overline{\lim_j} \|u_V^{\phi_j}(V-W)\|_{L_t^2 L_x^2} < C\sqrt{\varepsilon} \tag{45}$$

and hence (17) can be made arbitrarily small.

This completes the proof of Theorem 5.

Acknowledgements Research supported in part by NSF Grants DMS 1301619.

References

1. A. Anantharaman, F. Macia, Semi-classical measures for the Schrödinger equation on the torus. JEMS **16**(6), 1253–1288 (2014)
2. C. Bardos, G. Lebeau, J. Rauch, Sharp sufficient conditions for the observation, control and stabilization of waves from the boundary. SIAM J. Control Optim. **30**, 1024–1065 (1992)
3. J. Bourgain, Restriction phenomena for lattice subsets and applications to nonlinear evolution equations I, Schrödinger equations. GAFA **3**(2), 107–156 (1993)
4. J. Bourgain, *Green's Function Estimates for Lattice Schrödinger Equations and Applications*. Annals of Mathematics Studies, vol. 158 (Princeton University Press, Princeton, 2005)
5. J. Bourgain, Moment inequalities for trigonometric polynomials with spectrum in curved hypersurfaces. Israel J. Math. **193**, 441–450 (2013)
6. J. Bourgain, C. Demeter, New bounds for the discrete Fourier restriction to the sphere in four and five dimensions. IMRN (to appear)
7. J. Bourgain, N. Burq, M. Zworski, Control for Schrödinger operators on 2-tori: rough potentials. J. Eur. Math. Soc. **15**(5), 1597–1628 (2013)
8. N. Burq, M. Zworski, Control for Schrödinger equations on tori. Math. Res. Lett. **19**, 309–324 (2012)

Bounding the Norm of a Log-Concave Vector Via Thin-Shell Estimates

Ronen Eldan and Joseph Lehec

Abstract Chaining techniques show that if X is an isotropic log-concave random vector in $\mathbb{R}^n$ and Γ is a standard Gaussian vector then

$$\mathbb{E}\|X\| \leq C n^{1/4} \mathbb{E}\|\Gamma\|$$

for any norm $\|\cdot\|$, where C is a universal constant. Using a completely different argument we establish a similar inequality relying on the *thin-shell* constant

$$\sigma_n = \sup\left(\sqrt{\text{Var}(|X|)}; \ X \text{ isotropic and log-concave on } \mathbb{R}^n \right).$$

In particular, we show that if the thin-shell conjecture $\sigma_n = O(1)$ holds, then $n^{1/4}$ can be replaced by $\log(n)$ in the inequality. As a consequence, we obtain certain bounds for the mean-width, the dual mean-width and the isotropic constant of an isotropic convex body. In particular, we give an alternative proof of the fact that a positive answer to the thin-shell conjecture implies a positive answer to the slicing problem, up to a logarithmic factor.

1 Introduction

Given a stochastic process $(X_t)_{t \in T}$, the question of obtaining bounds for the quantity

$$\mathbb{E}\left(\sup_{t \in T} X_t \right)$$

R. Eldan (✉)
Microsoft Research, Microsoft Building 99, Redmond Microsoft Campus, One Microsoft Way, Redmond, WA 98052, USA
e-mail: roneneldan@gmail.com

J. Lehec
Ceremade (UMR CNRS 7534), Université Paris-Dauphine, place de Lattre de Tassigny, 75016 Paris, France
e-mail: lehec@ceremade.dauphine.fr

© Springer International Publishing Switzerland 2014
B. Klartag, E. Milman (eds.), *Geometric Aspects of Functional Analysis*,
Lecture Notes in Mathematics 2116, DOI 10.1007/978-3-319-09477-9_9

is a fundamental question in probability theory dating back to Kolmogorov, and the theory behind this type of question has applications in a variety of fields.

The case that $(X_t)_{t \in T}$ is a Gaussian process is perhaps the most important one. It has been studied intensively over the past 50 years, and numerous bounds on the supremum in terms of the geometry of the set T have been attained by Dudley, Fernique, Talagrand and many others.

The case of interest in this paper is a certain generalization of the Gaussian process. We consider the supremum of the process

$$(X_t = \langle X, t \rangle)_{t \in T}$$

where X is a log-concave random vector in $\mathbb{R}^n$ and $T \subset \mathbb{R}^n$ is a compact set. Throughout the article $\langle x, y \rangle$ denotes the inner product of $x, y \in \mathbb{R}^n$ and $|x| = \sqrt{\langle x, x \rangle}$ the Euclidean norm of x. Our aim is to obtain an upper bound on this supremum in terms of the supremum of a corresponding Gaussian process $Y_t = \langle \Gamma, t \rangle$ where Γ is a gaussian random vector having the same covariance structure as X.

Before we formulate the results, we begin with some notation. A probability density $\rho : \mathbb{R}^n \to [0, \infty)$ is called *log-concave* if it takes the form $\rho = \exp(-H)$ for a convex function $H : \mathbb{R}^n \to \mathbb{R} \cup \{+\infty\}$. A probability measure is log-concave if it has a log-concave density and a random vector taking values in $\mathbb{R}^n$ is said to be log-concave if its law is log-concave. Two canonical examples of log-concave measures are the uniform probability measure on a convex body and the Gaussian measure. It is a well-known fact that any log-concave probability density decays exponentially at infinity, and thus has moments of all orders. A log-concave random vector X is said to be *isotropic* if its expectation and covariance matrix satisfy

$$\mathbb{E}(X) = 0, \quad \mathrm{cov}(X) = \mathrm{id}.$$

Let σ_n be the so-called *thin-shell* constant:

$$\sigma_n = \sup_X \sqrt{\mathrm{Var}(|X|)} \tag{1}$$

where the supremum runs over all isotropic, log-concave random vectors X in $\mathbb{R}^n$. It is trivial that $\sigma_n \leq \sqrt{n}$ and it was proven initially by Klartag [9] that in fact

$$\sigma_n = o(\sqrt{n}).$$

Shortly afterwards, Fleury-Guédon-Paouris [5] gave an alternative proof of this fact. Several improvements on the bound have been established since then, and the current best estimate is $\sigma_n = O(n^{1/3})$ due to Guédon-Milman [6]. The *thin-shell* conjecture, which asserts that the sequence $(\sigma_n)_{n \geq 1}$ is bounded, is still open. Another related constant is:

$$\tau_n^2 = \sup_X \sup_{\theta \in \mathbb{S}^{n-1}} \sum_{i,j=1}^n \mathbb{E}\big(X_i X_j \langle X, \theta \rangle\big)^2, \tag{2}$$

where the supremum runs over all isotropic log-concave random vectors X in $\mathbb{R}^n$. Although it is not known whether $\tau_n = O(\sigma_n)$, we have the following estimate, proven in [3]

$$\tau_n^2 = O\left(\sum_{k=1}^{n} \frac{\sigma_k^2}{k}\right). \tag{3}$$

The estimate $\sigma_n = O(n^{1/3})$ thus gives $\tau_n = O(n^{1/3})$, whereas the *thin-shell* conjecture yields $\tau_n = O(\sqrt{\log n})$.

We denote by Γ the standard Gaussian vector in $\mathbb{R}^n$ (with identity covariance matrix). We are now ready to formulate our main theorem.

Theorem 1. *Let X be an isotropic log-concave random vector in $\mathbb{R}^n$ and let $\|\cdot\|$ be a norm. Then there is a universal constant C such that*

$$\mathbb{E}\|X\| \leq C\sqrt{\log n}\,\tau_n\,\mathbb{E}\|\Gamma\|. \tag{4}$$

Remark 1. It is well known that, as far as $\mathbb{E}\|X\|$ is concerned, there is no loss of generality assuming additionally that the support of X is contained in a ball of radius $C_0\sqrt{n}$ for some sufficiently large constant C_0 (see for instance Lemma 3 below). Then it is easily seen that X satisfies the following ψ_2 estimate

$$\mathbb{P}\big(|\langle X, \theta \rangle| \geq t\big) \leq Ce^{-ct^2/\sqrt{n}}, \quad \forall t \geq 0, \forall \theta \in \mathbb{S}^{n-1},$$

where C, c are universal constants. Combining this with chaining methods developed by Dudley-Fernique-Talagrand (more precisely, using Theorem 1.2.6. and Theorem 2.1.1. of [12]), one gets the inequality

$$\mathbb{E}\|X\| \leq C'n^{1/4}\mathbb{E}\|\Gamma\|,$$

we refer to [1] for more details. This means that using the current best-known bound for the thin-shell constant: $\sigma_n = O(n^{1/3})$, the above theorem does not give us anything new. On the other hand, under the thin-shell hypothesis we obtain using (3)

$$\mathbb{E}\|X\| \leq C\log n\,\mathbb{E}\|\Gamma\|.$$

As an application of Theorem 1, we derive several bounds related to the mean width and dual mean width of isotropic convex bodies and to the so-called *hyperplane conjecture*. We begin with a few definitions. A convex body $K \subset \mathbb{R}^n$ is a compact convex set whose interior contains the origin. For $x \in \mathbb{R}^n$, we define

$$\|x\|_K = \inf\{\lambda; \ x \in \lambda K\}$$

to be the gauge associated to K (it is a norm if K is symmetric about 0). The polar body of K is denoted by

$$K^\circ = \{y \in \mathbb{R}^n; \ \langle x, y \rangle \leq 1, \forall x \in K\}.$$

Next we define

$$M(K) = \int_{\mathbb{S}^{n-1}} \|x\|_K \, \sigma(\mathrm{d}x),$$

$$M^*(K) = \int_{\mathbb{S}^{n-1}} \|x\|_{K^\circ} \, \sigma(\mathrm{d}x),$$

where σ is the Haar measure on the sphere, normalized to be a probability measure. These two parameters play an important rôle in the asymptotic theory of convex bodies.

The convex body K is said to be isotropic if a random vector uniform on K is isotropic. When K is isotropic, the *isotropic constant* of K is then defined to be

$$L_K = |K|^{-1/n},$$

where $|K|$ denotes the Lebesgue measure of K. More generally, the isotropic constant of an isotropic log-concave random vector is $L_X = f(0)^{1/n}$ where f is the density of X. The *slicing* or *hyperplane* conjecture asserts that $L_K \leq C$ for some universal constant C. The current best estimate is $L_K \leq C n^{1/4}$ due to Klartag [8]. We are ready to formulate our corollary:

Corollary 1. *Let K be an isotropic convex body. Then one has,*

(i) $M^*(K) \geq c \sqrt{n}/(\sqrt{\log n} \, \tau_n)$,
(ii) $L_K \leq C \tau_n (\log n)^{3/2}$,

where $c, C > 0$ are universal constants.

Remark 2. Part (ii) of the corollary is nothing new. Indeed, in [4], it is shown that $L_K \leq C \sigma_n$ for a universal constant $C > 0$. Our proof uses different methods and could therefore shed some more light on this relation, which is the reason why we provide it.

Using similar methods, we attain an alternative proof of the following correlation inequality proven initially by Hargé in [7].

Proposition 1 (Hargé). *Let X be a random vector on $\mathbb{R}^n$. Assume that $\mathbb{E}(X) = 0$ and that X is more log-concave than Γ, i.e. the density of X has the form*

$$x \mapsto \exp\left(-V(x) - \frac{1}{2}|x|^2\right)$$

for some convex function $V : \mathbb{R}^n \to (-\infty, +\infty]$. Then for every convex function $\varphi : \mathbb{R}^n \to \mathbb{R}$ we have

$$\mathbb{E}\varphi(X) \leq \mathbb{E}\varphi(\Gamma).$$

The structure of the paper is as follows: in Sect. 2 we recall some properties of a stochastic process constructed in [3], which will serve as one of the central ingredients in the proof of Theorem 1, as well as establish some new facts about this process. In Sect. 3 we prove the main theorem and Proposition 1. Finally, in Sect. 4 we prove Corollary 1.

In this note, the letters $c, \tilde{c}, c', C, \tilde{C}, C', C''$ will denote positive universal constants, whose value is not necessarily the same in different appearances. Further notation used throughout the text: id will denote the identity $n \times n$ matrix. The Euclidean unit sphere is denoted by $\mathbb{S}^{n-1} = \{x \in \mathbb{R}^n; |x| = 1\}$. The operator norm and the trace of a matrix A are denoted by $\|A\|_{op}$ and $\mathrm{Tr}(A)$, respectively. For two probability measures μ, ν on $\mathbb{R}^n$, we let $T_2(\mu, \nu)$ be their *transportation cost* for the Euclidean distance squared:

$$T_2(\mu, \nu) = \inf_{\xi} \int_{\mathbb{R}^n \times \mathbb{R}^n} |x - y|^2 \, \xi(dx, dy)$$

where the infimum is taken over all measures ξ on $\mathbb{R}^{2n}$ whose marginals onto the first and last n coordinates are the measures μ and ν respectively. Finally, given a continuous martingale $(X_t)_{t \geq 0}$, we denote by $[X]_t$ its quadratic variation. If X is $\mathbb{R}^n$-valued, then $[X]_t$ is a non-negative matrix whose i, j coefficient is the quadratic covariation of the i-th and j-th coordinates of X at time t.

2   The Stochastic Construction

We make use of the construction described in [3]. There it is shown that, given a probability measure μ having compact support and whose density with respect to the Lebesgue measure is f, and given a standard Brownian motion $(W_t)_{t \geq 0}$ on $\mathbb{R}^n$; there exists an adapted random process $(\mu_t)_{t \geq 0}$ taking values in the space of absolutely continuous probability measures such that $\mu_0 = \mu$ and such that the density f_t of μ_t satisfies

$$\mathrm{d} f_t(x) = f_t(x) \langle A_t^{-1/2}(x - a_t), \mathrm{d}W_t \rangle, \quad \forall t \geq 0, \tag{5}$$

for every $x \in \mathbb{R}^n$, where

$$a_t = \int_{\mathbb{R}^n} x \, \mu_t(dx),$$

$$A_t = \int_{\mathbb{R}^n} (x - a_t) \otimes (x - a_t) \, \mu_t(dx)$$

are the barycenter and the covariance matrix of μ_t, respectively.

Let us give now the main properties of this process. Some of these properties have already been established in [3], in this case we will only give the general idea of the proof. We refer the reader to [3, Sect. 2,3] for complete proofs. Firstly, for every test function ϕ the process

$$\left(\int_{\mathbb{R}^n} \phi \, d\mu_t\right)_{t \geq 0}$$

is a martingale. In particular

$$\mathbb{E} \int_{\mathbb{R}^n} \phi \, d\mu_t = \int_{\mathbb{R}^n} \phi \, d\mu, \quad \forall t \geq 0. \tag{6}$$

The Itô differentials of a_t and A_t read

$$da_t = A_t^{1/2} dW_t \tag{7}$$

$$dA_t = -A_t \, dt + \int_{\mathbb{R}^n} (x - a_t) \otimes (x - a_t) \langle A_t^{-1/2}(x - a_t), dW_t \rangle \, \mu_t(dx). \tag{8}$$

Recall that μ is assumed to have compact support, and observe that the support of μ_t is included in that of μ, almost surely. This shows that the processes (a_t) and (A_t) as well as the process involved in the local martingale part of Eq. (8) are uniformly bounded. In particular the local martingales of the last two equations are actually genuine martingales. Thus we get from (8)

$$\frac{d}{dt} \mathbb{E}\mathrm{Tr}(A_t) = -\mathbb{E}\mathrm{Tr}(A_t).$$

Integrating this differential equation we obtain

$$\mathbb{E}\mathrm{Tr}(A_t) = e^{-t} \mathrm{Tr}(A_0), \quad t \geq 0. \tag{9}$$

Combining this with (7) we obtain

$$\mathbb{E}|a_t|^2 = |a_0|^2 + \int_0^t \mathbb{E}\mathrm{Tr}(A_s) \, ds = |a_0|^2 + (1 - e^{-t})\mathrm{Tr}(A_0).$$

The process $(a_t)_{t \geq 0}$ is thus a martingale bounded in L^2. By Doob's theorem, it converges almost surely and in L^2 to some random vector a_∞.

Proposition 2. *The random vector a_∞ has law μ.*

Proof. Let ϕ, ψ be functions on $\mathbb{R}^n$ satisfying

$$\phi(x) + \psi(y) \leq |x - y|^2, \quad x, y \in \mathbb{R}^n. \tag{10}$$

Then

$$\phi(a_t) + \int_{\mathbb{R}^n} \psi(y)\,\mu_t(dy) \le \int_{\mathbb{R}^n} |a_t - y|^2\,dy = \mathrm{Tr}(A_t).$$

Taking expectation and using (6) and (9) we obtain

$$\int_{\mathbb{R}^n} \phi\,d\nu_t + \int_{\mathbb{R}^n} \psi\,d\mu \le \mathrm{Tr}(A_0)e^{-t},$$

where ν_t is the law of a_t. This holds for every pair of functions satisfying the constraint (10). By the Monge-Kantorovich duality (see for instance [13, Theorem 5.10]) we obtain

$$T_2(\nu_t, \mu) \le e^{-t}\,\mathrm{Tr}(A_0)$$

where T_2 is the transport cost associated to the Euclidean distance squared, defined in the introduction. Thus $\nu_t \to \mu$ in the T_2 sense, which implies that $a_t \to \mu$ in law, hence the result.   □

Let us move on to properties of the operator norm of A_t. We shall use the following lemma which follows for instance from a theorem of Brascamp-Lieb [2, Theorem 4.1.]. We provide an elementary proof using the Prékopa-Leindler inequality.

Lemma 1. *Let X be a random vector on $\mathbb{R}^n$ whose density ρ has the form*

$$\rho(x) = \exp\left(-\frac{1}{2}\langle Bx, x\rangle - V(x)\right)$$

where B is a positive definite matrix, and $V \colon \mathbb{R}^n \to (-\infty + \infty]$ is a convex function. Then one has,

$$\mathrm{cov}(X) \le B^{-1}.$$

In other words, if a random vector X is more log-concave than a Gaussian vector Y, then $\mathrm{cov}(X) \le \mathrm{cov}(Y)$.

Proof. There is no loss of generality assuming that $B = \mathrm{id}$ (replace X by $B^{1/2}X$ otherwise). Let

$$\Lambda \colon x \mapsto \log \mathbb{E}(e^{\langle x, X\rangle}).$$

Since log-concave vectors have exponential moment Λ is C^∞ in a neighborhood of 0 and it is easily seen that

$$\nabla^2 \Lambda(0) = \mathrm{cov}(X). \tag{11}$$

Fix $a \in \mathbb{R}^n$ and define

$$f : x \mapsto \langle a, x \rangle - \frac{1}{2}|x|^2 - V(x),$$

$$g : y \mapsto -\langle a, y \rangle - \frac{1}{2}|y|^2 - V(y),$$

$$h : z \mapsto -\frac{1}{2}|z|^2 - V(z).$$

Using the inequality

$$\frac{1}{2}\langle a, x - y \rangle - \frac{1}{4}|x|^2 - \frac{1}{4}|y|^2 \leq \frac{1}{2}|a|^2 - \frac{1}{8}|x + y|^2,$$

and the convexity of V we obtain

$$\frac{1}{2}f(x) + \frac{1}{2}g(y) \leq \frac{1}{2}|a|^2 + h\left(\frac{x + y}{2}\right), \quad \forall x, y \in \mathbb{R}^n.$$

Hence by Prékopa-Leindler

$$\left(\int_{\mathbb{R}^n} e^{f(x)} \, dx\right)^{1/2} \left(\int_{\mathbb{R}^n} e^{g(y)} \, dy\right)^{1/2} \leq e^{|a|^2/2} \int_{\mathbb{R}^n} e^{h(z)} \, dz.$$

This can be rewritten as

$$\frac{1}{2}\Lambda(a) + \frac{1}{2}\Lambda(-a) - \Lambda(0) \leq \frac{1}{2}|a|^2.$$

Letting a tend to 0 we obtain $\langle \nabla^2 \Lambda(0)a, a \rangle \leq |a|^2$ which, together with (11), yields the result. $\qquad\square$

Recalling (5) and applying Itô's formula to $\log(f_t)$ yields

$$d\log(f_t)(x) = \langle A_t^{-1/2}(x - a_t), dW_t \rangle - \frac{1}{2}\langle A_t^{-1}(x - a_t), x - a_t \rangle \, dt$$

This shows that the density of the measure μ_t satisfies

$$f_t(x) = f(x) \exp\left(c_t + \langle b_t, x \rangle - \frac{1}{2}\langle B_t x, x \rangle\right) \tag{12}$$

where c_t, b_t are some random processes, and

$$B_t = \int_0^t A_s^{-1} \, ds. \tag{13}$$

Lemma 2. *If the initial measure μ is more-log-concave than the standard Gaussian measure, then almost surely*

$$\|A_t\|_{op} \leq e^{-t}, \quad \forall t \geq 0.$$

Proof. Since μ is more log-concave than the Gaussian, Eq. (12) shows that the density f_t of μ_t satisfies

$$f_t(x) = \exp\left(-\frac{1}{2}|x|^2 - \frac{1}{2}\langle B_t x, x\rangle - V_t(x)\right)$$

for some convex function V_t. By the previous lemma, the covariance matrix A_t of μ_t satisfies

$$A_t \leq (B_t + \mathrm{id})^{-1} \leq \frac{1}{\lambda_t + 1}\, \mathrm{id},$$

where λ_t is the lowest eigenvalue of B_t. Therefore

$$\|A_t\|_{op} \leq \frac{1}{\lambda_t + 1}.$$

On the other hand, the equality (13) yields

$$\lambda_t \geq \int_0^t \|A_s\|_{op}^{-1}\, ds,$$

showing that

$$\int_0^t \|A_s\|_{op}^{-1}\, ds + 1 \leq \|A_t\|_{op}^{-1}.$$

Integrating this differential inequality yields the result. □

The following proposition will be crucial for the proof of our main theorem. Its proof is more involved than the proof of previous estimate, and we refer to [3, Sect. 3].

Proposition 3. *If the initial measure μ is log-concave then*

$$\mathbb{E}\|A_t\|_{op} \leq C_0 \|A_0\|_{op} \tau_n^2 \log(n)\, e^{-t}, \quad \forall t \geq 0,$$

where C_0 is a universal constant.

3 Proof of the Main Theorem

We start with an elementary lemma.

Lemma 3. *Let X be a log-concave random vector in $\mathbb{R}^n$ and let $\|\cdot\|$ be a norm. Then for any event F*

$$\mathbb{E}(\|X\|;\ F) \leq C_1 \sqrt{\mathbb{P}(F)}\,\mathbb{E}(\|X\|),$$

where C_1 is a universal constant. In particular, if $\mathbb{P}(F) \leq (2C_1)^{-2}$, one has

$$\mathbb{E}(\|X\|) \leq 2\mathbb{E}(\|X\|;\ F^c), \tag{14}$$

where F^c is the complement of F.

Proof. This is an easy consequence of Borell's lemma, which states as follows. There exist universal constants $C, c > 0$ such that,

$$\mathbb{P}\left(\|X\| > t\,\mathbb{E}(\|X\|)\right) \leq C\,\mathrm{e}^{-ct}.$$

By Fubini's theorem and the Cauchy-Schwarz inequality

$$\mathbb{E}(\|X\|;\ F) = \int_0^\infty \mathbb{P}(\|X\| > t,\ F)\,\mathrm{d}t \leq \left(\int_0^\infty \sqrt{\mathbb{P}(\|X\| > t)}\,\mathrm{d}t\right) \times \sqrt{\mathbb{P}(F)}.$$

Plugging in Borell's inequality yields the result, with constant $C_1 = 2C/c$.   □

The next ingredient we will need is the following proposition, which we learnt from B. Maurey. The authors are not aware of any published similar result.

Proposition 4. *Let $(M_t)_{t\geq 0}$ be a continuous martingale taking values in $\mathbb{R}^n$. Assume that $M_0 = 0$ and that the quadratic variation of M satisfies*

$$\forall t > 0, \quad [M]_t \leq \mathrm{id},$$

almost surely. Then $(M_t)_{t\geq 0}$ converges almost surely, and the limit satisfies the following inequality. Letting Γ be a standard Gaussian vector, we have for every convex function $\varphi \colon \mathbb{R}^n \to \mathbb{R} \cup \{+\infty\}$

$$\mathbb{E}\varphi(M_\infty) \leq \mathbb{E}\varphi(\Gamma).$$

Proof. The hypothesis implies that M is bounded in L^2, hence convergent by Doob's theorem. Let X be a standard Gaussian vector on $\mathbb{R}^n$ independent of $(M_t)_{t\geq 0}$. We claim that

$$Y = M_\infty + (\mathrm{id} - [M]_\infty)^{1/2} X$$

is also a standard Gaussian vector. Indeed, for a fixed $x \in \mathbb{R}^n$ one has

$$\mathbb{E}\left(e^{i\langle x, Y\rangle} \mid (M_t)_{t\geq 0}\right) = \exp\left(i\langle x, M_\infty\rangle + \frac{1}{2}\langle [M]_\infty x, x\rangle - \frac{1}{2}|x|^2\right)$$

$$= \exp\left(iL_\infty + \frac{1}{2}[L]_\infty - \frac{1}{2}|x|^2\right),$$

where L is the real martingale defined by $L_t = \langle M_t, x\rangle$. Itô's formula shows that

$$D_t = \exp\left(iL_t + \frac{1}{2}[L]_t\right)$$

is a local martingale. On the other hand the hypothesis yields

$$|D_t| = \exp\left(\frac{1}{2}\langle [M]_t x, x\rangle\right) \leq \exp\left(\frac{1}{2}|x|^2\right)$$

almost surely. This shows that $(D_t)_{t\geq 0}$ is a bounded martingale; in particular

$$\mathbb{E}(D_\infty) = \mathbb{E}(D_0) = 1,$$

since $M_0 = 0$. Therefore

$$\mathbb{E}\left(e^{i\langle x, Y\rangle}\right) = e^{-|x|^2/2},$$

proving the claim. Similarly (just replace X by $-X$)

$$Z = M_\infty - (\mathrm{id} - [M]_\infty)^{1/2}X$$

is also standard Gaussian vector. Now, given a convex function ϕ, we have

$$\mathbb{E}\varphi(M_\infty) = \mathbb{E}\varphi\left(\frac{Y+Z}{2}\right) \leq \frac{1}{2}\mathbb{E}\left(\varphi(Y) + \varphi(Z)\right) = \mathbb{E}\varphi(Y),$$

which is the result. □

We are now ready to prove the main theorem.

Proof (of Theorem 1). Let us prove that given a norm $\|\cdot\|$ and a log-concave vector X satisfying $\mathbb{E}(X) = 0$ we have

$$\mathbb{E}\|X\| \leq C\tau_n (\log n)^{1/2} \|\mathrm{cov}(X)\|_{op}^{1/2} \mathbb{E}\|\Gamma\|, \tag{15}$$

for some universal constant C. If X is assumed to be isotropic, then $\mathrm{cov}(X) = \mathrm{id}$ and we end up with the desired inequality (4).

Our first step is to reduce the proof to the case that X has a compact support. Assume that (15) holds for such vectors, and for $r > 0$, let Y_r be a random vector distributed according to the conditional law of X given the event $\{|X| \le r\}$. Then Y_r is a compactly supported log-concave vector, and by our assumption,

$$\mathbb{E}\|Y_r - \mathbb{E}(Y_r)\| \le C\tau_n (\log n)^{1/2} \|\operatorname{cov}(Y_r)\|_{op}^{1/2} \, \mathbb{E}\|\Gamma\|. \tag{16}$$

Besides, it is easily seen by dominated convergence that

$$\lim_{r \to +\infty} \mathbb{E}\|Y_r - \mathbb{E}Y_r\| = \mathbb{E}\|X\|,$$

$$\lim_{r \to +\infty} \operatorname{cov}(Y_r) = \operatorname{cov}(X).$$

So letting r tend to $+\infty$ in (16) yields (15). Therefore, we may continue the proof under the assumption that X is compactly supported.

We use the stochastic process $(\mu_t)_{t \ge 0}$ defined in the beginning of the previous section, with the starting law μ being the law of X.

Let T be the following stopping time:

$$T = \inf\left(t \ge 0, \ \int_0^t A_s \, ds > C^2 \tau_n^2 \log n \, \|A_0\|_{op}\right),$$

where C is a positive constant to be fixed later and with the usual convention that $\inf(\emptyset) = +\infty$. Define the stopped process a^T by

$$(a^T)_t = a_{\min(t,T)}.$$

By the optional stopping theorem, this process is also a martingale and by definition of T its quadratic variation satisfies

$$[a^T]_t \le C^2 \tau_n^2 \log n \, \|A_0\|_{op}, \quad \forall t \ge 0.$$

Also $(a^T)_0 = a_0 = \mathbb{E}(X) = 0$. Applying Proposition 4 we get

$$\mathbb{E}\|a_T\| = \mathbb{E}\|(a^T)_\infty\| \le C\tau_n (\log n)^{1/2} \|A_0\|_{op}^{1/2} \, \mathbb{E}\|\Gamma\|. \tag{17}$$

On the other hand, using Proposition 3 and Markov inequality we get

$$\mathbb{P}(T < +\infty) = \mathbb{P}\left(\int_0^\infty \|A_s\|_{op} \, ds > C^2 \tau_n^2 \log n \, \|A_0\|_{op}\right) \le \frac{C_0}{C^2}.$$

So $\mathbb{P}(T < +\infty)$ can be rendered arbitrarily small by choosing C large enough. By Proposition 2 we have $a_\infty = X$ in law; in particular a_∞ is log-concave. If $\mathbb{P}(T < +\infty)$ is small enough, we get using Lemma 3

$$\mathbb{E}\|X\| = \mathbb{E}\|a_\infty\| \leq 2\mathbb{E}(\|a_\infty\|; T = \infty)$$
$$= 2\mathbb{E}(\|a_T\|; T = \infty) \leq 2\mathbb{E}\|a_T\|.$$

Combining this with (17) and recalling that $A_0 = \mathrm{cov}(X)$ we obtain the result (15).
□

The proof of Proposition 1 follows the same lines. The main difference is that Proposition 2 is used in lieu of Proposition 3.

Proof (of Proposition 1). Let Y_r b a random vector distributed according to the conditional law of X given $|X| \leq r$. Then Y_r is also more log-concave than Γ and

$$\mathbb{E}\varphi(Y_r) \to \mathbb{E}\varphi(X)$$

as $r \to +\infty$. So again we can assume that X is compactly supported, and consider the process $(\mu_t)_{t \geq 0}$ starting from the law of X.

By Lemma 2, the process $(a_t)_{t \geq 0}$ is a martingale whose quadratic variation satisfies

$$[a]_t = \int_0^t A_s\, ds \leq \mathrm{id}, \quad \forall t \geq 0,$$

almost surely. Since again $a_0 = \mathbb{E}(X) = 0$, Proposition 4 yields the result. □

Remark 3. This proof is essentially due to Maurey; although his (unpublished) argument relied on a different stochastic construction.

4 Application to Mean Width and to the Isotropic Constant

In this section, we prove Corollary 1.

Let Γ be a standard Gaussian vector in $\mathbb{R}^n$ and let Θ be a point uniformly distributed in $\mathbb{S}^{n-1}$. Integration in polar coordinates shows that for any norm $\|\cdot\|$,

$$\mathbb{E}\|\Gamma\| = c_n \mathbb{E}\|\Theta\|,$$

where

$$c_n = \mathbb{E}|\Gamma| = \sqrt{n} + O(1),$$

since Γ has the thin-shell property. Theorem 1 can thus be restated as follows. If Y is an isotropic log-concave random vector and K is a convex body containing 0 in its interior then

$$\mathbb{E}\|Y\|_K \leq C\sqrt{n \log n}\, \tau_n\, M(K). \tag{18}$$

Now let K be an isotropic convex body and let X be a random vector uniform on K. Since $X \in K$ almost surely, we have $\|X\|_{K^\circ} \geq |X|^2$, hence

$$\mathbb{E}\|X\|_{K^\circ} \geq \mathbb{E}|X|^2 = n.$$

Applying (18) to K° and to X thus gives

$$M^*(K) \geq \frac{c\sqrt{n}}{\sqrt{\log n}\, \tau_n},$$

which is *(i)*.

In [1], Bourgain combined the inequality

$$\mathbb{E}\|X\| \leq Cn^{1/4}\mathbb{E}\|\Gamma\| \tag{19}$$

with a theorem of Pisier to get the estimate

$$L_K \leq Cn^{1/4}\log n.$$

Part *(ii)* of the corollary is obtained along the same lines, replacing (19) by our main theorem. We sketch the argument for completeness.

Recall that K is assumed to be isotropic and that X is uniform on K. Let T be a positive linear map of determinant 1. Then by the arithmetic-geometric inequality

$$\mathbb{E}\|X\|_{(TK)^\circ} \geq \mathbb{E}\langle X, TX\rangle = \text{Tr}(T) \geq n.$$

Applying (18) to the random vector X and the convex body $(TK)^\circ$ we get

$$M^*(TK) \geq \frac{c\sqrt{n}}{\sqrt{\log n}\, \tau_n}. \tag{20}$$

Now we claim that given a convex body K containing 0 in its in interior, there exists a positive linear map T of determinant 1 such that

$$M^*(TK) \leq C|K|^{1/n}\sqrt{n}\log n. \tag{21}$$

Taking this for granted and combining it with (20) we obtain

$$|K|^{-1/n} \leq C'(\log n)^{3/2}\tau_n.$$

which is part (ii) of the corollary.

It remains to prove the claim (21). Clearly

$$M^*(K) \leq M^*(K - K),$$

and by the Rogers-Shephard inequality (see [11])

$$|K - K| \leq 4^n |K|.$$

This shows that it is enough to prove the claim when K is symmetric about the origin. Now if K is a symmetric convex body in $\mathbb{R}^n$, Pisier's Rademacher-projection estimate together with a result of Figiel and Tomczak-Jaegermann (see e.g. [10, Theorems 2.5 and 3.11]) guarantee the existence of T such that

$$M(TK)M^*(TK) \leq C \log(n),$$

where C is a universal constant. On the other hand, using Jensen's inequality and integrating in polar coordinate we get

$$M(TK) = \int_{\mathbb{S}^{n-1}} \|\theta\|_{TK}\, \sigma(\mathrm{d}\theta) \geq \left(\int_{\mathbb{S}^{n-1}} \|\theta\|_{TK}^{-n}\, \sigma(\mathrm{d}\theta) \right)^{-1/n}$$

$$= \left(\frac{|B_2^n|}{|TK|} \right)^{1/n} \geq \frac{c}{\sqrt{n}\, |K|^{1/n}},$$

finishing the proof of (21).

Acknowledgements The authors wish to thank Bo'az Klartag for a fruitful discussion and Bernard Maurey for allowing them to use an unpublished result of his.

References

1. J. Bourgain, On the distribution of polynomials on high dimensional convex sets, in *Geometric Aspects of Functional Analysis*, ed. by J. Lindenstrauss, V.D. Milman. Lecture Notes in Mathematics, vol. 1469 (Springer, Berlin, 1991), pp. 127–137
2. H.J. Brascamp, E.H. Lieb, On extensions of the Brunn-Minkowski and Prékopa Leindler theorems, including inequalities for log concave functions, and with an application to the diffusion equation. J. Funct. Anal. **22**(4), 366–389 (1976)
3. R. Eldan, Thin shell implies spectral gap up to polylog via a stochastic localization scheme. Geom. Funct. Anal. **23**(2), 532–569 (2013)
4. R. Eldan, B. Klartag, Approximately gaussian marginals and the hyperplane conjecture, in *Concentration, Functional Inequalities and Isoperimetry*. Contemporary Mathematics, vol. 545 (American Mathematical Society, Providence, 2011), pp. 55–68.
5. B. Fleury, O. Guédon, G. Paouris, A stability result for mean width of Lp-centroid bodies. Adv. Math. **214**(2), 865–877 (2007)
6. O. Guédon, E. Milman, Interpolating thin-shell and sharp large-deviation estimates for isotropic log-concave measures. Geom. Funct. Anal. **21**(5), 1043–1068 (2011)
7. G. Hargé, A convex/log-concave correlation inequality for Gaussian measure and an application to abstract Wiener spaces. Probab. Theory Relat. Fields **130**(3), 415–440 (2004)
8. B. Klartag, On convex perturbations with a bounded isotropic constant. Geom. Funct. Anal. **16**(6), 1274–1290 (2006)
9. B. Klartag, A central limit theorem for convex sets. Invent. Math. **168**, 91–131 (2007)

10. G. Pisier, *The Volume of Convex Bodies and Banach Space Geometry*. Cambridge Tracts in Mathematics, vol. 94 (Cambridge University Press, Cambridge, 1989)
11. C.A. Rogers, G.C. Shephard, The difference body of a convex body. Arch. Math. (Basel) **8**, 220–233 (1957)
12. M. Talagrand, *The Generic Chaining. Upper and Lower Bounds of Stochastic Processes*. Springer Monographs in Mathematics (Springer, Berlin, 2005)
13. C. Villani, *Optimal Transport. Old and New*. Grundlehren der Mathematischen Wissenschaften, vol. 338 (Springer, Berlin, 2009)

On the Oscillation Rigidity of a Lipschitz Function on a High-Dimensional Flat Torus

Dmitry Faifman, Bo'az Klartag, and Vitali Milman

Abstract Given an arbitrary 1-Lipschitz function f on the torus $\mathbb{T}^n$, we find a k-dimensional subtorus $M \subseteq \mathbb{T}^n$, parallel to the axes, such that the restriction of f to the subtorus M is nearly a constant function. The k-dimensional subtorus M is selected randomly and uniformly. We show that when $k \leq c \log n / (\log \log n + \log 1/\varepsilon)$, the maximum and the minimum of f on this random subtorus M differ by at most ε, with high probability.

1 Introduction

A uniformly continuous function f on an n-dimensional space X of finite volume tends to concentrate near a single value as n approaches infinity, in the sense that the ε-extension of some level set has nearly full measure. This phenomenon, which is called the *concentration of measure in high dimension*, is frequently related to a transitive group of symmetries acting on X. The prototypical example is the case of a 1-Lipschitz function on the unit sphere S^n, see [3, 4, 8].

One of the most important consequences of the concentration of measure is the emergence of *spectrum*, as was discovered in the 1970s by the third named author, see [5–7]. The idea is that not only does the distinguished level set have a large ε-extension in a sense of measure, but one may actually find structured subsets on which the function is nearly constant. When we have a group G acting transitively on X, this structured subset belongs to the orbit $\{gM_0 ; g \in G\}$ where $M_0 \subseteq X$ is a fixed subspace. The third named author also noted some connections with Ramsey theory, which were developed in two different directions: by Gromov [2] in the direction of metric geometry, and by Pestov [9, 10] in the unexpected direction of dynamical systems.

The phenomenon of spectrum thus follows from concentration, and it comes as no surprise that most of the results in Analysis which establish spectrum, have appeared as a consequence of concentration. In this note we demonstrate an instance

D. Faifman (✉) • B. Klartag (✉) • V. Milman (✉)
School of Mathematical Sciences, Tel Aviv University, Tel Aviv 69978, Israel
e-mail: dfaifman@gmail.com; klartagb@tau.ac.il; milman@tau.ac.il

© Springer International Publishing Switzerland 2014
B. Klartag, E. Milman (eds.), *Geometric Aspects of Functional Analysis*,
Lecture Notes in Mathematics 2116, DOI 10.1007/978-3-319-09477-9_10

where no concentration of measure is available, but nevertheless a geometrically structured level set arises.

To state our result, consider the standard flat torus $\mathbb{T}^n = \mathbb{R}^n/\mathbb{Z}^n = (\mathbb{R}/\mathbb{Z})^n$, which inherits its Riemannian structure from $\mathbb{R}^n$. We say that $M \subseteq \mathbb{T}^n$ is a *coordinate subtorus of dimension* k if it is the collection of all n-tuples $(\theta_j)_{j=1}^n \in \mathbb{T}^n$ with fixed $n - k$ coordinates. Given a manifold X and $f : X \to \mathbb{R}$ we denote the oscillation of f along X by

$$\text{Osc}(f; X) = \sup_X f - \inf_X f.$$

Theorem 1. *There is a universal constant $c > 0$, such that for any $n \geq 1, 0 < \varepsilon \leq 1$ and a function $f : \mathbb{T}^n \to \mathbb{R}$ which is 1-Lipschitz, there exists a k-dimensional coordinate subtorus $M \subseteq \mathbb{T}^n$ with $k = \left\lfloor c \frac{\log n}{\log \log(5n) + \log |\varepsilon|} \right\rfloor$, such that $\text{Osc}(f; M) \leq \varepsilon$.*

Note that the collection of all coordinate subtori equals the orbit $\{gM_0 : g \in G\}$ where $M_0 \subseteq \mathbb{T}^n$ is any fixed k-dimensional coordinate subtorus, and the group $G = \mathbb{R}^n \rtimes S_n$ acts on $\mathbb{T}^n$ by translations and permutations of the coordinates. Theorem 1 is a manifestation of *spectrum*, yet its proof below is inspired by proofs of the Morrey embedding theorem, and the argument does not follow the usual concentration paradigm. We think that the spectrum phenomenon should be much more widespread, perhaps even more than the concentration phenomenon, and we hope that this note will be a small step towards its recognition.

2 Proof of the Theorem

We write $|\cdot|$ for the standard Euclidean norm in $\mathbb{R}^n$ and we write log for the natural logarithm. The standard vector fields $\partial/\partial x_1, \ldots, \partial/\partial x_n$ on $\mathbb{R}^n$ are well-defined also on the quotient $\mathbb{T}^n = \mathbb{R}^n/\mathbb{Z}^n$. These n vector fields are the *coordinate directions* on the unit torus $\mathbb{T}^n$. Thus, the partial derivatives $\partial_1 f, \ldots, \partial_n f$ are well-defined for any smooth function $f : \mathbb{T}^n \to \mathbb{R}$, and we have $|\nabla f|^2 = \sum_{i=1}^n (\partial_i f)^2$. A k-dimensional subspace $E \subseteq T_x\mathbb{T}^n$ is a *coordinate subspace* if it is spanned by k coordinate directions. For $f : \mathbb{T}^n \to \mathbb{R}$ and $M \subseteq \mathbb{T}^n$ a submanifold, we write $\nabla_M f$ for the gradient of the restriction $f|_M : M \to \mathbb{R}$.

Throughout the proof, c, C will always denote universal constants, not necessarily having the same value at each appearance. Since the Riemannian volume of $\mathbb{T}^n$ equals one, Theorem 1 follows from the case $\alpha = 1$ of the following:

Theorem 2. *There is a universal constant $c > 0$ with the following property: Let $n \geq 1, 0 < \varepsilon \leq 1, 0 < \alpha \leq 1$ and $1 \leq k \leq c \frac{\log n}{\log \log(5n) + |\log \varepsilon| + |\log \alpha|}$. Let $f : \mathbb{T}^n \to \mathbb{R}$ be a locally-Lipschitz function such that, for $p = k(1 + \alpha)$,*

$$\int_{\mathbb{T}^n} |\nabla f|^p \le 1. \tag{1}$$

Then there exists a k-dimensional coordinate subtorus $M \subseteq \mathbb{T}^n$ with Osc $(f; M) \le \varepsilon$.

The essence of the proof is as follows. First, for some large k we find a k-dimensional coordinate subtorus M where the derivative is small on average, in the sense that $\left(\int_M |\nabla_M f|^p \right)^{1/p}$ is small. The existence of such a subtorus is a consequence of the observation that at every point, most of the partial derivatives in the coordinate directions are small. We then restrict our attention to this subtorus and take any two points $\tilde{x}, \tilde{y} \in M$. Our goal is to show that $f(\tilde{x}) - f(\tilde{y}) < \varepsilon$.

To this end we construct a polygonal line from $\tilde{x}$ to $\tilde{y}$ which consists of intervals of length $1/2$. For every such interval $[x, y]$ we randomly select a point Z in a $(k-1)$-dimensional ball which is orthogonal to the interval $[x, y]$ and is centered at its midpoint. We then show that $|f(x) - f(Z)|$ and $|f(y) - f(Z)|$ are typically small, since $|\nabla_M f|$ is small on average along the intervals $[x, Z]$ and $[y, Z]$.

We proceed with a formal proof of Theorem 2, beginning with the following computation:

Lemma 3. *For any $n \ge 1, 0 < \varepsilon \le 1, 0 < \alpha \le 1$ and $1 \le k \le c \frac{\log n}{\log \log(5n) + |\log \varepsilon| + |\log \alpha|}$, we have that $k \le n/2$ and*

$$\left(\frac{2k}{\delta^2 n} \right)^{1/p} \le \sqrt{k} \cdot \delta \tag{2}$$

where $p = (1 + \alpha)k$ and

$$\delta = \frac{\alpha}{16(1 + \alpha)} \cdot \frac{\varepsilon}{k^{3/2}}. \tag{3}$$

Proof. Take $c = 1/200$. The desired conclusion (2) is equivalent to $4k^{2-p} \le \delta^{2p+4} n^2$, which in turn is equivalent to

$$2^{8p+18} \cdot \left(\frac{\alpha + 1}{\alpha} \right)^{2p+4} \cdot k^{2p+8} \le \varepsilon^{2p+4} n^2. \tag{4}$$

Since $c \le 1/12$ and $\alpha \le 1$ we have that $6p \le 12k \le \log n / |\log \varepsilon|$ and hence $\varepsilon^{2p+4} n^2 \ge \varepsilon^{6p} n^2 \ge n$. Since $\alpha + 1 \le 2$ then in order to obtain (4) it suffices to prove

$$\left(\frac{32}{\alpha} \cdot k \right)^{2p+8} \le n. \tag{5}$$

Since $c \leq 1/200$ and $k \leq c \log n/(\log \log(5n))$ then $24k \log k \leq \log n$. Since $k \leq c \frac{\log n}{|\log \alpha| + \log(\log 5)}$ then $24k \log\left(\frac{32}{\alpha}\right) \leq \log n$. We conclude that $12k \log\left(\frac{32}{\alpha} \cdot k\right) \leq \log n$, and hence

$$\left(\frac{32}{\alpha} \cdot k\right)^{12k} \leq n. \tag{6}$$

However, $p = (1 + \alpha)k$ and hence $2p + 8 \leq 12k$. Therefore the desired bound (5) follows from (6). Since $k \leq \frac{1}{2} \log n \leq n/2$, the lemma is proven. $\square$

Our standing assumptions for the remainder of the proof of Theorem 2 are that $n \geq 1, 0 < \varepsilon \leq 1, 0 < \alpha \leq 1$ and that

$$1 \leq k \leq c \frac{\log n}{\log \log(5n) + |\log \varepsilon| + |\log \alpha|} \tag{7}$$

where $c > 0$ is the constant from Lemma 3. We also denote

$$p = (1 + \alpha)k \tag{8}$$

and we write $e_1, \ldots, e_n$ for the standard n unit vectors in $\mathbb{R}^n$.

Lemma 4. *Let $v \in \mathbb{R}^n$ and let $J \subseteq \{1, \ldots, n\}$ be a random subset of size k, selected uniformly from the collection of all $\binom{n}{k}$ subsets. Consider the k-dimensional subspace $E \subseteq \mathbb{R}^n$ spanned by $\{e_j; j \in J\}$ and let P_E be the orthogonal projection operator onto E in $\mathbb{R}^n$. Then,*

$$\left(\mathbb{E}|P_E v|^p\right)^{1/p} \leq \frac{\alpha}{8(1 + \alpha)} \cdot \frac{\varepsilon}{k} \cdot |v|.$$

Proof. We may assume that $v = (v_1, \ldots, v_n) \in \mathbb{R}^n$ satisfies $|v| = 1$. Let $\delta > 0$ be defined as in (3). Denote $I = \{i; |v_i| \geq \delta\}$. Since $|v| = 1$, we must have $|I| \leq 1/\delta^2$. We claim that

$$\mathbb{P}(I \cap J = \emptyset) \geq 1 - \frac{2k}{\delta^2 n}. \tag{9}$$

Indeed, if $\frac{2k}{\delta^2 n} \geq 1$ then (9) is obvious. Otherwise, $|I| \leq \delta^{-2} \leq n/2 \leq n - k$ and

$$\mathbb{P}(I \cap J = \emptyset) = \prod_{j=0}^{k-1} \frac{n - |I| - j}{n - j} \geq \left(1 - \frac{|I|}{n - k + 1}\right)^k \geq \left(1 - \frac{2}{\delta^2 n}\right)^k \geq 1 - \frac{2k}{\delta^2 n}.$$

Thus (9) is proven. Consequently,

$$\mathbb{E}|P_E v|^p = \mathbb{E}\left(\sum_{j \in J} v_j^2\right)^{p/2} \le \frac{2k}{\delta^2 n} + \mathbb{E}\left[1_{\{I \cap J = \emptyset\}} \cdot \left(\sum_{j \in J} v_j^2\right)^{p/2}\right] \le \frac{2k}{\delta^2 n} + \left(k \cdot \delta^2\right)^{p/2},$$

where 1_A equals one if the event A holds true and it vanishes otherwise. By using the inequality $(a + b)^{1/p} \le a^{1/p} + b^{1/p}$ we obtain

$$\left(\mathbb{E}|P_E v|^p\right)^{1/p} \le \left(\frac{2k}{\delta^2 n}\right)^{1/p} + \sqrt{k} \cdot \delta \le 2\sqrt{k} \cdot \delta = \frac{\alpha}{8(1+\alpha)} \cdot \frac{\varepsilon}{k},$$

where we utilized (3) and Lemma 3.   □

Corollary 5. *Let* $f : \mathbb{T}^n \to \mathbb{R}$ *be a locally-Lipschitz function with* $\int_{\mathbb{T}^n} |\nabla f|^p \le 1$. *Then there exists a k-dimensional coordinate subtorus* $M \subseteq \mathbb{T}^n$ *such that*

$$\left(\int_M |\nabla_M f|^p\right)^{1/p} \le \frac{\alpha}{8(1+\alpha)} \cdot \frac{\varepsilon}{k}. \tag{10}$$

Proof. The set of all coordinate k-dimensional subtori admits a unique probability measure, invariant under translations and coordinate permutations. Let M be a random coordinate k-subtorus, chosen with respect to the uniform distribution. All the tangent spaces $T_x \mathbb{T}^n$ are canonically identified with $\mathbb{R}^n$, and we let $E \subseteq \mathbb{R}^n$ denote a random, uniformly chosen k-dimensional coordinate subspace. Then we may write

$$\mathbb{E}_M \int_M |\nabla_M f|^p = \int_{\mathbb{T}^n} \mathbb{E}_E |P_E \nabla f|^p \le A^p \int_{\mathbb{T}^n} |\nabla f|^p \le A^p,$$

where $A = \frac{\alpha}{8(1+\alpha)} \cdot \frac{\varepsilon}{k}$ and we used Lemma 4. It follows that there exists a subtorus M which satisfies (10).   □

The following lemma is essentially Morrey's inequality (see [1, Sect. 4.5]).

Lemma 6. *Consider the k-dimensional Euclidean ball* $B(0, R) = \{x \in \mathbb{R}^k ; |x| \le R\}$. *Let* $f : B(0, R) \to \mathbb{R}$ *be a locally-Lipschitz function, and let* $x, y \in B(0, R)$ *satisfy* $|x - y| = 2R$. *Recall that* $p = (1 + \alpha)k$. *Then,*

$$|f(x) - f(y)| \le 4\frac{1+\alpha}{\alpha} \cdot k^{\frac{1}{2(1+\alpha)}} \cdot R^{1-\frac{k}{p}} \left(\int_{B(0,R)} |\nabla f(x)|^p dx\right)^{1/p}. \tag{11}$$

Proof. We may reduce matters to the case $R = 1$ by replacing $f(x)$ by $f(Rx)$; note that the right-hand side of (11) is invariant under such replacement. Thus x is a unit vector, and $y = -x$. Let Z be a random point, distributed uniformly in the $(k-1)$-dimensional unit ball

$$B(0,1) \cap x^{\perp} = \{v \in \mathbb{R}^k \; ; \; |v| \leq 1, \, v \cdot x = 0\},$$

where $v \cdot x$ is the standard scalar product of $x, v \in \mathbb{R}^k$. Let us write

$$\mathbb{E}|f(x) - f(Z)| \leq \mathbb{E}|x - Z| \int_0^1 |\nabla f((1-t)x + tZ)| \, dt \tag{12}$$

$$\leq 2\mathbb{E}|\nabla f((1-T)x + TZ)| = 2 \int_{B(0,1)} |\nabla f(z)| \rho(z) dz,$$

where T is a random variable uniformly distributed in $[0, 1]$, independent of Z, and where ρ is the probability density of the random variable $(1-T)x + TZ$. Then,

$$\rho((1-r)x + rz) = \frac{c_k}{r^{k-1}}$$

when $z \in B(0,1) \cap x^{\perp}, 0 < r < 1$. We may compute c_k as follows:

$$1 = c_k \int_0^1 \frac{1}{r^{k-1}} V_{k-1}(r) dr = c_k V_{k-1}(1) = c_k \frac{\pi^{k-1}}{\Gamma\left(\frac{k+1}{2}\right)},$$

where $V_{k-1}(r)$ is the $(k-1)$-dimensional volume of a $(k-1)$-dimensional Euclidean ball of radius r. Denote $q = p/(p-1)$. Then,

$$\int_{B(0,1)} \rho^q = \int_0^1 \left(\frac{c_k}{r^{k-1}}\right)^q V_{k-1}(r) dr = \frac{c_k^q V_{k-1}(1)}{(k-1)(1-q)+1} = \frac{p-1}{p-k}\left(\frac{\Gamma\left(\frac{k+1}{2}\right)}{\pi^{k-1}}\right)^{q-1},$$

and hence

$$\left(\int_{B(0,1)} \rho^q\right)^{1/q} = \left(\frac{p-1}{p-k}\right)^{1/q} \left(\frac{\Gamma\left(\frac{k+1}{2}\right)}{\pi^{k-1}}\right)^{1/p} \tag{13}$$

$$\leq \left(\frac{1+\alpha}{\alpha}\right)^{1/q} \left(\frac{k^{k/2}}{\pi^{k-1}}\right)^{1/p} \leq \frac{1+\alpha}{\alpha} \cdot k^{\frac{1}{2(1+\alpha)}}.$$

Denote $C_{\alpha,k} = \frac{1+\alpha}{\alpha} \cdot k^{\frac{1}{2(1+\alpha)}}$. From (12), (13) and the Hölder inequality,

$$\mathbb{E}|f(x) - f(Z)| \leq 2 \left(\int_{B(0,1)} |\nabla f|^p\right)^{\frac{1}{p}} \left(\int_{B(0,1)} \rho^q\right)^{\frac{1}{q}} \leq 2C_{\alpha,k} \left(\int_{B(0,1)} |\nabla f|^p\right)^{\frac{1}{p}}. \tag{14}$$

A bound similar to (14) also holds for $\mathbb{E}|f(y) - f(Z)|$, since $y = -x$. By the triangle inequality,

$$|f(x) - f(y)| \le \mathbb{E}|f(y) - f(Z)| + \mathbb{E}|f(Z) - f(x)| \le 4C_{\alpha,k} \left(\int_{B(0,1)} |\nabla f|^p \right)^{1/p}.$$

$\square$

Proof of Theorem 2. According to Corollary 5 we may select a coordinate subtorus $M = \mathbb{T}^k$ so that

$$\left(\int_M |\nabla_M f|^p \right)^{1/p} \le \frac{\alpha}{8(1+\alpha)} \cdot \frac{\varepsilon}{k}. \tag{15}$$

Given any two points $x, y \in M$, let us show that

$$|f(x) - f(y)| \le \varepsilon. \tag{16}$$

The distance between x and y is at most $\sqrt{k}/2$. Let us construct a curve, in fact a polygonal line, starting at x and ending at y which consists of at most $\sqrt{k} + 1$ intervals of length $1/2$. For instance, we may take all but the last two intervals to be intervals of length $1/2$ lying on a minimizing geodesic between x to y. The last two intervals need to connect two points whose distance is at most $1/2$, and this is easy to do by drawing an isosceles triangle whose base is the segment between these two points.

Let $[x_j, x_{j+1}]$ be any of the intervals appearing in the polygonal line constructed above. Let $B \subset \mathbb{T}^k = M$ be a geodesic ball of radius $R = 1/4$ centered at the midpoint of $[x_j, x_{j+1}]$. This geodesic ball on the torus is isometric to a Euclidean ball of radius $R = 1/4$ in $\mathbb{R}^k$. Lemma 6 applies, and implies that

$$|f(x_j) - f(x_{j+1})| \le 4 \frac{1+\alpha}{\alpha} \cdot k^{\frac{1}{2(1+\alpha)}} \left(\frac{1}{4} \right)^{1-\frac{k}{p}} \left(\int_B |\nabla_M f|^p \right)^{\frac{1}{p}}$$

$$\le 4 \frac{1+\alpha}{\alpha} \cdot \sqrt{k} \left(\int_M |\nabla_M f|^p \right)^{\frac{1}{p}}.$$

Since the number of intervals in the polygonal line is at most $\sqrt{k} + 1 \le 2\sqrt{k}$, then

$$|f(x) - f(y)| \le \sum_j |f(x_j) - f(x_{j+1})| \le 8 \frac{1+\alpha}{\alpha} \cdot k \left(\int_M |\nabla_M f|^p \right)^{1/p} \le \varepsilon,$$

where we used (15) in the last passage. The points $x, y \in M$ were arbitrary, and hence $\text{Osc}(f; M) \le \varepsilon$.

$\square$

Remarks. 1. It is evident from the proof of Theorem 2 that the subtorus M is selected randomly and uniformly over the collection of all k-dimensional coordinate subtori. It is easy to obtain that with probability at least $9/10$, we have that $Osc(M; f) \le \varepsilon$.

2. The assumption that f is locally-Lipschitz in Theorem 2 is only used to justify the use of the fundamental theorem of calculus in (12). It is possible to significantly weaken this assumption; it suffices to know that f admits weak derivatives $\partial_1 f, \ldots, \partial_n f$ and that (1) holds true, see [1, Chap. 4] for more information.

It is quite surprising that the conclusion of the theorem also holds for non-continuous, unbounded functions, with many singular points, as long as (1) is satisfied in the sense of weak derivatives. The singularities are necessarily of a rather mild type, and a variant of our proof yields a subtorus M on which the function f is necessarily continuous with $Osc(f; M) \le \varepsilon$.

3. Another possible approach to the problem would be along the lines of the proof of the classical concentration theorems—namely, finding an ε-net of points in a subtorus, where all the coordinate partial derivatives of the function are small. However, this approach requires some additional a-priori data about the function, such as a uniform bound on the Hessian.

4. We do not know whether the dependence on the dimension in Theorem 1 is optimal. Better estimates may be obtained if the subtorus $M \subseteq \mathbb{T}^n$ is permitted to be an arbitrary k-dimensional *rational* subtorus, which is not necessarily a coordinate subtorus. Recall that a rational torus is a quotient of $\mathbb{R}^n$ by a lattice which is spanned by n vectors with rational coordinates.

Acknowledgements We would like to thank Vladimir Pestov for his interest in this work. The first-named author was partially supported by ISF grants 701/08 and 1447/12. The second-named author was supported by a grant from the European Research Council (ERC). The third-named author was supported by ISF grant 387/09 and by BSF grant 2006079.

References

1. L.C. Evans, R.F. Gariepy, *Measure Theory and Fine Properties of Functions.* Studies in Advanced Mathematics (CRC Press, Boca Raton, FL, 1992)
2. M. Gromov, Filling Riemannian manifolds. J. Differ. Geom. **18**(1), 1–147 (1983)
3. M. Gromov, Isoperimetry of waists and concentration of maps. Geom. Funct. Anal. (GAFA) **13**(1), 178–215 (2003)
4. M. Ledoux, *The Concentration of Measure Phenomenon.* Mathematical Surveys and Monographs, vol. 89 (American Mathematical Society, Providence, RI, 2001)
5. V.D. Milman, The spectrum of bounded continuous functions which are given on the unit sphere of a B-space. Funkcional. Anal. i Priložen. **3**(2), 67–79 (1969) (Russian)
6. V.D. Milman, Geometric theory of Banach spaces. II. Geometry of the unit ball. Uspehi Mat. Nauk **26**(6)(162), 73–149 (1971) (Russian)
7. V.D. Milman, Asymptotic properties of functions of several variables that are defined on homogeneous spaces. Soviet Math. Dokl. **12**, 1277–1281 (1971). Translated from Dokl. Akad. Nauk SSSR **199**, 1247–1250 (1971) (Russian)

8. V.D. Milman, G. Schechtman, *Asymptotic Theory of Finite-Dimensional Normed Spaces. With an Appendix by M. Gromov.* Lecture Notes in Mathematics, vol. 1200 (Springer, Berlin, 1986)
9. V. Pestov, Ramsey-Milman phenomenon, Urysohn metric spaces, and extremely amenable groups. Israel J. Math. **127**, 317–357 (2002)
10. V. Pestov, *Dynamics of Infinite-Dimensional Groups. The Ramsey-Dvoretzky-Milman Phenomenon.* University Lecture Series, vol. 40 (American Mathematical Society, Providence, RI, 2006)

Identifying Set Inclusion by Projective Positions and Mixed Volumes

Dan Florentin, Vitali Milman, and Alexander Segal

Abstract We study a few approaches to identify inclusion (up to a shift) between two convex bodies in $\mathbb{R}^n$. To this goal we use mixed volumes and fractional linear maps. We prove that inclusion may be identified by comparing volume or surface area of all projective positions of the sets. We prove similar results for Minkowski sums of the sets.

1 Introduction and Results

Set inclusion $A \subseteq B$ of two convex bodies (elements of $\mathcal{K}^n$, namely compact convex non degenerate sets), implies that for every monotone functional $f : \mathcal{K}^n \to \mathbb{R}$, one has by definition, $f(A) \leq f(B)$. For example, for the volume functional we have $|A| \leq |B|$. Our goal is to achieve a reverse implication: describing a family $\mathcal{F}$ of such functionals, with the property of *identifying inclusion*, that is, given $A, B \in \mathcal{K}^n$, if $f(A) \leq f(B)$ for all $f \in \mathcal{F}$, then $A \subseteq B$ (or more generally, B contains a translate of A). Note that if a family of functionals $\mathcal{F}$ identifies inclusion, then it *separates elements* in $\mathcal{K}^n$, that is, if $f(A) = f(B)$ for all $f \in \mathcal{F}$, then $A = B$ (or more generally, B is a translate of A). The converse, however, is not true in general. That is, some families separate points but do not identify inclusion. For example, see the theorem by Chakerian and Lutwak below.

R. Schneider showed in [10] that a convex body is determined, up to translation, by the value of its mixed volumes with some relatively small family of convex bodies. An extension to this fact was given by W. Weil in the same year:

Theorem (Weil [12]). *Let* $A, B \in \mathcal{K}^n$. *Then* B *contains a translate of* A *if and only if for all* $K_2, \ldots, K_n \in \mathcal{K}^n$:

D.I. Florentin was partially supported by European Research Council grand Dimension 305629.

D.I. Florentin • V.D. Milman (✉) • A. Segal
Tel Aviv University, Tel Aviv, Israel
e-mail: milman@post.tau.ac.il

© Springer International Publishing Switzerland 2014   133
B. Klartag, E. Milman (eds.), *Geometric Aspects of Functional Analysis*,
Lecture Notes in Mathematics 2116, DOI 10.1007/978-3-319-09477-9_11

$$V(A, K_2, \ldots, K_n) \leq V(B, K_2, \ldots, K_n),$$

where $V(K_1, \ldots, K_n)$ denotes the n-dimensional mixed volume.

Actually, it is possible to reduce even further the information on the bodies A and B, as follows from the result of Lutwak [7] which we discuss in Sect. 3.

We will investigate a few approaches to achieve the same goal. First we use a family of transformations on $\mathbb{R}^n$. We examine two such families in this note: the group of affine transformations AF_n, and the group of the far less explored fractional linear (or projective) transformations FL_n. For example we may consider $\mathcal{F}_L = \{f \circ T : T \in AF_n\}$, where f is the volume functional, and T is considered as a map on $\mathcal{K}^n$. Unfortunately, the affine structure respects volume too well, that is, if $|A| \leq |B|$, then for every $T \in AF_n$ we have $|TA| \leq |TB|$ as well. In other words, the action of the affine group is not rich enough to describe inclusion. However, it turns out that the larger family $\mathcal{F}_P = \{f \circ T : T \in FL_n\}$ is identifying inclusion. We will consider replacing the volume functional by a mixed volume, as well as replacing the family of transformations by different operations (such as Minkowski sums with arbitrary bodies). We would like to emphasize that the proofs we present in this note are not very sophisticated. However, they point to some directions which Convexity Theory did not explore enough, and lead to new and intriguing questions.

Let us introduce a few standard notations. First, for convenience we will fix an Euclidean structure and some orthonormal basis $\{e_i\}_1^n$. For a vector $x \in \mathbb{R}^n$ we will often write $x = (x_1, \ldots, x_n) = \sum_1^n x_i e_i$. Given a subspace $E \subset \mathbb{R}^n$, P_E will denote the orthogonal projection onto E. We denote by D_n the n-dimensional Euclidean ball and, for $1 \leq i \leq n$, by $W_{n-i}(K)$ the quermassintegral:

$$W_{n-i}(K) = V(K[i], D_n[n - i]).$$

For further definitions and well known properties of mixed volumes and quermass-integrals see [11]. We denote the support function of K by:

$$h_K(u) = \sup_{x \in K} \langle x, u \rangle.$$

Let us recall the definition of fractional linear maps. We identify $\mathbb{R}^n$ with a subset of the projective space RP^n, by fixing some point $z \in \mathbb{R}^{n+1} \setminus \{0\}$, and considering the affine subspace $E_n = \{x | \langle z, x \rangle = 1\} \subset \mathbb{R}^{n+1}$. Every point in E_n corresponds to a unique line in R^{n+1} passing through the origin. A regular linear transformation $\tilde{L} : \mathbb{R}^{n+1} \to \mathbb{R}^{n+1}$ induces an injective map L on RP^n. A fractional linear map is the restriction of such a map to $E_n \cap L^{-1}(E_n)$. The maximal (open) domain $Dom(F)$, of a non-affine fractional linear map F is $\mathbb{R}^n \setminus H$, for some affine hyperplane H. Since our interest is in convex sets, we usually consider just one side of H as the domain, i.e. our maps are defined on half spaces. They are the homomorphisms of convexity, in the sense that there are no other injective maps that preserve convexity of every set in their domain. The big difference, compared to linear maps, is that the

Jacobian matrix is not constant, and its determinant is not bounded (on the maximal domain). In Sect. 2 we show that $\mathcal{F}_P$ is an identifying family:

Theorem 1.1. *Let $n \geq 1$, and $A, B \in \mathcal{K}^n$. If for every admissible $F \in FL_n$, one has*

$$Vol(FA) \leq Vol(FB),$$

then $A \subseteq B$.

When we say that F is admissible we mean that $A, B \subset Dom(F)$. Theorem 1.1 is a particular case of the following fact, where the volume is replaced by any of the quermassintegrals:

Theorem 1.2. *Let $n \geq 2$, $A, B \in \mathcal{K}^n$, and fix $1 \leq i \leq n$. If for every admissible $F \in FL_n$ one has*

$$W_{n-i}(FA) \leq W_{n-i}(FB),$$

then $A \subseteq B$.

The proof of Theorem 1.2 is based on the non boundedness of the Jacobians of admissible fractional linear maps. That is, if $A \setminus B$ is of positive volume, we may choose an admissible $F \in FL_n$ such that FA exceeds FB in volume or, say, surface area, regardless of how small $|A \setminus B|$ is. The exact formulation is given in Lemma 2.3.

Had we considered only $F \in FL_n$ which are affine in Theorem 1.1, clearly the conclusion could not have been reproduced (since $|FA| = \det(F)|A|$, we in fact only assume that $|A| \leq |B|$). One may ask the same question in the case of surface area, namely $i = n - 1$. Since the surface area is not a linear invariant, not even up to the determinant (as in the case of volume), the answer is not trivial, but it does follow immediately (along with the restriction to $n \geq 3$) from the negative answer to Shephard's problem. In [8, 9] Petty and Schneider showed:

Theorem (Petty, Schneider). *Let $n \geq 3$. Then there exist centrally symmetric bodies $A, B \in \mathcal{K}^n$, such that $Vol(A) > Vol(B)$, and yet for every $n - 1$ dimensional space E, $Vol_{n-1}(P_E A) \leq Vol_{n-1}(P_E B)$. In particular, $A \not\subseteq B$.*

We say that $K \in \mathcal{K}^n$ is centrally symmetric (or symmetric), if $K = -K$ (i.e. its center is 0). Chakerian and Lutwak [3] showed that the bodies from the previous theorem satisfy a surface area inequality in every position. For the sake of completeness, we append the proof.

Theorem (Chakerian, Lutwak). *Let $n \geq 3$. Then there exist centrally symmetric bodies $A, B \in \mathcal{K}^n$, such that $A \not\subseteq B$, and yet for every $L \in AF_n$ we have*

$$Vol_{n-1}(\partial L A) \leq Vol_{n-1}(\partial L B).$$

Proof. Let A, B be the sets whose existence is assured by the previous theorem. Recall the definition of the projection body ΠK of a convex body K:

$$h_{\Pi K}(u) = Vol_{n-1}(P_{u^\perp}(K)).$$

Using this notion, the assumption on A and B can be reformulated as $\Pi A \subseteq \Pi B$. By Kubota's formula $W_1(\Pi K) = c_n|\partial K|$, which implies that $|\partial A| \leq |\partial B|$. Let $L \in AF_n$. There exists $\tilde{L} \in GL_n$ such that for all $K \in \mathcal{K}^n$, $\Pi LK = \tilde{L}\Pi K$. Thus, the inclusion $\Pi LA \subseteq \Pi LB$ holds as before, and we get for all $L \in AF_n$:

$$|\partial LA| \leq |\partial LB|.$$

$\square$

In Sect. 2 we recall some of the properties of fractional linear maps and gather some technical lemmas, to deduce Theorem 1.2.

In Sect. 3 we consider a different type of an identifying family of functionals. The volume of projective (or linear) transformations of a body A is replaced by the volume of Minkowski sums of A with arbitrary bodies. First let us mention the following curious yet easy facts, which, in the same time, demonstrate well our intention.

Theorem 1.3. *Let $n \geq 1$, and let $A, B \in \mathcal{K}^n$ be centrally symmetric bodies. If for all $K \in \mathcal{K}^n$, one has $|A + K| \leq |B + K|$, then $A \subseteq B$.*
Moreover:
Let $n \geq 1$, $K_0 \in \mathcal{K}^n$, and let $A, B \in \mathcal{K}^n$ be centrally symmetric bodies. If for every linear image $K = LK_0$, one has $|A + K| \leq |B + K|$, then $A \subseteq B$.

Theorem 1.4. *Let $n \geq 1$, and let $A, B \in \mathcal{K}^n$. If for every $K \in \mathcal{K}^n$, one has $|A + K| \leq |B + K|$, then there exists $x_0 \in \mathbb{R}^n$ such that $A \subseteq B + x_0$.*
Moreover:
Let $n \geq 1$, and let $A, B \in \mathcal{K}^n$. If for every simplex $\Delta \in \mathcal{K}^n$, one has $|A + \Delta| \leq |B + \Delta|$, then there exists $x_0 \in \mathbb{R}^n$ such that $A \subseteq B + x_0$.

We include a direct proof of the symmetric case, since in this case the argument is far simpler. To prove the general case, we first obtain the inequality:

$$\forall K \in \mathcal{K}^n: \qquad V(A, K[n-1]) \leq V(B, K[n-1]),$$

where V is the n dimensional mixed volume, and $K[n-1]$ stands for $n-1$ copies of the body K. Finally, the proof may be completed by applying a beautiful result of Lutwak to the last inequality (see Sect. 3). In the rest of Sect. 3 we investigate the situation where we only assume that the $n-1$ dimensional volume of every section of $A + K$ is smaller than that of $B + K$ (for every K).

In the last two sections, we formulate conditions to achieve inclusions between two n-tuples of convex bodies, $K_1, \ldots, K_n$, $L_1, \ldots, L_n$, in terms of the mixed volume of their affine or projective positions.

We would like to thank Rolf Schneider for his remarks and attention.

2 Fractional Linear Maps

Let us recall the definition and some properties of fractional linear maps. A fractional linear map is a map $F : \mathbb{R}^n \setminus H \to \mathbb{R}^n$, of the form

$$F(x) = \frac{Ax + b}{\langle x, c \rangle + d},$$

where A is a linear operator in $\mathbb{R}^n$, $b, c \in \mathbb{R}^n$ and $d \in \mathbb{R}$, such that the matrix

$$\hat{A} = \begin{pmatrix} A & b \\ c & d \end{pmatrix}$$

is invertible (in $\mathbb{R}^{n+1}$). The affine hyperplane $H = \{x | \langle x, c \rangle + d = 0\}$ is called the defining hyperplane of F. Since such maps are traces of linear maps in $\mathbb{R}^{n+1}$, we sometimes call the image $F(K)$ of a convex body K under such a map a *projective position of the body* K. These are the only transformations (on, say, open convex domains), which map any interval to an interval. Note that affine maps are a subgroup of fractional linear maps, and that any projective position of a closed ellipsoid is again an ellipsoid. The same is true for a simplex. For more details, including proofs of the following useful facts, see [1].

Proposition. *Denote by H^+ the half space $\{x_1 > 1\}$ (where x_1 is the first coordinate of x) and let the map $F_0 : H^+ \to H^+$ (called the* canonical form *of a fractional linear map) be given by*

$$F_0(x) = \frac{x}{x_1 - 1}.$$

For any $x_0, y_0 \in \mathbb{R}^n$ and a non-affine fractional linear map F with $F(x_0) = y_0$, there exist $B, C \in GL_n$ such that for every $x \in \mathbb{R}^n$,

$$B(F(Cx + x_0) - y_0) = F_0(x).$$

Fractional linear maps turn up naturally in convexity. For example, they are strongly connected to the polarity map, as can be seen in the following, easily verified proposition.

Proposition. *Let $K \subseteq \{x_1 < 1\} \subset \mathbb{R}^n$ be a closed convex set containing 0. Then for the canonical form $F_0(x) = \frac{x}{x_1 - 1}$ the following holds:*

$$F_0(K) = (e_1 - K^\circ)^\circ,$$

where $e_1 = (1, 0, \ldots, 0) \in \mathbb{R}^n$.

Remark 2.1. Let $K, T \in \mathcal{K}^n$. It is useful to note that when $K \nsubseteq T$, there exist disjoint closed balls D_K, D_T of positive radius, and a hyperplane H (which divides $\mathbb{R}^n$ to two open half spaces H^+, H^-), such that:

$$D_K \subset K \cap H^-, \qquad T \subset D_T \subset H^+.$$

This trivial observation means it is sufficient in many cases to consider the action of fractional linear maps on dilations of the ball. To this end we write the following easily verified fact, resulting from a direct computation:

Fact 2.2 *Let $\mathcal{E}_{R,r,\delta}$ stand for the image of the Euclidean ball D_n under the diagonal linear map $A_{R,r} = \mathrm{diag}\{R, r, \ldots, r\}$, shifted by $(1 - \delta - R)e_1$, so that the distance of $\mathcal{E}_{R,r,\delta}$ from the hyperplane $H_0 = \{x_1 = 1\}$ is δ. That is:*

$$\mathcal{E}_{R,r,\delta} = A_{R,r} D_n + (1 - \delta - R)e_1.$$

Then for the canonical form F_0 one has:

$$F_0(\mathcal{E}_{R,R,\delta}) = \mathcal{E}_{\frac{R}{\delta(\delta+2R)}, \frac{R}{\delta\sqrt{\delta+2R}}, \frac{1}{\delta+2R}}.$$

In particular, $F_0(\mathcal{E}_{R,R,\delta})$ contains a translate of $m D_n$ and is contained in a translate of $M D_n$, where $m = \frac{R}{\delta} \min\{\frac{1}{\delta+2R}, \frac{1}{\sqrt{\delta+2R}}\}$, and $M = \frac{R}{\delta} \max\{\frac{1}{\delta+2R}, \frac{1}{\sqrt{\delta+2R}}\}$.

We shall now prove the main Lemma required for Theorem 1.2.

Lemma 2.3. *Let $K, T \in \mathcal{K}^n$ satisfy $K \nsubseteq T$. For every $\varepsilon > 0$ there exist $x_0 \in \mathbb{R}^n$ and a fractional linear map F such that:*

$$F(T) \subseteq \varepsilon D_n, \qquad D_n + x_0 \subseteq F(K).$$

Proof. Let $\varepsilon > 0$, and let D_K, D_T be balls satisfying the inclusions in Remark 2.1. It suffices to find a fractional linear map F such that $F(D_K)$ contains a translate of D_n, and $F(D_T) \subseteq \varepsilon D_n$. Without loss of generality (by applying an affine map), we may assume that the centers of D_K and D_T both lie on the x_1 coordinate axis, and that:

$$D_K = \mathcal{E}_{1,1,\delta}, \qquad D_T = \mathcal{E}_{R,R,d},$$

for some $d > 2$, and $R, \delta > 0$. From 2.2 it follows that $F_0(D_K)$ contains a translate of $\frac{1}{\delta(\delta+2)} D_n$, and $F_0(D_T)$ is contained in a translate of $\frac{R}{d\sqrt{d+2R}} D_n \subset R D_n$. Since δ is arbitrarily small, the result follows. $\square$

Theorem 1.2 now follows in an obvious way.

3 Comparing Convex Bodies Via Minkowski Sums

We will begin with proving the symmetric case.

Proof of Theorem 1.3. We assume that the symmetric sets A, B satisfy $|A + K| \leq |B + K|$, for all $K \in \mathcal{K}^n$. The case $n = 1$ is trivial. Assume $n \geq 2$ and let $u \in S^{n-1}$. Note that the inequality $|A + K| \leq |B + K|$ holds also for a compact convex set K with empty interior, and let $K \subset u^{\perp}$ be an $n - 1$ dimensional Euclidean ball of ($n - 1$ dimensional) volume 1. The leading coefficient in $|A + rK|$ for $r \to \infty$ is the width $w_A(u) = 2h_A(u)$, and since $|A + rK| \leq |B + rK|$ we have:

$$\forall u \in S^{n-1}, \qquad h_A(u) \leq h_B(u),$$

as required.   □

In the case of general bodies, the previous inequality on the widths:

$$\forall u \in S^{n-1}, \qquad w_A(u) \leq w_B(u),$$

does not imply the desired inclusion $A \subseteq B$. However, an inequality on the integrals $\int h_A d\sigma_K \leq \int h_B d\sigma_K$, against any surface area measure σ_K of a convex body K, may be obtained by an argument similar to that from the proof of Theorem 1.3. It turns out to be sufficient, by the following theorem from [7]:

Theorem (Lutwak). *Let $A, B \in \mathcal{K}^n$. Assume that for every simplex $\Delta \in \mathcal{K}^n$ one has*

$$V(A, \Delta[n - 1]) \leq V(B, \Delta[n - 1]).$$

Then there exists $x_0 \in \mathbb{R}^n$ such that $A + x_0 \subseteq B$.

Proof of Theorem 1.4. It follows from the assumption that for every $K \in \mathcal{K}^n$ and every $\varepsilon > 0$, we have $|K + \varepsilon A| \leq |K + \varepsilon B|$. Comparing derivatives at $\varepsilon = 0$ yields an inequality between mixed volumes:

$$\forall K \in \mathcal{K}^n, \qquad V(A, K[n - 1]) \leq V(B, K[n - 1]).$$

The conclusion follows from Lutwak's Theorem.   □

Clearly, these families of functionals are by no means minimal. In the proof of Theorem 1.4 we have in fact used only the functionals $\{A \mapsto |A + \Delta|\}$. In the symmetric case, we only used dilates of a flat ball. This leads to the formally stronger formulations of Theorems 1.3, 1.4.

Let us use these facts to add some information to the well known Busemann-Petty problem. The Busemann-Petty problem is concerned with comparisons of volume of central sections. That is, given two centrally symmetric sets $A, B \in \mathcal{K}^n$, satisfying $|A \cap E| \leq |B \cap E|$ for every $n - 1$ dimensional subspace E, does it imply that

$|A| \leq |B|$? As shown in [5, 6, 13], the answer is negative for all $n \geq 5$ and positive for $n \leq 4$. However, if we combine intersections with Minkowski sums, we may use Theorem 1.3 to get:

Corollary 3.1. *Let $n \geq 2$, and let $A, B \in \mathcal{K}^n$ be centrally symmetric bodies. If for all $K \in \mathcal{K}^n$ and $E \in G_{n,n-1}$ one has*

$$|E \cap (A + K)| \leq |E \cap (B + K)|,$$

then $A \subseteq B$. As in Theorem 1.3, it suffices to fix a body K_0, and check the condition only for linear images of K_0.

The non symmetric case is not as simple. In this case we know that for every $E \in G_{n,n-1}$ there exists a point $x_E \in E$, such that $E \cap A \subseteq E \cap B + x_E$. However, this does not imply that there exists a point $x \in \mathbb{R}^n$ such that $A \subseteq B + x$, as shown in the following example:

Example 3.2. There exist $A, B \in \mathcal{K}^2$, such that for every line E passing through the origin, the interval $B \cap E$ is longer than the interval $A \cap E$, and yet no translation of B contains A.

The construction is based on (the dual to) the Reuleaux triangle R, a planar body of constant width: $\forall u \in S^1$, $w_R(u) = h_R(u) + h_R(-u) = 2$. In other words, the projection of R to $u^\perp$ is an interval of length 2, say $P_{u^\perp}(R) = [\alpha - 2, \alpha]$ for some $\alpha = \alpha(u) \in (0, 2)$. Then:

$$|R^\circ \cap u^\perp| = |(P_{u^\perp}(R))^\circ| = |[1/(\alpha - 2), 1/\alpha]| \geq 2 = |D_2 \cap u^\perp|.$$

However, no translation of R° contains D_2.

Although Corollary 3.1 may not be extended to non symmetric bodies, one can show the following fact, in the same spirit:

Theorem 3.3. *Let $n \geq 2$ and $A, B \in \mathcal{K}^n$. If for all $K \in \mathcal{K}^n$ and $E \in G_{n,n-1}$ one has*

$$|E \cap (A + K)| \leq |E \cap (B + K)|, \tag{1}$$

then $A - A \subset B - B$. In particular, there exist $x_A, x_B \in \mathbb{R}^n$ such that:

$$A + x_A \subseteq B - B \subset (n + 1)(B + x_B).$$

Proof. Apply condition (1) for convex sets of the form $-A + K$ to get:

$$|E \cap (A - A + K)| \leq |E \cap (B - A + K)|.$$

Then apply condition (1) for convex sets of the form $-B + K$:

$$|E \cap (A - B + K)| \leq |E \cap (B - B + K)|.$$

Since $|E \cap (B - A + K)| = |E \cap (A - B - K)|$, we get for all *symmetric* $K \in \mathcal{K}^n$:

$$|E \cap (A - A + K)| \leq |E \cap (B - B + K)|,$$

which by Corollary 3.1 implies that $A - A \subset B - B$. The theorem now follows, since $A - A$ contains a translate of A, and nB contains a translate of $-B$ (as shown by Minkowski. See, e.g., Bonnesen and Fenchel [2], Sect. 7, 34, pp. 57–58 for a proof). □

We also consider the projection version of Theorem 3.3:

Theorem 3.4. *Let $n \geq 2$ and $A, B \in \mathcal{K}^n$. If for all $K \in \mathcal{K}^n$ and $E \in G_{n,n-1}$ one has*

$$|P_E(A + K)| \leq |P_E(B + K)|, \tag{2}$$

then there exists $x_0 \in \mathbb{R}^n$ such that $A + x_0 \subset 2B$.

Proof. Since projection is a linear operator, $P_E(A + K) = P_E(A) + P_E(K)$. Due to surjectivity we may rewrite condition (2) as follows. For every $n - 1$ dimensional subspace E and for every convex body $K' \subset E$ we have:

$$|P_E(A) + K'| \leq |P_E(B) + K'|.$$

Theorem 1.4 implies that there exists a shift $x_E \in E$ such that $P_E(A) \subseteq P_E(B) + x_E$. By a result of Chen, Khovanova, and Klain (see [4]) there exists $x_0 \in \mathbb{R}^n$ such that:

$$A + x_0 \subseteq \frac{n}{n-1} B \subseteq 2B.$$

Note that, while the dimension free constant equals 2, we have in fact seen the dimension dependent constant $\frac{n}{n-1}$, which tends to 1 when $n \to \infty$. □

Let us formulate a problem which arises from Theorem 3.3:

Problem A: Let $A, B \in \mathcal{K}^n$ such that $0 \in int(A \cap B)$. Assume that for every $n - 1$ dimensional subspace $E \in G_{n,n-1}$ there exists $x_E \in \mathbb{R}^n$ such that $A \cap E + x_E \subset B \cap E$. Does there exist a universal constant $C > 0$ such that $A + x_0 \subset CB$ for some $x_0 \in \mathbb{R}^n$? Of course, we are interested in C independent of the dimension. We suspect that $C = 4$ suffices.

4 Comparing n-Tuples: Affine Case

Although very little may be said about inclusion of the convex body in another one using only affine images of those bodies, some information is anyway available through the use of mixed volumes.

Theorem 4.1. *Let* $A, B, K_2, \ldots, K_n \in \mathcal{K}^n$ *be centrally symmetric bodies. If for all* $u \in SL_n$ *one has*

$$V(uA, K_2, \ldots, K_n) \le V(uB, K_2, \ldots, K_n), \tag{3}$$

then $A \subseteq B$.

Proof. Due to multilinearity of mixed volume, (3) holds for any $u \in GL_n$ as well. Since $V(uK_1, \ldots, uK_n) = |\det(u)| V(K_1, \ldots, K_n)$, we get for every $u \in GL_n$:

$$V(A, uK_2, \ldots, uK_n) \le V(B, uK_2, \ldots, uK_n).$$

Fix a direction $v \in S^{n-1}$, and let $E = v^\perp$. We may choose a sequence $\{u_n\} \subset GL_n$ such that for all $K \in \mathcal{K}^n$, $u_n(K) \to P_E(K)$ in the Hausdorff metric. By continuity of mixed volume with respect to the Hausdorff metric we get that:

$$V(A, P_E K_2, \ldots, P_E K_n) \le V(B, P_E K_2, \ldots, P_E K_n), \text{ that is:}$$

$$w_A(v) \cdot V(P_E K_2, \ldots, P_E K_n) \le w_B(v) \cdot V(P_E K_2, \ldots, P_E K_n).$$

In the last inequality, V stands for the $n-1$ dimensional mixed volume, and since K_i have non empty interior, it does not vanish. Thus for all $v \in S^{n-1}$, $w_A(v) \le w_B(v)$, which implies inclusion, since the bodies are symmetric. $\qquad \square$

Theorem 4.2. *Let* $A_1, B_1, \ldots, A_n, B_n \in \mathcal{K}^n$ *be centrally symmetric bodies. If for all* $u_1, \ldots, u_n \in SL_n$ *one has*

$$V(u_1 A_1, \ldots, u_n A_n) \le V(u_1 B_1, \ldots, u_n B_n), \tag{4}$$

then there exist positive constants $t_1, \ldots, t_n$ *such that* $\prod_1^n t_i = 1$ *and* $A_i \subset t_i B_i$.

Proof. For $2 \le i \le n$, denote:

$$t_i = \max_{t>0}\{t : \quad t A_i \subseteq B_i\},$$

and let $v_i \in S^{n-1}$ be such that $t_i h_{A_i}(v_i) = h_{B_i}(v_i)$. Since we may rotate the bodies A_i separately as desired, we may assume without loss of generality that $v_i = e_i$. Let $v \in S^{n-1}$, let g be a rotation such that $g(v) = e_1$, and denote $t_1 = (t_2 \ldots t_n)^{-1}$. Due to continuity of mixed volume, (4) holds also for degenerate u_i. Applying it to $g \circ P_v, P_{v_2}, \ldots, P_{v_n}$ (where P_w denotes the orthogonal projection onto the 1-dimensional subspace spanned by w) yields:

$$2^n h_{A_1}(v) \cdot h_{A_2}(v_2) \cdot \ldots \cdot h_{A_n}(v_n) = V(g P_v A_1, P_{v_2} A_2, \ldots, P_{v_n} A_n)$$
$$\le V(g P_v B_1, P_{v_2} B_2, \ldots, P_{v_n} B_n)$$

$$= 2^n h_{B_1}(v) \cdot h_{B_2}(v_2) \cdot \ldots \cdot h_{B_n}(v_n)$$

$$= \frac{2^n}{t_1} h_{B_1}(v) \cdot h_{A_2}(v_2) \cdot \ldots \cdot h_{A_n}(v_n).$$

This implies $h_{t_1 A_1}(v) \leq h_{B_1}(v)$. Since v was arbitrary, $t_1 A_1 \subseteq B_1$. This completes the proof, since for $i \geq 2$, $t_i A_i \subseteq B_i$ by the definition of t_i, and $\Pi_{i=1}^n t_i = 1$. $\quad\square$

Corollary 4.2. *Let* $A, B \in \mathcal{K}^n$ *be centrally symmetric bodies. If for all* $u_i \in SL_n$ *one has*

$$V(u_1 A, \ldots, u_n A) \leq V(u_1 B, \ldots, u_n B),$$

then $A \subseteq B$. $\hfill\square$

Remark 4.4. Theorem 4.2 is not true if the bodies are not assumed to be centrally symmetric. One example in $\mathbb{R}^2$ is given by the bodies $K_1 = R$, the Reuleaux triangle, and $K_2 = L_1 = L_2 = D_2$, the Euclidean unit ball. In fact for any $n \geq 2$, the choice $K_1 = -L_1 = A$ and $L_i = K_i = S_i$ yields a counter example, for trivial reasons, provided that $A \neq -A$, and $S_i = -S_i$, for $i \geq 2$.

Let us mention a problem inspired by Corollary 4.2.

Problem B: Let K, L be convex bodies with barycenter at the origin. Denote

$$a(K, L) = \sup_{x \neq 0} \frac{h_K(x)}{h_L(x)}.$$

Assume that for any $n - 1$ dimensional subspace E we have

$$a(K, L)|P_E(K)| \leq |P_E(L)|.$$

Does this imply that $|K| \leq |L|$?

5 Comparing n-Tuples: Fractional Linear Case

Let us denote by $\mathcal{K}_{(0)}^n$ the subset of $\mathcal{K}^n$ of bodies with 0 in their interior. We have:

Theorem 5.3. *Let* $n \geq 1$ *and let* $K_1, \ldots, K_n, L_1, \ldots, L_n \in \mathcal{K}_{(0)}^n$ *such that for every* $\lambda_1, \ldots, \lambda_n > 0$ *and every admissible (for* $\lambda_1 K_1, \ldots, \lambda_n K_n$) $F \in FL_n$ *one has*

$$V(F(\lambda_1 K_1), \ldots, F(\lambda_n K_n)) \leq V(F(\lambda_1 L_1), \ldots, F(\lambda_n L_n)).$$

Then for every $1 \leq i \leq n$, $K_i \subseteq L_i$.

Proof. Assume the claim is false, for example $K_1 \nsubseteq L_1$. Apply Remark 2.1 for K_1, L_1 and select H^+ to contain 0. Let D_{K_1}, D_{L_1} be the balls from Remark 2.1 (note again that these balls are not necessarily centered at 0), and take a ball D centered at the origin such that $D \subset D_{L_1}$. Without loss of generality (by a correct rescaling) we may assume that D is the Euclidean unit ball D_n. Fix $\lambda_2, \ldots, \lambda_n > 0$ such that $\lambda_i K_i, \lambda_i L_i$ are contained in the unit ball, and choose ε_0 such that for every $2 \le i \le n$:

$$\varepsilon_0 D_n \subseteq \lambda_i K_i \subseteq D_n, \qquad \varepsilon_0 D_n \subseteq \lambda_i L_i \subseteq D_n.$$

It follows that for every admissible F we have:

$$V(F(D_{K_1}), F(\varepsilon_0 D_n)[n-1]) \le V(F(D_{L_1})[n]). \tag{5}$$

Without loss of generality, we may assume that the centers of the balls D_{K_1}, $\varepsilon_0 D_n$, and D_{L_1} are collinear (for example, we may replace D_{L_1} by a larger ball containing it, while keeping the intersection $D_{L_1} \cap D_{K_1}$ empty). By applying an affine map A we may further assume that:

$$A(D_{K_1}) = \mathcal{E}_{1,1,\delta}, \quad A(D_{L_1}) = \mathcal{E}_{R,R,d}, \quad A(\varepsilon_0 D_n) = \mathcal{E}_{\varepsilon_1,\varepsilon_1,d_1},$$

where $R, \varepsilon_1 > 0, d, d_1 > 2$, and $\delta > 0$ is arbitrarily small. By 2.2, we have:

$$\frac{1}{\delta(\delta + 2)} D_n \subset F_0 A D_{K_1}, \quad \frac{\varepsilon_1}{d_1(d_1 + 2\varepsilon_1)} \subset F_0 A \varepsilon_0 D_n, \quad F_0 A D_L \subset \frac{R}{d\sqrt{d + 2R}} D_n.$$

Substituting $F = F_0 A$ in (5), and using multilinearity of mixed volumes, we get:

$$\frac{1}{\delta(\delta + 2)} \left(\frac{\varepsilon_1}{d_1(d_1 + 2\varepsilon_1)} \right)^{n-1} \le \left(\frac{R}{d\sqrt{d + 2R}} \right)^n,$$

which is false for sufficiently small $\delta > 0$. Thus $K_1 \subseteq L_1$, as required. $\square$

References

1. S. Artstein-Avidan, D. Florentin , V.D. Milman, *Order Isomorphisms on Convex Functions in Windows*. GAFA Lecture Notes in Mathematics, vol. 2050 (Springer, Berlin, 2012), pp. 61–122
2. T. Bonnesen, W. Fenchel, *Theory of Convex Bodies* (BCS Associates, Moscow, 1987). Translated from the German and edited by L. Boron, C. Christenson and B. Smith
3. G.D. Chakerian, E. Lutwak, On the Petty-Schneider theorem. Contemp. Math. **140**, 31–37 (1992)
4. C. Chen, T. Khovanova, D. Klain, Volume bounds for shadow covering. Trans. Am. Math. Soc. **366**(3), 1161–1177 (2014)

5. R.J. Gardner, A. Koldobsky , T. Schlumprecht, An analytic solution to the Busemann-Petty problem on sections of convex bodies. Ann. Math. Second Ser. **149**(2), 691–703 (1999)
6. A. Koldobsky, Intersection bodies, positive definite distributions, and the Busemann-Petty problem. Am. J. Math. **120**(4), 827–840 (1998)
7. E. Lutwak, Containment and circumscribing simplices. Discrete Comput. Geom. **19**, 229–235 (1998)
8. C.M. Petty, Projection bodies, in *Proc. Colloquium on Convexity*, Copenhagen, 1965 (Københavns Univ. Mat. Inst., Copenhagen, Denmark, 1967), pp. 234–241
9. R. Schneider, Zur einem Problem von Shephard über die Projektionen Konvexer Körper. Math. Z. **101**, 71–82 (1967)
10. R. Schneider, Additive Transformationen Konvexer Körper. Geom. Dedicata **3**, 221–228 (1974)
11. R. Schneider, *Convex Bodies: The Brunn Minkowski Theory. Second Expanded Edition.* Encyclopedia of Mathematics and its Applications, vol. 151 (Cambridge University Press, Cambridge, 2014).
12. W. Weil, Decomposition of convex bodies. Mathematika **21**, 19–25 (1974)
13. G. Zhang, A positive solution to the Busemann-Petty problem in $\mathbb{R}^4$. Ann. Math. Second Ser. **149**(2), 535–543 (1999)

Vitushkin-Type Theorems

Omer Friedland and Yosef Yomdin

Abstract It is shown that for a subset $A \subset \mathbb{R}^n$ that has the global Gabrielov property, a Vitushkin-type estimate holds. Concrete examples are given for sub-level sets of certain classes of functions.

2010 *Mathematics Subject Classification:* 14P10, 26B15 (primary), and 14R10, 26D05, 42105 (secondary)

1 Introduction

The metric entropy of a subset $A \subset \mathbb{R}^n$ can be bounded in terms of the i-dimensional "size" of A. Indeed, the theory of multi-dimensional variations, developed by Vitushkin [6, 7], Ivanov [2], and other, provides a bound by measuring the i-dimensional "size" of A in terms of its variations.

Let us recall a general definition of the metric entropy of a set. Let X be a metric space, $A \subset X$ a relatively compact subset. For every $\varepsilon > 0$, denote by $M(\varepsilon, A)$ the minimal number of closed balls of radius ε in X, covering A (note that this number does exist because A is relatively compact). The real number $H_\varepsilon(A) = \log M(\varepsilon, A)$ is called the ε-entropy of the set A. In our setting, we assume $X = \mathbb{R}^n$, and it will be convenient to modify slightly this definition, and consider coverings by the ε-cubes Q_ε, which are translations of the standard ε-cube, $Q_\varepsilon^n = [0, \varepsilon]^n$, that is, the $\frac{\varepsilon}{2}$-ball in the ℓ_∞ norm.

O. Friedland (✉)
Institut de Mathématiques de Jussieu, Université Pierre et Marie Curie (Paris 6), 4 Place Jussieu, 75005 Paris, France
e-mail: omer.friedland@imj-prg.fr

Y. Yomdin
Department of Mathematics, The Weizmann Institute of Science, Rehovot 76100, Israel
e-mail: yosef.yomdin@weizmann.ac.il

© Springer International Publishing Switzerland 2014
B. Klartag, E. Milman (eds.), *Geometric Aspects of Functional Analysis*,
Lecture Notes in Mathematics 2116, DOI 10.1007/978-3-319-09477-9__12

The following inequality, which we refer to as Vitushkin's bound, bounds the metric entropy of a set A in terms of its multi-dimensional variations, that is, for every $A \subset \mathbb{R}^n$ it holds

$$M(\varepsilon, A) \leq c(n) \sum_{i=0}^{n} V_i(A)/\varepsilon^i, \tag{1}$$

where the i-th variation of A, $V_i(A)$, is the average of the number of connected components of the section $A \cap P$ over all $(n - i)$-affine planes P in $\mathbb{R}^n$. Note that if the components are points, the Vitushkin invariants are essentially the volumes, via the Crofton formula, but when the components are of higher dimensions, the geometric interpretation of these invariants is less straightforward.

In particular, our definition of the metric entropy implies that the last term in (1) has the form $\mu_n(A)/\varepsilon^n$, where $\mu_n(A)$ denotes the n-dimensional Lebesgue measure (or the volume) of the set A.

Vitushkin's bound is sometimes considered as a difficult result, mainly because of the so-called multi-dimensional variations which are used. However, in some cases (cf. [1, 8, 9]) the proof is indeed very short and transparent. In this note, we present a Vitushkin-type theorem, which works in situations where we can control the number of connected components of the sections $A \cap P$ over all $(n - i)$-affine planes P. In Sect. 2 we present our main observation that we can replace the i-th variation $V_i(A)$ of A, with an upper bound on the number of connected components of the section $A \cap P$ over all $(n - i)$-affine planes P. As in some cases, it is easier to compute this upper bound rather than V_i which is the average. In Sect. 3, we extend Vitushkin's bound for semi-algebraic sets, to sub-level sets of functions, for which we have a certain replacement of the polynomial Bézout theorem. These results can be proved using a general result of Vitushkin in [6, 7] through the use of multi-dimensional variations. However, in our specific case the results below are much simpler and shorter and produce explicit ("in one step") constants.

2 Gabrielov Property and Vitushkin's Bound

In this section we establish a relation between Vitushkin's bound and the global Gabrielov property of a set A. We show that an a priori knowledge about the maximal number of connected components of a set A, intersected with every ℓ-affine plane in $\mathbb{R}^n$, allows us to estimate the metric entropy of A.

More precisely, we say that a subset $A \subset \mathbb{R}^n$ has the local Gabrielov property if for $a \in A$ there exist a neighborhood U of a and an integer $\hat{C}_\ell$ such that for every ℓ-affine plane P, the number of connected components of $U \cap A \cap P$ is bounded by $\hat{C}_\ell$. If we can take $U = \mathbb{R}^n$, we say that A has the global Gabrielov property. For example, every tame set has the local Gabrielov property (for more details see [9]).

The following theorem is applicable to arbitrary subsets $A \subset Q_1^n$. The boundary ∂A of A is defined as the intersection of the closures of A and of $Q_1^n \setminus A$.

Theorem 1 (Vitushkin-Type Theorem). *Let $A \subset Q_1^n$ and let $0 < \varepsilon \le 1$. Assume that the boundary ∂A of A has the global Gabrielov property, with explicit bound $\hat{C}_\ell$ for $0 \le \ell \le n$. Then*

$$M(\varepsilon, A) \le C_0 + C_1/\varepsilon + \cdots + C_{n-1}/\varepsilon^{n-1} + \mu_n(A)/\varepsilon^n,$$

where $C_t := \hat{C}_{n-t} 2^t \binom{n}{t}$.

Proof. Let us subdivide Q_1^n into adjacent ε-cubes Q_ε^n, with respect to the standard Cartesian coordinate system. Each Q_ε^n, having a non-empty intersection with A, is either entirely contained in A, or it intersects the boundary ∂A of A. Certainly, the number of those cubes Q_ε^n, which are entirely contained in A, is bounded by $\mu_n(A)/\mu_n(Q_\varepsilon) = \mu_n(A)/\varepsilon^n$. In the other case, in which Q_ε^n intersects ∂A, it means that there exist faces of Q_ε^n that have a non-empty intersection with ∂A. Among all these faces, let us take one with the smallest dimension s, and denote it by F. In other words, there exists an s-face F of the smallest dimension s that intersects ∂A, for some $s = 0, 1, \ldots, n$. Let us fix an s-affine plane V, which corresponds to F. Then, by the minimality of s, F contains completely some of the connected components of $\partial A \cap V$, otherwise ∂A would intersect a face of Q_ε^n of a dimension strictly less than s. By our assumption, the number of connected components with respect to an s-affine plane is bounded by $\hat{C}_s$. According to the subdivision of Q_1^n to Q_ε cubes, we have at most $\left(\frac{1}{\varepsilon} + 1\right)^{n-s}$ s-affine planes with respect to the same s coordinates, and the number of different choices of s coordinates is $\binom{n}{s}$. It means that the number of cubes, that have an s-face F which contains completely some connected component of $A \cap V$, is at most

$$\hat{C}_s \binom{n}{s} \left(\frac{1}{\varepsilon} + 1\right)^{n-s} \le \hat{C}_s 2^{n-s} \binom{n}{s} / \varepsilon^{n-s}.$$

Let us define the constant

$$C_{n-s} := \hat{C}_s 2^{n-s} \binom{n}{s}.$$

Note that C_0 is the bound on the number of cubes that contain completely some of the connected components of A. Thus, we have

$$M(\varepsilon, A) \le C_0 + C_1/\varepsilon + \cdots + C_{n-1}/\varepsilon^{n-1} + \mu_n(A)/\varepsilon^n.$$

This completes the proof. $\qquad\square$

3 Entropy Estimates of Sub-level Sets

In this section we extend Vitushkin's bound to sub-level sets of certain natural classes of functions, beyond polynomials. We do so by "counting" the singularities of these functions, and then bounding the number of the connected components of their sub-level set through the number of singularities.

We start with a general simple "meta"-lemma, which implies, together with a specific computation of the bound on the number of singularities, all our specific results below. Consider a class of functions $\mathcal{F}$ on $\mathbb{R}^n$. We assume that $\mathcal{F}$ is closed with respect to taking partial derivatives, restrictions to affine subspaces of $\mathbb{R}^n$, and with respect to sufficiently rich perturbations. There are many classes of functions that comply with this condition, for example, we may speak about the class of real polynomials of n variables and degree d, and the classes considered below in this section. Assume that for each $f_1, \ldots, f_n \in \mathcal{F}$ the number of non-degenerate solutions of the system

$$f_1 = f_2 = \cdots = f_n = 0,$$

is bounded by the constant $C(D(f_1, \ldots, f_n))$, where $D(f_1, \ldots, f_n)$ is a collection of "combinatorial" data of f_i, like degrees, which we call a "Diagram" of $f_1, \ldots, f_n$. We assume that the diagram is stable with respect to the deformations we use. In each of the examples below we define the appropriate diagram specifically.

Let $f \in \mathcal{F}$. Denote by

$$W = W(f, \rho) = \{x \in Q_1^n : f(x) \le \rho\}, \tag{2}$$

the ρ-sub-level set of f, and let $\hat{C}_s = C(D(\frac{\partial f(x)}{\partial x_1}, \ldots, \frac{\partial f(x)}{\partial x_s}))$, $s = 1, \ldots, n$.

Lemma 2. *The boundary ∂W has the global Gabrielov property, i.e. the number of connected components of $\partial W \cap P$, where P is an s-affine plane in $\mathbb{R}^n$, is bounded by $\hat{C}_s$.*

Proof. We may assume that P is a parallel translation of the coordinate plane in $\mathbb{R}^n$ generated by $x_{j_1}, \ldots, x_{j_s}$. Inside each connected component of $\partial W \cap P$ there is a critical point of f restricted to P (its local maximum or minimum), which is defined by the system of equations

$$\frac{\partial f(x)}{\partial x_{j_1}} = \cdots = \frac{\partial f(x)}{\partial x_{j_s}} = 0.$$

After a small perturbation of f, we can always assume that all such critical points are non-degenerate. Hence by the assumptions above, the number of these points, and therefore the number of connected components, is bounded by $\hat{C}_s$. $\square$

In other words, W has the global Gabrielov property, with explicit bound $\hat{C}_\ell$. Therefore, Theorem 1 can be applied to this set, and under the assumptions above, we have

Corollary 3. *Let $0 < \varepsilon \le 1$. Then*

$$M(\varepsilon, W) \le C_0 + C_1/\varepsilon + \cdots + C_{n-1}/\varepsilon^{n-1} + \mu_n(W)/\varepsilon^n,$$

where $C_t := \hat{C}_{n-t} 2^t \binom{n}{t}$.

4 Concrete Bounds on $\hat{C}_s$

In view of Corollary 3, our main goal now is to give concrete bounds on constants $\hat{C}_s$ in specific situations.

4.1 Polynomials and Bézout's Theorem

Let $p(x) = p(x_1, \ldots, x_n)$ be a polynomial in $\mathbb{R}^n$ of degree d. We consider the sub-level set $W(p, \rho)$ as defined in (2). Clearly, in this situation, by Bézout's theorem, we have

$$\hat{C}_s(W(p, \rho)) \le (d - 1)^s.$$

4.2 Laurent Polynomials and Newton Polytypes

Let $\alpha \in \mathbb{Z}^n$. A Laurent monomial in the variables $x_1, \ldots, x_n$ is $x^\alpha = x_1^{\alpha_1} \cdots x_n^{\alpha_n}$. A Laurent polynomial is a finite sum of Laurent monomials,

$$p(x) = p(x_1, \ldots, x_n) = \sum_{\alpha \in A \subset \mathbb{Z}^n} a_\alpha x^\alpha.$$

The Newton polytope of p is the polytope

$$N(p) = \text{conv}\{\alpha \in \mathbb{Z}^n \mid a_\alpha \ne 0\}.$$

A natural generalization of the Bézout bound above is the following Kušhnirenko bound for polynomial systems with the prescribed Newton polytope.

Theorem 4 ([4]). *Let* $f_1, \ldots, f_n$ *be Laurent polynomials with the Newton polytope* $N \subset \mathbb{R}^n$. *Then the number of non-degenerate solutions of the system*

$$f_1 = f_2 = \cdots = f_n = 0,$$

is at most $n! \operatorname{Vol}_n(N)$.

The Newton polytope of a general polynomial of degree d is the simplex

$$\Delta_d = \{\alpha \in \mathbb{R}^n, |\alpha| \le d\}.$$

Its volume is $d^n/n!$, and the bound of Theorem 4 coincides with the Bézout's bound d^n. The notion of the Newton polytope is connected to a representation of the polynomial p in a fixed coordinate system $x_1, \ldots, x_n$ in $\mathbb{R}^n$. As we perform a coordinate changes (which may be necessary when we restrict a polynomial p to a certain affine subspace P of $\mathbb{R}^n$), $N(p)$ may change strongly. However, in Theorem 1 we restrict our functions only to affine subspaces P spanned by a part of the standard basis vectors in a fixed coordinate system. The following lemma describes the behavior of the Newton polytope of p under such restrictions, and under partial differentiation.

Lemma 5. *The Newton polytope* $N(\frac{\partial p}{\partial x_i})$ *is* $N_i(p)$ *obtained by a translation of* $N(p)$ *to the vector* $-e_i$, *where* $e_1, \ldots, e_n$ *are the vectors of the standard basis in* $\mathbb{R}^d$. *The Newton polytope* $N(p|P)$ *of a restriction of* p *to* P, *where* P *is a translation of a certain coordinate subspace, is contained in the projection* $\pi_P(N(p))$ *of* $N(p)$ *on* P.

Proof. The proof of the first claim is immediate. In a restriction of p to P, we substitute some of the x_i's for their specific values. The degrees of the free variables remain the same.   □

For a Newton polytope $N \subset \mathbb{R}^n$ define

$$C_s(N) := \max\{\operatorname{Vol}_s(N_{P_s}) \mid s\text{-dimensional coordinate subspaces } P_s\},$$

where N_{P_s} is the convex hull of the sets $\pi_{P_s}(N_i(p))$, for all the coordinate directions in P_s. Here, $\pi_{P_s}(N_i(p))$ is the projection of N to P_s, shifted by -1 in one of coordinate directions x_i in P_s.

Theorem 6. *Let* p *be a Laurent polynomial with the Newton polytope* N. *Then for* $s = 1, \ldots, n$

$$\hat{C}_s(W(p, \rho)) \le \frac{C_s(N)}{s!}.$$

Proof. According to Lemma 2, the constants $\hat{C}_s(W(p, \rho))$ do not exceed the number of solutions in Q_1^n of the system $\frac{\partial p(x)}{\partial x_{j_1}} = \cdots = \frac{\partial p(x)}{\partial x_{j_s}} = 0$. Now application of Theorem 4, Lemma 5, and of the definition of $C_s(N)$ above, completes the proof.
□

An important example is provided by "multi-degree d" polynomials. A polynomial $p(x) = p(x_1, \ldots, x_n)$ is called multi-degree d, if each of its variable enters p with degrees at most d. The total degree of such p may be nd. In particular, multi-linear polynomials contain each variable with the degree at most one. Multi-linear polynomials appear in various problems in Mathematics and Computer Science, in particular, since the determinant of a matrix is a multi-linear polynomial in its entries.

Theorem 7. *Let p be a multi-degree d polynomial. Then for $s = 1, \ldots, n$*

$$\hat{C}_s(W(p, \rho)) \le \frac{d^s}{s!}.$$

Proof. The Newton polytope $N(p)$ of a multi-degree d polynomial p is contained in a cube Q_d^n with the edge d. Its projection to each P_s is Q_d^s. After the shift by -1 in one of the coordinate directions in P_s, it remains in Q_d^s. So Q_d^s can be taken as the common Newton polytope of $\frac{\partial p(x)}{\partial x_{j_1}}, \cdots, \frac{\partial p(x)}{\partial x_{j_s}}$. Application of Theorem 6 completes the proof.
□

4.3   Quasi-Polynomials and Khovanskii's Theorem

Let $f_1, \ldots, f_k \in (\mathbb{C}^n)^*$ be a pairwise different set of complex linear functionals f_j which we identify with the scalar products $f_j \cdot z$, $z = (z_1, \ldots, z_n) \in \mathbb{C}^n$. We shall write $f_j = a_j + ib_j$. A quasi-polynomial is a finite sum

$$p(z) = \sum_{j=1}^{k} p_j(z) e^{f_j \cdot z},$$

where $p_j \in \mathbb{C}[z_1, \ldots, z_n]$ are polynomials in z of degrees d_j. The degree of p is $m = \deg p = \sum_{j=1}^{k}(d_j + 1)$.

Below we consider $p(x)$ for the real variables $x = (x_1, \ldots, x_n) \in \mathbb{R}^n$, and we are interested in the following sub-level set of p which is defined as $\{x \in Q_1^n : |p(x)| \le \rho\}$. Denote $q(x) = |p(x)|^2$, then this sub-level set is also defined by $W(q, \rho^2)$.

A simple observation that $q(x) = |p(x)|^2 = p(x)\bar{p}(x)$ tells us that we can rewrite q as follows

Lemma 8. $q(x)$ *is a real exponential trigonometric quasi-polynomial with* P_{ij}, Q_{ij} *real polynomials in* x *of degree* $d_i + d_j$, *and at most* $\kappa := k(k+1)/2$ *exponents, sinus and cosinus elements. Moreover,*

$$q(x) = \sum_{0 \leq i \leq j \leq k} e^{\langle a_i + a_j, x \rangle} \big[P_{ij}(x) \sin\langle b_{ij}, x \rangle + Q_{ij}(x) \cos\langle b_{ij}, x \rangle \big],$$

where $b_{ij} = b_i - b_j \in \mathbb{R}^n$.

Now, we need to bound the singularities of q. This can be done using the following theorem due to Khovanskii, which gives an estimate of the number of solutions of a system of real exponential trigonometric quasi-polynomials.

Theorem 9 (Khovanskii Bound [3], Sect. 1.4). *Let* $P_1 = \cdots = P_n = 0$ *be a system of n equations with n real unknowns* $x = x_1, \ldots, x_n$, *where* P_i *is polynomial of degree* m_i *in* $n+k+2p$ *real variables* $x, y_1, \ldots, y_k, u_1, \ldots, u_p, v_1, \ldots, v_p$, *where* $y_i = \exp\langle a_j, x \rangle$, $j = 1, \ldots, k$ *and* $u_q = \sin\langle b_q, x \rangle$, $v_q = \cos\langle b_q, x \rangle$, $q = 1, \ldots, p$. *Then the number of non-degenerate solutions of this system in the region bounded by the inequalities* $|\langle b_q, x \rangle| < \pi/2$, $q = 1, \ldots, p$, *is finite and less than*

$$m_1 \cdots m_n \left(\sum m_i + p + 1 \right)^{p+k} 2^{p+(p+k)(p+k-1)/2}.$$

Clearly, all the partial derivatives $\frac{\partial q(x)}{\partial x_j}$ have exactly the same form as q. Therefore, Khovanskii's theorem gives the following bound on the number of critical points of q. More precisely, we have

Lemma 10. *Let* V *be a parallel translation of the coordinate subspace in* $\mathbb{R}^n$ *generated by* $x_{j_1}, \ldots, x_{j_s}$. *Then the number of non-degenerate real solutions in* $V \cap Q_\rho^n$ *of the system*

$$\frac{\partial q(x)}{\partial x_{j_1}} = \cdots = \frac{\partial q(x)}{\partial x_{j_s}} = 0,$$

is at most

$$\left(\frac{2}{\pi} \sqrt{s} \rho \lambda \right)^s \prod_{r=1}^s (d_{j_r} + d_{i_r}) \left(\sum_{r=1}^s (d_{j_r} + d_{i_r}) + 2\kappa + 1 \right)^{2\kappa} 2^{\kappa + (2\kappa)(2\kappa-1)/2},$$

where $\lambda := \max \|b_{ij}\|$ *is the maximal frequency in* q.

Proof. The following geometric construction is required by the Khovanskii bound: Let $Q_{ij} = \{x \in \mathbb{R}^n, |\langle b_{ij}, x \rangle| \leq \frac{\pi}{2}\}$ and let $Q = \bigcap_{0 \leq i \leq j \leq k} Q_{ij}$. For every $B \subset \mathbb{R}^n$ we define $M(B)$ as the minimal number of translations of Q covering B. For an affine subspace V of $\mathbb{R}^n$ we define $M(B \cap V)$ as the minimal number of translations of $Q \cap V$ covering $B \cap V$. Notice that for $B = Q_r^n$, a cube of size r, we have

$M(Q_r^n) \leq (\frac{2}{\pi}\sqrt{nr}\lambda)^n$. Indeed, Q always contains a ball of radius $\frac{\pi}{2\lambda}$. Now, applying the Khovanskii's theorem on the system

$$\frac{\partial q(x)}{\partial x_{j_1}} = \cdots = \frac{\partial q(x)}{\partial x_{j_s}} = 0,$$

we get that the number of non-degenerate real solutions in $V \cap Q_\rho^n$ is at most

$$(\frac{2}{\pi}\sqrt{s}\rho\lambda)^s \prod_{r=1}^{s}(d_{j_r} + d_{i_r}) \left(\sum_{r=1}^{s}(d_{j_r} + d_{i_r}) + 2\kappa + 1 \right)^{2\kappa} 2^{\kappa+(2\kappa)(2\kappa-1)/2}.$$

Note that this bound is given in term of the "diagram" of q, and therefore of p.

$\square$

Theorem 11. *Let $p(x)$ a real quasi-polynomial (as described above). Then for $s = 1, \ldots, n$*

$$\hat{C}_s(W(q, \rho^2))$$

$$\leq (\frac{2}{\pi}\sqrt{s}\rho\lambda)^s \prod_{r=1}^{s}(d_{j_r} + d_{i_r}) \left(\sum_{r=1}^{s}(d_{j_r} + d_{i_r}) + 2\kappa + 1 \right)^{2\kappa} 2^{\kappa+(2\kappa)(2\kappa-1)/2},$$

where $q(x) = |p(x)|^2$.

4.4 Exponential Polynomials and Nazarov's Lemma

In a particular case where p is an exponential polynomial, that is,

$$p(t) = \sum_{k=0}^{m} c_k e^{\lambda_k t},$$

where $c_k, \lambda_k \in \mathbb{C}$, $t \in \mathbb{R}$, we can avoid Khovanskii's theorem and instead use the following result of Nazarov [5, Lemma 4.2], which gives a bound on the local distribution of zeroes of an exponential polynomial.

Lemma 12. *The number of zeroes of $p(z)$ inside each disk of radius $r > 0$ does not exceed*

$$4m + 7\hat{\lambda}r,$$

where $\hat{\lambda} = \max |\lambda_k|$.

Note that this result is applicable only in dimension 1. Let $B \subset \mathbb{R}$ be an interval. Therefore, the number of real solutions of $p(t) \leq \rho$ inside the interval B does not exceed Nazarov's bound, that is, $4m + 7\hat{\lambda}\mu_1(B)$, which gives us

$$\hat{C}_1(W(p,\rho)) \leq 4m + 7\hat{\lambda}.$$

For the case of a real exponential polynomial $p(t) = \sum_{k=0}^{m} c_k e^{\lambda_k t}$, $c_k, \lambda_k \in \mathbb{R}$, we get an especially simple and sharp result, as the number of zeroes of a real exponential polynomial is always bounded by its degree m (indeed, the "monomials" $e^{\lambda_k t}$ form a Chebyshev system on each real interval).

4.5 Semialgebraic and Tame Sets

We conclude with a remark about even more general settings for which Theorem 1 is applicable.

A set $A \subset \mathbb{R}^n$ is called semialgebraic, if it is defined by a finite sequence of polynomial equations and inequalities, or any finite union of such sets. More precisely, A can be represented in a form $A = \bigcup_{i=1}^{k} A_i$ with $A_i = \bigcap_{j=1}^{j_i} A_{ij}$, where each A_{ij} has the form

$$\{x \in \mathbb{R}^n : p_{ij}(x) > 0\} \text{ or } \{x \in \mathbb{R}^n : p_{ij}(x) \geq 0\},$$

where p_{ij} is a polynomial of degree d_{ij}. The diagram $D(A)$ of A is the collective data

$$D(A) = (n, k, j_1, \ldots, j_k, (d_{ij})_{i=1,\ldots,k, \ j=1,\ldots,j_i}).$$

A classical result tells us that the number of connected components of a plane section $A \cap P$ is uniformly bounded. More precisely, we have

Theorem 13 ([9]). *Let $A \subset \mathbb{R}^n$ be a semialgebraic set with diagram $D(A)$. Then the number of connected components of $A \cap P$, where P is an ℓ-affine plane of $\mathbb{R}^n$, is bounded by*

$$\hat{C}_\ell \leq \frac{1}{2} \sum_{i=1}^{k} (d_i + 2)(d_i + 1)^{\ell-1},$$

where $d_i = \sum_{j=1}^{j_i} d_{ij}$.

In other words, Theorem 13 says that any semialgebraic set has the global Gabrielov property.

However, not only semialgebraic sets, but a very large class of sets has the Gabrielov property. These sets are called *tame sets*. The precise definition of these sets and the fact that they satisfy the Gabrielov property can be found, in particular, in [9].

Acknowledgements This research was partially supported by ISF grant No. 639/09 and by the Minerva foundation.

References

1. O. Friedland, Y. Yomdin, An observation on the Turán-Nazarov inequality. Studia Math. **218**(1), 27–39 (2013)
2. L.D. Ivanov, in *Variatsii mnozhestv i funktsii* (Russian) [Variations of Sets and Functions], ed. by A.G. Vitushkin (Izdat. "Nauka", Moscow, 1975), 352 pp.
3. A.G. Khovanskii, *Fewnomials*. Translated from the Russian by Smilka Zdravkovska. Translations of Mathematical Monographs, vol. 88 (American Mathematical Society, Providence, RI, 1991), viii+139 pp.
4. A.G. Kušnirenko, Newton polyhedra and Bezout's theorem (Russian). Funkcional. Anal. i Priložen. **10**(3), 82–83 (1976)
5. F.L. Nazarov, Local estimates for exponential polynomials and their applications to inequalities of the uncertainty principle type. Algebra i Analiz **5**(4), 3–66 (1993). Translation in St. Petersburg Math. J. **5**(4), 663–717 (1994)
6. A.G. Vitushkin, *O mnogomernyh Variaziyah* (Gostehisdat, Moskow, 1955)
7. A.G. Vitushkin, *Ozenka sloznosti zadachi tabulirovaniya* (Fizmatgiz, Moskow, 1959)
8. Y. Yomdin, Remez-type inequality for discrete sets. Israel J. Math. **186**, 45–60 (2011)
9. Y. Yomdin, G. Comte, *Tame Geometry with Application in Smooth Analysis*. Lecture Notes in Mathematics, vol. 1834 (Springer, Berlin, 2004), viii+186 pp.

M-Estimates for Isotropic Convex Bodies and Their L_q-Centroid Bodies

Apostolos Giannopoulos and Emanuel Milman

Abstract Let K be a centrally-symmetric convex body in $\mathbb{R}^n$ and let $\|\cdot\|$ be its induced norm on $\mathbb{R}^n$. We show that if $K \supseteq rB_2^n$ then:

$$\sqrt{n}M(K) \leq C \sum_{k=1}^{n} \frac{1}{\sqrt{k}} \min\left(\frac{1}{r}, \frac{n}{k}\log\left(e + \frac{n}{k}\right)\frac{1}{v_k^-(K)}\right)$$

where $M(K) = \int_{S^{n-1}} \|x\| \, d\sigma(x)$ is the mean-norm, $C > 0$ is a universal constant, and $v_k^-(K)$ denotes the minimal volume-radius of a k-dimensional orthogonal projection of K. We apply this result to the study of the mean-norm of an isotropic convex body K in $\mathbb{R}^n$ and its L_q-centroid bodies. In particular, we show that if K has isotropic constant L_K then:

$$M(K) \leq \frac{C \log^{2/5}(e+n)}{\sqrt[10]{n}L_K}.$$

1 Introduction

Let K be a centrally-symmetric convex compact set with non-empty interior ("body") in Euclidean space $(\mathbb{R}^n, \langle \cdot, \cdot \rangle)$. We write $\|\cdot\|$ for the norm induced on $\mathbb{R}^n$ by K and h_K for the support function of K; this is precisely the dual norm $\|\cdot\|^*$. The parameters:

$$M(K) = \int_{S^{n-1}} \|x\| \, d\sigma(x) \quad \text{and} \quad M^*(K) = \int_{S^{n-1}} h_K(x) \, d\sigma(x), \tag{1}$$

A. Giannopoulos
Department of Mathematics, University of Athens, Panepistimioupolis 157-84, Athens, Greece
e-mail: apgiannop@math.uoa.gr

E. Milman (✉)
Department of Mathematics, Technion - Israel Institute of Technology, Haifa 32000, Israel
e-mail: emilman@tx.technion.ac.il

© Springer International Publishing Switzerland 2014   159
B. Klartag, E. Milman (eds.), *Geometric Aspects of Functional Analysis*,
Lecture Notes in Mathematics 2116, DOI 10.1007/978-3-319-09477-9_13

where σ denotes the rotationally invariant probability measure on the unit Euclidean sphere S^{n-1}, play a central role in the asymptotic theory of finite dimensional normed spaces.

Let $\mathrm{vrad}(K) := \left(|K|/|B_2^n|\right)^{1/n}$ denote the volume-radius of K, where $|A|$ denotes Lebesgue measure in the linear hull of A and B_2^n denotes the unit Euclidean ball. It is easy to check that:

$$M(K)^{-1} \leq \mathrm{vrad}(K) \leq M^*(K) = M(K^\circ),   \tag{2}$$

where $K^\circ = \{y \in \mathbb{R}^n : \langle x, y \rangle \leq 1 \text{ for all } x \in K\}$ is the polar body to K, i.e. the unit-ball of the dual norm $\|\cdot\|^*$. Indeed, the left-hand side is a simple consequence of Jensen's inequality after we express the volume of K as an integral in polar coordinates, while the right-hand side is the classical Urysohn inequality. In particular, one always has $M(K)M^*(K) \geq 1$.

In the other direction, it is known from results of Figiel–Tomczak-Jaegermann [11], Lewis [18] and Pisier's estimate [29] on the norm of the Rademacher projection, that for any centrally-symmetric convex body K, there exists $T \in GL(n)$ such that:

$$M(TK)M^*(TK) \leq C \log n,   \tag{3}$$

where $C > 0$ is a universal constant. Throughout this note, unless otherwise stated, all constants $c, c', C, \ldots$ denote universal numeric constants, independent of any other parameter, whose value may change from one occurrence to the next. We write $A \simeq B$ if there exist absolute constants $c_1, c_2 > 0$ such that $c_1 A \leq B \leq c_2 A$.

The role of the linear map T in (3) is to put the body in a good "position", since without it $M(K)M^*(K)$ can be arbitrarily large. The purpose of this note is to obtain good upper bounds on the parameter $M(K)$, when K is already assumed to be in a good position—the isotropic position. A convex body K in $\mathbb{R}^n$ is called isotropic if it has volume 1, its barycenter is at the origin, and there exists a constant $L_K > 0$ such that:

$$\int_K \langle x, \theta \rangle^2 dx = L_K^2, \text{ for all } \theta \in S^{n-1}.   \tag{4}$$

It is not hard to check that every convex body K has an isotropic affine image which is uniquely determined up to orthogonal transformations [24]. Consequently, the isotropic constant L_K is an affine invariant of K. A central question in asymptotic convex geometry going back to Bourgain [5] asks if there exists an absolute constant $C > 0$ such that $L_K \leq C$ for every (isotropic) convex body K in $\mathbb{R}^n$ and every $n \geq 1$. Bourgain [6] proved that $L_K \leq C \sqrt[4]{n} \log n$ for every centrally-symmetric convex body K in $\mathbb{R}^n$. The currently best-known general estimate, $L_K \leq C \sqrt[4]{n}$, is due to Klartag [14] (see also the work of Klartag and Milman [16] and a further refinement of their approach by Vritsiou [32]).

It is known that if K is a centrally-symmetric isotropic convex body in $\mathbb{R}^n$ then $K \supseteq L_K B_2^n$, and hence trivially $M(K) \leq 1/L_K$. It seems that, until recently, the problem of bounding $M(K)$ in isotropic position had not been studied and there were no other estimates besides the trivial one. The example of the normalized ℓ_∞^n ball shows that the best one could hope is $M(K) \leqslant C\sqrt{\log n}/\sqrt{n}$. Note that obtaining a bound of the form $M(K) \leqslant n^{-\delta}L_K^{-1}$ immediately provides a non-trivial upper bound on L_K, since $M(K) \geqslant \mathrm{vrad}(K)^{-1} \simeq 1/\sqrt{n}$, and hence $L_K \leqslant c^{-1}n^{\frac{1}{2}-\delta}$. The current best-known upper bound on L_K suggests that $M(K) \leqslant C(n^{1/4}L_K)^{-1}$ might be a plausible goal.

Paouris and Valettas (unpublished) proved that for every isotropic centrally-symmetric convex body K in $\mathbb{R}^n$ one has:

$$M(K) \leqslant \frac{C\sqrt[3]{\log(e+n)}}{\sqrt[12]{n}L_K}. \tag{5}$$

Subsequently, this was extended by Giannopoulos, Stavrakakis, Tsolomitis and Vritsiou in [12] to the case of the L_q-centroid bodies $Z_q(\mu)$ of an isotropic log-concave probability measure μ on $\mathbb{R}^n$ (see Sect. 5 for the necessary definitions). The approach of [12] was based on a number of observations regarding the local structure of $Z_q(\mu)$; more precisely, lower bounds for the in-radius of their proportional projections and estimates for their dual covering numbers (we briefly sketch an improved version of this approach in Sect. 7).

In this work we present a different method, applicable to general centrally-symmetric convex bodies, which yields better quantitative estimates. As always, our starting point is Dudley's entropy estimate (see e.g. [31, Theorem 5.5]):

$$\sqrt{n}M^*(K) \leqslant C\sum_{k\geqslant 1}\frac{1}{\sqrt{k}}e_k(K, B_2^n), \tag{6}$$

where $e_k(K, B_2^n)$ are the entropy numbers of K. Recall that the covering number $N(K, L)$ is defined to be the minimal number of translates of L whose union covers K, and that $e_k(K, L) := \inf\{t > 0 : N(K, tL) \leqslant 2^k\}$.

Our results depend on the following natural volumetric parameters associated with K for each $k = 1, \ldots, n$:

$$w_k(K) := \sup\{\mathrm{vrad}(K \cap E) : E \in G_{n,k}\}, \quad v_k^-(K) := \inf\{\mathrm{vrad}(P_E(K)) : E \in G_{n,k}\},$$

where $G_{n,k}$ denotes the Grassmann manifold of all k-dimensional linear subspaces of $\mathbb{R}^n$, and P_E denotes orthogonal projection onto $E \in G_{n,k}$. Note that by the Blaschke–Sanataló inequality and its reverse form due to Bourgain and V. Milman (see Sect. 2), it is immediate to verify that $w_k(K^\circ) \simeq \frac{1}{v_k^-(K)}$.

Theorem 1. *For every centrally-symmetric convex body K in $\mathbb{R}^n$ and $k \geqslant 1$:*

$$e_k(K, B_2^n) \leqslant C\frac{n}{k}\log\left(e + \frac{n}{k}\right)\sup_{1\leqslant m\leqslant \min(k,n)}\left\{2^{-\frac{k}{3m}}w_m(K)\right\}.$$

By invoking Carl's theorem (see Sect. 2), a slightly weaker version of Theorem 1 may be deduced from the following stronger statement:

Theorem 2. *Let K be a centrally-symmetric convex body in $\mathbb{R}^n$. Then for any $k = 1, \ldots, \lfloor n/2 \rfloor$ there exists $F \in G_{n,n-2k}$ so that:*

$$K \cap F \subseteq C \frac{n}{k} \log \left(e + \frac{n}{k} \right) w_k(K) B_2^n \cap F, \tag{7}$$

and dually, there exists $F \in G_{n,n-2k}$ so that:

$$P_F(K) \supseteq \frac{1}{C \frac{n}{k} \log(e + \frac{n}{k})} v_k^-(K) P_F(B_2^n). \tag{8}$$

A weaker version of Theorem 2, with the parameters $w_k(K)$, $v_k^-(K)$ above replaced by:

$$v_k(K) := \sup \{ \mathrm{vrad}(P_E(K)) : E \in G_{n,k} \} , \ w_k^-(K) := \inf \{ \mathrm{vrad}(K \cap E) : E \in G_{n,k} \},$$

respectively, was obtained by V. Milman and G. Pisier in [25] (see Theorem 7). Our improved version is crucial for properly exploiting the corresponding properties of isotropic convex bodies.

By (essentially) inserting the estimates of Theorem 1 into (6) (with K replaced by $K°$), we obtain that if K is a centrally-symmetric convex body in $\mathbb{R}^n$ with $K \supseteq r B_2^n$ then:

$$\sqrt{n} M(K) \leq C \sum_{k=1}^{n} \frac{1}{\sqrt{k}} \min \left(\frac{1}{r}, \frac{n}{k} \log \left(e + \frac{n}{k} \right) \frac{1}{v_k^-(K)} \right). \tag{9}$$

In the case of the centroid bodies $Z_q(\mu)$ of an isotropic log-concave probability measure μ on $\mathbb{R}^n$, one can obtain precise information on the growth of the parameters $v_k^-(Z_q(\mu))$. We recall the relevant definitions in Sect. 5, and use (9) to deduce in Sect. 6 that:

$$2 \leq q \leq q_0 := (n \log n)^{2/5} \implies M(Z_q(\mu)) \leq C \frac{\sqrt{\log q}}{\sqrt[4]{q}}. \tag{10}$$

In particular, since $Z_n(\mu) \supseteq Z_{q_0}(\mu)$ and $M(K) \simeq M(Z_n(\lambda_{K/L_K}))/L_K$, where λ_A denotes the uniform probability measure on A, we immediately obtain:

Theorem 3. *If K is a centrally-symmetric isotropic convex body in $\mathbb{R}^n$ then:*

$$M(K) \leq \frac{C \log^{2/5}(e + n)}{\sqrt[10]{n} L_K}. \tag{11}$$

It is clear that (11) is not optimal. Note that if (10) were to remain valid until $q_0 = n$, we would obtain the bound $M(K) \leq C \frac{\sqrt{\log(e+n)}}{n^{1/4} L_K}$, which as previously explained would in turn imply that $L_K \leqslant C \sqrt{\log(e+n)} \, n^{1/4}$, in consistency with the best-known upper bound on the isotropic constant. We believe that it is an interesting question to extend the range where (10) remains valid. In Sect. 6, we obtain such an extension when μ is in addition assumed to be Ψ_α (see Sect. 6 for definitions).

Our entire method is based on Pisier's regular versions of V. Milman's M-ellipsoids associated to a given centrally-symmetric convex body K, comparing between volumes of sections and projections of K and those of its associated regular ellipsoids. This expands on an approach already employed in [7, 12, 15, 17, 31].

We conclude the introduction by remarking that the dual question of providing an upper bound for the mean-width $M^*(K)$ of an isotropic convex body K has attracted more attention in recent years. Until recently, the best known estimate was $M^*(K) \leqslant Cn^{3/4}L_K$, where $C > 0$ is an absolute constant (see [9, Chap. 9] for a number of proofs of this inequality). The second named author has recently obtained in [21] an essentially optimal answer to this question—for every isotropic convex body K in $\mathbb{R}^n$ one has $M^*(K) \leqslant C \sqrt{n} \log^2 n \, L_K$.

2 Preliminaries and Notation from the Local Theory

Let us introduce some further notation. Given $F \in G_{n,k}$, we denote $B_F = B_2^n \cap F$ and $S_F = S^{n-1} \cap F$. A centrally-symmetric convex body K in $\mathbb{R}^n$ is a compact convex set with non-empty interior so that $K = -K$. The norm induced by K on $\mathbb{R}^n$ is given by $\|x\|_K = \min\{t \geqslant 0 : x \in tK\}$. The support function of K is defined by $h_K(y) := \|y\|_K^* = \max\{\langle y, x \rangle : x \in K\}$, with K° denoting the unit-ball of the dual-norm. By the Blaschke–Santaló inequality (the right-hand side below) and its reverse form due to Bourgain and V. Milman [8] (the left-hand side), it is known that:

$$0 < c \leqslant \mathrm{vrad}(K)\mathrm{vrad}(K^\circ) \leqslant 1. \tag{12}$$

Recall that the k-th entropy number is defined as

$$e_k(K, L) := \inf\left\{t > 0 \ : \ N(K, tL) \leqslant 2^k\right\}.$$

A deep and very useful fact about entropy numbers is the Artstein–Milman–Szarek duality of entropy theorem [1], which states that:

$$e_k(B_2^n, K) \leqslant Ce_{ck}(K^\circ, B_2^n) \tag{13}$$

for every centrally-symmetric convex body K and $k \geqslant 1$.

In what follows, a crucial role is played by G. Pisier's regular version of V. Milman's M-ellipsoids. It was shown by Pisier (see [30] or [31, Chap. 7]) that for any centrally-symmetric convex body K in $\mathbb{R}^n$ and $\alpha \in (0, 2)$, there exists an ellipsoid $\mathscr{E} = \mathscr{E}_{K,\alpha}$ so that:

$$\max\{e_k(K,\mathscr{E}), e_k(K^\circ,\mathscr{E}^\circ), e_k(\mathscr{E}, K), e_k(\mathscr{E}^\circ, K^\circ)\} \leqslant P_\alpha \left(\frac{n}{k}\right)^{1/\alpha}, \tag{14}$$

where $P_\alpha \leqslant C \left(\frac{\alpha}{2-\alpha}\right)^{1/2}$ is a positive constant depending only on α.

Given a pair of centrally-symmetric convex bodies K, L in $\mathbb{R}^n$, the Gelfand numbers $c_k(K, L)$ are defined as:

$$c_k(K, L) := \begin{cases} \inf\{\text{diam}_{L\cap F}(K \cap F) : F \in G_{n,n-k}\} & k = 0, \dots, n-1 \\ 0 & \text{otherwise} \end{cases},$$

where $\text{diam}_A(B) := \inf\{R > 0 : B \subseteq RA\}$. We denote $c_k(K) = c_k(K, B_2^n)$ and $e_k(K) = e_k(K, B_2^n)$.

Carl's theorem [10] relates any reasonable Lorentz norm of the sequence of entropy numbers $\{e_m(K, L)\}$ with that of the Gelfand numbers $\{c_m(K, L)\}$. In particular, for any $\alpha > 0$, there exist constants $C_\alpha, C'_\alpha > 0$ such that for any $k \geqslant 1$:

$$\sup_{m=1,\dots,k} m^\alpha e_m(K, L) \leqslant C_\alpha \sup_{m=1,\dots,k} m^\alpha c_m(K, L), \tag{15}$$

and:

$$\sum_{m=1}^{k} m^{-1+\alpha} e_m(K, L) \leqslant C'_\alpha \sum_{m=1}^{k} m^{-1+\alpha} c_m(K, L). \tag{16}$$

In fact, Pisier deduces the covering estimates of (14) from an application of Carl's theorem, after establishing the following estimates:

$$\max\{c_k(K, \mathscr{E}), c_k(K^\circ, \mathscr{E}^\circ)\} \leqslant P_\alpha \left(\frac{n}{k}\right)^{1/\alpha} \quad \text{for all } k \in \{1, \dots, n\}. \tag{17}$$

Our estimates depend on a number of volumetric parameters of K, already defined in the Introduction, which we now recall:

$$w_k(K) := \sup\{\text{vrad}(K \cap E) : E \in G_{n,k}\}, v_k(K) := \sup\{\text{vrad}(P_E(K)) : E \in G_{n,k}\},$$

and

$$w_k^-(K) := \inf\{\text{vrad}(K \cap E) : E \in G_{n,k}\}, v_k^-(K) := \inf\{\text{vrad}(P_E(K)) : E \in G_{n,k}\}.$$

Note that $0 < c \leqslant w_k^-(K)v_k(K^\circ), v_k^-(K)w_k(K^\circ) \leqslant 1$ by (12). Also observe that $k \mapsto v_k(K)$ is non-increasing by the Alexandrov inequalities and Kubota's formula, and that $k \mapsto w_k^-(K)$ is non-decreasing by polar-integration and Jensen's inequality.

We refer to the books [26] and [31] for additional basic facts from the local theory of normed spaces.

3 New Covering Estimates

The main result of this section provides a general upper bound for the entropy numbers $e_k(K, B_2^n)$.

Theorem 4. *Let K be a centrally-symmetric convex body in $\mathbb{R}^n$, and let $k \geq 1$. Then:*

$$e_k(K, B_2^n) \leq C \frac{n}{k} \log \left(e + \frac{n}{k} \right) \sup_{1 \leq m \leq \min(k,n)} \left\{ 2^{-\frac{k}{3m}} w_m(K) \right\}.$$

We combine this fact with Dudley's entropy estimate

$$\sqrt{n} M^*(K) \leq C \sum_{k \geq 1} \frac{1}{\sqrt{k}} e_k(K, B_2^n). \tag{18}$$

(see [31, Theorem 5.5] for this formulation). As an immediate consequence, we obtain:

Corollary 5. *Let K be a centrally-symmetric convex body in $\mathbb{R}^n$ with $K \subseteq RB_2^n$. Then:*

$$\sqrt{n} M^*(K) \leq C \sum_{k \geq 1} \frac{1}{\sqrt{k}} \min \left\{ R, \frac{n}{k} \log \left(e + \frac{n}{k} \right) \sup_{1 \leq m \leq \min(k,n)} \left\{ 2^{-\frac{k}{3m}} w_m(K) \right\} \right\}.$$

Dually, let K be a centrally-symmetric convex body in $\mathbb{R}^n$ with $K \supseteq r B_2^n$. Then:

$$\sqrt{n} M(K) \leq C \sum_{k \geq 1} \frac{1}{\sqrt{k}} \min \left\{ \frac{1}{r}, \frac{n}{k} \log \left(e + \frac{n}{k} \right) \sup_{1 \leq m \leq \min(k,n)} \left\{ 2^{-\frac{k}{3m}} \frac{1}{v_m^-(K)} \right\} \right\}$$

Proof. The first claim follows by a direct application of (18) if we estimate $e_k(K, B_2^n)$ using Theorem 4 and the observation that $e_k(K, B_2^n) \leq R$ for all $k \geq 1$ (recall that $K \subseteq RB_2^n$). Then, the second claim follows by duality since $w_m(K^\circ) \simeq \frac{1}{v_m^-(K)}$. □

We will see in the next section that the supremum over m above is unnecessary and that one may always use $m = k$, only summing over $k = 1, \ldots, n$. But we proceed with the proof of Theorem 4, as it is a simpler approach.

Proof of Theorem 4. Assume without loss of generality that k is divisible by 3, and use the estimate:

$$e_k(K, B_2^n) \leq e_{k/3}(K, \mathcal{E}) e_{2k/3}(\mathcal{E}, B_2^n),$$

where $\mathcal{E} = \mathcal{E}_{K,\alpha_k}$ is Pisier's α_k-regular M-ellipsoid associated to K, with $\alpha_k \in [1, 2)$ to be determined. The first term is controlled directly by Pisier's regular covering estimate (14). For the second term we use the following simple fact about covering numbers of ellipsoids (see e.g. [31, Remark 5.15]):

$$e_j(\mathscr{E}, B_2^n) \simeq \sup_{1 \le m \le n} 2^{-j/m} w_m(\mathscr{E}) \simeq \sup_{1 \le m \le \min(j,n)} 2^{-j/m} w_m(\mathscr{E});$$

the latter equivalence follows since $w_m(\mathscr{E})$ is the geometric average of the m largest principal radii of $\mathscr{E}$, and so $m \mapsto w_m(\mathscr{E})$ is non-increasing. Now recall that

$$w_m(\mathscr{E}) \simeq 1/v_m^-(\mathscr{E}^\circ). \tag{19}$$

To estimate $v_m^-(\mathscr{E}^\circ)$, we use a trivial volumetric bound: for any $E \in G_{n,m}$,

$$\frac{\operatorname{vrad}(P_E(K^\circ))}{\operatorname{vrad}(P_E(\mathscr{E}^\circ)) e_s(K^\circ, \mathscr{E}^\circ)} \le N(P_E(K^\circ), e_s(K^\circ, \mathscr{E}^\circ) P_E(\mathscr{E}^\circ))^{1/m}$$

$$\le N(K^\circ, e_s(K^\circ, E^\circ)\mathscr{E}^\circ)^{1/m} \le 2^{s/m},$$

for $s \ge 1$ to be determined. Consequently:

$$v_m^-(\mathscr{E}^\circ) \ge \frac{1}{2^{s/m} e_s(K^\circ, \mathscr{E}^\circ)} v_m^-(K^\circ),$$

and plugging this back into (19), we deduce:

$$w_m(\mathscr{E}) \le C 2^{s/m} e_s(K^\circ, \mathscr{E}^\circ) w_m(K),$$

and hence:

$$e_{2k/3}(\mathscr{E}, B_2^n) \le C \sup_{1 \le m \le \min(k,n)} 2^{\frac{s-2k/3}{m}} e_s(K^\circ, \mathscr{E}^\circ) w_m(K).$$

Setting $s = k/3$, we conclude that:

$$e_{2k/3}(\mathscr{E}, B_2^n) \le C e_{k/3}(K^\circ, \mathscr{E}^\circ) \sup_{1 \le m \le \min(k,n)} 2^{-\frac{k}{3m}} w_m(K).$$

Combining everything, we obtain:

$$e_k(K, \mathscr{E}) \le C e_{k/3}(K, \mathscr{E}) e_{k/3}(K^\circ, \mathscr{E}^\circ) \sup_{1 \le m \le \min(k,n)} 2^{-\frac{k}{3m}} w_m(K)$$

$$\le \frac{C'}{2 - \alpha_k} \left(\frac{n}{k}\right)^{\frac{2}{\alpha_k}} \sup_{1 \le m \le \min(k,n)} 2^{-\frac{k}{3m}} w_m(K).$$

Setting $\alpha_k = 2 - \frac{1}{\log(e+n/k)}$, the assertion follows. $\square$

Remark 6. Theorem 4 implies the following dual covering estimate:

$$e_k(B_2^n, K) \leq C \frac{n}{k} \log \left(e + \frac{n}{k} \right) \sup_{1 \leq m \leq \min(k,n)} \left\{ 2^{-\frac{k}{3m}} \frac{1}{v_m^-(K)} \right\}. \tag{20}$$

Indeed, this is immediate from the duality of entropy theorem (13) and the fact that $w_m(K^\circ) \simeq \frac{1}{v_m^-(K)}$. Alternatively, one may simply repeat the proof of Theorem 4 with the roles of K and B_2^n exchanged.

4 New Diameter Estimates

This section may be read independently of the rest of this work, and contains a refinement of the following result of V. Milman and G. Pisier from [25], as exposed in [31, Lemma 9.2]:

Theorem 7 (Milman–Pisier). *Let K be a centrally-symmetric convex body in $\mathbb{R}^n$. Then, for any $k = 1, \ldots, n/2$:*

$$c_{2k}(K) \leq C \frac{n}{k} \log \left(e + \frac{n}{k} \right) v_k(K).$$

In other words, there exists $F \in G_{n,n-2k}$ so that:

$$K \cap F \subseteq C \frac{n}{k} \log \left(e + \frac{n}{k} \right) v_k(K) B_F, \tag{21}$$

and dually, there exists $F \in G_{n,n-2k}$ so that:

$$P_F(K) \supseteq \frac{1}{C \frac{n}{k} \log(e + \frac{n}{k})} w_k^-(K) B_F. \tag{22}$$

Our version refines these estimates by replacing $v_k(K)$ and $w_k^-(K)$ above by the stronger $w_k(K)$ and $v_k^-(K)$ parameters, respectively; this refinement is crucial for our application in this paper.

Theorem 8. *Let K be a centrally-symmetric convex body in $\mathbb{R}^n$. Then for any $k = 1, \ldots, n/2$:*

$$c_{2k}(K) \leq C \frac{n}{k} \log \left(e + \frac{n}{k} \right) w_k(K).$$

In other words, there exists $F \in G_{n,n-2k}$ so that:

$$K \cap F \subseteq C \frac{n}{k} \log \left(e + \frac{n}{k} \right) w_k(K) B_F, \tag{23}$$

and dually, there exists $F \in G_{n,n-2k}$ so that:

$$P_F(K) \supseteq \frac{1}{C\frac{n}{k}\log(e + \frac{n}{k})} v_k^-(K) B_F. \tag{24}$$

Our refinement will come from exploiting the full strength of Pisier's result on the existence of regular M-ellipsoids. In contrast, the Milman–Pisier result is based on V. Milman's quotient-of-subspace theorem, from which it seems harder to obtain enough regularity to deduce our proposed refinement.

Proof of Theorem 8. Given $k = 1, \ldots, n/2$, let $\mathscr{E} = \mathscr{E}_{K,\alpha_k}$ denote Pisier's α_k-regular M-ellipsoid, for some $\alpha_k \in [1, 2)$ to be determined. By the second estimate in (17), we know that there exists $E \in G_{n,n-k}$ so that:

$$P_E(K) \supseteq \frac{1}{P_{\alpha_k}} \left(\frac{k}{n}\right)^{1/\alpha_k} P_E(\mathscr{E}).$$

For the ellipsoid $\mathscr{E}' := P_E(\mathscr{E}) \subseteq E$, we may always find a linear subspace $F \subseteq E$ of codimension m in E so that:

$$P_F(\mathscr{E}') \supseteq \inf_{H \in G_m(E)} \sup_{H' \subseteq H} \{\mathrm{vrad}(P_{H'}(\mathscr{E}'))\} B_F,$$

where $G_m(E)$ is the Grassmannian of all m-dimensional linear subspaces of E. Indeed, this is immediate by choosing H to be the subspace spanned by the m shortest axes of $\mathscr{E}'$, and setting F to be its orthogonal complement. Consequently, there exists a subspace $F \in G_{n,n-(k+m)}$ so that:

$$P_F(K) \supseteq \frac{1}{P_{\alpha_k}} \left(\frac{k}{n}\right)^{1/\alpha_k} \inf_{H \in G_{n,m}} \sup_{H' \subseteq H} \{\mathrm{vrad}(P_{H'}(\mathscr{E}))\} B_F. \tag{25}$$

We now deviate from the proof of our refined version, to show how one may recover the Milman–Pisier estimate; the reader solely interested in the proof of our refinement may safely skip this paragraph. Assume for simplicity that $k < n/3$. By the first estimate in (17), we know that there exists $J \in G_{n,n-k}$ so that:

$$K \cap J \subseteq P_{\alpha_k} \left(\frac{n}{k}\right)^{1/\alpha_k} \mathscr{E} \cap J.$$

Given $H \in G_{n,m}$ and denoting $H' := H \cap J \in G_{m'}(H)$ with $m' \in [m - k, m]$, it follows that:

$$P_{H'}(\mathscr{E}) \supseteq \mathscr{E} \cap H' \supseteq \frac{1}{P_{\alpha_k}} \left(\frac{k}{n}\right)^{1/\alpha_k} K \cap H'.$$

Setting $m = 2k$, it follows from (25) that there exists $F \in G_{n,n-3k}$ so that:

$$P_F(K) \geq \frac{1}{P_{\alpha_k}^2} \left(\frac{k}{n}\right)^{2/\alpha_k} \inf\left\{\mathrm{vrad}(K \cap H') : H' \in G_{n,m'}, m' \in [k, 2k]\right\} B_F.$$

Noting that the sequence $m' \mapsto w_{m'}^-(K)$ is non-decreasing, and setting $\alpha_k = 2 - \frac{1}{\log(e+n/k)}$, we have found $F \in G_{n,n-3k}$ such that

$$P_F(K) \geq \frac{c}{\frac{n}{k}\log(e + \frac{n}{k})} w_k^-(K),$$

as asserted in (22) (with perhaps an immaterial constant 3 instead of 2). The assertion of (21) follows by duality.

To obtain our refinement, we will use instead of the first estimate in (17), the covering estimate (14) (which Pisier obtains from (17) by an application of Carl's theorem, requiring the entire sequence of c_k estimates, not just the one for our specific k). Setting $m = k$, we use a trivial volumetric estimate to control $\mathrm{vrad}(P_H(\mathscr{E}))$, exactly as in the proof of Theorem 4: for any $H \in G_{n,k}$,

$$\frac{\mathrm{vrad}(P_H(K))}{\mathrm{vrad}(P_H(\mathscr{E}))e_k(K,\mathscr{E})} \leq N(P_H(K), e_k(K,\mathscr{E})P_H(\mathscr{E}))^{1/k}$$

$$\leq N(K, e_k(K, E)\mathscr{E})^{1/k} \leq 2.$$

Together with (14), we obtain:

$$\mathrm{vrad}(P_H(\mathscr{E})) \geq \frac{1}{2e_k(K,\mathscr{E})}\mathrm{vrad}(P_H(K)) \geq \frac{1}{2P_{\alpha_k}}\left(\frac{k}{n}\right)^{1/\alpha_k}\mathrm{vrad}(P_H(K)).$$

Plugging this into (25) and setting as usual $\alpha_k = 2 - \frac{1}{\log(e+n/k)}$, the asserted estimate (24) follows. The other estimate (23) follows by duality. $\qquad\square$

As immediate corollaries, we have:

Corollary 9. *For every centrally-symmetric convex body K in $\mathbb{R}^n$, $k = 1,\ldots,n$ and $\alpha > 0$:*

$$e_k(K, B_2^n) \leq C_\alpha \sup_{m=1,\ldots,k} \left(\frac{m}{k}\right)^\alpha \frac{n}{m}\log\left(e + \frac{n}{m}\right)w_m(K),$$

where $C_\alpha > 0$ is a constant depending only on α.

Proof. This is immediate from Theorem 8 and Carl's theorem (15). Note that $k \mapsto c_k(K, B_2^n)$ is non-increasing, and so there is no difference whether we take the supremum on the right-hand-side just on the even integers. $\qquad\square$

Corollary 10. *For every centrally-symmetric convex body K in $\mathbb{R}^n$ so that $K \subseteq RB_2^n$, we have:*

$$\sqrt{n}M^*(K) \le C \sum_{k=1}^{n} \frac{1}{\sqrt{k}} \min\left(R, \frac{n}{k}\log\left(e + \frac{n}{k}\right)w_k(K)\right).$$

Dually, for every centrally-symmetric convex body K in $\mathbb{R}^n$ so that $K \supseteq rB_2^n$, we have:

$$\sqrt{n}M(K) \le C \sum_{k=1}^{n} \frac{1}{\sqrt{k}} \min\left(\frac{1}{r}, \frac{n}{k}\log\left(e + \frac{n}{k}\right)\frac{1}{v_k^-(K)}\right).$$

Proof. Let us verify the first claim, the second follows by duality. Indeed, this is immediate from Dudley's entropy estimate (6) coupled with Carl's theorem (16):

$$\sqrt{n}M^*(K) \le C \sum_{k=1}^{n} \frac{1}{\sqrt{k}}e_k(K) \le C' \sum_{k=1}^{n} \frac{1}{\sqrt{k}}c_k(K).$$

Obviously $c_k(K) \le R$ for all k, and so the assertion follows from the estimates of Theorem 8. $\qquad\square$

Both Corollaries should be compared with the results of the previous section.

Remark 11. It may be insightful to compare Theorem 8 to some other known estimates on diameters of k-codimensional sections, besides the Milman–Pisier Theorem 7. One sharp estimate is the Pajor–Tomczak-Jaegermann refinement [27] of V. Milman's low-M^* estimate [22]:

$$c_k(L) \le C\sqrt{\frac{n}{k}}M^*(L), \tag{26}$$

for any origin-symmetric convex L and $k = 1, \ldots, n$. However, for our application, we cannot use this to control $c_k(K^\circ)$ since we do not a-priori know $M^*(K^\circ) = M(K)$. A type of dual low-M estimate was observed by Klartag [13]:

$$c_k(L) \le C^{\frac{n}{k}}\mathrm{vrad}(L)^{\frac{n}{k}}M(L)^{\frac{n-k}{k}}.$$

Since $M(K^\circ) = M^*(K)$ is now well understood for an isotropic origin-symmetric convex body [21], this would give good estimates for low-dimensional sections (large codimension k), but unfortunately this is not enough for controlling $M(K)$. Klartag obtains the latter estimate from the following one, which is more in the spirit of the estimates we obtain in this work:

$$c_k(L) \le C^{\frac{n}{k}}\frac{\mathrm{vrad}(L)^{\frac{n}{k}}}{w_{n-k}(L)^{\frac{n-k}{k}}}.$$

Again, this seems too rough for controlling the diameter of high-dimensional sections.

5 Preliminaries from Asymptotic Convex Geometry

An absolutely continuous Borel probability measure μ on $\mathbb{R}^n$ is called log-concave if its density f_μ is of the form $\exp(-\varphi)$ with $\varphi : \mathbb{R}^n \to \mathbb{R} \cup \{+\infty\}$ convex. Note that the uniform probability measure on K, denoted λ_K, is log-concave for any convex body K.

The barycenter of μ is denoted by $\operatorname{bar}(\mu) := \int_{\mathbb{R}^n} x d\mu(x)$. The isotropic constant of μ, denoted L_μ, is the following affine invariant quantity:

$$L_\mu := (\sup_{x \in \mathbb{R}^n} f_\mu(x))^{\frac{1}{n}} \det \operatorname{Cov}(\mu)^{\frac{1}{2n}}, \tag{27}$$

where $\operatorname{Cov}(\mu) := \int x \otimes x d\mu(x) - \int x d\mu(x) \otimes \int x d\mu(x)$ denotes the covariance matrix of μ. We say that a log-concave probability measure μ on $\mathbb{R}^n$ is isotropic if $\operatorname{bar}(\mu) = 0$ and $\operatorname{Cov}(\mu)$ is the identity matrix. Note that a convex body K of volume 1 is isotropic if and only if the log-concave probability measure λ_{K/L_K} is isotropic, and that L_{λ_K} indeed coincides with L_K. It was shown by K. Ball [2, 3] that given $n \geqslant 1$:

$$\sup_\mu L_\mu \leqslant C \sup_K L_K,$$

where the suprema are taken over all log-concave probability measures μ and convex bodies K in $\mathbb{R}^n$, respectively (see e.g. [14] for the non-even case). Klartag's bound on the isotropic constant [14] thus reads $L_\mu \leqslant Cn^{1/4}$ for all log-concave probability measures μ on $\mathbb{R}^n$.

Given $E \in G_{n,k}$, we denote by $\pi_E \mu := \mu \circ P_E^{-1}$ the push-forward of μ via P_E. Obviously, if μ is centered or isotropic then so is $\pi_E \mu$, and by the Prékopa–Leindler theorem, the same also holds for log-concavity.

Given a log-concave probability measure μ on $\mathbb{R}^n$ and $q \geqslant 1$, the L_q-centroid body of μ, denoted $Z_q(\mu)$, is the centrally-symmetric convex body with support function:

$$h_{Z_q(\mu)}(y) := \left(\int_{\mathbb{R}^n} |\langle x, y \rangle|^q d\mu(x) \right)^{1/q}. \tag{28}$$

Observe that μ is isotropic if and only if it is centered and $Z_2(\mu) = B_2^n$. By Jensen's inequality $Z_1(\mu) \subseteq Z_p(\mu) \subseteq Z_q(\mu)$ for all $1 \leqslant p \leqslant q < \infty$. Conversely, it follows from work of Berwald [4] or by employing Borell's lemma (see [26, Appendix III]), that:

$$1 \leqslant p \leqslant q \implies Z_q(\mu) \subseteq C \frac{q}{p} Z_p(\mu).$$

When $\mu = \lambda_K$ is the uniform probability measure on a centrally-symmetric convex body K in $\mathbb{R}^n$, it is easy to check (e.g. [9]) using the Brunn–Minkowski inequality that:

$$cK \subseteq Z_n(\lambda_K) \subseteq K.$$

Let μ denote an isotropic log-concave probability measure μ on $\mathbb{R}^n$. It was shown by Paouris [28] that

$$1 \leq q \leq \sqrt{n} \implies M^*(Z_q(\mu)) \simeq \sqrt{q}, \tag{29}$$

and that:

$$1 \leq q \leq n \implies \mathrm{vrad}(Z_q(\mu)) \leq C\sqrt{q}. \tag{30}$$

Conversely, it was shown by Klartag and E. Milman in [16] that:

$$1 \leq q \leq \sqrt{n} \implies \mathrm{vrad}(Z_q(\mu)) \geq c_1\sqrt{q}. \tag{31}$$

This determines the volume radius of $Z_q(\mu)$ for all $1 \leq q \leq \sqrt{n}$. For larger values of q one can still use the lower bound:

$$1 \leq q \leq n \implies \mathrm{vrad}(Z_q(\mu)) \geq c_2\sqrt{q}\, L_\mu^{-1}, \tag{32}$$

obtained by Lutwak et al. [20] via symmetrization.

We refer to the book [9] for further information on isotropic convex bodies and log-concave measures.

6 M-Estimates for Isotropic Convex Bodies and Their L_q-Centroid Bodies

Let μ denote an isotropic log-concave probability measure on $\mathbb{R}^n$, and fix $H \in G_{n,k}$. A very useful observation is that:

$$P_H(Z_q(\mu)) = Z_q(\pi_H(\mu)).$$

It follows from (31) that:

$$1 \leq q \leq \sqrt{k} \implies \mathrm{vrad}(P_H(Z_q(\mu))) \geq c\sqrt{q}. \tag{33}$$

Furthermore, using (32), we see that:

$$q \geq \sqrt{k} \implies \mathrm{vrad}(P_H(Z_q(\mu))) \geq c' \max\left(\sqrt[4]{k}, \frac{\sqrt{\min(q,k)}}{L_{\pi_H\mu}}\right). \tag{34}$$

Unfortunately, we can only say in general that $\sup\{L_{\pi_H\mu} : H \in G_{n,k}\} \leq C\sqrt[4]{k}$, and so the estimate (34) is not very useful, unless we have some additional information on μ. Recalling the definition of $v_k^-(Z_q(\mu))$, we summarize this (somewhat sloppily) in:

Lemma 12. *Let μ be an isotropic log-concave probability measure on $\mathbb{R}^n$. For any $q \geq 1$ and $k = 1, \ldots, n$ we have:*

$$v_k^-(Z_q(\mu)) \geq c\sqrt{\min(q, \sqrt{k})}.$$

Assuming that $\sup\{L_{\pi_H\mu} : H \in G_{n,k}\} \leq A_k$ we have:

$$v_k^-(Z_q(\mu)) \geq \frac{c'}{A_k}\sqrt{\min(q, k)}.$$

6.1 Estimates for $Z_q(\mu)$

Plugging these lower bounds for $v_k^-(Z_q(\mu))$ into either Theorem 4 or Corollary 9 coupled with Remark 6, we immediately obtain estimates on the entropy numbers $e_k(B_2^n, Z_q(\mu))$. Similar estimates on the maximal (with respect to $F \in G_{n,n-k}$) in-radius of $P_F(Z_q(\mu))$ are obtained by invoking Theorem 8.

Theorem 13. *Given $q \geq 2$ and an integer $k = 1, \ldots, n$, denote:*

$$R_{k,q} := \min\left\{1, C\frac{1}{\min(\sqrt{q}, \sqrt[4]{k})}\frac{n}{k}\log\left(e + \frac{n}{k}\right)\right\}.$$

Then, for any isotropic log-concave probability measure μ on $\mathbb{R}^n$:

$$e_k(B_2^n, Z_q(\mu)) \leq R_{k,q},$$

and there exists $F \in G_{n,n-k}$ so that:

$$P_F(Z_q(\mu)) \supseteq \frac{1}{R_{k,q}}B_F.$$

Proof. From (20) and Lemma 12 we have:

$$e_k(B_2^n, Z_q(\mu)) \leq C\frac{n}{k}\log\left(e + \frac{n}{k}\right)\sup_{1 \leq m \leq k}\left\{2^{-\frac{k}{3m}}\frac{1}{\min(\sqrt{q}, \sqrt[4]{m})}\right\}.$$

Then, it suffices to observe that:

$$\sup_{1 \le m \le k} \left\{ 2^{-\frac{k}{3m}} \frac{1}{\min(\sqrt{q}, \sqrt[4]{m})} \right\} \simeq \sup_{1 \le m \le k} \left\{ 2^{-\frac{k}{3m}} \left(\frac{1}{\sqrt{q}} + \frac{1}{\sqrt[4]{m}} \right) \right\}$$

$$\le C \left(\frac{1}{\sqrt{q}} + \frac{1}{\sqrt[4]{k}} \right) \simeq \frac{1}{\min(\sqrt{q}, \sqrt[4]{k})},$$

because $2^{-\frac{k}{3m}}/\sqrt{q} \le 1/\sqrt{q}$ for all $1 \le m \le k$, and $m \mapsto 2^{\frac{k}{3m}}\sqrt[4]{m}$ attains its minimum at $m \simeq k$, so that $\sup_{1 \le m \le k}(2^{-\frac{k}{3m}}/\sqrt[4]{m}) \le C/\sqrt[4]{k}$. We also use the fact that in a certain range of values for $q \ge 2$ and $k \ge 1$, we might as well use the trivial estimates:

$$e_k(B_2^n, Z_q(\mu)) \le 1 , \quad P_F(Z_q(\mu)) \supseteq B_F, \tag{35}$$

which hold since $Z_q(\mu) \supseteq Z_2(\mu) = B_2^n$. $\qquad\square$

An elementary computation based on Corollary 10 then yields a non-trivial estimate for $M(Z_q(\mu))$. It is interesting to note that without using the trivial information that $Z_q(\mu) \supseteq B_2^n$ (or equivalently, the trivial estimates in (35)), Corollary 10 would not yield anything meaningful.

Theorem 14. *For any isotropic log-concave probability measure μ on $\mathbb{R}^n$:*

$$2 \le q \le q_0 := (n \log(e + n))^{2/5} \implies M(Z_q(\mu)) \le C \frac{\sqrt{\log q}}{\sqrt[4]{q}}.$$

Proof. We use the estimate:

$$\sqrt{n} M(Z_q(\mu)) \le C \sum_{k=1}^{n} \frac{1}{\sqrt{k}} \min\left\{ 1, C \frac{1}{\min(\sqrt{q}, \sqrt[4]{k})} \frac{n}{k} \log\left(e + \frac{n}{k} \right) \right\},$$

which follows from Corollary 10 combined with Theorem 13. We set $k(n, q) = (n \log q)/\sqrt{q}$. Note that if $k \ge k(n, q)$ then $k \ge cq^2$. Therefore, we may write:

$$\sqrt{n} M(Z_q(\mu)) \le C \sum_{k=1}^{k(n,q)} \frac{1}{\sqrt{k}} + \frac{Cn}{\sqrt{q}} \sum_{k=k(n,q)}^{n} \frac{1}{k^{3/2}} \log\left(e + \frac{n}{k} \right)$$

$$\le C_1 \sqrt{k(n,q)} + C_2 \frac{n \log q}{\sqrt{q k(n,q)}} \le C_3 \frac{\sqrt{n \log q}}{\sqrt[4]{q}}.$$

The result follows. $\qquad\square$

For larger values of q, we obtain no additional information beyond the trivial monotonicity:

$$q_0 \le q \quad \Longrightarrow \quad M(Z_q(\mu)) \le M(Z_{q_0}(\mu)) \le C \frac{\log^{2/5}(e+n)}{n^{1/10}}.$$

If K is an isotropic centrally-symmetric convex body in $\mathbb{R}^n$, using that λ_{K/L_K} is isotropic log-concave and that $Z_n(\lambda_{K/L_K})$ is isomorphic to K/L_K, one immediately translates the above results to corresponding estimates for K.

Theorem 15. *Given $k = 1, \ldots, n$, set:*

$$R_k := \min \left\{ 1, C \frac{1}{\sqrt[4]{k}} \frac{n}{k} \log\left(e + \frac{n}{k}\right) \right\}.$$

Then, for any isotropic centrally-symmetric convex body K in $\mathbb{R}^n$:

$$e_k(B_2^n, K) \le \frac{R_k}{L_K},$$

and there exists $F \in G_{n,n-k}$ so that:

$$P_F(K) \supseteq \frac{L_K}{R_k} B_F.$$

Moreover:

$$M(K) \le \frac{C}{L_K} \frac{\log^{2/5}(e+n)}{n^{1/10}}.$$

6.2   Assuming that the Isotropic Constant is Bounded

It is interesting to perform the same calculations under the assumption that $L_\mu \le C$ for any log-concave probability measure μ (regardless of dimension). In that case:

$$v_k^-(Z_q(\mu)) \ge c\sqrt{\min(q,k)}.$$

This would yield the following conditional result:

Theorem 16. *Given $q \ge 2$ and an integer $k = 1, \ldots, n$, denote:*

$$R_{k,q} := \min \left\{ 1, C \frac{1}{\sqrt{\min(q,k)}} \frac{n}{k} \log\left(e + \frac{n}{k}\right) \right\}.$$

Assuming that $L_\mu \le C$ for any log-concave probability measure (regardless of dimension), then for any isotropic log-concave probability measure μ on $\mathbb{R}^n$:

$$e_k(B_2^n, Z_q(\mu)) \leq R_{k,q},$$

and there exists $F \in G_{n,n-k}$ so that:

$$P_F(Z_q(\mu)) \supseteq \frac{1}{R_{k,q}} B_F.$$

Furthermore:

$$M(Z_q(\mu)) \leq C \frac{\sqrt{\log q}}{\sqrt[4]{q}} \text{ for all } 2 \leq q \leq (n \log n)^{2/3}.$$

Consequently, for every isotropic convex body K in $\mathbb{R}^n$ one would have:

$$M(K) \leq C \frac{\log^{1/3}(e+n)}{n^{1/6}}.$$

6.3 ψ_α-Measures

Finally, rather than assuming that L_μ is always bounded, we repeat the calculations for a log-concave measure μ which is assumed to be ψ_α-regular. Recall that μ is called ψ_α with constant b_α ($\alpha \in [1, 2]$) if:

$$Z_q(\mu) \subseteq b_\alpha q^{1/\alpha} Z_2(\mu) \quad \text{for all } q \geq 2.$$

Note that this property is inherited by all marginals of μ, and that any log-concave measure is ψ_1 with $b_1 = C$ a universal constant.

It was shown by Klartag and E. Milman [16] that when μ is a ψ_α log-concave probability measure on $\mathbb{R}^n$ with constant b_α, then:

$$1 \leq q \leq C \frac{n^{\frac{\alpha}{2}}}{b_\alpha^\alpha} \implies \mathrm{vrad}(Z_q(\mu)) \geq c\sqrt{q},$$

and:

$$L_\mu \leq C\sqrt{b_\alpha^\alpha n^{1-\alpha/2}}.$$

This implies that for such a measure, for any $H \in G_{n,k}$:

$$1 \leq q \leq C \frac{k^{\frac{\alpha}{2}}}{b_\alpha^\alpha} \implies \mathrm{vrad}(P_H(Z_q(\mu))) \geq c\sqrt{q}.$$

By (32), we know that:

$$q \geq q_0 := C \frac{k^{\frac{\alpha}{2}}}{b_\alpha^\alpha} \quad \Longrightarrow \quad \text{vrad}(P_H(Z_q(\mu))) \geq c' \max\left(\sqrt{q_0}, \frac{\sqrt{\min(q,k)}}{L_{\pi_H\mu}}\right).$$
$$(36)$$

Unfortunately, since we only know that:

$$L_{\pi_H\mu} \leq C \sqrt{b_\alpha^\alpha k^{1-\alpha/2}},$$

we again see that the maximum in (36) is always attained by the $\sqrt{q_0}$ term. Summarizing, we have:

Lemma 17. *Let μ be an isotropic log-concave probability measure on $\mathbb{R}^n$ which is ψ_α with constant b_α for some $\alpha \in [1, 2]$. Then for any $q \geq 1$ and $k = 1, \ldots, n$ we have:*

$$v_k^-(Z_q(\mu)) \geq c \sqrt{\min\left(q, \frac{k^{\alpha/2}}{b_\alpha^\alpha}\right)}.$$

Plugging this estimate into the general results of Sects. 3 and 4, we obtain:

Theorem 18. *Let μ denote an isotropic log-concave probability measure on $\mathbb{R}^n$ which is ψ_α with constant b_α for some $\alpha \in [1, 2]$. Given $q \geq 2$ and an integer $k = 1, \ldots, n$, denote:*

$$R_{k,q} := \min\left\{1, C \frac{1}{\sqrt{\min\left(q, \frac{k^{\alpha/2}}{b_\alpha^\alpha}\right)}} \frac{n}{k} \log\left(e + \frac{n}{k}\right)\right\}.$$

Then:

$$e_k(B_2^n, Z_q(\mu)) \leq R_{k,q},$$

and there exists $F \in G_{n,n-k}$ so that:

$$P_F(Z_q(\mu)) \supseteq \frac{1}{R_{k,q}} B_F.$$

Furthermore:

$$M(Z_q(\mu)) \leq C \frac{\sqrt{\log q}}{\sqrt[4]{q}} \text{ for all } 2 \leq q \leq c \frac{(n \log(e+n))^{\frac{2\alpha}{\alpha+4}}}{b_\alpha^{\frac{4\alpha}{\alpha+4}}}.$$

Consequently, for every isotropic convex body K in $\mathbb{R}^n$ so that λ_K is ψ_α with constant b_α, one has:

$$M(K) \leqslant \frac{C}{L_K} b_\alpha^{\frac{\alpha}{\alpha+4}} \frac{\log^{\frac{2}{\alpha+4}}(e+n)}{n^{\frac{\alpha}{2(\alpha+4)}}}.$$

Remark 19. Better estimates for the entropy-numbers $e_k(B_2^n, Z_q(\mu))$ and Gelfand numbers $c_k(Z_q(\mu)^\circ)$ may be obtained for various ranges of k by employing the alternative known estimates mentioned in Remark 11. However, these do not result in improved estimates on $M(Z_q(\mu))$, which was our ultimate goal. We therefore leave these improved estimates on the entropy and Gelfand numbers to the interested reader. We only remark that even the classical low-M^* estimate (26) coupled with our estimate on $M(Z_q(\mu))$ yield improved estimates for e_k and c_k in a certain range—a type of "bootstrap" phenomenon.

7   Concluding Remarks

In this section we briefly describe an improved and simplified version of the arguments from [12] and compare the resulting improved estimates to the ones from the previous section. Following the general approach we employ in this work, the arguments are presented for general centrally-symmetric convex bodies, and this in fact further simplifies the exposition of [12].

We mainly concentrate on presenting an alternative proof of the following slightly weaker variant of Theorem 13:

Theorem 20. *Let K be a centrally-symmetric convex body in $\mathbb{R}^n$. For any $k = 1, \ldots, \lfloor n/2 \rfloor$ there exists $F \in G_{n,n-2k}$ such that:*

$$P_F(K) \supseteq \frac{c}{\frac{n}{k} \log^2\left(e + \frac{n}{k}\right)} v_k^-(K) B_F$$

where $c > 0$ is an absolute constant.

For the proof of Theorem 20, we use a sort of converse to Carl's theorem (15) on the diameter of sections of a convex body satisfying 2-regular entropy estimates, which is due to V. Milman [23] (see also [9, Chap. 9]).

Lemma 21. *Let L be a symmetric convex body in $\mathbb{R}^n$. Then:*

$$\sqrt{k}\, c_k(L, B_2^n) \leqslant C \log(e + n/k) \sup_{k \leqslant m \leqslant n} \sqrt{m}\, e_m(L, B_2^n).$$

Remark 22. Clearly, by applying a linear transformation, the statement equally holds with B_2^n replaced by an arbitrary ellipsoid.

Proof of Theorem 20. Given $k = 1, \ldots, \lfloor n/2 \rfloor$, let $\mathscr{E} = \mathscr{E}_{K,\alpha_k}$ denote Pisier's α_k-regular M-ellipsoid, for some $\alpha_k \in [1, 2)$ to be determined. Instead of directly using Pisier's estimate (17) on the Gelfand numbers as in the proof of Theorem 8 to deduce the existence of $E \in G_{n,n-k}$ so that:

$$P_E(K) \supseteq \frac{1}{P_{\alpha_k}} \left(\frac{k}{n} \right)^{1/\alpha_k} P_E(\mathscr{E}), \qquad (37)$$

the starting point in [12] are the more traditional covering estimates (14):

$$\max\{e_k(K, \mathscr{E}), e_k(K^\circ, \mathscr{E}^\circ), e_k(\mathscr{E}, K), e_k(\mathscr{E}^\circ, K^\circ)\} \leq P_\alpha \left(\frac{n}{k} \right)^{1/\alpha_k}. \qquad (38)$$

In [12], the following estimate was used (see [31, Theorem 5.14]):

$$c_k(K^\circ, \mathscr{E}^\circ) \leq C \sqrt{\frac{n}{k}} e_k(K^\circ, \mathscr{E}^\circ).$$

However, this estimate does not take into account the regularity of the covering. Consequently, a significantly improved estimate is obtained by employing Lemma 21 (and the subsequent remark) which exploits this regularity:

$$\sqrt{k} \, c_k(K^\circ, \mathscr{E}^\circ) \leq C \log(e + n/k) \sup_{k \leq m \leq n} \sqrt{m} e_m(K^\circ, \mathscr{E}^\circ)$$

$$\leq C \log(e + n/k) \sup_{k \leq m \leq n} \sqrt{m} \, P_{\alpha_k} \left(\frac{n}{m} \right)^{1/\alpha_k}.$$

Even with this improvement, note that this is where the current approach incurs some unnecessary logarithmic price with respect to the approach in the previous sections: instead of using (37) directly, one uses (38) which Pisier obtains from (37) by applying Carl's theorem, and then uses the converse to Carl's theorem (Lemma 21) to pass back to Gelfand number estimates.

Using $\alpha_k = 2 - \frac{1}{\log(e+n/k)}$, we deduce that:

$$c_k(K^\circ, \mathscr{E}^\circ) \leq C \sqrt{\frac{n}{k}} \log^{3/2}(e + n/k),$$

or in other words, the existence of $E \in G_{n,n-k}$ such that:

$$P_E(K) \supseteq \frac{1}{C \sqrt{\frac{n}{k}} \log^{3/2}(e + n/k)} P_E(\mathscr{E}).$$

The rest of the proof is identical to that of Theorem 8. For the ellipsoid $\mathscr{E}' := P_E(\mathscr{E})$ we may always find a linear subspace $F \subseteq E$ of codimension k in E so that:

$$P_F(\mathscr{E}') \supseteq \inf_{H \in G_k(E)} \{\mathrm{vrad}(P_H(\mathscr{E}'))\} B_F.$$

Estimating $\mathrm{vrad}(P_H(\mathscr{E}')) = \mathrm{vrad}(P_H(\mathscr{E}))$ by comparing to $\mathrm{vrad}(P_H(K))$ via the dual covering estimate on $e_k(K, \mathscr{E})$ (note that there is no need to use the duality of entropy theorem here), we obtain:

$$\mathrm{vrad}(P_H(\mathscr{E}')) \geq \frac{1}{2e_k(K,\mathscr{E})}\mathrm{vrad}(P_H(K)) \geq \frac{1}{2C\sqrt{\frac{n}{k}}\log^{1/2}(e+n/k)}\mathrm{vrad}(P_H(K)).$$

Combining all of the above, we deduce the existence of $F \in G_{n,n-2k}$ so that:

$$P_F(K) \supseteq \frac{1}{C'\frac{n}{k}\log^2(e+n/k)}\mathrm{vrad}(P_H(K))B_F.$$

This concludes the proof. $\square$

Having obtained a rather regular estimate on the Gelfand numbers, the next goal is to obtain an entropy estimate. To this end, one can use Carl's theorem (15) or (16), as we do in Sect. 4. The approach in [12] proceeds by employing an entropy extension theorem of Litvak et al. [19]. We remark that this too may be avoided, by employing the following elementary covering estimate (see e.g. [9, Chap. 9]):

Lemma 23. *Let K be a symmetric convex body in $\mathbb{R}^n$ and assume that $B_2^n \subseteq \rho K$ for some $\rho \geq 1$. Let W be a subspace of $\mathbb{R}^n$ with $\dim W = m$ and $P_{W^\perp}(K) \supseteq B_{W^\perp}$. Then, we have*

$$N(B_2^n, 4K) \leq (3\rho)^m.$$

Finally, having a covering estimate at hand, the estimate on $M(K)$ is obtained by Dudley's entropy bound (6). Plugging in the lower bounds on $v_k^-(Z_q(\mu))$ given in Sect. 6, the results of [12] are recovered and improved.

As the reader may wish to check, the improved approach of this section over the arguments of [12] yields estimates which are almost as good as the ones obtained in Sect. 6, and only lose by logarithmic terms.

Acknowledgements The first named author acknowledges support from the programme "ΑΡΙΣΤΕΙΑ II" of the General Secretariat for Research and Technology of Greece. The second named author is supported by ISF (grant no. 900/10), BSF (grant no. 2010288), Marie-Curie Actions (grant no. PCIG10-GA-2011-304066) and the E. and J. Bishop Research Fund.

References

1. S. Artstein, V.D. Milman, S.J. Szarek, Duality of metric entropy. Ann. Math. **159**, 1313–1328 (2004)
2. K.M. Ball, Isometric problems in ℓ_p and sections of convex sets. Ph.D. dissertation, Trinity College, Cambridge, 1986
3. K.M. Ball, Logarithmically concave functions and sections of convex sets in $\mathbb{R}^n$. Stud. Math. **88**, 69–84 (1988)

4. L. Berwald, Verallgemeinerung eines Mittelwertsatzes von J. Favard für positive konkave Funktionen. Acta Math. **79**, 17–37 (1947)
5. J. Bourgain, On high-dimensional maximal functions associated to convex bodies, Am. J. Math. **108**, 1467–1476 (1986)
6. J. Bourgain, On the distribution of polynomials on high dimensional convex sets, in *Geometric Aspects of Functional Analysis*. Lecture Notes in Mathematics, vol. 1469 (Springer, Berlin, 1991), pp. 127–137
7. J. Bourgain, B. Klartag, V.D. Milman, Symmetrization and isotropic constants of convex bodies, in *Geometric Aspects of Functional Analysis*. Lecture Notes in Mathematics, vol. 1850 (Springer, Berlin, 2002–2003), pp. 101–115
8. J. Bourgain, V.D. Milman, New volume ratio properties for convex symmetric bodies in $\mathbb{R}^n$. Invent. Math. **88**, 319–340 (1987)
9. S. Brazitikos, A. Giannopoulos, P. Valettas, B.-H. Vritsiou, *Geometry of Isotropic Convex Bodies*. Mathematical Surveys and Monographs, vol. 196 (American Mathematical Society, Providence, 2014)
10. B. Carl, Entropy numbers, *s*-numbers, and eigenvalue problems. J. Funct. Anal. **41**, 290–306 (1981)
11. T. Figiel, N. Tomczak-Jaegermann, Projections onto Hilbertian subspaces of Banach spaces. Israel J. Math. **33**, 155–171 (1979)
12. A. Giannopoulos, P. Stavrakakis, A. Tsolomitis, B.-H. Vritsiou, Geometry of the L_q-centroid bodies of an isotropic log-concave measure. Trans. Am. Math. Soc. (to appear)
13. B. Klartag, A geometric inequality and a low *M*-estimate. Proc. Am. Math. Soc. **132**, 2619–2628 (2004)
14. B. Klartag, On convex perturbations with a bounded isotropic constant. Geom. Funct. Anal. **16**, 1274–1290 (2006)
15. B. Klartag, Uniform almost sub-Gaussian estimates for linear functionals on convex sets. Algebra i Analiz **19**, 109–148 (2007)
16. B. Klartag, E. Milman, Centroid bodies and the logarithmic laplace transform – a unified approach. J. Funct. Anal. **262**, 10–34 (2012)
17. B. Klartag, V.D. Milman, Rapid Steiner symmetrization of most of a convex body and the slicing problem. Comb. Probab. Comput. **14**, 829–843 (2005)
18. D.R. Lewis, Ellipsoids defined by Banach ideal norms. Mathematika **26**, 18–29 (1979)
19. A. Litvak, V.D. Milman, A. Pajor, N. Tomczak-Jaegermann, Entropy extension. Funct. Anal. Appl. **40**, 298–303 (2006).
20. E. Lutwak, D. Yang, G. Zhang, L^p affine isoperimetric inequalities. J. Differ. Geom. **56**, 111–132 (2000)
21. E. Milman, On the mean width of isotropic convex bodies and their associated L_p-centroid bodies. Int. Math. Res. Not. doi: 10.1093/imrn/rnu040
22. V.D. Milman, Geometrical inequalities and mixed volumes in the local theory of Banach spaces. Astérisque **131**, 373–400 (1985)
23. V.D. Milman, A note on a low M^*-estimate, in *Geometry of Banach Spaces, Proceedings of a Conference Held in Strobl, Austria, 1989*, ed. by P.F. Müller, W. Schachermayer. LMS Lecture Note Series, vol. 158 (Cambridge University Press, Cambridge, 1990), pp. 219–229
24. V.D. Milman, A. Pajor, Isotropic position and interia ellipsoids and zonoids of the unit ball of a normed *n*-dimensional space, in *Geometric Aspects of Functional Analysis*. Lecture Notes in Mathematics, vol. 1376 (Springer, Berlin, 1987–1988), pp. 64–104
25. V.D. Milman, G. Pisier, Gaussian processes and mixed volumes. Ann. Probab. **15**, 292–304 (1987)
26. V.D. Milman, G. Schechtman, *Asymptotic Theory of Finite Dimensional Normed Spaces*. Lecture Notes in Mathematics, vol. 1200 (Springer, Berlin, 1986)
27. A. Pajor, N. Tomczak-Jaegermann, Subspaces of small codimension of finite-dimensional Banach spaces. Proc. Am. Math. Soc. **97**, 637–642 (1986)
28. G. Paouris, Concentration of mass in convex bodies. Geom. Funct. Anal. **16**, 1021–1049 (2006)

29. G. Pisier, Holomorphic semi-groups and the geometry of Banach spaces. Ann. Math. **115**, 375–392 (1982)
30. G. Pisier, A new approach to several results of V. Milman. J. Reine Angew. Math. **393**, 115–131 (1989)
31. G. Pisier, *The Volume of Convex Bodies and Banach Space Geometry*. Cambridge Tracts in Mathematics, vol. 94 (Cambridge University Press, Cambridge, 1989)
32. B.-H. Vritsiou, Further unifying two approaches to the hyperplane conjecture. Int. Math. Res. Not. doi: 10.1093/imrn/rns263

Remarks on the Central Limit Theorem for Non-convex Bodies

Uri Grupel

Abstract In this note, we study possible extensions of the Central Limit Theorem for non-convex bodies. First, we prove a Berry-Esseen type theorem for a certain class of unconditional bodies that are not necessarily convex. Then, we consider a widely-known class of non-convex bodies, the so-called p-convex bodies, and construct a counter-example for this class.

1 Introduction

Let $X_1, \ldots, X_n$ be random variables with $\mathbb{E}X_i = 0$ and $\mathbb{E}X_i X_j = \delta_{i,j}$ for $i, j = 1, 2, \ldots, n$. Let $\theta \in S^{n-1}$, where $S^{n-1} \subseteq \mathbb{R}^n$ is the unit sphere centered at 0, and let G be a standard Gaussian random variable, that is G has density function $1/\sqrt{2\pi}e^{-x^2/2}$. We denote $X = (X_1, \ldots, X_n)$. In this paper we examine different conditions on X under which $X \cdot \theta$ is close to G in distribution. The classical central limit theorem states that if $X_1, \ldots, X_n$ are independent then for most $\theta \in S^{n-1}$ the marginal $X \cdot \theta$ is close to G. It was conjectured by Anttila et al. [1] and by Brehm and Voigt [6] that if X is distributed uniformly in a convex body $K \subseteq \mathbb{R}^n$, then for most $\theta \in S^{n-1}$ the marginal $X \cdot \theta$ is close to G. This is known as the central limit theorem for convex sets and was first proved by Klartag [12].

In this note we examine extensions of the above theorem to non-convex settings. Our study was motivated by the following observation on the unit balls of l_p spaces for $0 < p < 1$:

We denote by $B_p^n = \{x \in \mathbb{R}^n; |x_1|^p + \cdots + |x_n|^p \leq 1\}$ the unit ball of the space l_p^n. For $X = (X_1, \ldots, X_n)$ that is distributed uniformly on $c_{p,n} B_p^n$, $p > 0$, $\theta \in S^{n-1}$, and G a standard Gaussian, one can show that

$$|\mathbb{P}(\theta \cdot X \leq t) - \mathbb{P}(G \leq t)| \leq C_p \sum_{k=1}^{n} |\theta_k|^3$$

U. Grupel (✉)
Tel Aviv University, Tel Aviv, Israel
e-mail: urigrupe@post.tau.ac.il

© Springer International Publishing Switzerland 2014
B. Klartag, E. Milman (eds.), *Geometric Aspects of Functional Analysis*,
Lecture Notes in Mathematics 2116, DOI 10.1007/978-3-319-09477-9__14

where $c_{p,n}$ is chosen such that $\mathbb{E}X_i = 0$ and $\mathbb{E}X_i X_j = \delta_{i,j}$ for $i, j = 1, 2, \ldots, n$, and $C_p > 0$ does not depend on n.

In order to formulate our results we use the following definitions: Let $X = (X_1, \ldots, X_n)$ be a random vector in $\mathbb{R}^n$. A random vector X is called isotropic if $\mathbb{E}X_i = 0$ and $\mathbb{E}X_i X_j = \delta_{i,j}$ for $i, j = 1, 2, \ldots, n$. A random vector X is called unconditional if the distribution of $(\varepsilon_1 X_1, \ldots, \varepsilon_n X_n)$ is the same as the distribution of X for any $\varepsilon_i = \pm 1$, $i = 1, \ldots, n$.

The first class of densities we define is based on Klartag's recent work [14] and includes the uniform distribution over B_p^n for $0 < p < 1$.

Theorem 1. *Let X be an unconditional, isotropic random vector with density $e^{-u(x)}$, where the function $u\left(x_1^\kappa, \ldots, x_n^\kappa\right)$ is convex in $\mathbb{R}_+^n = \left\{x \in \mathbb{R}^n; \ x_i \geq 0 \ \forall i \in \{1, 2, \ldots, n\}\right\}$ for $\kappa > 1$. Let G be a standard Gaussian random variable and $\theta \in S^{n-1}$. Then*

$$\left|\mathbb{P}\left(\theta \cdot X \geq t\right) - \mathbb{P}(G \geq t)\right| \leq C_\kappa \sum_{k=1}^n |\theta_k|^3,$$

where $C_\kappa > 0$ depends on κ only, and does not depend on n.

In order to see that Theorem 1 includes the uniform distribution over B_p^n for $0 < p < 1$ take

$$u(x) = \begin{cases} 0, & x_1^p + \cdots + x_n^p \leq 1 \\ \infty, & \text{otherwise} \end{cases},$$

and set $\kappa = 1/p$.

The error rate in Theorem 1 is the same as in the classical Central Limit Theorem. For example, by choosing $\theta = \left(1/\sqrt{n}, \ldots, 1/\sqrt{n}\right)$, we get an error rate of $O\left(1/\sqrt{n}\right)$.

The symmetry conditions in Theorem 1 are highly restrictive. Hence, we are led to study p-convex bodies, which satisfy fewer symmetry conditions and are shown to share some of the properties of convex bodies.

We say that $K \subset \mathbb{R}^n$ is *p-convex* with $0 < p < 1$ if $K = -K$ and for all $x, y \in K$ and $0 < \lambda < 1$, we have

$$\lambda^{1/p} x + (1 - \lambda)^{1/p} y \in K.$$

These bodies are related to unit balls of *p-norms* and were studied in relation to local theory of Banach spaces by Gordon and Lewis [10], Gordon and Kalton [9], Litvak et al. [16] and others (see [2, 7, 11, 15, 17]).

The following discussion explains why the class of p-convex bodies does not give the desired result.

Theorem 2. *Set $N = n + n^{5/2} \log^2 n$. There exists a random vector X distributed uniformly in a $1/2$-convex body $K \subseteq \mathbb{R}^N$, and a subspace E with $\dim(E) = n$,* *such that for any* $\theta \in S^{N-1} \cap E$, *the random variable* $\theta \cdot Proj_E X$ *is not close to a Gaussian random variable in any reasonable sense (Kolmogorov distance, Wasserstein distance and others).*

A similar construction can be made for any fixed parameter $0 < p < 1$. Since $\dim(E)$ tends to infinity with n, a similar theorem is not true in the convex case. Hence, the central limit theorem for convex sets cannot be extended for the p-convex case. Thus, we need to look for a new class of bodies (densities) that includes the l_p^n unit balls, with a weaker condition than the unconditional one.

Remark 1. In [15] Litvak constructed an example of a *p*-convex body for which the volume distribution is very different from the convex case. Litvak's work studies the large deviations regime for *p*-convex distributions, while our work is focused on the central limit theorem.

Throughout the text the letters c, C, c', C' will denote universal positive constants that do not depend on the dimension n. The value of the constant may change from one instance to another. We use $C_\alpha, C(\alpha)$ for constants that depend on a parameter α and nothing else. σ_{n-1} will denote the Haar probability measure on S^{n-1}. $f(n) = O(g(n))$ is the big O notation, i.e. there exists a constant $C > 0$ such that $|f(n)| \leq Cg(n), \; \forall n \in \mathbb{N}$.

2 A Class of Densities with Symmetries

In this section we use Klartag's recent work [14] in order to exhibit a family of functions, which includes the indicator functions of l_p^n unit balls, for $0 < p < 1$, having almost Gaussian marginals.

A special case of Theorem 1.1 in [14] gives us the following lemma.

Lemma 1. *Let* $\kappa > 1$ *and let* $\phi : \mathbb{R}^n \to \mathbb{R}$ *be an unconditional function such that* $e^{-\phi(x)}$ *is a probability density function and* $\phi(x_1^\kappa, \ldots, x_n^\kappa)$ *is convex on* $\mathbb{R}_+^n$. *Let* X *be a random vector with density* $e^{-\phi(x)}$. *Then*

$$Var|X|^2 \leq c_\kappa \sum_{j=1}^n \mathbb{E}|X_j|^4,$$

where c_κ *depends only on* κ.

Lemma 2. *Let* $\kappa \geq 1$ *and let* $\phi : \mathbb{R}^n \to \mathbb{R}$ *be an unconditional function such that* $e^{-\phi(x)}$ *is a probability density function and* $\phi(x_1^\kappa, \ldots, x_n^\kappa)$ *is convex on* $\mathbb{R}_+^n$. *Let* X *be a random vector with density* $e^{-\phi(x)}$. *Then for any* $p \geq 1$ *and* $i = 1, \ldots, n$,

$$\mathbb{E}|X_i|^p \leq c_{p,\kappa} \left(\mathbb{E}|X_i|^2\right)^{p/2}.$$

Proof. If $p \leq 2$ then, by Hölder's inequality, we have $c_{p,\kappa} = 1$. Assume that $p \geq 2$. Define $\pi : \mathbb{R}_+^n \to \mathbb{R}_+^n$ by $\pi(x) = (|x_1|^\kappa, \dots, |x_n|^\kappa)$. The Jacobian of π is $\prod_{j=1}^n \kappa |x_j|^{\kappa-1}$. Using the symmetry of ϕ we obtain

$$\int_{\mathbb{R}^n} |x_i|^p e^{-\phi(x)} dx = 2^n \int_{\mathbb{R}_+^n} |x_i|^p e^{-\phi(x)} dx$$

$$= 2^n \int_{\mathbb{R}_+^n} |x_i|^{p\kappa} \left(\prod_{j=1}^n \kappa |x_j|^{\kappa-1} \right) e^{-\phi(\pi(x))} dx$$

Now set $u(x) = \phi(\pi(x)) - (\kappa - 1) \sum_{j=1}^n \log |x_j|$. The function $e^{-u(x)}$ is log-concave on $\mathbb{R}_+^n$, with $\kappa^n \int_{\mathbb{R}_+^n} e^{-u(x)} = 1/2^n$, and

$$\int_{\mathbb{R}_+^n} |x_i|^p e^{-\phi(x)} dx = \kappa^n \int_{\mathbb{R}_+^n} |x_i|^{p\kappa} e^{-u(x)} dx.$$

By Borell's Lemma (see [4, 5, 18]) we obtain

$$(2\kappa)^n \int_{\mathbb{R}_+^n} |x_i|^{p\kappa} e^{-u(x)} dx \leq C_{\kappa,p} \left((2\kappa)^n \int_{\mathbb{R}_+^n} |x_i|^{2\kappa} e^{-u(x)} dx \right)^{p/2}$$

$$= C_{\kappa,p} \left(\int_{\mathbb{R}^n} |x_i|^2 e^{-\phi(x)} dx \right)^{p/2}$$

Lemma 3. *Let $\kappa > 1$ and let $\phi : \mathbb{R}^n \to \mathbb{R}$ be an unconditional function such that $e^{-\phi(x)}$ is an isotropic probability density and $\phi(x_1^\kappa, \dots, x_n^\kappa)$ is convex on $\mathbb{R}_+^n$. Let X be a random vector with density $e^{-\phi(x)}$. Then, for any $a \in \mathbb{R}^n$*

$$\mathrm{Var}(a_1^2 X_1^2 + \cdots + a_n^2 X_n^2) \leq C_\kappa \sum_{j=1}^n |a_j|^4.$$

Proof. By applying a linear transformation, Lemma 1 gives

$$\mathrm{Var}(a_1^2 X_1^2 + \cdots + a_n^2 X_n^2) \leq C_\kappa' \sum_{j=1}^n \mathbb{E} a_j^4 |X_j|^4.$$

By Lemma 2, we obtain

$$\text{Var}(a_1^2 X_1^2 + \cdots + a_n^2 X_n^2) \le C_\kappa' \sum_{j=1}^n \mathbb{E} a_j^4 |X_j|^4 \le C_\kappa \sum_{j=1}^n a_j^4 \left(\mathbb{E}|X_j|^2\right)^2 = C_\kappa \sum_{j=1}^n |a_j|^4.$$

We are now ready to prove Theorem 1.

Proof. Since X is unconditional,

$$\mathbb{P}\left(\theta \cdot X \ge t\right) = \mathbb{P}\left(\sum_{k=1}^n \theta_k X_k \varepsilon_k \ge t\right),$$

where $\varepsilon_1, \ldots, \varepsilon_n$ are i.i.d. random variables distributed uniformly on $\{\pm 1\}$ that are independent of X. By the triangle inequality,

$$\left| \mathbb{P}\left(\sum_{k=1}^n \theta_k X_k \varepsilon_k \ge t\right) - \mathbb{P}(G \ge t)\right| \le \mathbb{E}_X \left| \mathbb{P}(G \ge t) - \mathbb{P}_G\left(G \ge \frac{t}{\sqrt{\sum_{k=1}^n \theta_k^2 X_k^2}}\right)\right|$$

$$+\mathbb{E}_X \left| \mathbb{P}_\varepsilon\left(\sum_{k=1}^n \varepsilon_k \theta_k X_k \ge t\right) - \mathbb{P}_G\left(G \ge \frac{t}{\sqrt{\sum_{k=1}^n \theta_k^2 X_k^2}}\right)\right|.$$

We estimate each term separately. Denote $Y_n = \sum_{k=1}^n \theta_k^2 X_k^2$. By the Berry-Esseen Theorem (see [8]),

$$\mathbb{E}_X \left| \mathbb{P}_\varepsilon\left(\sum_{k=1}^n \varepsilon_k \theta_k X_k \ge t\right) - \mathbb{P}_G\left(G \ge \frac{t}{\sqrt{Y_n}}\right)\right|$$

$$\le C \left(\mathbb{E}_X \sum_{k=1}^n \frac{|\theta_k|^3 |X_k|^3}{(Y_n)^{3/2}} 1_{[1/2,\infty)}(Y_n) + 2\mathbb{P}\left(Y_n < \frac{1}{2}\right)\right)$$

$$\le C \left(10 \sum_{k=1}^n \mathbb{E}_X |\theta_k|^3 |X_k|^3 + 2\mathbb{P}\left(Y_n < \frac{1}{2}\right)\right) \le C_\kappa \sum_{k=1}^n |\theta_k|^3 + C\mathbb{P}\left(Y_n < \frac{1}{2}\right)$$

Here we used Lemma 2 to estimate $\mathbb{E}|X_k|^3$. Note that

$$\mathbb{E}_X Y_n = \mathbb{E}_X \sum_{j=1}^n \theta_j^2 X_j^2 = \sum_{j=1}^n \theta_j^2 \mathbb{E}_X X_j^2 = \sum_{j=1}^n \theta_j^2 = 1,$$

so by Chebyshev's inequality and Lemma 3

$$\mathbb{P}\left(|Y_n - 1| \ge \frac{1}{2}\right) \le \frac{\text{Var}(Y_n)}{1/4} \le 4C_\kappa \sum_{j=1}^n |\theta_j|^4 \tag{1}$$

Hence, since $|\theta_i| \leq 1$ for all $i = 1, \ldots, n$,

$$\mathbb{E}_X \left| \mathbb{P}_\varepsilon \left(\sum_{k=1}^n \varepsilon_k \theta_k X_k \geq t \right) - \mathbb{P} \left(G \geq \frac{t}{\sqrt{Y_n}} \right) \right| \leq C_\kappa \sum_{k=1}^n |\theta_k|^3.$$

Now, in order to estimate $\mathbb{E}_X \left| \mathbb{P}(G \geq t) - \mathbb{P} \left(G \geq \frac{t}{\sqrt{Y_n}} \right) \right|$ we use (1) and Klartag's argument in [13] (Sect. 6, Lemma 7) and conclude that it is enough to show that

$$\mathbb{E} \left((Y_n - 1)^2 \,\Big|\, Y_n \geq \frac{1}{2} \right) \leq C \left(\sum_{j=1}^n |\theta_j|^3 \right)$$

By Lemma 3 we get

$$\mathbb{E} \, (Y_n - 1)^2 = \text{Var} \, (Y_n) \leq C_\kappa \sum_{j=1}^n |\theta_j|^4$$

Hence,

$$\mathbb{E} \left((Y_n - 1)^2 \,\Big|\, Y_n \geq \frac{1}{2} \right) \leq \mathbb{E} \, (Y_n - 1)^2 \, \mathbb{P} \left(Y_n \geq \frac{1}{2} \right)^{-1}$$

$$\leq C_\kappa \left(\sum_{j=1}^n |\theta_j|^4 \right) \mathbb{P} \left(Y_n \geq \frac{1}{2} \right)^{-1}$$

From inequality (1) it follows that

$$\left(\mathbb{P} \left(Y_n \geq \frac{1}{2} \right) \right)^{-1} = \mathbb{P} \left(\sum_{j=1}^n \theta_j^2 X_j^2 \geq \frac{1}{2} \right)^{-1} \leq \frac{1}{1 - C_\kappa \sum_{j=1}^n |\theta_j|^4}.$$

We may assume that $\sum_{j=1}^n |\theta_j|^4$ is bounded by some small positive constant depending on κ, since otherwise the result is trivial, and obtain

$$\frac{1}{1 - C_\kappa \sum_{j=1}^n |\theta_j|^4} \leq 1 + C_\kappa \sum_{j=1}^n |\theta_j|^4$$

which completes our proof.

3 The p-Convex Case

In this section we construct a random vector X, distributed uniformly in a $1/2$-convex body K, such that for a large subspace $E \subseteq \mathbb{R}^n$ the random vector $\mathrm{Proj}_E X$ has no single approximately Gaussian marginal. We define a function $f : \mathbb{R}_+ \to \mathbb{R}_+$ such that the radial density $r^{n-1}e^{-f(r)}$ is spread across an interval of length proportional to $\sqrt{n}$; that is, we want $r^{n-1}e^{-f(r)}$ to be constant (or close to constant) on such an interval. Such densities have marginals that are far from Gaussian. We use the density function introduced above and an approximation argument to construct the desired body K.

In order to construct a p-convex body from a function f, we restrict ourselves to *p-convex functions*.

Definition 1. A function $f : \mathbb{R}^n \to \mathbb{R} \cup \{\infty\}$ is called *p-convex* if for any $x, y \in \mathbb{R}^n$ and $t \in [0, 1]$,

$$f\left(t^{1/p}x + (1-t)^{1/p}y\right) \le tf(x) + (1-t)f(y). \tag{2}$$

The following proposition allows us to construct a p-convex body with $0 < p < 1$ from a p-convex function.

Proposition 1. *For $\psi : \mathbb{R}^n \to \mathbb{R}_+$ p-convex function with $0 < p < 1$ and fixed $N > 0$, define $f_N(x) = (1 - \psi(x)/N)_+^N$. Then the set*

$$K_N(\psi) = \left\{(x, y); \; x \in \mathbb{R}^n, \; y \in \mathbb{R}^N, \; |y| < f_N^{1/N}(x)\right\}$$

is p-convex.

Proof. Let $(x_1, y_1), (x_2, y_2) \in K_N(\psi)$. Since $(x_i, y_i) \in K_N(\psi)$ we have $f_N(x_i) > 0$. Therefore,

$$f_N^{1/N}(x_i) = 1 - \frac{\psi(x_i)}{N}.$$

Let $0 \le t \le 1$ we get

$$f_N^{1/N}(t^{1/p}x_1 + (1-t)^{1/p}x_2) \ge 1 - \frac{1}{N}\psi(t^{1/p}x_1 + (1-t)^{1/p}x_2)$$

$$\ge 1 - \frac{1}{N}(t\psi(x_1) + (1-t)\psi(x_2))$$

$$= tf_N^{1/N}(x_1) + (1-t)f_N^{1/N}(x_2) > t|y_1| + (1-t)|y_2|$$

$$\ge |t^{1/p}y_1| + |(1-t)^{1/p}y_2| \ge |t^{1/p}y_1 + (1-t)^{1/p}y_2|.$$

Hence, $t^{1/p}(x_1, y_1) + (1-t)^{1/p}(x_2, y_2) \in K_N(\psi)$, as needed.

Proposition 2. *There exists a universal constant $C > 0$ such that, for $a \geq C$ the function*

$$f(x) = \begin{cases} \log a, & \text{if } 0 \leq x \leq a \\ \log x, & \text{if } a \leq x \leq 2a \\ \sqrt{x} - \sqrt{2a} + \log 2a, & \text{if } 2a \leq x \end{cases}$$

is $1/2$-convex.

Proof. We begin by verifying that the function f is $1/2$-convex for each interval $[0, a]$, $[a, 2a]$, $[2a, \infty)$. Then we need to check that condition (2) holds when x and y are from different intervals. By symmetry, we may assume that $x < y$. The cases $x, y \in [0, a]$ and $x, y \in [2a, \infty)$ are straightforward. In order for condition (2) to hold for the function $\log x$ on an interval $[a, b]$ we must show that for any $x, y \in [a, b]$

$$\log((1 - t)^2 x + t^2 y) \leq (1 - t) \log(x) + t \log(y) = \log \left(x^{1-t} y^t \right). \tag{3}$$

This is equivalent to

$$(1 - t)^2 x + t^2 y - x^{1-t} y^t \leq 0.$$

Setting here $y = cx$, we obtain

$$(1 - t)^2 + t^2 c - c^t \leq 0.$$

This inequality holds for every $1 \leq c \leq 4$ and $0 \leq t \leq 1$. To see that note that $g(t, c) = (1-t)^2 + t^2 c - c^t$ is a convex function in c (as a sum of convex functions). Hence, it is enough to verify that $g(t, 1) \leq 0$ and $g(t, 4) \leq 0$ for any $0 \leq t \leq 1$. Indeed,

$$g(t, 1) = (1 - t)^2 + t^2 - 1 = 2t(t - 1) \leq 0$$

and

$$g(t, 4) = (1 - t)^2 + t^2 4 - 4^t \implies \frac{\partial^2 g(t, 4)}{\partial t^2} = 2 + 8 - (\log 4)^2 4^t \geq 2.$$

Hence, $g(t, 4)$ is convex in t. Since $g(0, 4) = g(1, 4) = 0$, we obtain, $g(t, 4) \leq 0$ for all $0 \leq t \leq 1$.

Consequently (3) holds for any interval of the form $[a, b] \subseteq [a, 4a]$.

Next, we verify condition (2) for f when $x \in [a, 2a]$, $y \in [2a, \infty)$, and $t^2 x + (1 - t)^2 y \in [a, 2a]$. We consider two cases

1. $y \in [2a, 4a]$. By inequality (3),

$$f(t^2x + (1-t)^2y) = \log(t^2x + (1-t)^2y) \leq t\log(x) + (1-t)\log(y)$$
$$\leq \log(x) + (1-t)(\log(2a) + \sqrt{y} - \sqrt{2a})$$
$$= tf(x) + (1-t)f(y).$$

The second inequality holds thanks to the elementary inequality $\log(y) - \log(2a) \leq \sqrt{y} - \sqrt{2a}$. Since for $y = 2a$ we have equality, and $(\sqrt{y})' = 1/\sqrt{4y} \geq 1/y = (\log(y))'$ for $y \geq 4$, the inequality holds if $2a \geq 4$.

2. $y \geq 4a$. Define

$$g(t) = \log(t^2x + (1-t)^2y) - t\log(x) - (1-t)(\sqrt{y} - \sqrt{2a} + \log(2a)).$$

We need to show that $g(t) \leq 0$ for all $t \in [0, 1]$. Since $g(1) = 0$, it is enough to show that $g'(t) \geq 0$ for all $0 \leq t \leq 1$. We have,

$$g'(t) = \frac{2tx - 2(1-t)y}{t^2x + (1-t)^2y} - \log(x) + \sqrt{y} - \sqrt{2a} + \log(2a)$$
$$\geq \frac{2tx - 2(1-t)y}{t^2x + (1-t)^2y} + \left(1 - \frac{1}{\sqrt{2}}\right)\sqrt{y}$$

Hence, if $2tx - 2(1-t)y + \left(1 - 1/\sqrt{2}\right)\sqrt{y}(t^2x + (1-t)^2y) \geq 0$, then $g'(t) \geq 0$. Recalling that $t^2x + (1-t)^2y \geq a$, it suffices to prove that

$$2tx - 2(1-t)y + \left(1 - \frac{1}{\sqrt{2}}\right)\sqrt{y}a \geq 0.$$

Using the fact that $(1-t)^2y \leq t^2x + (1-t)^2y \leq 2a$, we obtain $(1-t)\sqrt{y} \leq \sqrt{2a}$. Hence,

$$2tx - 2(1-t)y + \left(1 - \frac{1}{\sqrt{2}}\right)a\sqrt{y} \geq 2ta - 2\sqrt{2a}\sqrt{y} + \left(1 - \frac{1}{\sqrt{2}}\right)a\sqrt{y}$$
$$\geq \sqrt{y}\left(\left(1 - \frac{1}{\sqrt{2}}\right)a - 2\sqrt{2a}\right).$$

This gives the condition

$$\left(1 - \frac{1}{\sqrt{2}}\right)a - 2\sqrt{2a} \geq 0,$$

Which is satisfied for $a \geq 100$.

When $x \in [a, 2a]$ and $y \in [2a, \infty)$ and $t^2 x + (1-t)^2 y \geq 2a$, we have

$$f(t^2 x + (1-t)^2 y) = \sqrt{t^2 x + (1-t)^2 y} - \sqrt{2a} + \log 2a$$
$$\leq t\sqrt{x} + (1-t)\sqrt{y} - \sqrt{2a} + \log 2a$$

and

$$tf(x) + (1-t)f(y) = t \log x + (1-t)(\sqrt{y} - \sqrt{2a} + \log 2a).$$

Hence, (2) holds thanks to the elementary inequality $\log 2a - \log x + \sqrt{x} - \sqrt{2a} \leq 0$, which holds for $a \geq 4$.

If $x \in [0, a]$, then $f(x) = f(a)$ and $f(t^2 x + (1-t)^2 y) \leq f(t^2 a + (1-t)^2 y)$. Hence, for $x \in [0, a]$ and $y \in [a, \infty)$ we have

$$f(t^2 x + (1-t)^2 y) \leq f(t^2 a + (1-t)^2 y) \leq tf(a) + (1-t)f(y) = tf(x) + (1-t)f(y).$$

Proposition 3. *Let $f : \mathbb{R}_+ \to \mathbb{R}_+$ be a p-convex function with parameter $0 < p < 1$. Then $x \mapsto f(|x|)$ is a p-convex function on $\mathbb{R}^n$.*

Proof. First, we prove that f is non-decreasing. Let $0 < x < y$. There exists some $k \geq 1$ such that $2^{-k(1/p-1)} y \leq x$. We proceed by induction on k. For $k = 1$, note that $h(t) = t^{1/p} y + (1-t)^{1/p} y$ is continuous, $h(0) = y$, and $h(1/2) = 2^{-(1/p-1)} y$. Hence, there exists some $0 \leq t_0 \leq 1$ for which $h(t_0) = x$, and so

$$f(x) = f(t_0^{1/p} y + (1-t_0)^{1/p} y) \leq t_0 f(y) + (1-t_0)f(y) = f(y)$$

For $k \geq 2$, $f(2^{-(k-1)(1/p-1)} y) \leq f(y)$ by the induction hypothesis, and by the same argument as above

$$f(x) \leq f(2^{-(k-1)(1/p-1)} y) \leq f(y).$$

We thus showed that f is monotone non-decreasing. Now, by the triangle inequality, for any $x, y \in \mathbb{R}^n$ and $0 < t < 1$ we have

$$f(|t^{1/p} x + (1-t)^{1/p} y|) \leq f(t^{1/p}|x| + (1-t)^{1/p}|y|) \leq tf(|x|) + (1-t)f(|y|)$$

Using the function from Proposition 2, we are ready to construct the $1/2$-convex body K and prove Theorem 2.

Definition 2. A sequence of probability measures $\{\mu_n\}$ on $\mathbb{R}^n$ is called essentially isotropic if $\int x d\mu_n(x) = 0$ and $\int x_i x_j d\mu_n(x) = (1 + \varepsilon_n)\delta_{ij}$ for all $i, j = 1, \ldots, n$, when $\varepsilon_n \underset{n \to \infty}{\longrightarrow} 0$.

Proposition 4. *The probability measure $d\mu = C_n e^{-(n-1)f(|x|)} dx$, where f is defined as in Proposition 2, with $a = \sqrt{3n/7}$, is essentially isotropic. That is,*

$$\int x_i x_j \, d\mu(x) = (1 + \varepsilon_n) \delta_{ij}$$

for all $i, j = 1, 2, \ldots, n$, when $|\varepsilon_n| \leq C/n$.

Proof. The density μ is spherically symmetric, hence

$$\int_{\mathbb{R}^n} x_i x_j \, d\mu(x) = 0,$$

for $i \neq j$, and

$$\int_{\mathbb{R}^n} x_i^2 \, d\mu(x) = \frac{1}{n} \int_{\mathbb{R}^n} |x|^2 \, d\mu(x),$$

for $i = 1, 2, \ldots, n$. Integration in spherical coordinates and using Laplace asymptotic method yields

$$\int |x|^2 \, d\mu(x) = \frac{\left[\begin{array}{c} \displaystyle\int_0^{\sqrt{3n/7}} \frac{r^{n+1}}{\left(\sqrt{3n/7}\right)^{n-1}} \, dr + \int_{\sqrt{3n/7}}^{2\sqrt{3n/7}} r^2 \, dr \\[2em] + \left(\frac{e^{\sqrt{2\sqrt{3n/7}}}}{2\sqrt{3n/7}}\right)^{n-1} \displaystyle\int_{2\sqrt{3n/7}}^{\infty} r^{n+1} e^{-(n-1)\sqrt{r}} \, dr \end{array} \right]}{\left[\begin{array}{c} \displaystyle\int_0^{\sqrt{3n/7}} \left(\frac{r}{\sqrt{3n/7}}\right)^{n-1} dr + \int_{\sqrt{3n/7}}^{2\sqrt{3n/7}} dr \\[2em] + \left(\frac{e^{\sqrt{2\sqrt{3n/7}}}}{2\sqrt{3n/7}}\right)^{n-1} \displaystyle\int_{2\sqrt{3n/7}}^{\infty} r^{n-1} e^{-(n-1)\sqrt{r}} \, dr \end{array} \right]}$$

$$= \frac{\sqrt{3/7} \, n^{3/2} + O\left(\sqrt{n}\right)}{\sqrt{\frac{3}{7}n} + O\left(1/\sqrt{n}\right)} = n + O(1).$$

Proposition 5. *Let X be a random vector in $\mathbb{R}^n$ distributed according to μ from Proposition 4. Then,*

$$\mathbb{P}\left(\sqrt{\frac{3}{7}n} \leq |X| \leq 2\sqrt{\frac{3}{7}n} \right) \geq 1 - \frac{C}{n}.$$

Proof. By the same arguments as in Proposition 4

$$\mathbb{P}\left(\sqrt{3n/7} \leq |X| \leq 2\sqrt{3n/7} \right) = \frac{\displaystyle\int_{\sqrt{3n/7}}^{2\sqrt{3n/7}} dr}{\sqrt{3n/7} + O\left(1/\sqrt{n}\right)} = 1 + O\left(1/n\right).$$

Proposition 6. *Let X be a random vector in $\mathbb{R}^n$ distributed according to μ from Proposition 4, and let $\widetilde{X}$ be a random variable distributed according to $d\widetilde{\mu} = \widetilde{C_n} \left(1 - (n-1)f(|x|)/N\right)_+^N$. Then for $N \geq n^{5/2} \log^2 n$, $\widetilde{X}$ is essentially isotropic, namely*

$$\int x_i x_j \, d\widetilde{\mu}(x) = (1 + \varepsilon'_n)\delta_{ij}$$

for all $i, j = 1, 2, \ldots, n$, when $|\varepsilon'_n| \leq C/\sqrt{n}$. Also

$$\forall t, \quad \left| \mathbb{P}(|X| \leq t) - \mathbb{P}(|\widetilde{X}| \leq t) \right| \leq \frac{C}{\sqrt{n}}.$$

Proof. The random vector $\widetilde{X}$ is spherically symmetric. Hence

$$\int_{\mathbb{R}^n} x_i x_j \, d\widetilde{\mu}(x) = 0,$$

for $i \neq j$, and

$$\int_{\mathbb{R}^n} x_i^2 \, d\widetilde{\mu}(x) = \frac{1}{n} \int_{\mathbb{R}^n} |x|^2 \, d\widetilde{\mu}(x),$$

for $i = 1, 2, \ldots, n$. Since both densities are spherically symmetric, we need to estimate the one-dimensional integrals

$$I_k = \int_0^\infty r^k \left(e^{-(n-1)f(r)} - \left(1 - \frac{(n-1)f(r)}{N}\right)_+^N \right) dr$$

for $k = n - 1, n + 1$. Define α by the equation

$$\left(\sqrt{\alpha} - \sqrt{2\sqrt{\frac{3}{7}n}} + \log\left(2\sqrt{\frac{3}{7}n}\right) \right)(n-1) = \frac{N}{2}.$$

That is, for any $r \leq \alpha$ we have $(n-1)f(r)/N \leq 1/2$. By Taylor's Theorem, for any $r \leq \alpha$,

$$\left| \log\left(1 - \frac{(n-1)f(r)}{N}\right)_+^N - (-(n-1)f(r)) \right| \leq C \frac{(n-1)^2}{N} f^2(r).$$

Hence, for any $r \leq \alpha$

$$\left| e^{-(n-1)f(r)} - \left(1 - \frac{(n-1)}{N} f(r)\right)_+^N \right|$$

$$= e^{-(n-1)f(r)} \left| 1 - \exp\left((n-1)f(r) - \log\left(1 - \frac{(n-1)}{N} f(r)\right)_+^N\right) \right|$$

$$\leq C \frac{n^2}{N} e^{-(n-1)f(r)} f^2(r).$$

Note that

$$\left| \int_\alpha^\infty \left(e^{-(n-1)f(r)} - \left(1 - \frac{(n-1)}{N} f(r)\right)_+^N \right) dr \right| \leq C \int_\alpha^\infty e^{-(n-1)f(r)} dr \leq C e^{-n}.$$

Combining the above inequalities, we obtain

$$|I_k| \leq C_1 \frac{n^2}{N} \int_0^\alpha r^k e^{-(n-1)f(r)} f^2(r) dr + C_2 e^{-n} \leq C \frac{n^2}{N} \int_0^\infty r^k e^{-(n-1)f(r)} f^2(r) dr.$$

Hence,

$$|I_{n-1}| \leq C \frac{n^2}{N} \left(\sqrt{n} \log^2 n + O\left(\frac{\log^2 n}{\sqrt{n}}\right) \right) \leq C_1,$$

$$|I_{n+1}| \leq C \frac{n^2}{N} \left(n^{\frac{3}{2}} \log^2 n + O\left(\log^2 n \sqrt{n}\right) \right) \leq C_2 n.$$

By the estimation on I_{n-1}, and the calculations in Proposition 4 we obtain

$$\left| \int_0^\infty \left(1 - \frac{(n-1)}{N} f(r)\right)_+^N dr - \sqrt{\frac{3}{7}n} \right|$$

$$\leq \left| \int_0^\infty \left(1 - \frac{(n-1)}{N} f(r)\right)_+^N dr - \int_0^\infty e^{-(n-1)f(r)} dr \right| + O\left(\frac{1}{\sqrt{n}}\right)$$

$$= |I_{n-1}| + O\left(\frac{1}{\sqrt{n}}\right) \leq C_1.$$

Hence,

- $$\left(\int_0^\infty \left(1 - \frac{(n-1)}{N} f(r)\right)_+^N dr \right)^{-1} = \sqrt{\frac{1}{\frac{2}{7}n}} \left(1 + O\left(\frac{1}{\sqrt{n}}\right)\right);$$
- $\forall t, \quad \left| \mathbb{P}(|X| \leq t) - \mathbb{P}(|\tilde{X}| \leq t) \right| \leq \frac{C}{\sqrt{n}}.$

By the estimation of I_{n+1} we obtain,

$$\left| \mathbb{E} X_i^2 - \mathbb{E} \tilde{X}_i^2 \right| = \frac{1}{n} \left| \mathbb{E} |X|^2 - \mathbb{E} |\tilde{X}|^2 \right| \leq C \frac{1}{\sqrt{n}} \frac{1}{n} |I_{n+1}| \leq \frac{C}{\sqrt{n}}.$$

Remark 2. It is possible to take $a \approx \sqrt{3n/7}$ in the definition of f, such that $\tilde{X}$ is isotropic.

We use the following estimation in our proof of Theorem 2.

Proposition 7. *Let $Z_1, .., Z_n$ be independent standard Gaussian random variables, and let $0 < \delta < 1/2$. Then,*

$$\mathbb{P} \left(\left| \sqrt{Z_1^2 + \ldots + Z_n^2} - \sqrt{n} \right| \leq n^\delta \right) \geq 1 - C e^{-cn^{2\delta}},$$

where $c, C > 0$ are constants.

Proof. Note that

$$\left| Z_1^2 + \cdots + Z_n^2 - n \right| = \left| \sqrt{Z_1^2 + \cdots + Z_n^2} - \sqrt{n} \right| \left| \sqrt{Z_1^2 + \cdots + Z_n^2} + \sqrt{n} \right|$$

$$\geq \left| \sqrt{Z_1^2 + \cdots + Z_n^2} - \sqrt{n} \right| \sqrt{n}.$$

Therefore it is enough to show that

$$\mathbb{P} \left(\left| Z_1^2 + \cdots + Z_n^2 - n \right| \leq n^{\delta + \frac{1}{2}} \right) \geq 1 - C e^{-cn^{2\delta}}.$$

Note that for all $m \geq 1$ and for all $i = 1, \ldots, n$, we have

$$\mathbb{E} |Z_i^2 - 1|^m \leq \sum_{k=1}^m \binom{m}{k} \mathbb{E} Z_i^{2k} \leq 2^m (2m)!! \leq 4^m m!$$

where $(2m)!! = 1 \cdot 3 \cdot 5 \cdots (2m - 1)$. Hence, by Bernstein's inequality [3] we obtain

$$\mathbb{P} \left(\left| (Z_1^2 - 1) + \cdots + (Z_n^2 - 1) \right| > n^{\frac{1}{2} + \delta} \right) \leq C e^{-cn^{2\delta}}.$$

We are now ready to prove Theorem 2.

Proof. By Proposition 2, the function $(n - 1) f(|x|)$ is $1/2$-convex. Proposition 1 with $N = n^{5/2} \log^2 n$ yields a $1/2$-convex body K. Let X be a random vector distributed uniformly in K. By the definition of K the marginal of X with respect to the first n coordinates has density proportional to $(1 - (n - 1) f(|x|)/N)_+^N$. Denote this subspace by E. By Proposition 6, $\text{Proj}_E X$ is essentially isotropic. Let G be a standard Gaussian random variable. In order to show that $Y = \mathrm{Proj}_E X$ has no approximately Gaussian marginals, we examine $\mathbb{P}(|\theta_0 \cdot Y| \le t)$, for any $\theta_0 \in S^{n-1}$. Using the symmetry of Y and the rotation invariance of σ_{n-1}, we obtain,

$$\mathbb{P}(|\theta_0 \cdot Y| \le t) = \mathbb{E}1_{[0,t]}(|\theta_0 \cdot Y|) = \int_{S^{n-1}} \mathbb{E}1_{[0,t]}(|\theta \cdot Y|)d\sigma_{n-1}(\theta)$$

$$= \mathbb{E}\int_{S^{n-1}} 1_{[0,t]}(\theta_1|Y|)d\sigma_{n-1}(\theta),$$

where $\theta = (\theta_1, \ldots, \theta_n)$. Let $Z = (Z_1, \ldots, Z_n)$, where Z_i are independent standard Gaussian random variable. Since Z is invariant under rotations, $Z/|Z|$ is distributed uniformly on S^{n-1}. Hence,

$$\mathbb{P}(|\theta_0 \cdot Y| \le t) = \mathbb{P}\left(|Z_1||Y| \le t\sqrt{Z_1^2 + \cdots + Z_n^2}\right).$$

By Proposition 7, $\mathbb{P}(|\sqrt{Z_1^2 + \cdots + Z_n^2} - \sqrt{n}| \le n^{1/100}) \ge 1 - Ce^{-cn^{1/50}}$. Hence,

$$\mathbb{P}(|\theta_0 \cdot Y| \le t) = \mathbb{P}\left(|Z_1||Y| \le t\sqrt{n}\left(1 + O\left(n^{-1/2+1/100}\right)\right)\right) + O\left(e^{-cn^{1/50}}\right).$$
$$(4)$$

By Propositions 6 and 5, there exists a random vector Y' such that

$$\forall t \;\; \left|\mathbb{P}\left(|Y'| \le t\right) - \mathbb{P}(|Y| \le t)\right| \le \frac{C}{\sqrt{n}}, \;\; \mathbb{P}\left(\sqrt{\frac{3}{7}n} \le |Y'| \le 2\sqrt{\frac{3}{7}n}\right) \ge 1 - \frac{C}{n},$$

and $|Y'|$ has constant density function on $\left[\sqrt{3n/7}, 2\sqrt{3n/7}\right]$. By the triangle inequality, for W distributed uniformly on $\left[\sqrt{3/7}, 2\sqrt{3/7}\right]$ and any $\sqrt{3/7} \le \alpha \le \beta \le 2\sqrt{3/7}$ we have

$$|\mathbb{P}(\sqrt{n}\alpha \le |Y| \le \sqrt{n}\beta) - \mathbb{P}(\alpha \le W \le \beta)| \le \frac{C}{\sqrt{n}}.$$

Combining with (4),

$$\mathbb{P}(|\theta_0 \cdot Y| \le t) = \mathbb{P}\left(|G|W \le t(1 + O(n^{-1/2+1/100}))\right) + O\left(\frac{1}{\sqrt{n}}\right).$$

We conclude that $|Y \cdot \theta_0|$ is very close to a distribution which is the product of a Gaussian with a uniform random variable, and the latter distribution is far from Gaussian.

Acknowledgements This paper is part of the authors M.Sc. thesis written under the supervision of Professor Bo'az Klartag whose guidance, support and patience were invaluable. In addition, I would like to thank Andrei Iacob for his helpful editorial comments. Supported by the European Research Council (ERC).

References

1. M. Anttila, K. Ball, I. Perissinaki, The central limit problem for convex bodies. Trans. Am. Math. Soc. **355**, 4723–4735 (2003)
2. J. Bastero, J.Bernues, A.Pena, An extension of Milman's reverse Brunn-Minkowski inequality. Geom. Funct. Anal. **5**(3), 572–581 (1995)
3. G. Bennet, Probability inequalities for the sum of independent random variables. J. Am. Stat. Assoc. **57**, 33–45 (1962)
4. L. Berwald, Verallgemeinerung eines Mittelwertsatzes von J. Favard für positive konkave Funktionen. Acta Math. **79**, 17–37 (1947)
5. C. Borell, Convex measures on locally convex spaces. Arkiv för Matematik **12**(1–2), 239–252 (1974)
6. U. Brehm, J. Voigt, Asymptotics of cross sections for convex bodies. Beiträge Algebra Geom. **41**, 437–454 (2000)
7. S.J. Dilworth, The dimension of Euclidean subspaces of quasinormed spaces. Math. Proc. Camb. Philos. Soc. **97**(2), 311–320 (1985)
8. W. Feller, *An Introduction to Probability Theory and its Applications*, vol. II, Sect. XVI.5 (Wiley, New York, 1971)
9. Y. Gordon, N.J. Kalton, Local structure theory for quasi-normed spaces. Bull. Sci. Math. **118**, 441–453 (1994)
10. Y. Gordon, D.R. Lewis, Dvoretzky's theorem for quasi-normed space. Illinois J. Math. **35**(2), 250–259 (1991)
11. N.J. Kalton, Convexity, type and the three space problem. Stud. Math. **69**, 247–287 (1981)
12. B. Klartag, A central limit theorem for convex sets. Invent. Math. **168**, 91–131 (2007)
13. B. Klartag, A Berry-Esseen type inequality for convex bodies with an unconditional basis. Probab. Theory Relat. Fields **45**(1), 1–33 (2009)
14. B. Klartag, Poincaré inequalities and moment maps. Ann. Fac. Sci. Toulouse Math. **22**(1), 1–41 (2013)
15. A.E. Litvak, Kahane-Khinchin's inequality for the quasi-norms. Can. Math. Bull. **43**, 368–379 (2000)
16. A.E. Litvak, V.D. Milman, N. Tomczak-Jaegermann, Isomorphic random subspaces and quotients of convex and quasi-convex bodies, in *GAFA*. Lecture Notes in Mathematics, vol. 1850 (Springer, Berlin, 2004), pp. 159–178
17. V. Milman, Isomorphic Euclidean regularization of quasi-norms in Rn. C. R. Acad. Sci. Paris **321**(7), 879–884 (1995)
18. V.D. Milman, A. Pajor, Isotropic position and inertia ellipsoids and zonoids of the unit ball of a normed n-dimensional space, in *Geometric Aspects of Functional Analysis - Israel Seminar*. Lecture Notes in Mathematics, vol. 1376 (Springer, Berlin, 1989), pp. 64–104

Reflectionless Measures and the Mattila-Melnikov-Verdera Uniform Rectifiability Theorem

Benjamin Jaye and Fedor Nazarov

Abstract The aim of these notes is to provide a new proof of the Mattila-Melnikov-Verdera theorem on the uniform rectifiability of an Ahlfors-David regular measure whose associated Cauchy transform operator is bounded. They are based on lectures given by the second author in the analysis seminars at Kent State University and Tel-Aviv University.

1   Introduction

The purpose of these notes is to provide a new proof of Mattila, Melnikov, and Verdera's theorem. The exposition is self-contained, relying only on a knowledge of basic real analysis.

Theorem 1.1 ([8]). *An Ahlfors-David regular measure μ whose associated Cauchy transform operator is bounded in $L^2(\mu)$ is uniformly rectifiable.*

The precise statement of this theorem is given in Sect. 3. The scheme employed to prove Theorem 1.1 in these notes is quite different from that in [8], and relies upon a characterization of *reflectionless measures*. In this regard, one may compare the proof to that of Mattila's theorem [5]: *Suppose that μ is a finite Borel measure satisfying* $\liminf_{r \to 0} \frac{\mu(B(z,r))}{r} \in (0, \infty)$ *for μ-a.e. $z \in \mathbb{C}$. If the Cauchy transform of μ exists μ-a.e. in the sense of principal value, then μ is rectifiable.* Mattila's proof of this theorem uses a characterization of *symmetric measures*, the reader may consult Chap. 14 of the book [4] for more information.

Subsequently, Mattila's theorem was generalized to the case of singular integrals in higher dimensions by Mattila and Preiss in [6]. To find the analogous generalization of the proof we carry out here would answer a longstanding problem of

B. Jaye • F. Nazarov (✉)
Department of Mathematics, Kent State University, Kent, OH 44240, USA
e-mail: bjaye@kent.edu; nazarov@math.kent.edu

© Springer International Publishing Switzerland 2014
B. Klartag, E. Milman (eds.), *Geometric Aspects of Functional Analysis*,
Lecture Notes in Mathematics 2116, DOI 10.1007/978-3-319-09477-9_15

David and Semmes [1]. Very recently, Nazarov et al. [10] completed the solution of the problem of David and Semmes in the case of singular integral operators of co-dimension 1. They proved that if μ is a d-dimensional Ahlfors-David regular measure in $\mathbb{R}^{d+1}$, then the boundedness of the d-dimensional Riesz transform in $L^2(\mu)$ implies that one of the criteria for uniform rectifiability given in [1] is satisfied. See [10] for more details, and for further references and history about this problem.

Throughout this paper, we shall only consider Ahlfors-David regular measures. For closely related results without this assumption, see the paper of Léger [3].

2 Notation

We shall adopt the following notation:

- $B(z, r)$ denotes the open disc centred at z with radius $r > 0$.
- For a square Q, we set z_Q to be the centre of Q, and $\ell(Q)$ to denote the side-length of Q.
- We shall denote by $\mathcal{D}$ the standard lattice of dyadic squares in the complex plane. A dyadic square is any square of the form $[k2^j, (k + 1)2^j) \times [\ell 2^j, (\ell + 1)2^j)$ for j, k and ℓ in $\mathbb{Z}$.
- We define the Lipschitz norm of a function f by

$$\|f\|_{\text{Lip}} = \sup_{z, \xi \in \mathbb{C}, z \neq \xi} \frac{|f(z) - f(\xi)|}{|z - \xi|}.$$

- We denote by $\text{Lip}_0(\mathbb{C})$ the space of compactly supported functions with finite Lipschitz norm. The continuous functions with compact support are denoted by $C_0(\mathbb{C})$.
- For $f : \mathbb{C} \to \mathbb{C}$, we set $\|f\|_\infty = \sup_{z \in \mathbb{C}} |f(z)|$. In particular, note that we are taking the *pointwise everywhere* supremum here.
- The closure of a set E is denoted by $\overline{E}$
- The support of a measure μ is denoted by $\text{supp}(\mu)$.
- For a line L, we write the one dimensional Hausdorff measure restricted to L by $\mathcal{H}^1_L$. If $L = \mathbb{R}$, we instead write m_1.
- We will denote by C and c various positive absolute constants. These constants may change from line to line within an intermediate argument. The constant C is thought of as large (at the very least greater than 1), while c is thought of as small (certainly smaller than 1). We shall usually make any dependence of a constant on a parameter explicit, unless it is clear from the context what the dependencies are.

3   The Precise Statement of Theorem 1.1

3.1   The Cauchy Transform of a Measure μ

Let $K(z) = \frac{1}{z}$ for $z \in \mathbb{C}\setminus\{0\}$. For a measure ν, the *Cauchy transform* of ν is formally defined by

$$\mathcal{C}(\nu)(z) = \int_{\mathbb{C}} K(z - \xi)d\nu(\xi), \text{ for } z \in \mathbb{C}.$$

In general, the singularity in the kernel is too strong to expect the integral to converge absolutely on supp(ν). It is therefore usual to introduce a regularized Cauchy kernel. For $\delta > 0$, define

$$K_\delta(z) = \frac{\bar{z}}{\max(\delta, |z|)^2}.$$

Then the δ-regularized Cauchy transform of ν is defined by

$$\mathcal{C}_\delta(\nu)(z) = \int_{\mathbb{C}} K_\delta(z - \xi)d\nu(\xi), \text{ for } z \in \mathbb{C}.$$

Before we continue, let us introduce a very natural condition to place upon μ. A measure μ is called C_0-*nice* if $\mu(B(z, r)) \leq C_0 r$ for any disc $B(z, r) \subset \mathbb{C}$.

If μ is a C_0-nice measure, then for any $f \in L^2(\mu)$ and $z \in \mathbb{C}$, we have that $\mathcal{C}_{\mu,\delta}(f)(z) := \mathcal{C}_\delta(f\mu)(z)$ is bounded in absolute value in terms of δ, C_0, and $\|f\|_{L^2(\mu)}$. To see this, we shall need an elementary tail estimate, which we shall refer to quite frequently in what follows:

Lemma 3.1. *Suppose μ is C_0-nice measure. For every $\varepsilon > 0$ and $r > 0$, we have*

$$\int_{\mathbb{C}\setminus B(0,r)} \frac{1}{|\xi|^{1+\varepsilon}}d\mu(\xi) \leq \frac{C_0(1 + \varepsilon)}{\varepsilon}r^{-\varepsilon}.$$

The proof of Lemma 3.1 is a standard exercise, and is left to the reader. With this lemma in hand, we return to our claim that $\mathcal{C}_{\mu,\delta}(f)$ is bounded. First apply the Cauchy-Schwarz inequality to estimate

$$|\mathcal{C}_{\mu,\delta}(f)(z)| \leq \left(\int_{\mathbb{C}} |K_\delta(z - \xi)|^2 d\mu(\xi)\right)^{1/2}\|f\|_{L^2(\mu)}.$$

But now $\int_{\mathbb{C}} |K_\delta(z-\xi)|^2 d\mu(\xi) \leq \int_{B(z,\delta)} \frac{|\xi-z|^2}{\delta^4}d\mu(\xi) + \int_{\mathbb{C}\setminus B(z,\delta)} \frac{1}{|\xi-z|^2}d\mu(\xi)$. The first term on the right hand side of this inequality is at most $\frac{\mu(B(z,\delta))}{\delta^2} \leq \frac{C_0}{\delta}$, and the second term is no greater than $\frac{2C_0}{\delta}$ by Lemma 3.1. We therefore see that $|\mathcal{C}_{\mu,\delta}(f)(z)| \le$ $\left(\frac{3C_0}{\delta}\right)^{1/2}\|f\|_{L^2(\mu)}$. In particular, we have $\mathcal{C}_{\mu,\delta}(f)(z) \in L^2_{\text{loc}}(\mu)$.

One conclusion of this discussion is that for any nice measure μ, it makes sense to ask if $\mathcal{C}_{\mu,\delta}$ is a bounded operator from $L^2(\mu)$ to $L^2(\mu)$.

Definition 3.2. We say that μ is C_0-*good* if it is C_0-nice and

$$\sup_{\delta > 0} \|\mathcal{C}_{\mu,\delta}\|_{L^2(\mu) \to L^2(\mu)} \le C_0.$$

By definition, the Cauchy transform operator associated with μ is bounded in $L^2(\mu)$ if μ is good.

The two dimensional Lebesgue measure restricted to the unit disc is good. However, this measure is not supported on a 1-rectifiable set and so such measures should be ruled out in a statement such as Theorem 1.1. To this end, we shall deal with Ahlfors-David (AD) regular measures.

Definition 3.3. A nice measure μ is called AD-regular, with regularity constant $c_0 > 0$, if $\mu(B(z,r)) \ge c_0 r$ for any disc $B(z,r) \subset \mathbb{C}$ with $z \in \text{supp}(\mu)$.

3.2 Uniform Rectifiability

A set $E \subset \mathbb{C}$ is called *uniformly rectifiable* if there exists $M > 0$ such that for any dyadic square $Q \in \mathcal{D}$, there exists a Lipschitz mapping $F : [0,1] \to \mathbb{C}$ with $\|F\|_{\text{Lip}} \le M\ell(Q)$ and $E \cap Q \subset F([0,1])$.

We can alternatively say that E is uniformly rectifiable if there exists $M > 0$ such that for any dyadic square $Q \in \mathcal{D}$, there is a rectifiable curve containing $E \cap Q$ of length no greater than $M\ell(Q)$.

A measure μ is uniformly rectifiable if the set $E = \text{supp}(\mu)$ is uniformly rectifiable.

We may now restate Theorem 1.1 in a more precise way.

Theorem 3.4. *A good AD-regular measure μ is uniformly rectifiable.*

4 Making Sense of the Cauchy Transform on supp(μ)

The definition of a good measure does not immediately provide us with a workable definition of the Cauchy transform on the support of μ. In this section, we rectify this matter by defining an operator $\mathcal{C}_\mu$ as a weak limit of the operators $\mathcal{C}_{\mu,\delta}$ as $\delta \to 0$. This idea goes back to Mattila and Verdera [7]. We fix a C_0-good measure μ.

Note that if $f \in \text{Lip}_0(\mathbb{C})$, then f is bounded in absolute value by $\|f\|_{\text{Lip}} \cdot \text{diam}(\text{supp}(f))$.

Fix $f, g \in \mathrm{Lip}_0(\mathbb{C})$. Then for any $\delta > 0$, we may write

$$\langle \mathcal{C}_{\mu,\delta}(f), g \rangle_\mu = \frac{1}{2} \iint\limits_{\mathbb{C} \times \mathbb{C}} K_\delta(z - \xi)[f(\xi)g(z) - f(z)g(\xi)]d\mu(z)d\mu(\xi).$$

Let $H(z, \xi) = \frac{1}{2}[f(\xi)g(z) - f(z)g(\xi)]$. It will be useful to denote by $I_\delta(f, g)$ the expression

$$I_\delta(f, g) = I_{\delta,\mu}(f, g) = \iint\limits_{\mathbb{C} \times \mathbb{C}} K_\delta(z - \xi)H(z, \xi)d\mu(\xi)d\mu(z).$$

Now, note that if $S = \mathrm{supp}(f) \cup \mathrm{supp}(g)$, it is clear that $\mathrm{supp}(H) \subset S \times S$.

In addition H is Lipschitz in $\mathbb{C}^2$ with Lipschitz norm no greater than $\frac{1}{\sqrt{2}}(\|f\|_\infty\|g\|_{\mathrm{Lip}} + \|g\|_\infty\|f\|_{\mathrm{Lip}})$. To see this, first observe that $|H(z, \xi) - H(\omega, \xi)| \leq \frac{1}{2}(\|f\|_\infty\|g\|_{\mathrm{Lip}} + \|g\|_\infty\|f\|_{\mathrm{Lip}})|z - \omega|$, whenever $z, \omega, \xi \in \mathbb{C}$. By using this inequality twice, we see that

$$|H(z_1, z_2) - H(\xi_1, \xi_2)| \leq \frac{1}{2}(\|f\|_\infty\|g\|_{\mathrm{Lip}} + \|g\|_\infty\|f\|_{\mathrm{Lip}})[|z_1 - \xi_1| + |z_2 - \xi_2|],$$

and the claim follows since $|z_1 - \xi_1| + |z_2 - \xi_2| \leq \sqrt{2}\sqrt{|z_1 - \xi_1|^2 + |z_2 - \xi_2|^2}$. Since $H(z, z) = 0$, this Lipschitz bound immediately yields

$$|H(z, \xi)| \leq \frac{1}{\sqrt{2}}(\|f\|_\infty\|g\|_{\mathrm{Lip}} + \|g\|_\infty\|f\|_{\mathrm{Lip}})|z - \xi| \text{ for any } z \neq \xi.$$

As a result of this bound on the absolute value of H, there exists a constant $C(f, g) > 0$ such that

$$|K_\delta(z - \xi)||H(z, \xi)| \leq C(f, g)\chi_{S \times S}(z, \xi).$$

On the other hand, since μ is a nice measure, the set $\{(z, \xi) \in \mathbb{C} \times \mathbb{C} : z = \xi\}$ is $\mu \times \mu$ null, and so for $\mu \times \mu$ almost every (z, ξ), the limit as $\delta \to 0$ of $K_\delta(z - \xi)$ is equal to $K(z - \xi)$. As a result, the Dominated Convergence Theorem applies to yield

$$\lim_{\delta \to 0} I_\delta(f, g) = \iint\limits_{\mathbb{C} \times \mathbb{C}} K(z - \xi)H(z, \xi)d\mu(z)d\mu(\xi).$$

This limit will be denoted by $I(f, g) = I_\mu(f, g)$. Moreover, there is a quantitative estimate on the speed of convergence:

$$|I(f, g) - I_\delta(f, g)| \leq \iint\limits_{\substack{(z,\xi) \in S \times S: \\ |z-\xi| < \delta}} C(f, g)d\mu(z)d\mu(\xi) \leq C(f, g)\delta\mu(S).$$

Since μ is C_0-nice, $\mu(S)$ can be bounded in terms of the diameters of the supports of f and g, and we see that $|I(f, g) - I_\delta(f, g)| \leq C(f, g)\delta$.

We have now justified the existence of an operator C_μ acting from the space of compactly supported Lipschitz functions to its dual with respect to the standard pairing $\langle f, g \rangle_\mu = \int_{\mathbb{C}} fg d\mu$.

Since μ is C_0-good, for any $\delta > 0$ we have

$$|I_\delta(f, g)| \leq C_0 \|f\|_{L^2(\mu)} \|g\|_{L^2(\mu)}, \text{ for any } f, g \in L^2(\mu), \tag{1}$$

and this inequality allows us to extend the definition of $I(f, g)$ to the case when f and g are $L^2(\mu)$ functions. To do this, we first pick functions f and g in $L^2(\mu)$. Let $\varepsilon > 0$. Using the density of $\text{Lip}_0(\mathbb{C})$ in $L^2(\mu)$, we write $f = f_1 + f_2$ and $g = g_1 + g_2$, where f_1 and g_1 are compactly supported Lipschitz functions, and the norms of f_2 and g_2 in $L^2(\mu)$ are as small as we wish (say, less than ε). We know that $I_\delta(f_1, g_1) \to I(f_1, g_1)$ as $\delta \to 0$. Consequently, for each $\varepsilon > 0$, $I_\delta(f, g)$ can be written as a sum of two terms, the first of which (namely $I_\delta(f_1, g_1)$) has a finite limit, and the second term (which is $I_\delta(f_1, g_2) + I_\delta(f_2, g_1) + I_\delta(f_2, g_2)$) has absolute value no greater than $C_0 \varepsilon (3\varepsilon + \|f\|_{L^2(\mu)} + \|g\|_{L^2(\mu)})$. It follows that the limit as $\delta \to 0$ of $I_\delta(f, g)$ exists. We define this limit to be $I(f, g) = I_\mu(f, g)$.

From (1), we see that $|I(f, g)| \leq C_0 \|f\|_{L^2(\mu)} \|g\|_{L^2(\mu)}$. Therefore we may apply the Riesz-Fisher theorem to deduce the existence of a (unique) bounded linear operator $C_\mu : L^2(\mu) \to L^2(\mu)$ such that

$$\langle C_\mu(f), g \rangle_\mu = I(f, g) \text{ for all } f, g \in L^2(\mu).$$

Having defined an operator C_μ for any good measure μ, we now want to see what weak continuity properties this operator has.

Definition 4.1. We say that the sequence μ_k tends to μ weakly if, for any $\varphi \in C_0(\mathbb{C})$,

$$\int_{\mathbb{C}} \varphi \, d\mu_k \to \int_{\mathbb{C}} \varphi \, d\mu \text{ as } k \to \infty.$$

We now recall a standard weak compactness result, which can be found in Chap. 1 of [4] (or any other book in real analysis).

Lemma 4.2. *Let $\{\mu_k\}_k$ be a sequence of measures. Suppose that for each compact set $E \subset \mathbb{C}$, $\sup_k \mu_k(E) < \infty$. Then there exists a subsequence $\{\mu_{k_j}\}_{k_j}$ and a measure μ such that μ_{k_j} converges to μ weakly.*

An immediate consequence of this lemma is that any sequence $\{\mu_k\}_k$ of C_0-nice measures has a subsequence that converges weakly to a measure μ. The next lemma shows that the various regularity properties of measures that we consider are inherited by weak limits.

Lemma 4.3. *Suppose that μ_k converges to μ weakly. If each measure μ_k is C_0-good with AD regularity constant c_0, then the limit measure μ is also C_0-good with AD regularity constant c_0.*

Proof. We shall first check that μ is AD regular. Let $x \in \mathrm{supp}(\mu)$, $r > 0$, and choose $\varepsilon \in (0, r/2)$. Consider a smooth non-negative function f, supported in the disc $B(x, \varepsilon)$, with $f \equiv 1$ on $B(x, \frac{\varepsilon}{2})$. Then $\int_{\mathbb{C}} f d\mu > 0$. Hence, for all sufficiently large k, $\int_{\mathbb{C}} f d\mu_k > 0$. For all such k, $B(x, \varepsilon) \cap \mathrm{supp}(\mu_k) \neq \varnothing$, and so there exists $x_k \in B(x, \varepsilon)$ satisfying $\mu_k(B(x_k, r - 2\varepsilon)) \geq c_0(r - 2\varepsilon)$. As a result, $\mu_k(B(x, r - \varepsilon)) > c_0(r - 2\varepsilon)$.

Now let $\varphi \in C_0(\mathbb{C})$ be nonnegative and supported in $B(x, r)$, satisfying $\|\varphi\|_\infty \leq 1$ and $\varphi \equiv 1$ on $B(x, r - \varepsilon)$. Then

$$\mu(B(x, r)) \geq \int_{\mathbb{C}} \varphi d\mu = \lim_{k \to \infty} \int_{\mathbb{C}} \varphi d\mu_k \geq c_0(r - 2\varepsilon).$$

Letting $\varepsilon \to 0$, we arrive at the desired AD regularity. The property that μ is C_0-nice is easier and left to the reader (it also follows from standard lower-semicontinuity properties of the weak limit).

It remains to show that μ is C_0-good. Fix $f, g \in \mathrm{Lip}_0(\mathbb{C})$ and define H and S as before. Note that $K_\delta(z - \xi)H(z, \xi)$ is a Lipschitz function in $\mathbb{C}^2$, and has support contained in $S \times S$. Let $U \supset S$ be an open set with $\mu(U) \leq \mu(S) + 1$. The (complex valued) Stone-Weierstrass theorem for a locally compact space tells us that the algebra of finite linear combinations of functions in $C_0(U) \times C_0(U)$ is dense in $C_0(U \times U)$ (with respect to the uniform norm in $\mathbb{C}^2$). Let $\varepsilon > 0$. There are functions $\varphi_1, \ldots, \varphi_n$ and $\psi_1, \ldots, \psi_n$, all belonging to $C_0(U)$, such that $|K_\delta(z - \xi)H(z, \xi) - \sum_{j=1}^n \varphi(z)\psi(\xi)| < \varepsilon$ for any $(z, \xi) \in U \times U$. For each $j = 1, \ldots, n$, we have

$$\lim_{k \to \infty} \iint_{\mathbb{C} \times \mathbb{C}} \varphi_j(z)\psi_j(\xi)d\mu_k(z)d\mu_k(\xi) = \iint_{\mathbb{C} \times \mathbb{C}} \varphi_j(z)\psi_j(\xi)d\mu(z)d\mu(\xi).$$

It therefore follows that

$$\limsup_{k \to \infty} |I_{\delta, \mu_k}(f, g) - I_{\delta, \mu}(f, g)| \leq \varepsilon(\limsup_{k \to \infty} \mu_k(U)^2 + \mu(U)^2).$$

On the other hand, μ_k is C_0 nice, and so $\mu_k(U) \leq C(f, g)$. Since $\varepsilon > 0$ was arbitrary, we conclude that $I_{\delta, \mu_k}(f, g) \to I_{\delta, \mu}(f, g)$ as $k \to \infty$.

As a result of this convergence, we have that

$$|I_{\mu, \delta}(f, g)| \leq C_0 \liminf_{k \to \infty} (\|f\|_{L^2(\mu_k)}\|g\|_{L^2(\mu_k)}).$$

But since both $|f|^2$ and $|g|^2$ are in $C_0(\mathbb{C})$, the right hand side of this inequality equals $\|f\|_{L^2(\mu)}\|g\|_{L^2(\mu)}$. We now wish to appeal to the density of $\mathrm{Lip}_0(\mathbb{C})$ in $L^2(\mu)$

to extend this inequality to all $f, g \in L^2(\mu)$. Let $R > 0$. As μ is C_0-nice, we saw in Sect. 3 that $\mathcal{C}_{\mu,\delta} : L^2(\mu) \to L^2(B(0, R), \mu)$. But then, since the space of Lipschitz function compactly supported in $B(0, R)$ is dense in $L^2(B(0, R), \mu)$, we conclude that $\|\mathcal{C}_{\mu,\delta}\|_{L^2(\mu) \to L^2(B(0,R),\mu)} \leq C_0$. Finally, taking the limit as $R \to \infty$, the monotone convergence theorem guarantees that $\|\mathcal{C}_{\mu,\delta}\|_{L^2(\mu) \to L^2(\mu)} \leq C_0$, and hence μ is C_0-good. $\qquad\square$

The proof of the next lemma is left as an exercise.

Lemma 4.4. *Suppose that μ_k is a sequence of c_0 AD-regular measures converging weakly to a measure μ. If $z_k \in \mathrm{supp}(\mu_k)$ with $z_k \to z$ as $k \to \infty$, then $z \in \mathrm{supp}(\mu)$.*

Our last task is to check that the bilinear form I_{μ_k} has nice weak convergence properties. For this, let $f, g \in \mathrm{Lip}_0(\mathbb{C})$. For $\delta > 0$, we write

$$|I_{\mu_k}(f, g) - I_{\mu}(f, g)| \leq |I_{\mu_k}(f, g) - I_{\delta,\mu_k}(f, g)| + |I_{\delta,\mu_k}(f, g) - I_{\delta,\mu}(f, g)|$$
$$+ |I_{\delta,\mu}(f, g) - I_{\mu}(f, g)|.$$

The first and third terms are bounded by $C(f, g)\delta$. The second term converges to 0 as $k \to \infty$. Therefore

$$\limsup_{k \to \infty} |I_{\mu_k}(f, g) - I_{\mu}(f, g)| \leq C(f, g)\delta.$$

But $\delta > 0$ was arbitrary, and so $I_{\mu_k}(f, g)$ converges to $I_{\mu}(f, g)$ as $k \to \infty$.

5 Riesz Systems

Throughout this section we fix a C_0-nice measure μ.

A system of functions ψ_Q $(Q \in \mathcal{D})$ is called a C-Riesz system if $\psi_Q \in L^2(\mu)$ for each Q, and

$$\left\| \sum_{Q \in \mathcal{D}} a_Q \psi_Q \right\|_{L^2(\mu)}^2 \leq C \sum_{Q \in \mathcal{D}} |a_Q|^2, \tag{2}$$

for every sequence $\{a_Q\}_{Q \in \mathcal{D}}$. By a simple duality argument, we see that if ψ_Q is a C-Riesz system, then

$$\sum_{Q \in \mathcal{D}} |\langle f, \psi_Q \rangle_{\mu}|^2 \leq C \|f\|_{L^2(\mu)}^2, \text{ for any } f \in L^2(\mu).$$

Suppose now that with each square $Q \in \mathcal{D}$, we associate a set Ψ_Q of $L^2(\mu)$ functions. We say that Ψ_Q $(Q \in \mathcal{D})$ is a C-Riesz family if, for any choice of functions $\psi_Q \in \Psi_Q$, the system ψ_Q forms a C-Riesz system.

We now introduce a particularly useful Riesz family. Suppose that μ is a C_0-nice measure. Fix $A > 1$, and define

$$\Psi^\mu_{Q,A} = \left\{ \psi : \operatorname{supp}(\psi) \subset B(z_Q, A\ell(Q)), \|\psi\|_{\operatorname{Lip}} \leq \ell(Q)^{-3/2}, \int_{\mathbb{C}} \psi \, d\mu = 0 \right\}.$$

Lemma 5.1. *For any $A > 1$, $\Psi^\mu_{Q,A}$ is a C-Riesz family, with constant $C = C(C_0, A)$.*

Proof. For each $Q \in \mathcal{D}$, pick a function $\psi_Q \in \Psi^\mu_{Q,A}$. Then we have

$$\|\psi_Q\|_\infty \leq \|\psi_Q\|_{\operatorname{Lip}} \cdot \operatorname{diam}(\operatorname{supp}(\psi_Q)) \leq \ell(Q)^{-3/2} \cdot 2A\ell(Q) \leq CA\ell(Q)^{-1/2},$$

and

$$\|\psi_Q\|^2_{L^2(\mu)} \leq \|\psi_Q\|^2_{L^\infty} \mu(B(z_Q, A\ell(Q))) \leq CA^3.$$

Now, if $Q', Q'' \in \mathcal{D}$ with $\ell(Q') \leq \ell(Q'')$, then $\langle \psi_{Q'}, \psi_{Q''} \rangle_\mu = 0$ provided that $B(z_{Q'}, A\ell(Q')) \cap B(z_{Q''}, A\ell(Q'')) = \varnothing$. If $B(z_{Q'}, A\ell(Q'))$ intersects $B(z_{Q''}, A\ell(Q''))$, we instead have the bound

$$|\langle \psi_{Q'}, \psi_{Q''} \rangle_\mu| \leq CA^3 \left(\frac{\ell(Q')}{\ell(Q'')} \right)^{3/2}.$$

Indeed, note that $\|\psi_{Q'}\|_{L^1(\mu)} \leq \|\psi_{Q'}\|_{L^\infty} \mu(B(z_{Q'}, A\ell(Q'))) \leq CA^2 \ell(Q')^{1/2}$, while the oscillation of $\psi_{Q''}$ on the set $B(z_{Q'}, A\ell(Q'))$ (which contains the support of $\psi_{Q'}$) is no greater than $\frac{A\ell(Q')}{\ell(Q'')^{3/2}}$. By multiplying these two estimates we arrive at the desired bound on the absolute value of the inner product.

Consider a sequence $\{a_Q\}_Q \in \ell^2(\mathcal{D})$. Then

$$\left\| \sum_{Q \in \mathcal{D}} a_Q \psi_Q \right\|^2_{L^2(\mu)} \leq 2 \sum_{\substack{Q', Q'' \in \mathcal{D}: \\ \ell(Q') \leq \ell(Q'')}} |a_{Q'}| |a_{Q''}| |\langle \psi_{Q'}, \psi_{Q''} \rangle_\mu|.$$

Inserting our bounds on the inner products into this sum, we see that we need to bound the sum

$$CA^3 \sum_{\substack{\ell(Q') \leq \ell(Q''), \\ B(z_{Q'}, A\ell(Q')) \cap B(z_{Q''}, A\ell(Q'')) \neq \varnothing}} |a_{Q'}| |a_{Q''}| \left(\frac{\ell(Q')}{\ell(Q'')} \right)^{3/2}.$$

(Since all sums involving squares will be taken over the lattice $\mathcal{D}$, we will not write this explicitly from now on.) Using Cauchy's inequality, we estimate

$$|a_{Q'}| |a_{Q''}| \left(\frac{\ell(Q')}{\ell(Q'')} \right)^{3/2} \leq \frac{|a_{Q'}|^2}{2} \left(\frac{\ell(Q')}{\ell(Q'')} \right)^{1/2} + \frac{|a_{Q''}|^2}{2} \left(\frac{\ell(Q')}{\ell(Q'')} \right)^{5/2}.$$

It therefore suffices to estimate two double sums:

$$I = \sum_{Q'} |a_{Q'}|^2 \sum_{\substack{Q'':\ell(Q')\leq\ell(Q''), \\ B(z_{Q'},A\ell(Q'))\cap B(z_{Q''},A\ell(Q''))\neq\varnothing}} \left(\frac{\ell(Q')}{\ell(Q'')}\right)^{1/2},$$

and

$$II = \sum_{Q''} |a_{Q''}|^2 \sum_{\substack{Q':\ell(Q')\leq\ell(Q''), \\ B(z_{Q'},A\ell(Q'))\cap B(z_{Q''},A\ell(Q''))\neq\varnothing}} \left(\frac{\ell(Q')}{\ell(Q'')}\right)^{5/2}.$$

For each dyadic length ℓ greater than $\ell(Q')$, there are at most CA^2 squares Q'' of side length ℓ for which $B(z_{Q''}, A\ell)$ has non-empty intersection with $B(z_{Q'}, A\ell(Q'))$. Hence

$$I \leq \sum_{Q'} |a_{Q'}|^2 \sum_{k\geq 0} CA^2 2^{-k/2} \leq CA^2 \sum_Q |a_Q|^2.$$

Concerning II, all the relevant squares Q' in the inner sum are contained in the disc $B(z_{Q''}, 3A\ell(Q''))$. Therefore, at scale ℓ there are at most $CA^2\left(\frac{\ell(Q'')}{\ell}\right)^2$ squares Q' of side length ℓ that can contribute to the inner sum. As a result,

$$II \leq CA^2 \sum_{Q''} |a_{Q''}|^2 \sum_{k\geq 0} 2^{-k/2} \leq CA^2 \sum_Q |a_Q|^2.$$

Combining our bounds, we see that the $\Psi_{Q,A}^\mu$ is a Riesz family, with Riesz constant $C(C_0)A^5$. $\qquad\qquad\qquad\qquad\qquad\qquad\qquad\qquad\qquad\qquad\qquad\qquad\qquad\qquad\square$

6 Bad Squares and Uniform Rectifiability

In this section we identify a local property of the support of a measure, which ensures that the measure is uniformly rectifiable. The mathematics in this section is largely due to David, Jones, and Semmes, see [1, Chap. 2.1], and is simpler than Jones' geometric Traveling Salesman theory [2], which was used in [8].

Fix a C_0-nice measure μ, which is AD-regular with regularity constant c_0. Set $E = \text{supp}(\mu)$.

6.1 The Construction of a Lipschitz Mapping

We will begin by constructing a certain graph. For our purposes, a *graph* $\Gamma = (\mathcal{N}, \mathcal{E})$ is a set of points $\mathcal{N}$ (the vertices), endowed with a collection of line

Fig. 1 An example of a
graph consisting of two
connected components

segments $\mathcal{E}$ (the edges) where each segment has its end-points at vertices. A
connected component of the graph is a maximal subset of vertices that can be
connected through the edges. For example, the graph depicted in Fig. 1 below has
two connected components.

The *distance between connected components* of a graph is measured as the
distance between the relevant sets of vertices. Therefore, the distance between the
components of the graph depicted in Fig. 1 is the distance between the vertices
labeled p and q.

Definition 6.1. For a graph $\Gamma = (\mathcal{N}, \mathcal{E})$, and a square Q, we define Γ_Q to be the
subgraph with vertex set $\mathcal{N} \cap 7Q$, endowed with the edges from $\mathcal{E}$ connecting those
vertices in $7Q$.

Let $\tau \in (0, 1)$. Fix $P \in \mathcal{D}$ (this square is to be considered as the viewing window
in the definition of uniform rectifiability). Choose a (small) dyadic fraction ℓ_0 with
$\ell_0 < \ell(P)$.

We shall construct a graph adapted to P inductively. Set $\mathcal{N}$ to be a maximal $\tau \ell_0$
separated subset of E. Note that $\mathcal{N}$ forms a $\tau \ell_0$ net of E.

The Base Step. For each square $Q \in \mathcal{D}$ with $\ell(Q) = \ell_0$ and $3Q \cap \mathcal{N} \neq \varnothing$, fix a
point which lies in $3Q \cap \mathcal{N}$. Then join together every point of $\mathcal{N} \cap 3Q$ to this fixed
point by line segments, as illustrated in the figure below.

In $3Q$, there are at most $C\tau^{-2}$ points of $\mathcal{N}$, and so the total length of the line
segments in $3Q$ is $C\tau^{-2}\ell_0$ (Fig. 2).

We thereby form the graph $\Gamma_{\ell_0}(\ell_0)$ comprised of the vertex set $\mathcal{N}$, and the set of
line segments $\mathcal{E}_{\ell_0}(\ell_0)$ obtained by carrying out the above procedure for all squares
$Q \in \mathcal{D}$ with $\ell(Q) = \ell_0$.

This is the base step of the construction.

The Inductive Step. Let ℓ be a dyadic fraction no smaller than ℓ_0. Suppose that we
have constructed the graph $\Gamma_{\ell_0}(\ell) = (\mathcal{N}, \mathcal{E}_{\ell_0}(\ell))$.

The graph $\Gamma_{\ell_0}(2\ell)$ is set to be the pair $(\mathcal{N}, \mathcal{E}_{\ell_0}(2\ell))$, where $\mathcal{E}_{\ell_0}(2\ell)$ is obtained
by taking the union of $\mathcal{E}_{\ell_0}(\ell)$ with the collection of line segments obtained by
performing the following algorithm (Fig. 3):

For every square $Q \in \mathcal{D}$ with $\ell(Q) = 2\ell$, consider the graph $\Gamma = (\Gamma_{\ell_0}(\ell))_Q$.
If Γ has at least two components that intersect $3Q$, then for each such component,

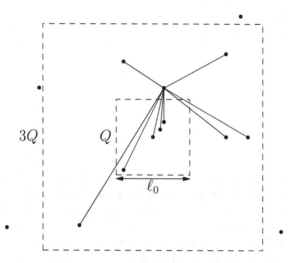

Fig. 2 The base step in the construction applied in $3Q$

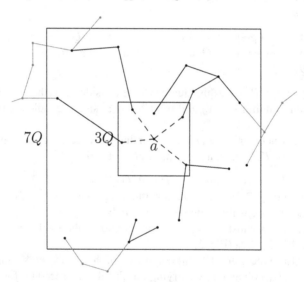

Fig. 3 The induction algorithm applied to a square Q. The *grey* edges indicate the edges of $\Gamma_{\ell_0}(\ell)$ not included in the subgraph $\Gamma = (\Gamma_{\ell_0}(\ell))_Q$. The *dashed lines* indicate the edges added by applying the algorithm. Note that in this case the graph Γ has seven components, four of which intersect $3Q$. The fixed point in $3Q \cap \mathcal{N}$ is denoted by a

choose a vertex that lies in its intersection with $\mathcal{N} \cap 3Q$. Fix a point in $3Q \cap \mathcal{N}$, and join each of the chosen points to this fixed point with an edge.

We carry out the inductive procedure for $\ell = \ell_0, \ldots, \frac{\ell(P)}{2}$, and thereby obtain the graph $\Gamma_{\ell_0}(\ell(P))$. To continue our analysis, first note the following elementary fact:

Lemma 6.2. *Let $Q \in \mathcal{D}$ with $\ell(Q) = 2\ell$. For any two points $z_1, z_2 \in 4Q$ with $|z_1 - z_2| < \ell$, there is a dyadic square Q' of sidelength ℓ, such that $7Q' \subset 7Q$, and $z_1, z_2 \in 3Q'$.*

Proof. Pick the square Q' to be the dyadic square of side length ℓ containing z_1. Then $\text{dist}(Q', \mathbb{C} \backslash 3Q') = \ell$, so $z_2 \in 3Q'$.

Since $\ell(Q) = 2\ell$, we have that $4Q$ is a union of dyadic squares of side-length ℓ. Therefore Q' is contained in $4Q$. As the square annulus $7Q \backslash 4Q$ is of width $\frac{3}{2}\ell(Q) = 3\ell$, we conclude that $7Q' \subset 7Q$.   □

We shall use this lemma (or rather a weaker statement with $4Q$ replaced by $3Q$) to deduce the following statement.

Claim 6.3. For each $\ell \geq \ell_0$, and $Q \in \mathcal{D}$ with $\ell(Q) = 2\ell$, any two connected components of $(\Gamma_{\ell_0}(\ell))_Q$ which intersect $3Q$ are ℓ-separated in $3Q$.

Proof. First suppose $\ell = \ell_0$. Let $z_1, z_2 \in 3Q$ with $|z_1 - z_2| < \ell_0$. Choose Q' as in Lemma 6.2. We have that $z_1, z_2 \in 3Q'$, and $3Q' \subset 7Q$. But then z_1 and z_2 must have been joined when the base step rule was applied to the square Q', and so they lie in the same component of $(\Gamma_{\ell_0}(\ell))_Q$.

Now suppose that $\ell > \ell_0$, and $z_1, z_2 \in 3Q$ with $|z_1 - z_2| < \ell$. Again, let Q' be the square of Lemma 6.2. The induction step applied at level $\frac{\ell}{2}$ to the square Q' ensures that z_1 and z_2 are joined by edges in $\mathcal{E}_{\ell_0}(\ell)$ that are contained in $7Q' \subset 7Q$. Therefore, z_1 and z_2 lie in the same component of $(\Gamma_{\ell_0}(\ell))_Q$.   □

Claim 6.4. There exists a constant $C > 0$, such that for each $\ell \geq \ell_0$, and for every $Q \in \mathcal{D}$ with $\ell(Q) = 2\ell$, the iterative procedure applied to Q increases the length of $\Gamma_{\ell_0}(2\ell)$ by at most $C\ell$.

Proof. From Claim 6.3, we see that the graph $(\Gamma_{\ell_0}(\ell))_Q$ can have at most C components which have non-empty intersection with $3Q$. Consequently, the application of the inductive procedure can generate at most C new edges, each of which having length no greater than $\sqrt{2} \cdot 6\ell$. The claim follows.   □

Adapting the Graph to P. We begin with another observation about the induction algorithm. Note that any two vertices in $3P \cap \mathcal{N}$ can be joined by edges in $\mathcal{E}_{\ell_0}(\ell(P))$ that are contained in $7P$. Thus, the graph $(\Gamma_{\ell_0}(\ell(P)))_P$ has only one connected component which intersects $3P$, and we denote this component by Γ. Let us denote by $L = L(\ell_0)$ the total length of Γ.

By Euler's theorem, there is a walk through Γ which visits each vertex of Γ at least once, and travels along each edge at most twice. By a suitable parametrization of this walk, we arrive at the following lemma:

Lemma 6.5. *There exists $F : [0, 1] \to \mathbb{C}$, with $\|F\|_{\text{Lip}} \leq 2L$, and such that $F([0, 1]) \supset \mathcal{N} \cap 3P$.*

If we have a suitable control over $L(\ell_0)$ independently of ℓ_0, then $E \cap P$ is contained in the image of a Lipschitz graph:

Lemma 6.6. *Suppose that there exists $M > 0$ such that $L(\ell_0) \le M\ell(P)$ for every $\ell_0 > 0$. Then there exists $F : [0, 1] \to \mathbb{C}$, such that $\|F\|_{\mathrm{Lip}} \le 2M\ell(P)$, and $E \cap P \subset F([0, 1])$.*

Proof. Let $\ell_0 = 2^{-k}$. Let F_k denote the function of Lemma 6.5. Then $F_k([0, 1])$ is a $\tau 2^{-k}$-net of $E \cap P$.

By appealing to the Arzela-Ascoli theorem, we see that there is a subsequence of the F_k (which we again denote by F_k), converging uniformly to some limit function F. The function F is Lipschitz continuous with Lipschitz norm no greater than $2M\ell(P)$.

Now, for any $x \in E \cap P$, there exists a sequence $\{x_k\}_k$ where $x_k \in F_k([0, 1])$, and $|x - x_k| < \tau 2^{-k}$. Take $t_k \in [0, 1]$ with $F_k(t_k) = x_k$. There is a convergent subsequence of $\{t_k\}_k$ which converges to some $t \in [0, 1]$. But then $F(t) = x$, and the proof is complete. $\square$

We shall now estimate $L(\ell_0)$ in terms of the total side length of squares where the induction step has been carried out. Note that only the base and inductive steps applied to the dyadic squares Q contained in $7P$ can contribute to the length.

We shall first estimate the contribution to the length by the base step.

Claim 6.7. *The contribution to the length of Γ from the base step is no greater than $C\tau^{-2}\ell(P)$.*

Proof. Let N denote the number of dyadic sub-squares of $7P$ with side length ℓ_0 where the base step has been carried out. For any such square Q, we must have $3Q \cap \mathrm{supp}(\mu) \ne \varnothing$. From the AD-regularity of μ, it follows that $\mu(4Q) \ge c_0\ell(Q) = c_0\ell_0$. Hence

$$c_0\ell_0 N \le \sum_{Q \in \mathcal{D}:\, Q \subset 7P,\, \ell(Q) = \ell_0} \mu(4Q) \le C\mu(CP) \le C\ell(P),$$

and therefore $N \le C\frac{\ell(P)}{\ell_0}$. Consequently, the contribution to the length of Γ from the base step is no greater than $C\tau^{-2}\ell_0\frac{\ell(P)}{\ell_0}$, as required. $\square$

We now denote by $\mathcal{Q}(P, \ell_0)$ the collection of dyadic squares $Q \in \mathcal{D}$ such that $\ell(Q) \in [\ell_0, \ell(P)]$, $Q \subset 7P$, and the inductive step has been carried out non-vacuously in Q at scale $\frac{\ell(Q)}{2}$.

Claim 6.4 guarantees that for each $Q \in \mathcal{Q}(P, \ell_0)$, an application of the inductive procedure increases the length $L(\ell_0)$ by no more than $C\ell(Q)$. Combining this observation with Claim 6.7, we infer the following bound:

$$L(\ell_0) \le C\tau^{-2}\ell(P) + C \sum_{Q \in \mathcal{Q}(P, \ell_0)} \ell(Q). \tag{3}$$

6.2 Bad Squares

Given the construction above, we would like to find a convenient way of identifying whether a square has been used in the inductive procedure at some scale. Since we don't want these squares to occur very often, we call them *bad squares*.

Definition 6.8. We say that $Q \in \mathcal{D}$ is a (μ)-bad square if there exist $\zeta, \xi \in B(z_Q, 10\ell(Q)) \cap \text{supp}(\mu)$, such that $|\zeta - \xi| \geq \ell(Q)/2$, and there exists $z \in [\zeta, \xi]$ such that $B(z, \tau\ell(Q)) \cap E = \varnothing$.

We now justify the use of this definition:

Lemma 6.9. *Suppose that* $\tau < \frac{1}{16}$. *Suppose that the inductive algorithm has been applied to* $Q \in \mathcal{D}$. *Then* Q *is a bad square.*

Proof. If the inductive algorithm has been applied, then there is a graph $\Gamma = (\mathcal{N}, \mathcal{E})^1$, with the following properties:

(1) The set $\mathcal{N}$ forms a $\frac{\tau\ell(Q)}{2}$ net of E.
(2) For every dyadic square Q' with $\ell(Q') < \ell(Q)$ and $7Q' \subset 7Q$, we have that if $z_1, z_2 \in 3Q' \cap \mathcal{N}$, then z_1 and z_2 lie in the same component of Γ_Q.
(3) The connected components of Γ_Q that intersect $3Q$ are at least $\frac{\ell(Q)}{2}$ separated in $3Q$.

(In fact, property (2) implies property (3), as was seen in Claim 6.3.)

By assumption, there exist two points ζ and ξ in $3Q \cap \mathcal{N}$ that lie in different components of Γ_Q. Then $|\zeta - \xi| \geq \frac{\ell(Q)}{2}$. Consider the line segment $[\zeta, \xi]$. Cover this segment with overlapping discs of radius $\tau\ell(Q)$, such that the centre of each disc lies in the line segment $[\zeta, \xi]$ (see Fig. 4).

Suppose that every disc has positive μ measure. If $\tau < \frac{1}{16}$, the concentric double of each disc is contained in $4Q$. Furthermore, in the concentric double of each disc, there must be a point from $\mathcal{N}$. We therefore form a chain of points in $\mathcal{N} \cap 4Q$, with every consecutive pair of points in the chain are separated by a distance of at most $8\tau\ell(Q) < \ell(Q)/2$. Furthermore, the first point in the chain is within a distance of $\ell(Q)/2$ of ζ, and the last point in the chain is no further than $\ell(Q)/2$ from ξ. Therefore, Lemma 6.2 ensures that each consecutive pair of points in the chain are contained in the concentric triple of some dyadic square Q' with $\ell(Q') < \ell(Q)$ and $7Q' \subset 7Q$. But then property (2) yields that each such pair lies in the same component of Γ_Q. As a result, ζ and ξ also lie in the same component of Γ_Q.

From this contradiction, we see that one of the discs of radius $\tau\ell(Q)$ has zero measure, which implies that Q is a bad square. $\square$

Now let $\mathcal{B}^\mu$ denote the set of those squares $Q \in \mathcal{D}$ that are bad. To prove Theorem 1.1, it suffices to prove the following proposition:

---

[1]In the notation of the previous section, $\Gamma = \Gamma_{\ell_0}(\frac{\ell(Q)}{2})$, for some $\ell_0 \leq \frac{\ell(Q)}{2}$.

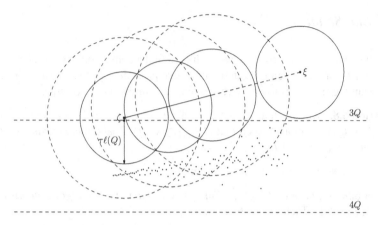

Fig. 4 The intersecting discs of radius $\tau\ell(Q)$, along with their concentric doubles. The cloud of points represents those points of $\mathcal{N}$

Proposition 6.10. *Suppose that μ is a C_0-good measure with AD regularity constant c_0. There is a constant $C = C(A, C_0, c_0) > 0$ such that for each $P \in \mathcal{D}$,*

$$\sum_{Q \in \mathcal{B}^\mu, Q \subset P} \ell(Q) \le C\ell(P). \tag{4}$$

Let us see how Theorem 1.1 follows from this proposition. Fix $P \in \mathcal{D}$, and construct $\Gamma_{\ell_0}(\ell(P))$ for $\ell_0 < \ell(P)$. From Proposition 6.10, the bound (3) for the length $L(\ell_0)$ is no more than $M\ell(P)$, where M can be chosen to depend on A, c_0, C_0, and τ (in particular, M can be chosen independently of P). But now Lemma 6.6 yields the existence of a function $F : [0, 1] \to \mathbb{C}$ with Lipschitz norm no greater than $M\ell(P)$, such that $E \cap P \subset F([0, 1])$. This is the required uniform rectifiability.

The condition (4) is very well known in harmonic analysis, and a family of squares $\mathcal{B}^\mu$ satisfying (4) is often referred to as a *Carleson family*. The best constant $C > 0$ such that (4) holds for all $P \in \mathcal{D}$ is called the *Carleson norm* of $\mathcal{B}^\mu$.

7 Bad Squares and the Riesz Family $\{\Psi_{Q,A}^\mu\}_Q$

Fix a C_0-good measure μ with AD-regularity constant $c_0 > 0$.

Choose $A' > 1$, with $A' \ge A$. Recall the Riesz family $\Psi_{Q,A}^\mu$ introduced in Sect. 5. For each $Q \in \mathcal{D}$, we define

$$\Theta_{A,A'}(Q) = \Theta_{A,A'}^\mu(Q) = \inf_{F \supset B(z_Q, A'\ell(Q))} \sup_{\psi \in \Psi_{Q,A}^\mu} \ell(Q)^{-1/2} |\langle \mathcal{C}_\mu(\chi_F), \psi \rangle_\mu|.$$

Consider a fixed $P \in \mathcal{D}$. Then for each $Q \subset P$ there exists a function $\psi_Q \in \Psi^\mu_{Q,A}$ such that $\Theta_{A,A'}(Q)^2 \ell(Q) \leq 2|\langle \mathcal{C}_\mu(\chi_{B(z_P,2A'\ell(P))}), \psi_Q \rangle_\mu|^2$ (note here that $B(z_Q, A'\ell(Q)) \subset B(z_P, 2A'\ell(P))$ whenever $Q \subset P$). Hence

$$\sum_{Q \subset P} \Theta_{A,A'}(Q)^2 \ell(Q) \leq 2 \sum_{Q \subset P} |\langle \mathcal{C}_\mu(\chi_{B(z_P,2A'\ell(P))}), \psi_Q \rangle_\mu|^2.$$

Since ψ_Q ($Q \in \mathcal{D}$) forms a $C(C_0, A)$-Riesz system, the right hand side of this inequality is bounded by $C(C_0, A) \|\mathcal{C}_\mu(\chi_{B(z_P,2A'\ell(P))})\|^2_{L^2(\mu)}$. As μ is C_0-good, this quantity is in turn bounded by $C(C_0, A)\mu(B(z_P, 2A'\ell(P)))$, which is at most $C(C_0, A, A')\ell(P)$. Therefore

$$\sum_{Q \subset P} \Theta_{A,A'}(Q)^2 \ell(Q) \leq C(C_0, A, A')\ell(P).$$

As an immediate corollary of this discussion, we arrive at the following result:

Lemma 7.1. *Let* $\gamma > 0$. *Consider the set* $\mathcal{F}_\gamma$ *of dyadic squares* Q *satisfying* $\Theta_{A,A'}(Q) > \gamma$. *Then* $\mathcal{F}_\gamma$ *is a Carleson family, with Carleson norm bounded by* $C(C_0, A, A')\gamma^{-2}$.

In order to prove Proposition 6.10 (from which Theorem 1.1 follows), it therefore suffices to prove the following proposition:

Proposition 7.2. *Suppose* μ *is a* C_0-*good measure with AD regularity constant* $c_0 > 0$. *There exist constants* $A, A' > 1$, *and* $\gamma > 0$, *such that for any square* $Q \in \mathcal{B}^\mu$,

$$\Theta^\mu_{A,A'}(Q) \geq \gamma.$$

We end this section with a simple remark about scaling.

Remark 7.3 (Scaling Remark). Fix a square Q, a function $\psi \in \Psi^\mu_{Q,A}$, and a compact set $F \subset \mathbb{C}$. For $z_0 \in \mathbb{C}$, set $\widetilde{\mu}(\cdot) = \frac{1}{\ell(Q)}\mu(\ell(Q) \cdot +z_0)$, $\widetilde{\psi}(\cdot) = \ell(Q)^{1/2}\psi(\ell(Q) \cdot +z_0)$ and $\widetilde{F} = \frac{1}{\ell(Q)}(F - z_0)$. Then $\|\widetilde{\psi}\|_{\mathrm{Lip}} \leq 1$, $\mathrm{supp}(\widetilde{\psi}) \subset B(\frac{z_Q-z_0}{\ell(Q)}, A)$, and

$$\langle \mathcal{C}_{\widetilde{\mu}}(\chi_{\widetilde{F}}), \widetilde{\psi} \rangle_{\widetilde{\mu}} = \ell(Q)^{-1/2}\langle \mathcal{C}_\mu(\chi_F), \psi \rangle_\mu.$$

8 Reflectionless Measures

In this section, we explore what happens if Proposition 7.2 fails. To do this, we shall need a workable definition of the Cauchy transform operator of a good measure acting on the constant function 1. Suppose that ν is a C_0-good measure with $0 \notin \mathrm{supp}(\nu)$.

8.1   The Function $\widetilde{C}_\nu(1)$

Let us begin with an elementary lemma.

Lemma 8.1. *Suppose that σ is a C_0-nice measure with $0 \notin \mathrm{supp}(\sigma)$. Let $z \in \mathbb{C}$ with $z \notin \mathrm{supp}(\sigma)$. Set $d_0 = \mathrm{dist}(\{0, z\}, \mathrm{supp}(\sigma))$. Then*

$$\int_{\mathbb{C}} \left| \frac{1}{z - \xi} + \frac{1}{\xi} \right| d\sigma(\xi) \le \frac{C(C_0)|z|}{d_0}.$$

Proof. Note the estimate

$$\int_{\mathbb{C}} \left| \frac{1}{z - \xi} + \frac{1}{\xi} \right| d\sigma(\xi) \le \frac{2}{d_0} \sigma(B(0, 2|z|)) + 2 \int_{\mathbb{C} \backslash B(0, 2|z|)} \frac{|z|}{|\xi|^2} d\sigma(\xi).$$

The first term on the right hand side has size no greater than $\frac{2C_0|z|}{d_0}$. Since the domain of integration in the second term can be replaced by $\mathbb{C} \backslash B(0, \max(d_0, 2|z|))$, Lemma 3.1 guarantees that the second integral is bounded by $\frac{C|z|}{\max(2|z|, d_0)}$. $\square$

For $z \notin \mathrm{supp}(\nu)$, define

$$\widetilde{C}_\nu(1)(z) = \int_{\mathbb{C}} \left[\frac{1}{z - \xi} + \frac{1}{\xi} \right] d\mu(\xi) = \int_{\mathbb{C}} [K(z - \xi) - K(-\xi)] d\nu(\xi). \qquad (5)$$

Lemma 8.1 guarantees that this integral converges absolutely.

To extend the definition to the support of ν, we shall follow a rather standard path. We shall initially define $\widetilde{C}_\nu(1)$ as a distribution, before showing it is a well defined function ν-almost everywhere. Recall from Sect. 4 how we interpret C_ν as a bounded operator in $L^2(\nu)$.

Fix $\psi \in \mathrm{Lip}_0(\mathbb{C})$. Choose $\varphi \in \mathrm{Lip}_0(\mathbb{C})$ satisfying $\varphi \equiv 1$ on a neighbourhood of the support of ψ. Then define

$$\langle \widetilde{C}_\nu(1), \psi \rangle_\nu = \langle C_\nu(\varphi), \psi \rangle_\nu - C_\nu(\varphi)(0) \cdot \int_{\mathbb{C}} \psi \, d\nu$$
$$+ \int_{\mathbb{C}} \psi(z) \int_{\mathbb{C}} (1 - \varphi(\xi)) [K(z - \xi) - K(-\xi)] d\nu(\xi) d\nu(z). \qquad (6)$$

Note that Lemma 8.1, applied with $\sigma = |1 - \varphi| \cdot \nu$, yields that

$$\sup_{z \in \mathrm{supp}(\psi)} \int_{\mathbb{C}} |(1 - \varphi(\xi))| \cdot |K(z - \xi) - K(-\xi)| d\nu(\xi) < \infty.$$

Therefore the inner product in (6) is well defined. We now claim that this inner product is independent of the choice of φ. To see this, let φ_1 and φ_2 be two compactly supported Lipschitz continuous functions, that are both identically equal to 1 on some neighbourhood of supp(ψ). If $z \in$ supp(ψ), then

$$\int_{\mathbb{C}} (1 - \varphi_1(\xi)) \Big[\frac{1}{z - \xi} + \frac{1}{\xi} \Big] d\nu(\xi) - \int_{\mathbb{C}} (1 - \varphi_2(\xi)) \Big[\frac{1}{z - \xi} + \frac{1}{\xi} \Big] d\nu(\xi)$$

$$= \int_{\mathbb{C}} (\varphi_2(\xi) - \varphi_1(\xi)) \Big[\frac{1}{z - \xi} + \frac{1}{\xi} \Big] d\nu(\xi)$$

$$= \int_{\mathbb{C}} (\varphi_2(\xi) - \varphi_1(\xi)) K(z - \xi) d\nu(\xi) + \mathcal{C}_\nu(\varphi_1)(0) - \mathcal{C}_\nu(\varphi_2)(0).$$

(All integrals in this chain of equalities converge absolutely.) Now consider

$$\int_{\mathbb{C}} \psi(z) \int_{\mathbb{C}} (\varphi_2(\xi) - \varphi_1(\xi)) K(z - \xi) d\nu(\xi) d\nu(z).$$

As a result of the anti-symmetry of K, this equals

$$\frac{1}{2} \int_{\mathbb{C}} \int_{\mathbb{C}} K(z - \xi) \big[\psi(z) [\varphi_2(\xi) - \varphi_1(\xi)] - \psi(\xi) [\varphi_2(z) - \varphi_1(z)] \big] d\nu(\xi) d\nu(z).$$

However, as we saw in Sect. 4, $K(z - \xi)(\psi(z)\varphi_j(\xi) - \psi(\xi)\varphi_j(z)) \in L^1(\nu \times \nu)$ for each $j = 1, 2$. Hence, by using the linearity of the integral, and applying Fubini's theorem, we see that the last line equals $I_\nu(\varphi_2, \psi) - I_\nu(\varphi_1, \psi)$. By definition, this is equal to $\langle \mathcal{C}_\nu(\varphi_2), \psi \rangle_\nu - \langle \mathcal{C}_\nu(\varphi_1), \psi \rangle_\nu$. The claim follows.

We have seen that $\widetilde{\mathcal{C}}_\nu(1)$ is well-defined as a distribution. For any bounded open set $U \subset \mathbb{C}$, if we choose φ to be identically equal to 1 on a neighbourhood of U, then $\widetilde{\mathcal{C}}_\nu(1) \in L^2(U, \nu)$. Since Lipschitz functions with compact support are dense in $L^2(U, \nu)$, we find that $\widetilde{\mathcal{C}}_\nu(1)$ is well-defined ν-almost everywhere.

Finally, we note that the smoothness of the function φ is not essential. If $\psi \in \text{Lip}_0(\mathbb{C})$, let U be a bounded open set containing supp(ψ). Then it is readily seen that $\langle \widetilde{\mathcal{C}}_\nu(1), \psi \rangle_\nu$ equals

$$\langle \mathcal{C}_\nu(\chi_U), \psi \rangle_\nu - \mathcal{C}_\nu(\chi_U)(0) \cdot \langle 1, \psi \rangle_\nu + \Big\langle \int_{\mathbb{C}\backslash U} [K(\cdot - \xi) + K(\xi)] d\nu(\xi), \psi \Big\rangle_\nu.$$

8.2 The Weak Continuity of $\widetilde{\mathcal{C}}_\nu(1)$

We shall introduce a couple more sets of functions. Φ_A^ν will denote those functions ψ with $\|\psi\|_{\text{Lip}} \leq 1$, that satisfy $\int_{\mathbb{C}} \psi \, d\nu = 0$ and supp(ψ) $\subset B(0, A)$. We define Φ^ν to be the set of compactly supported functions ψ with $\|\psi\|_{\text{Lip}} \leq 1$, satisfying $\int_{\mathbb{C}} \psi \, d\nu = 0$.

We start with another standard estimate.

Lemma 8.2. *Suppose that v is C_0-nice measure. For $R > 0$, suppose that $\psi \in \Phi_R^v$. Then $\|\psi\|_{L^1(v)} \leq C(C_0)R^2$.*

Proof. Simply note that

$$\int_{B(0,R)} |\psi| dv = \int_{B(0,R)} \left| \psi - \frac{1}{v(B(0,R))} \int_{B(0,R)} \psi dv \right| dv.$$

This quantity is no greater than $\operatorname{osc}_{B(0,R)}(\psi)v(B(0,R))$, which is less than or equal to $2R \cdot C_0 R$. $\square$

Our next lemma concerns a weak continuity property of $\widetilde{C}_v(1)$.

Lemma 8.3. *Let v_k be a sequence of C_0-good measures, with $0 \notin \operatorname{supp}(v_k)$. Suppose that v_k converge weakly to v (and so v is C_0-good), with $0 \notin \operatorname{supp}(v)$. Fix non-negative sequences $\widetilde{\gamma}_k$ and $\widetilde{A}_k$, satisfying $\widetilde{\gamma}_k \to 0$, and $\widetilde{A}_k \to \widetilde{A} \in (0, \infty]$. If $|\langle \widetilde{C}_{v_k}(1), \psi \rangle_{v_k}| \leq \widetilde{\gamma}_k$ for all $\psi \in \Phi_{\widetilde{A}_k}^{v_k}$, then*

$$|\langle \widetilde{C}_v(1), \psi \rangle_v| = 0 \text{ for all } \psi \in \Phi_{\widetilde{A}}^v.$$

(Here $\Phi_{\widetilde{A}}^v = \Phi^v$ if $\widetilde{A} = \infty$.)

Proof. If $v(B(0,\widetilde{A})) = 0$, then there is nothing to prove, so assume that $v(B(0,\widetilde{A})) > 0$. Let $\varepsilon > 0$. Pick $\psi \in \Phi_{\widetilde{A}}^v$. Then there exists $R \in (0, \infty)$ such that $\operatorname{supp}(\psi) \subset B(0, R) \subset B(0, \widetilde{A}_k)$ for all sufficiently large k, and $v(B(0, R)) > 0$.

Fix $\rho \in \operatorname{Lip}_0$ with $\operatorname{supp}(\rho) \subset B(0, R)$, such that $\int_{\mathbb{C}} \rho dv = c_\rho > 0$. Define

$$\psi_k = \psi - b_k \rho, \text{ with } b_k = \frac{1}{\int_{\mathbb{C}} \rho dv_k} \int_{\mathbb{C}} \psi dv_k.$$

Note that ψ_k is supported in $B(0, R)$, and has μ_k-mean zero. Since $b_k \to 0$, we have that $\|\psi_k\|_{\operatorname{Lip}} \leq 2$ for all sufficiently large k. Therefore, for these k, we have $|\langle \widetilde{C}_{v_k}(1), \psi_k \rangle_{v_k}| \leq 2\widetilde{\gamma}_k$.

Now pick $\varphi \in \operatorname{Lip}_0$ with $\varphi \equiv 1$ on $B(0, 2R)$ and $0 \leq \varphi \leq 1$ on $\mathbb{C}$, such that both

$$|\langle C_{v_k}(\varphi), \psi_k \rangle_{v_k} - \langle \widetilde{C}_{v_k}(1), \psi_k \rangle_{v_k}| < \varepsilon,$$

for all sufficiently large k, and

$$|\langle C_v(\varphi), \psi \rangle_v - \langle \widetilde{C}_v(1), \psi \rangle_v| < \varepsilon.$$

To see that such a choice is possible, note that if $\varphi \equiv 1$ on $B(0, R')$ for $R' > 2R$, then $|\langle C_{v_k}(\varphi), \psi_k \rangle_{v_k} - \langle \widetilde{C}_{v_k}(1), \psi_k \rangle_{v_k}|$ is bounded by

$$\int_{B(0,R)} |\psi_k(z)| \int_{\mathbb{C}} |1 - \varphi(\xi)| \left| \frac{1}{z - \xi} + \frac{1}{\xi} \right| dv_k(\xi) dv_k(z),$$

(recall here that ψ_k has ν_k mean zero). For any $z \in B(0, R)$, note that $\text{dist}(z, \text{supp}(1 - \varphi)) \geq \frac{R'}{2}$, and so by applying Lemma 8.1, we see that the above quantity is no greater than $C \|\psi_k\|_{L^1(\nu_k)} \frac{R}{R'}$. Applying Lemma 8.2, we see that $|\langle \mathcal{C}_{\nu_k}(\varphi), \psi_k \rangle_{\nu_k} - \langle \widetilde{\mathcal{C}}_{\nu_k}(1), \psi_k \rangle_{\nu_k}| \leq C \frac{R^3}{R'}$, which can be made smaller than ε with a reasonable choice of R'. The same reasoning shows that $|\langle \mathcal{C}_\nu(\varphi), \psi \rangle_\nu - \langle \widetilde{\mathcal{C}}_\nu(1), \psi \rangle_\nu| < \varepsilon$ provided R' is chosen suitably.

On the other hand, as ν_k is C_0-good, we have $|\langle \mathcal{C}_{\nu_k}(\varphi), \rho \rangle_{\nu_k}| \leq C_0 \|\varphi\|_{L^2(\nu_k)}$ $\|\rho\|_{L^2(\nu_k)}$. Since φ and ρ are compactly supported Lipschitz functions, the right hand side of this inequality converges to $C_0 \|\varphi\|_{L^2(\nu)} \|\rho\|_{L^2(\nu)}$, and so it is bounded independently of k.

Bringing together these observations, we see that $\langle \mathcal{C}_{\nu_k}(\varphi), \psi_k \rangle_{\nu_k}$ converges to $\langle \mathcal{C}_\nu(\varphi), \psi \rangle_\nu$ as $k \to \infty$. But since $|\langle \widetilde{\mathcal{C}}_{\nu_k}(1), \psi_k \rangle_{\nu_k}| \leq 2\widetilde{\gamma}_k$ for k large enough, we deduce from the triangle inequality that $|\langle \widetilde{\mathcal{C}}_\nu(1), \psi \rangle_\nu| \leq 4\varepsilon$. □

Let us now suppose that Proposition 7.2 is false. Fix $A \geq 100$. For each $k \in \mathbb{N}$, $k \geq 2A$, there is a C_0-good measure μ_k with AD-regularity constant $c_0 > 0$, a square $Q_k \in \mathcal{B}^{\mu_k}$, and a set $E_k \supset B(z_{Q_k}, k\ell(Q_k))$ such that

$$|\langle \mathcal{C}_{\mu_k}(\chi_{E_k}), \psi \rangle_{\mu_k}| \leq \frac{1}{k}, \text{ for all } \psi \in \Psi^{\mu_k}_{A, Q_k}. \tag{7}$$

In addition, by the scale invariance of the condition (7) (see Remark 7.3), we may dilate and translate the square Q_k so that it has side length 1, and so that there are $\zeta_k, \xi_k \in B(z_{Q_k}, 10) \cap \text{supp}(\mu_k)$ with $|\zeta_k - \xi_k| \geq 1/2$, such that $0 \in [\zeta_k, \xi_k]$ and $B(0, \tau) \cap \text{supp}(\mu_k) = \varnothing$. Note that the translated and dilated square is not necessarily dyadic.

By passing to a subsequence if necessary, we may assume that μ_k converge weakly to a measure $\mu^{(A)}$ (using the uniform niceness of the μ_k). This limit measure is C_0-good, with AD-regularity constant c_0, and $0 \notin \mu^{(A)}$. Furthermore, it is routine to check that $\mu^{(A)}$ satisfies the following property (recall Lemma 4.4):

$$\text{There exist } \xi, \zeta \in \overline{B(0, 20)} \cap \text{supp}(\mu^{(A)}), \text{ with } |\xi - \zeta| \geq \frac{1}{2},$$

$$\text{such that } 0 \in [\zeta, \xi] \text{ and } B(0, \tau) \cap \text{supp}(\mu^{(A)}) = \varnothing. \tag{8}$$

Now, for each k we have that $B(0, \frac{A}{2}) \subset B(z_{Q_k}, A)$ and $E_k \supset B(0, \frac{k}{2}) \supset B(0, A)$. We claim that

$$|\langle \widetilde{\mathcal{C}}_{\mu_k}(1), \psi \rangle_{\mu_k}| \leq \frac{1}{k} + \frac{CA^3}{k}, \text{ for all } \psi \in \Phi^{\mu_k}_{\frac{A}{2}}.$$

To see this, note that for any $\psi \in \Phi^{\mu_k}_{\frac{A}{2}}$, $\langle \widetilde{\mathcal{C}}_{\mu_k}(1), \psi \rangle_{\mu_k}$ is equal to

$$\langle \mathcal{C}_{\mu_k}(\chi_{E_k}), \psi \rangle_{\mu_k} + \int_{B(0,\frac{A}{2})} \psi(z) \int_{\mathbb{C}\setminus E_k} \left(\frac{1}{z-\xi} + \frac{1}{\xi} \right) d\mu_k(\xi) d\mu_k(z).$$

The first term is smaller than $\frac{1}{k}$ in absolute value. To bound the second term, note that for any $z \in B(0, \frac{A}{2})$, $\mathrm{dist}(z, \mathbb{C}\setminus E_k) \geq \frac{k}{2}$ so Lemma 8.1 yields that this second term is no larger than $\frac{CA}{k} \|\psi\|_{L^1(\mu_k)}$, and applying Lemma 8.2 yields the required estimate.

We now apply Lemma 8.3 with $\nu_k = \mu_k$, $\widetilde{\gamma}_k = \frac{1}{k} + \frac{CA^3}{k}$, and $\widetilde{A}_k = \frac{A}{2}$. Our conclusion is that $|\langle \widetilde{\mathcal{C}}_{\mu^{(A)}}(1), \psi \rangle_{\mu^{(A)}}| = 0$, for all $\psi \in \Phi_{\frac{A}{2}}^{\mu^{(A)}}$.

We now set $A = k$, for $k > 100$. The above argument yields a measure $\mu^{(k)}$ satisfying $|\langle \widetilde{\mathcal{C}}_{\mu^{(k)}}(1), \psi \rangle_{\mu^{(k)}}| = 0$, for all $\psi \in \Phi_{\frac{k}{2}}^{\mu^{(k)}}$. We now pass to a subsequence of $\{\mu^{(k)}\}_k$ so that $\mu^{(k)} \to \mu$ weakly as $k \to \infty$. The measure μ is C_0-good with AD-regularity constant c_0, and satisfies the property (8) with μ replacing $\mu^{(A)}$. By applying Lemma 8.3 with $\widetilde{\nu}_k = \mu^{(k)}$, $\widetilde{\nu} = \mu$, $\widetilde{A}_k = \frac{k}{2}$, and $\widetilde{\gamma}_k = 0$, we arrive at the following result:

Lemma 8.4. *Suppose that Proposition 7.2 fails. Then there exists a C_0-good measure μ with AD-regularity constant c_0, such that*

$$|\langle \widetilde{\mathcal{C}}_\mu(1), \psi \rangle_\mu| = 0, \text{ for all } \psi \in \Phi^\mu, \tag{9}$$

and there exist $\xi, \zeta \in \overline{B(0, 20)} \cap \mathrm{supp}(\mu)$, with $|\xi - \zeta| \geq \frac{1}{2}$, such that $0 \in [\zeta, \xi]$ and $B(0, \tau) \cap \mathrm{supp}(\mu) = \varnothing$.

We call any measure μ that satisfies (9) a *reflectionless measure*. It turns out that there aren't too many good AD-regular reflectionless measures.

Proposition 8.5. *Suppose that μ is a non-trivial reflectionless good AD-regular measure. Then $\mu = c\mathcal{H}_L^1$ for a line L, and a positive constant $c > 0$.*

Note that Proposition 8.5 contradicts the existence of the measure μ in Lemma 8.4. Therefore, once Proposition 8.5 is proved, we will have asserted Proposition 7.2, and Theorem 1.1 will follow. Hence it remains to prove the proposition. It is at this stage where the precise structure of the Cauchy transform is used.

9   The Cauchy Transform of a Reflectionless Good Measure μ is Constant in Each Component of $\mathbb{C}\setminus \mathrm{supp}(\mu)$

Our goal is now to prove Proposition 8.5. Suppose that μ is a reflectionless C_0-good measure. We may assume that $0 \notin \mathrm{supp}(\mu)$. All constants in this section may depend on C_0 without explicit mention.

Since $\widetilde{C}_\mu(1)$ is a well defined μ-almost everywhere function and satisfies (9), we conclude that it is a constant μ-almost everywhere in $\mathbb{C}$, say with value $\varkappa \in \mathbb{C}$.

Lemma 9.1. *Suppose that μ is a C_0-good reflectionless measure, and $0 \notin \mathrm{supp}(\mu)$. Then there exists $\varkappa \in \mathbb{C}$ such that $\widetilde{C}_\mu(1) = \varkappa$ μ-almost everywhere.*

Our considerations up to now have been quite general, but now our hand is forced to use the magic of the complex plane. The main difficulty is to obtain some information about the values of $\widetilde{C}_\mu(1)$ away from the support of μ in terms of the constant value $\varkappa$.

9.1 The Resolvent Identity

Lemma 9.2. *For every $z \notin \mathrm{supp}(\mu)$,*

$$[\widetilde{C}_\mu(1)(z)]^2 = 2\varkappa \cdot \widetilde{C}_\mu(1)(z).$$

An immediate consequence of Lemma 9.2 is that either $\widetilde{C}_\mu(1)(z) = 2\varkappa$ or $\widetilde{C}_\mu(1)(z) = 0$ for any $z \notin \mathrm{supp}(\mu)$. Since $\widetilde{C}_\mu(1)$ is a continuous function away from $\mathrm{supp}(\mu)$, it follows that $\widetilde{C}_\mu(1)$ is constant in each connected component of $\mathbb{C} \setminus \mathrm{supp}(\mu)$.

A variant of Lemma 9.2, where the Cauchy transform is considered in the sense of principal value, has previously appeared in work of Melnikov, Poltoratski, and Volberg, see Theorem 2.2 of [9]. We shall modify the proof from [9] in order to prove Lemma 9.2.

We shall first provide an incorrect proof of this lemma. Indeed, note the following regularized version of the resolvent identity: for any three distinct points $z, \xi, \omega \in \mathbb{C}$,

$$\left[\frac{1}{z-\xi} + \frac{1}{\xi}\right] \cdot \left[\frac{1}{\xi-\omega} + \frac{1}{\omega}\right] + \left[\frac{1}{z-\omega} + \frac{1}{\omega}\right] \cdot \left[\frac{1}{\omega-\xi} + \frac{1}{\xi}\right]$$

$$= \left[\frac{1}{z-\xi} + \frac{1}{\xi}\right] \cdot \left[\frac{1}{z-\omega} + \frac{1}{\omega}\right]. \tag{10}$$

Integrating both sides of this equality with respect to $d\mu(\xi)d\mu(\omega)$, we (only formally!) arrive at $2\widetilde{C}_\mu(\widetilde{C}_\mu(1))(z) = [\widetilde{C}_\mu(z)]^2$. Once this is established, Lemma 9.1 completes the proof. The proof that follows is a careful justification of this integration.

Proof. We shall define $\widetilde{C}_{\mu,\delta}(\varphi)(\omega) = \int_{\mathbb{C}}[K_\delta(\omega - \xi) + \frac{1}{\xi}]\varphi(\xi)d\mu(\xi)$ for $\varphi \in \mathrm{Lip}_0(\mathbb{C})$. In particular, as δ tends to 0, $\widetilde{C}_{\mu,\delta}(\varphi)$ converges to $\widetilde{C}_\mu(\varphi) = C_\mu(\varphi) - C_\mu(\varphi)(0)$ weakly in $L^2(\mu)$.

Set $d_0 = \mathrm{dist}(\{z, 0\}, \mathrm{supp}(\mu))$. Since $d_0 > 0$, Lemma 8.1 tells us that $[K(z - \cdot) + K(\cdot)] \in L^1(\mu)$.

For $N > 0$, define a bump function $\varphi_N \in \text{Lip}_0(\mathbb{C})$, satisfying $\varphi_N \equiv 1$ on $B(0, N)$, and $\text{supp}(\varphi) \subset B(0, 2N)$. Consider the identity (10), and multiply both sides by $\varphi_N(\xi)\varphi_N(\omega)$. After integration against $d\mu(\xi)d\mu(\omega)$, the right hand side of this equality becomes

$$\int_{\mathbb{C}}\Big[\frac{1}{z-\xi} + \frac{1}{\xi}\Big]\varphi_N(\xi)d\mu(\xi) \int_{\mathbb{C}}\Big[\frac{1}{z-\omega} + \frac{1}{\omega}\Big]\varphi_N(\omega)d\mu(\omega).$$

But since $\big[K(z - \cdot) + K(\cdot)\big] \in L^1(\mu)$, the dominated convergence theorem ensures that as $N \to \infty$, this expression converges to $[\widetilde{\mathcal{C}}_\mu(1)(z)]^2$.

Now, let $\delta > 0$, and note that

$$\frac{1}{\xi - \omega} = K_\delta(\xi - \omega) + \chi_{B(0,\delta)}(\xi - \omega) \cdot \Big[\frac{1}{\xi - \omega} - \frac{\overline{\xi - \omega}}{\delta^2}\Big].$$

Consider the integral

$$\int_{\mathbb{C}}\int_{\mathbb{C}}\chi_{B(0,\delta)}(\xi - \omega)\Big[\frac{1}{\xi - \omega} - \frac{\overline{\xi - \omega}}{\delta^2}\Big] \cdot \Big[\frac{1}{z-\xi} + \frac{1}{\xi} - \frac{1}{z-\omega} - \frac{1}{\omega}\Big]$$
$$\varphi_N(\xi)\varphi_N(\omega)d\mu(\xi)d\mu(\omega).$$

Note that $\big|\frac{1}{z-\xi} + \frac{1}{\xi} - \frac{1}{z-\omega} - \frac{1}{\omega}\big| \le \frac{2}{d_0^2}|\xi - \omega|$ for $\xi, \omega \in \text{supp}(\mu)$, and so this integral is bounded in absolute value by a constant multiple of $\int_{\mathbb{C}}\varphi_N(\xi)\mu(B(\xi,\delta))d\mu(\xi)$, which is bounded by $C\delta N$. This converges to zero as $\delta \to 0$.

Making reference to (10), we have thus far shown that

$$\lim_{N\to\infty}\lim_{\delta\to 0}\int_{\mathbb{C}}\int_{\mathbb{C}}\varphi_N(\xi)\varphi_N(\omega)\Big\{\Big[\frac{1}{z-\xi} + \frac{1}{\xi}\Big]\Big[K_\delta(\xi - \omega) + \frac{1}{\omega}\Big]$$
$$+ \Big[\frac{1}{z-\omega} + \frac{1}{\omega}\Big]\Big[K_\delta(\omega - \xi) + \frac{1}{\xi}\Big]\Big\}d\mu_N(\xi)d\mu_N(\omega) = [\widetilde{\mathcal{C}}_\mu(1)(z)]^2.$$

By Fubini's theorem, and the weak convergence of $\widetilde{\mathcal{C}}_{\mu,\delta}(\varphi)$ to $\widetilde{\mathcal{C}}_\mu(\varphi)$, the left hand side of this equality is equal to twice the following limit

$$\lim_{N\to\infty}\Big[\int_{\mathbb{C}}\varphi_N(\xi)\Big[\frac{1}{z-\xi} + \frac{1}{\xi}\Big]\widetilde{\mathcal{C}}_\mu(\varphi_N)(\xi)d\mu(\xi)\Big]. \tag{11}$$

Therefore, to prove the lemma, it suffices to show that this limit equals $\varkappa\widetilde{\mathcal{C}}_\mu(1)(z)$. To do this, let $\alpha \in (\frac{1}{2}, 1)$. First consider

$$I_N = \int_{B(0,N^\alpha)}\varphi_N(\xi)\Big|\frac{1}{z-\xi} + \frac{1}{\xi}\Big| \cdot |\widetilde{\mathcal{C}}_\mu(\varphi_N)(\xi) - \widetilde{\mathcal{C}}_\mu(1)(\xi)|d\mu(\xi).$$

Note that, for $|\xi| \le N^\alpha$, we have

$$|\widetilde{C}_\mu(\varphi_N)(\xi) - \widetilde{C}_\mu(1)(\xi)| \le \int_{\mathbb{C}} |1 - \varphi_N(\omega)| \left| \frac{1}{\xi - \omega} - \frac{1}{\omega} \right| d\mu(\omega).$$

Applying Lemma 8.1 yields an upper bound for the right hand side of $C \frac{|\xi|}{N} \le CN^{\alpha-1}$. But as $[K(z - \cdot) + K(\cdot)] \in L^1(\mu)$, we conclude that $I_N \to 0$ as $N \to \infty$. Next, note that

$$\int_{B(0,N^\alpha)} \varphi_N(\xi) \left[\frac{1}{z - \xi} + \frac{1}{\xi} \right] \widetilde{C}_\mu(1)(\xi) d\mu(\xi) = \varkappa \int_{B(0,N^\alpha)} \varphi_N(\xi) \left[\frac{1}{z - \xi} + \frac{1}{\xi} \right] d\mu(\xi),$$

which converges to $\varkappa \cdot \widetilde{C}_\mu(1)(z)$ as $N \to \infty$.

To complete the proof the lemma, it now remains to show that

$$\lim_{N \to \infty} \int_{B(0,2N) \setminus B(0,N^\alpha)} |\widetilde{C}_\mu(\varphi_N)(\xi)| \cdot \left| \frac{1}{z - \xi} - \frac{1}{\xi} \right| d\mu(\xi) = 0.$$

To do this, first note that $|\widetilde{C}_\mu(\varphi_N)(\xi)| \le |C_\mu(\varphi_N)(\xi)| + C \log \frac{N}{d_0}$ (this merely uses the C_0-niceness of μ). On the other hand, for sufficiently large N, $\left| \frac{1}{z-\xi} - \frac{1}{\xi} \right| \le \frac{8|z|}{N^{2\alpha}}$ for $|\xi| \ge N^\alpha$. Therefore, there is a constant $C = C(C_0, d_0) > 0$ such that

$$\int_{B(0,2N) \setminus B(0,N^\alpha)} |\widetilde{C}_\mu(1)(\xi)| \cdot \left| \frac{1}{z - \xi} - \frac{1}{\xi} \right| d\mu(\xi)$$

$$\le \frac{C|z| \log N}{N^{2\alpha}} \mu(B(0, 2N)) + \frac{C|z|}{N^{2\alpha}} \int_{B(0,2N)} |C_\mu(\varphi_N)(\xi)| d\mu(\xi).$$

Finally, since $\|C_\mu(\varphi_N)\|_{L^2(\mu)} \le C \sqrt{\mu(B(0, 2N))}$, and $\mu(B(0,2N)) \le CN$, we estimate the right hand side here by a constant multiple of $\frac{|z|N \log N}{N^{2\alpha}}$, which tends to zero as $N \to \infty$. $\qquad\square$

10 The Proof of Proposition 8.5

In this section we conclude our analysis by proving Proposition 8.5. To do this, we shall use the notion of a tangent measure, which was developed by Preiss [11]. Suppose that ν is a Borel measure on $\mathbb{C}$. The measure $\nu_{z,\lambda}(A) = \frac{\nu(\lambda A + z)}{\lambda}$ is called a λ-blowup of ν at z. A tangent measure of ν at z is any measure that can be obtained as a weak limit of a sequence of λ-blowups of ν at z with $\lambda \to 0$.

Now suppose that μ is a nontrivial C_0-good with AD regularity constant c_0. Then any λ-blowup measure of μ at $z \in \text{supp}(\mu)$ will again have these properties (C_0-goodness, and c_0-AD regularity). Therefore, both properties are inherited by any tangent measure of μ. In particular, every tangent measure of μ at $z \in$ supp(μ) is non-trivial, provided that μ is non-trivial. Lastly, we remark that if μ is reflectionless, then any tangent measure of μ is also reflectionless. This follows from a simple application of Lemma 8.3.

In what follows, it will often be notationally convenient to translate a point on supp(μ) to the origin. Whenever this is the case, the definition of $\widetilde{\mathcal{C}}_\mu(1)$ in (5) is translated with the support of the measure, and becomes

$$\widetilde{\mathcal{C}}_\mu(1)(z) = \int_{\mathbb{C}} [K(z - \xi) - K(z_0 - \xi)] d\mu(\xi), \tag{12}$$

for some $z_0 \notin$ supp(μ). If μ is reflectionless, then $\widetilde{\mathcal{C}}_\mu(1)$ is constant in each component of $\mathbb{C}\setminus$ supp(μ), and takes one of two values in $\mathbb{C}\setminus$ supp(μ).

10.1 Step 1

Suppose that supp(μ) $\subset L$, for some line L. Then by translation and rotation we may as well assume that $L = \mathbb{R}$. If the support is not the whole line, then there exists an interval (x, x') disjoint from the support of μ, with either x or x' in the support of μ. By rotating the support if necessary, we may assume that $x' \in$ supp(μ).

Denote by $\tilde{\mu}$ a non-zero tangent measure of μ at x'. Then $\tilde{\mu}$ has support contained in the segment $[x', \infty)$, and $x' \in$ supp($\tilde{\mu}$). Since $\tilde{\mu}$ is reflectionless, we may apply Lemma 9.2 to deduce that $\widetilde{\mathcal{C}}_{\tilde{\mu}}(1)(x' - t)$ is constant for all $t > 0$. Differentiating this function with respect to t, we arrive at $\int_{x'}^\infty \frac{1}{(x-t-y)^2} d\tilde{\mu}(y)$. This integral is strictly positive as $\tilde{\mu}$ is not identically zero. From this contradiction we see that supp(μ) $= \mathbb{R}$.

Consequently, we have that $d\mu(t) = h(t)dt$, where $c_0 \le h(t) \le C_0$. Now let $y > 0$ and consider, for $x \in \mathbb{R}$,

$$\widetilde{\mathcal{C}}_\mu(1)(x - yi) - \widetilde{\mathcal{C}}_\mu(1)(x + yi) = 2i \int_{\mathbb{R}} \frac{y}{(x - t)^2 + y^2} h(t)dt.$$

The expression on the left hand side is constant in $x \in \mathbb{R}$ and $y > 0$. On the other hand, the integral on the right hand side is a constant multiple of the harmonic extension of h to $\mathbb{R}_+^2$. The Poisson kernel is an approximate identity, and so by letting $y \to 0^+$ we conclude that h is a constant. Therefore $\mu = cm_1$, with $c > 0$.

10.2 Step 2

We now turn to the general case. We first introduce some notation. For $z \in \mathbb{C}$ and a unit vector e, $H_{z,e}$ denotes the (closed) half space containing z on the boundary, with inner unit normal e. With $\alpha \in (0, 1)$, we denote $C_{z,e}(\alpha) = \{\xi \in \mathbb{C} : \langle \xi - z, e \rangle > \alpha|\xi - z|\}$, where $\langle \cdot, \cdot \rangle$ is the standard inner product in $\mathbb{R}^2$.

Lemma 10.1. *Suppose that $z \notin \mathrm{supp}(\mu)$. Let $\tilde{z}$ be a closest point in $\mathrm{supp}(\mu)$ to z, and set $e = \frac{\tilde{z}-z}{|\tilde{z}-z|}$. For each $\alpha \in (0, 1)$, there is a radius $r_\alpha > 0$ such that $B(\tilde{z}, r_\alpha) \cap C_{\tilde{z},e}(\alpha)$ is disjoint from $\mathrm{supp}(\mu)$.*

Proof. We may suppose that $z = -ri$ for some $r > 0$ and $\tilde{z} = 0$, (and so $e = i$). We shall examine the imaginary part of the Cauchy transform evaluated at $-ti$ for $t \in (0, \frac{r}{2})$:

$$\Im[\widetilde{\mathcal{C}}_\mu(1)(-ti)] = \int_{\mathbb{C}} \left[\frac{\Im(\xi) + t}{|\xi + it|^2} - \frac{\Im(\xi - z_0)}{|\xi - z_0|^2} \right] d\mu(\xi).$$

Lemma 9.2 guarantees that $\Im[\widetilde{\mathcal{C}}_\mu(1)(-ti)] = \Im[\widetilde{\mathcal{C}}_\mu(1)(z)]$ for any $t > 0$. In particular, it is bounded independently of t.

Making reference to Fig. 5, we let $R > 3r$, and define three regions: $I = \mathbb{C} \backslash B(0, R)$, $II = \{\xi \in B(0, R) : \Im(\xi) < -t\}$, and $III = B(0, R) \backslash II$.

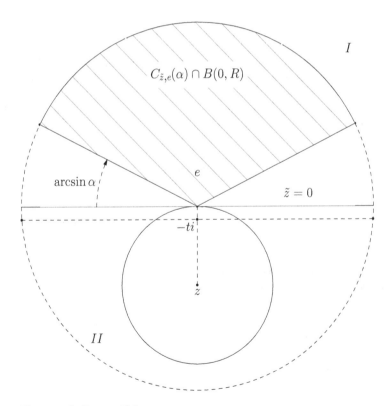

Fig. 5 The set-up for Lemma 10.1

Set $d_0 = \text{dist}(z_0, \text{supp}(\mu))$. First note that if $R > 3|z_0|$, and $\xi \in \mathbb{C} \setminus B(0, R)$, then

$$\left| \frac{\Im(\xi) + t}{|\xi + it|^2} - \frac{\Im(\xi - z_0)}{|\xi - z_0|^2} \right| \leq \frac{C}{|\xi|^2}.$$

Therefore,

$$\int_{|\xi| \geq R} \left| \frac{\Im(\xi) + t}{|\xi + it|^2} - \frac{\Im(\xi - z_0)}{|\xi - z_0|^2} \right| d\mu(\xi) + \int_{B(0,R)} \left| \frac{\Im(\xi - z_0)}{|\xi - z_0|^2} \right| d\mu(\xi)$$

$$\leq \int_{|\xi| \geq R} \frac{C}{|\xi|^2} d\mu(\xi) + \int_{B(0,R)} \frac{1}{|\xi - z_0|} d\mu(\xi).$$

The right hand side of this inequality if finite and independent of t.

Next, note that if $\xi \in II \cap \text{supp}(\mu)$, then $|\xi - it|^2 \geq -(\Im(\xi) + t)r$, provided that $|\Im(\xi)| < \frac{r}{2}$ and $t < \frac{r}{2}$. To see this, note that $|\xi - z| > r$, and so by elementary geometry, $|\xi - it|^2 \geq r^2 - (r + (\Im(\xi) + t))^2$. This is at least $-(\Im(\xi) + t)r$ under our assumptions on ξ and t. Therefore, if $t < \frac{r}{2}$, then

$$\int_{II} \frac{\Im(\xi) + t}{|\xi + it|^2} d\mu(\xi) \geq -\int_{II \cap B(0, \frac{r}{2})} \frac{1}{r} d\mu(\xi) - \left| \int_{II \setminus B(0, \frac{r}{2})} \frac{\Im(\xi) + t}{|\xi + it|^2} d\mu(\xi) \right|.$$

Both terms on the right hand side are bounded in absolute value by $C \frac{\mu(B(0,R))}{r} \leq \frac{CR}{r}$ (recall that $B(z, r) \cap \text{supp}(\mu) = \varnothing$). Note that the integral on the left hand side is at most zero.

Our conclusion thus far is that there is a constant Δ, depending on C_0, d_0, R, r and $\Im(\widetilde{C}_\mu(1)(z))$, such that for any $t < \frac{r}{2}$,

$$\left| \int_{III} \frac{\Im(\xi) + t}{|\xi + it|^2} d\mu(\xi) \right| \leq \Delta. \tag{13}$$

Note that the integrand in this integral is positive for any $\xi \in III$.

Suppose now that the statement of the lemma is false. Then there exists $\alpha > 0$, along with a sequence $z_j \in C_{0,e}(\alpha) \cap \text{supp}(\mu)$ with $z_j \to 0$ as $j \to \infty$. By passing to a subsequence, we may assume that $|z_j| \leq \frac{R}{2}$ for each j, and also that the balls $B_j = B(z_j, \frac{\alpha}{2}|z_j|)$ are pairwise disjoint.

Each ball $B_j \subset III$, and provided that $t \leq \frac{\alpha}{2}|z_j|$, we have

$$\frac{\Im(\xi) + t}{|\xi + ti|^2} \geq \frac{\alpha|z_j|}{8|z_j|^2} = \frac{\alpha}{8|z_j|}, \quad \text{for any } \xi \in B_j.$$

As a result, we see that

$$\int_{III} \frac{\Im(\xi) + t}{|\xi + it|^2} d\mu(\xi) \geq \sum_{j : t \leq |z_j|/2} \int_{B_j} \frac{\Im(\xi) + t}{|\xi + ti|^2} d\mu(\xi) \geq \sum_{j : t \leq |z_j|/2} \mu(B_j) \frac{\alpha}{8|z_j|}.$$

But $\mu(B_j) \geq \frac{c_0 \alpha}{2} |z_j|$, and so the previous integral over III has size at least $\frac{c_0 \alpha^2}{16}$ · card$\{j : t \leq \frac{1}{2}|z_j|\}$. However, if t is sufficiently small, then this quantity may be made larger than the constant Δ appearing in (13). This is absurd. □

We now pause to prove a simple convergence lemma.

Lemma 10.2. *Suppose that ν_k is a sequence of C_0-nice measures with AD-regularity constant c_0 that converges to ν weakly as $k \to \infty$ (and so ν is C_0-nice with AD-regularity constant c_0). If $z_0 \notin \operatorname{supp}(\nu)$, then for any $z \notin \operatorname{supp}(\nu)$, $\widetilde{C}_{\nu_k}(1)(z)$ is well defined (as in (12)) for sufficiently large k, and $\widetilde{C}_{\nu_k}(1)(z) \to \widetilde{C}_{\nu}(1)(z)$ as $k \to \infty$.*

Proof. First note that there exists $r > 0$ such that $\nu(B(z_0, r)) = 0 = \nu(B(z, r))$. But then, by the AD regularity of each ν_k, we must have that $\nu_k(B(z_0, \frac{r}{2})) = 0 = \nu_k(B(z, \frac{r}{2}))$ for sufficiently large k, and hence $\widetilde{C}_{\nu_k}(1)(z)$ is well defined. Let $N > 0$, and choose $\varphi_N \in \operatorname{Lip}_0(\mathbb{C})$ satisfying $\varphi_N \equiv 1$ on $B(0, N)$ and $0 \leq \varphi_N \leq 1$ in $\mathbb{C}$. For large enough k, $|\widetilde{C}_{\nu}(1)(z) - \widetilde{C}_{\nu_k}(1)(z)|$ is no greater than the sum of $|\int_{\mathbb{C}}[K(z - \xi) - K(z_0 - \xi)]\varphi_N(\xi)d(\nu - \nu_k)(\xi)|$ and $|\int_{\mathbb{C}}[K(z - \xi) - K(z_0 - \xi)][1 - \varphi_N(\xi)]d(\nu - \nu_k)(\xi)|$. The first of these two terms tends to zero as $k \to \infty$, while the second has size at most $\frac{C|z - z_0|}{N}$ (for sufficiently large N) due to Lemma 8.1. This establishes the required convergence. □

Lemma 10.3. *Suppose that $z \notin \operatorname{supp}(\mu)$, and $\tilde{z}$ is a closest point on the support of μ to z. Let $e = \frac{\tilde{z} - z}{|\tilde{z} - z|}$. Then $\operatorname{supp}(\mu) \subset H_{\tilde{z}, e}$.*

Proof. Write $e = e^{i\theta}$. By translation, we may assume that $\tilde{z} = 0$. To prove the lemma, it suffices to show that $B(-\rho e, \rho) \cap \operatorname{supp}(\mu) = \varnothing$ for all $\rho > 0$.

Fix t_0 small enough to ensure that $te \notin \operatorname{supp}(\mu)$ for any $0 < t \leq t_0$. The existence of $t_0 > 0$ is guaranteed by Lemma 10.1. Now set $\sigma = \widetilde{C}_{\mu}(1)(z) - \widetilde{C}_{\mu}(1)(t_0 e)$. Notice that the value of σ is independent of the choice of $z_0 \notin \operatorname{supp}(\mu)$ in (12), so we shall fix $z_0 = t_0 e$. Now, let μ^* denote a tangent measure to μ at 0. On account of Lemma 10.1, the support of μ^* is contained in the line L through 0 perpendicular to e. By Step 1, $\mu^* = c^* \mathcal{H}_L^1$ with $c^* \in [c_0, C_0]$. As a result, for any $y > 0$, we have that

$$\widetilde{C}_{\mu^*}(1)(-ye) - \widetilde{C}_{\mu^*}(1)(ye) = \int_{\mathbb{R}} \frac{-2e^{-i\theta}y}{t^2 + y^2} c^* dm_1(t) = -2\pi c^* e^{-i\theta}. \tag{14}$$

We claim that $\widetilde{C}_{\mu^*}(1)(-ye) - \widetilde{C}_{\mu^*}(1)(ye) = \sigma$. To see this, note that for $\lambda > 0$ small enough so that $y\lambda \leq t_0$, we have $\widetilde{C}_{\mu_{0,\lambda}}(1)(-ye) - \widetilde{C}_{\mu_{0,\lambda}}(1)(ye) = \widetilde{C}_{\mu}(1)(-\lambda ye) - \widetilde{C}_{\mu}(1)(\lambda ye)$. But this equals σ because $-\lambda ye$ and λye lie in the same connected components of $\mathbb{C} \setminus \operatorname{supp}(\mu)$ as z and $t_0 e$ respectively. Since μ^* is a weak limit of measures μ_{0,λ_k} for some sequence $\lambda_k \to 0$, applying Lemma 10.2 proves the claim. Consequently, we have that σ determines the direction of tangency from z to $\mathrm{supp}(\mu)$ (the angle θ).

The right hand side of (14) is non-zero, and so $t_0 e$ lies in a different component of $\mathbb{C} \backslash \mathrm{supp}(\mu)$ to z. As there are only two possible values that $\widetilde{C}_\mu(1)$ can take in $\mathbb{C} \backslash \mathrm{supp}(\mu)$, σ is determined by $\widetilde{C}_\mu(1)(z)$. Since $\widetilde{C}_\mu(1)$ is constant in each connected component of $\mathbb{C} \backslash \mathrm{supp}(\mu)$, the direction of tangency from any point in the connected component of $\mathbb{C} \backslash \mathrm{supp}(\mu)$ containing z to $\mathrm{supp}(\mu)$ is the same. Finally, set

$$\mathcal{I} = \{\rho > 0 : \{-te : t \in (0, \rho]\} \text{ lies in the same connected component}$$

$$\text{of } \mathbb{C} \backslash \mathrm{supp}(\mu) \text{ as } z\}.$$

We claim that if $\rho \in \mathcal{I}$, then $B(-\rho e, \rho) \cap \mathrm{supp}(\mu) = \varnothing$. Indeed, otherwise there is a point $\zeta \neq 0$ which is a closest point in $\mathrm{supp}(\mu)$ to $-\rho e$. But then it follows that $e = \frac{\zeta + \rho e}{|\zeta + \rho e|}$. Given that $\{-te : t \in (0, \rho]\} \cap \mathrm{supp}(\mu) = \varnothing$, this is a contradiction. From this claim, we see that if $\rho \in \mathcal{I}$, then $(0, 2\rho) \subset \mathcal{I}$. Since $|z| \in \mathcal{I}$, it follows that $\mathcal{I} = (0, \infty)$, so $B(-\rho e, \rho) \cap \mathrm{supp}(\mu) = \varnothing$ for any $\rho > 0$. $\qquad\square$

Proof of Proposition 8.5. An immediate corollary of Lemma 10.3 is the following statement: For each $z \notin \mathrm{supp}(\mu)$, there is a half space with z on its boundary which does not intersect $\mathrm{supp}(\mu)$.

Now, suppose that there are three points $z, \xi, \zeta \in \mathrm{supp}(\mu)$ which are not collinear. Then they form a triangle. Since μ is AD-regular, there is a point in the interior of this triangle outside of the support of μ. Let's call this point ω. But then there is a half space, with ω on its boundary, which is disjoint from $\mathrm{supp}(\mu)$. This half space must contain at least one of the points z, ξ or ζ. This is absurd. $\qquad\square$

References

1. G. David, S. Semmes, *Analysis of and on Uniformly Rectifiable Sets.* Mathematical Surveys and Monographs, vol. 38 (American Mathematical Society, Providence, 1993)
2. P.W. Jones, Rectifiable sets and the traveling salesman problem. Invent. Math. **102**(1), 1–15 (1990)
3. J.C. Léger, Menger curvature and rectifiability. Ann. Math. **149**(3), 831–869 (1999)
4. P. Mattila, *Geometry of Sets and Measures in Euclidean Spaces. Fractals and Rectifiability.* Cambridge Studies in Advanced Mathematics, vol. 44 (Cambridge University Press, Cambridge, 1995)
5. P. Mattila, Cauchy singular integrals and rectifiability in measures of the plane. Adv. Math. **115**(1), 1–34 (1995)
6. P. Mattila, D. Preiss, Rectifiable measures in $\mathbb{R}^n$ and existence of principal values for singular integrals. J. Lond. Math. Soc. (2) **52**(3), 482–496 (1995)
7. P. Mattila, J. Verdera, Convergence of singular integrals with general measures. J. Eur. Math. Soc. **11**(2), 257–271 (2009)
8. P. Mattila, M. Melnikov, J. Verdera, The Cauchy integral, analytic capacity, and uniform rectifiability. Ann. Math. **144**, 127–136 (1996)

9. M. Melnikov, A. Poltoratski, A. Volberg, Uniqueness theorems for Cauchy integrals. Publ. Mat. **52**(2), 289–314 (2008)
10. F. Nazarov, X. Tolsa, A. Volberg, On the uniform rectifiability of AD regular measures with bounded Riesz transform operator: the case of codimension 1. Preprint (2012). arXiv:1212.5229
11. D. Preiss, Geometry of measures in $\mathbb{R}^n$: distribution, rectifiability, and densities. Ann. Math. (2) **125**(3), 537–643 (1987)

Logarithmically-Concave Moment Measures I

Bo'az Klartag

Abstract We discuss a certain Riemannian metric, related to the toric Kähler-Einstein equation, that is associated in a linearly-invariant manner with a given log-concave measure in $\mathbb{R}^n$. We use this metric in order to bound the second derivatives of the solution to the toric Kähler-Einstein equation, and in order to obtain spectral-gap estimates similar to those of Payne and Weinberger.

1 Introduction

In this paper we explore a certain geometric structure related to the *moment measure* of a convex function. This geometric structure is well-known in the community of complex geometers, see, e.g., Donaldson [13] for a discussion from the perspective of Kähler geometry.

Our motivation stems from the Kannan-Lovász-Simonovits conjecture [17, Sect. 5], which is concerned with the isoperimetric problem for high-dimensional convex bodies. Essentially, our idea is to replace the standard Euclidean metric by a special Riemannian metric on the given convex body K. This Riemannian metric has many favorable properties, such as a Poincaré inequality with constant one, a positive Ricci tensor, the linear functions are eigenfunctions of the Laplacian, etc. Perhaps this alternative geometry does not deviate too much from the standard Euclidean geometry on K, and it is conceivable that the study of this Riemannian metric will turn out to be relevant to the Kannan-Lovász-Simonovits conjecture.

Let μ be an arbitrary Borel probability measure on $\mathbb{R}^n$ whose barycenter is at the origin. Assume furthermore that μ is not supported in a hyperplane. It was proven in [12] that there exists an essentially-continuous convex function $\psi : \mathbb{R}^n \to \mathbb{R} \cup \{+\infty\}$, uniquely determined up to translations, such that μ is the *moment measure* of ψ, i.e.,

$$\int_{\mathbb{R}^n} b(y)d\mu(y) = \int_{\mathbb{R}^n} b(\nabla\psi(x))e^{-\psi(x)}dx \tag{1}$$

B. Klartag (✉)
School of Mathematical Sciences, Tel Aviv University, Tel Aviv 69978, Israel
e-mail: klartagb@tau.ac.il

© Springer International Publishing Switzerland 2014

B. Klartag, E. Milman (eds.), *Geometric Aspects of Functional Analysis*,
Lecture Notes in Mathematics 2116, DOI 10.1007/978-3-319-09477-9_16

for any μ-integrable function $b : \mathbb{R}^n \to \mathbb{R}$. In other words, the gradient map $x \mapsto \nabla \psi(x)$ pushes the probability measure $e^{-\psi(x)} dx$ forward to μ. The argument in [12] closely follows the variational approach of Berman and Berndtsson [5], which succeeded the continuity methods of Wang and Zhu [29] and Donaldson [13].

Even in the case where μ is absolutely-continuous with a C^∞-smooth density, it is not guaranteed that ψ is differentiable. From the regularity theory of the Brenier map, developed by Caffarelli [9] and Urbas [28], we learn that in order to conclude that ψ is sufficiently smooth, one has to assume that the support of μ is convex.

An absolutely-continuous probability measure on $\mathbb{R}^n$ is called *log-concave* if it is supported on an open, convex set $K \subset \mathbb{R}^n$, and its density takes the form $\exp(-\rho)$ where the function $\rho : K \to \mathbb{R}$ is convex. An important example of a log-concave measure is the uniform probability measure on a convex body in $\mathbb{R}^n$. Here we assume that μ is log-concave and furthermore, we require that the following conditions are met:

The convex set $K \subset \mathbb{R}^n$ is bounded, the function ρ is C^∞-smooth, and $\quad$ (2)
ρ and its derivatives of all orders are bounded in K.

Under these regularity assumptions, we can assert that

The convex function ψ is finite and C^∞-smooth in the entire $\mathbb{R}^n$. $\quad$ (3)

The validity of (3) under the assumption (2) was proven by Wang and Zhu [29] and by Donaldson [13] via the continuity method. Berman and Berndtsson [5] explained how to deduce (3) from (2) by using Caffarelli's regularity theory [9]. In fact, the argument in [5] requires only the boundness of ρ, and not of its derivatives, see also the Appendix in Alesker et al. [2]. Since the function ψ is smooth, it follows from (1) that the transport equation

$$e^{-\rho(\nabla \psi(x))} \det \nabla^2 \psi(x) = e^{-\psi(x)} \qquad (4)$$

holds everywhere in $\mathbb{R}^n$, where $\nabla^2 \psi(x)$ is the Hessian matrix of ψ. In the case where $\rho \equiv Const$, Eq. (4) is called the *toric Kähler-Einstein equation*. We write $x \cdot y$ for the standard scalar product of $x, y \in \mathbb{R}^n$, and $|x| = \sqrt{x \cdot x}$.

Theorem 1. *Let μ be a log-concave probability measure on $\mathbb{R}^n$ with barycenter at the origin that satisfies the regularity conditions (2). Then, with the above notation, for any $x \in \mathbb{R}^n$,*

$$\Delta \psi(x) \leq 2R^2(K)$$

where $R(K) = \sup_{x \in K} |x|$ is the outer radius of K, and $\Delta \psi = \sum_i \partial^2 \psi / \partial x_i^2$ is the Laplacian of ψ.

Theorem 1 is proven by analyzing a certain *weighted Riemannian manifold*. A weighted Riemannian manifold, sometimes called a *Riemannian metric-measure space*, is a triple

$$X = (\Omega, g, \mu)$$

where Ω is a smooth manifold (usually an open set in $\mathbb{R}^n$), where g is a Riemannian metric on Ω, and μ is a measure on Ω with a smooth density with respect to the Riemannian volume measure. In this paper we study the weighted Riemannian manifold

$$M_\mu^* = \left(\mathbb{R}^n, \nabla^2 \psi, e^{-\psi(x)} dx \right). \tag{5}$$

That is, the measure associated with M_μ^* has density $e^{-\psi}$ with respect to the Lebesgue measure on $\mathbb{R}^n$, and the Riemannian tensor on $\mathbb{R}^n$ which is induced by the Hessian of ψ is

$$\sum_{i,j=1}^n \psi_{ij} dx^i dx^j, \tag{6}$$

where we abbreviate $\psi_{ij} = \partial^2 \psi / \partial x^i \partial x^j$. There is also a dual description of M_μ^*. Recall that the Legendre transform of $f : \mathbb{R}^n \to \mathbb{R} \cup \{+\infty\}$ is the convex function

$$f^*(x) = \sup_{\substack{y \in \mathbb{R}^n \\ f(y) < +\infty}} [x \cdot y - f(y)] \qquad (x \in \mathbb{R}^n).$$

We refer the reader to Rockafellar [26] for the basic properties of the Legendre transform. Denote $\varphi = \psi^*$. From (4) we see that the Hessian matrix of the convex function ψ is always invertible, hence it is positive-definite. Therefore φ is a smooth function in K whose Hessian is always positive-definite. Consequently, the map $\nabla \varphi : K \to \mathbb{R}^n$ is a diffeomorphism, and $\nabla \psi$ is its inverse map. One may directly verify that the weighted Riemannian manifold M_μ^* is canonically isomorphic to

$$M_\mu = \left(K, \nabla^2 \varphi, \mu \right),$$

with $x \mapsto \nabla \psi(x)$ being the isomorphism map. In differential geometry, the isomorphism between M_μ and M_μ^* is the passage from complex coordinates to action/angle coordinates, see, e.g., Abreu [1]. Here are some basic properties of our weighted Riemannian manifold:

(i) The space M_μ is stochastically complete. That is, the diffusion process associated with M_μ is well-defined, it has μ as a stationary measure and "it never reaches the boundary of K".

(ii) The Bakry-Émery-Ricci tensor of M_μ is positive. In fact, it is at least half of the Riemannian metric tensor.

(iii) The Laplacian associated with M_μ has an interesting spectrum: The first non-zero eigenvalue is -1, and the corresponding eigenspace contains all linear functions.

Property (ii) is a particular case of the results of Kolesnikov [23, Theorem 4.3] (the notation of Kolesnikov is related to ours via $V = \Phi = \psi$), and properties (i) and (iii) are discussed below.

It is important to note that the construction of M_μ does not rely on the Euclidean structure: Suppose that V is a real n-dimensional linear space and μ is a probability measure on V satisfying the assumptions of Theorem 1. Then the convex function $\psi : V^* \to \mathbb{R}$ whose moment measure is μ is well-defined up to translations, and it induces the weighted Riemannian manifolds M_μ and M_μ^* via the procedure described above. The fact that M_μ is well-defined without any reference to a Euclidean structure is in sharp contrast with the Riemannian metric-measure space $(\mathbb{R}^n, |\cdot|, \mu)$ that is frequently used for the analysis of the log-concave measure μ.

In the following sections we prove the assertions made in the Introduction, and as a sample of possible applications, we explain below how to recover the classical Payne-Weinberger spectral gap inequality [25], up to a constant factor:

Corollary 1. *Let μ be a log-concave probability measure on $\mathbb{R}^n$ with barycenter at the origin that satisfies the regularity conditions (2). Then, for any μ-integrable, smooth function $f : K \to \mathbb{R}$,*

$$\int_K f^2 d\mu - \left(\int_K f d\mu \right)^2 \leq 2R^2(K) \int_K |\nabla f|^2 d\mu. \tag{7}$$

The constant $2R^2(K)$ on the right-hand side of (7) is not optimal. In the case where μ is the uniform probability measure on a convex body $K \subset \mathbb{R}^n$ with a central symmetry (i.e., $K = -K$), the best possible constant is $4R^2(K)/\pi^2$, see Payne and Weinberger [25].

Throughout this note, a convex body in $\mathbb{R}^n$ is a bounded, open, convex set. We write log for the natural logarithm. A smooth function or a smooth manifold are C^∞-smooth. The unit sphere is $S^{n-1} = \{x \in \mathbb{R}^n; |x| = 1\}$. The five sections below use a variety of techniques, from Itô calculus to maximum principles. We tried to make each section as independent of the others as possible.

2   Continuity of the Moment Measure

This section is concerned with the continuity of the correspondence between convex functions and their moment measures. Our main result here is Proposition 1 below. We say that a convex function $\psi : \mathbb{R}^n \to \mathbb{R}$ is *centered* if

$$\int_{\mathbb{R}^n} e^{-\psi(x)} dx = 1, \qquad \int_{\mathbb{R}^n} x_i e^{-\psi(x)} dx = 0, \ i = 1, \ldots, n. \qquad (8)$$

The role of the barycenter condition in (8) is to prevent translations of ψ which result in the same moment measure. It is well-known that any convex function $\psi :$ $\mathbb{R}^n \to \mathbb{R}$ satisfying $\int e^{-\psi} = 1$ must tend to $+\infty$ at infinity. More precisely, for any such convex function ψ there exist $A, B > 0$ with

$$\psi(x) \geq A|x| - B \qquad (x \in \mathbb{R}^n), \qquad (9)$$

see, e.g., [19, Lemma 2.1]).

Proposition 1. *Let $\Omega \subset \mathbb{R}^n$ be a compact set, and let $\psi, \psi_1, \psi_2, \ldots : \mathbb{R}^n \to \mathbb{R}$ be centered, convex functions. Denote by $\mu, \mu_1, \mu_2, \ldots$ the corresponding moment measures, which are assumed to be supported in Ω. Then the following are equivalent:*

(i) $\psi_\ell \longrightarrow \psi$ pointwise in $\mathbb{R}^n$.
(ii) $\mu_\ell \longrightarrow \mu$ weakly (i.e., $\int b d\mu_\ell \to \int b d\mu$ for any continuous function $b : \Omega \to \mathbb{R}$).

Several lemmas are required for the proof of Proposition 1. For a centered, convex function $\psi : \mathbb{R}^n \to \mathbb{R}$ we define

$$K(\psi) = \left\{ x \in \mathbb{R}^n ; \psi(x) \leq 2n + \inf_{y \in \mathbb{R}^n} \psi(y) \right\},$$

a convex set in $\mathbb{R}^n$. Since the barycenter of $e^{-\psi(x)} dx$ lies at the origin, then $\psi(0) \leq n + \inf_{x \in \mathbb{R}^n} \psi(x)$, according to Fradelizi [14]. Hence the origin is necessarily in the interior of $K(\psi)$. For $x \in \mathbb{R}^n$ consider the Minkowski functional

$$\|x\|_\psi = \inf \{\lambda > 0; x/\lambda \in K(\psi)\}.$$

Since a convex function is continuous, then $\psi(x/\|x\|_\psi) = 2n + \inf \psi$ for $0 \neq x \in \mathbb{R}^n$.

Lemma 1. *Let $\psi : \mathbb{R}^n \to \mathbb{R}$ be a centered, convex function. Then,*

$$\psi(x) \geq n\|x\|_\psi + \psi(0) - 2n \qquad (x \in \mathbb{R}^n). \qquad (10)$$

Proof. Since the barycenter of $e^{-\psi(x)} dx$ lies at the origin, from Fradelizi [14],

$$\psi(0) \leq n + \inf_{x \in \mathbb{R}^n} \psi(x). \qquad (11)$$

Whenever $x \in K(\psi)$ we have $\|x\|_\psi \leq 1$. Therefore (10) follows from (11) for $x \in K(\psi)$. In order to prove (10) for $x \notin K(\psi)$, we observe that for such x we have $\|x\|_\psi \geq 1$ and hence

$$\psi(0) + n \leq \inf_{y \in \mathbb{R}^n} \psi(y) + 2n = \psi\left(\frac{x}{\|x\|_\psi}\right) \leq \left(1 - \frac{1}{\|x\|_\psi}\right) \cdot \psi(0) + \frac{1}{\|x\|_\psi} \cdot \psi(x),$$

due to the convexity of ψ. We conclude that $\psi(x) \geq \psi(0) + n\|x\|_\psi$ for any $x \notin K(\psi)$, and (10) is proven in all cases.   □

Proof of the direction (i) $\Rightarrow$ (ii) in Proposition 1. Denote

$$K = \{x \in \mathbb{R}^n; \psi(x) \leq 2n + 1 + \psi(0)\},$$

a convex set containing a neighborhood of the origin. Since $e^{-\psi}$ is integrable, then K must be of finite volume, hence bounded. According to Rockafellar [26, Theorem 10.8], the convergence of ψ_ℓ to ψ is locally uniform in $\mathbb{R}^n$. In particular, the convergence is uniform on K, and there exists $\ell_0 \geq 1$ such that $\psi_\ell(x) > 2n + \psi_\ell(0)$ for any $x \in \partial K$ and $\ell \geq \ell_0$. Setting $M = \psi(0) - 1$ we conclude that

$$K(\psi_\ell) \subseteq K, \quad \psi_\ell(0) \geq M \qquad \text{for all } \ell \geq \ell_0. \tag{12}$$

Denote $R = \sup_{x \in K} |x|$. From (12) and Lemma 1, for any $\ell \geq \ell_0$,

$$\psi_\ell(x) \geq n\|x\|_{\psi_\ell} + \psi_\ell(0) - 2n \geq \frac{n}{R}|x| + (M - 2n) \qquad (x \in \mathbb{R}^n). \tag{13}$$

According to our assumption (i) and [26, Theorem 24.5] we have that

$$\nabla\psi_\ell(x) \overset{\ell \to \infty}{\longrightarrow} \nabla\psi(x)$$

for any $x \in \mathbb{R}^n$ in which $\psi, \psi_1, \psi_2, \ldots$ are differentiable. Let $b : \Omega \to \mathbb{R}$ be a continuous function. Since a convex function is differentiable almost everywhere, we conclude that

$$b(\nabla\psi_\ell(x))e^{-\psi_\ell(x)} \overset{\ell \to \infty}{\longrightarrow} b(\nabla\psi(x))e^{-\psi(x)} \quad \text{for almost any } x \in \mathbb{R}^n.$$

The function b is bounded because Ω is compact. We may use the dominated convergence theorem, thanks to (13), and conclude that

$$\int_\Omega b\, d\mu_\ell = \int_{\mathbb{R}^n} b(\nabla\psi_\ell(x))e^{-\psi_\ell(x)}\, dx \overset{\ell \to \infty}{\longrightarrow} \int_{\mathbb{R}^n} b(\nabla\psi(x))e^{-\psi(x)}\, dx = \int_\Omega b\, d\mu.$$

Thus (ii) is proven.   □

It still remains to prove the direction (ii) $\Rightarrow$ (i) in Proposition 1. A function $f : \mathbb{R}^n \to \mathbb{R}$ is L-Lipschitz if $|f(x) - f(y)| \leq L|x - y|$ for any $x, y \in \mathbb{R}^n$.

Lemma 2. *Let $L, \varepsilon > 0$. Suppose that $\psi : \mathbb{R}^n \to \mathbb{R}$ is a centered, L-Lipschitz, convex function, such that*

$$\int_{\mathbb{R}^n} |\nabla\psi(x) \cdot \theta| e^{-\psi(x)} dx \geq \varepsilon \qquad \text{for all } \theta \in S^{n-1}. \qquad (14)$$

Then,

$$\alpha|x| - \beta \leq \psi(x) \leq L|x| + \gamma \qquad (x \in \mathbb{R}^n), \qquad (15)$$

where $\alpha, \beta, \gamma > 0$ are constants depending only on L, ε and n.

Proof. Fix $\theta \in S^{n-1}$ and set $H = \theta^{\perp}$, the hyperplane orthogonal to θ. The function

$$m_\theta(y) = \inf_{t \in \mathbb{R}} \psi(y + t\theta) \qquad (y \in H)$$

is convex. Furthermore, for any fixed $y \in H$, the function $t \mapsto \psi(y + t\theta)$ is convex, L-Lipschitz and tends to $+\infty$ as $t \to \pm\infty$. Hence the one-dimensional convex function $t \mapsto \psi(y + t\theta)$ attains its minimum at a certain point $t_0 \in \mathbb{R}$, is non-decreasing on $[t_0, +\infty)$ and non-increasing on $(-\infty, t_0]$. Therefore, for any $y \in H$,

$$\int_{-\infty}^{\infty} \left| \frac{\partial\psi(y + t\theta)}{\partial t} \right| e^{-\psi(y+t\theta)} dt = \int_{-\infty}^{\infty} \left| \frac{\partial}{\partial t} e^{-\psi(y+t\theta)} \right| dt = 2e^{-m_\theta(y)}.$$

We now integrate over $y \in H$ and use Fubini's theorem to conclude that

$$\int_{\mathbb{R}^n} |\nabla\psi(x) \cdot \theta| e^{-\psi(x)} dx = 2 \int_H e^{-m_\theta(y)} dy. \qquad (16)$$

Consider the interval

$$I_\theta = \{t \in \mathbb{R} \,;\, t\theta \in K(\psi)\}. \qquad (17)$$

Then,

$$\int_{-\infty}^{\infty} e^{-\psi(t\theta)/2} dt \geq \int_{I_\theta} e^{-\psi(t\theta)/2} dt \geq e^{-n-\frac{m_\theta(0)}{2}} |I_\theta| \qquad (18)$$

where $|I_\theta|$ is the length of the interval I_θ. Fix a point $y \in H$. Then there exists $t_0 \in \mathbb{R}$ for which $m_\theta(y) = \psi(y + t_0\theta)$. From (18) and from the convexity of ψ,

$$\int_{-\infty}^{\infty} e^{-\psi\left(\frac{y}{2} + t\theta\right)} dt = \frac{1}{2} \int_{-\infty}^{\infty} e^{-\psi\left(\frac{y + t_0\theta}{2} + \frac{t\theta}{2}\right)} dt \geq \frac{1}{2} e^{-\frac{m_\theta(y)}{2}} \int_{-\infty}^{\infty} e^{-\frac{\psi(t\theta)}{2}} dt$$

$$\geq \frac{1}{2} e^{-\frac{m_\theta(y) + m_\theta(0)}{2}} e^{-n} |I_\theta| \geq \frac{1}{2} e^{-m_\theta(y)} e^{-2n} |I_\theta|, \tag{19}$$

where in the last passage we used that $m_\theta(0) \leq \psi(0) \leq n + \inf \psi \leq n + m_\theta(y)$, because the barycenter of $e^{-\psi(x)} dx$ lies at the origin. Integrating (19) over $y \in H$, we see that

$$\int_H e^{-m_\theta(y)} dy \leq \frac{2e^{2n}}{|I_\theta|} \int_H \int_{-\infty}^{\infty} e^{-\psi(\frac{y}{2} + t\theta)} dt dy = \frac{2^n e^{2n}}{|I_\theta|} \int_{\mathbb{R}^n} e^{-\psi} = \frac{2^n e^{2n}}{|I_\theta|}.$$

Combine the last inequality with (14) and (16). This leads to the bound

$$|I_\theta| \leq C_n \left(\int_{\mathbb{R}^n} |\nabla \psi(x) \cdot \theta| e^{-\psi(x)} dx \right)^{-1} \leq \frac{C_n}{\varepsilon}, \tag{20}$$

for some constant C_n depending only on n. Recall that the origin belongs to $K(\psi)$ and hence $0 \in I_\theta$. By letting θ range over all of S^{n-1} and glancing at (17) and (20), we see that

$$K(\psi) \subseteq B(0, C_n/\varepsilon) \tag{21}$$

where $B(x, r) = \{y \in \mathbb{R}^n; |y - x| \leq r\}$. From (21) and from Lemma 1,

$$\psi(x) \geq \psi(0) - 2n + n\|x\|_\psi \geq \psi(0) - 2n + \frac{\varepsilon}{\tilde{C}_n} |x| \qquad (x \in \mathbb{R}^n), \tag{22}$$

for $\tilde{C}_n = C_n/n$. By integrating (22) we obtain

$$1 = \int_{\mathbb{R}^n} e^{-\psi} \leq e^{-(\psi(0) - 2n)} \int_{\mathbb{R}^n} e^{-\varepsilon |x|/\tilde{C}_n} dx.$$

Therefore, $\psi(0) \leq \gamma$ for $\gamma = 2n + \log(\int_{\mathbb{R}^n} e^{-\varepsilon |x|/\tilde{C}_n} dx)$. Since ψ is L-Lipschitz, then the right-hand side inequality of (15) follows. Next, observe that

$$1 = \int_{\mathbb{R}^n} e^{-\psi(x)} dx \geq \int_{\mathbb{R}^n} e^{-\psi(0) - L|x|} dx = e^{-\psi(0)} \int_{\mathbb{R}^n} e^{-L|x|} dx.$$

Hence $\psi(0) \geq \log(\int_{\mathbb{R}^n} e^{-L|x|} dx)$, and the left-hand side inequality of (15) follows from (22). □

Proof of the direction (ii) $\Rightarrow$ (i) in Proposition 1.

Step 1. We claim that

$$\liminf_{\ell \to \infty} \left(\inf_{\theta \in S^{n-1}} \int_\Omega |x \cdot \theta| d\mu_\ell(x) \right) > 0. \tag{23}$$

Assume that (23) fails. Then there exist sequences $\ell_j \in \mathbb{N}$ and $\theta_j \in S^{n-1}$ such that

$$\lim_{j \to \infty} \int_\Omega |x \cdot \theta_j| d\mu_{\ell_j}(x) = 0. \tag{24}$$

Passing to a subsequence, if necessary, we may assume that $\theta_j \longrightarrow \theta_0 \in S^{n-1}$. The sequence of functions $|x \cdot \theta_j|$ tends to $|x \cdot \theta_0|$ uniformly in $x \in \Omega$. Hence, from (ii) and (24),

$$\int_\Omega |x \cdot \theta_0| d\mu(x) = \lim_{j \to \infty} \int_\Omega |x \cdot \theta_0| d\mu_{\ell_j}(x) = \lim_{j \to \infty} \int_\Omega |x \cdot \theta_j| d\mu_{\ell_j}(x) = 0.$$

Therefore μ is supported in the hyperplane $\theta_0^\perp$. However, μ is the moment measure of the convex function $\psi : \mathbb{R}^n \to \mathbb{R}$, and according to [12, Proposition 1], it cannot be supported in a hyperplane. We have thus arrived at a contradiction, and (23) is proven.

Step 2. We will prove that there exist $\alpha, \beta, \gamma > 0$ and $\ell_0 \geq 1$ such that

$$\alpha|x| - \beta \leq \psi_\ell(x) \leq L|x| + \gamma \qquad (\ell \geq \ell_0, x \in \mathbb{R}^n). \tag{25}$$

Indeed, according to Step 1, there exists $\ell_0 \geq 1$ and $\varepsilon_0 > 0$ such that

$$\int_{\mathbb{R}^n} |\nabla \psi_\ell(x) \cdot \theta| e^{-\psi_\ell(x)} dx = \int_\Omega |x \cdot \theta| d\mu_\ell(x) > \varepsilon_0 \qquad (\ell \geq \ell_0, \theta \in S^{n-1}). \tag{26}$$

Denote $L = \sup_{x \in \Omega} |x|$. The function ψ_ℓ is centered and convex. Furthermore, for almost any $x \in \mathbb{R}^n$ we know that $\nabla \psi_\ell(x) \in \Omega$, because the moment measure of ψ_ℓ is supported in Ω. Hence, for $\ell \geq 1$,

$$|\nabla \psi_\ell(x)| \leq L \qquad \text{for almost any } x \in \mathbb{R}^n. \tag{27}$$

Since a convex function is always locally-Lipschitz, then (27) implies that ψ_ℓ is L-Lipschitz, for any ℓ. We may now apply Lemma 2, thanks to (26), and conclude (25).

Step 3. Assume by contradiction that there exists $x_0 \in \mathbb{R}^n$ for which $\psi_\ell(x_0)$ does not converge to $\psi(x_0)$. Then there exist $\varepsilon > 0$ and a subsequence ℓ_j such that

$$|\psi_{\ell_j}(x_0) - \psi(x_0)| \geq \varepsilon \qquad (j = 1, 2, \ldots). \tag{28}$$

From (25) we know that the sequence of functions $\{\psi_{\ell_j}\}_{j=1,2,\ldots}$ is uniformly bounded on any compact subset of $\mathbb{R}^n$. Furthermore, ψ_{ℓ_j} is L-Lipschitz for any j. According to the Arzelá-Ascoli theorem, we may pass to a subsequence and assume that ψ_{ℓ_j} converges locally uniformly in $\mathbb{R}^n$, to a certain function F.

The function F is convex and L-Lipschitz, as it is the limit of convex and L-Lipschitz functions. Furthermore, thanks to (25) we may apply the dominated convergence theorem and conclude that F is centered.

To summarize, the functions $F, \psi_{\ell_1}, \psi_{\ell_2}, \ldots$ are L-Lipschitz, centered and convex. We know that $\psi_{\ell_j} \longrightarrow F$ locally uniformly in $\mathbb{R}^n$. According to the implication (i) $\Rightarrow$ (ii) proven above, the sequence of measures $\{\mu_{\ell_j}\}_{j=1,2,\ldots}$ converges weakly to the moment measure of F. But we assumed that μ_{ℓ_j} converges weakly to μ, and hence μ is the moment measure of F. Thus $\psi, F : \mathbb{R}^n \to \mathbb{R}$ are two centered, convex functions with the same moment measure μ. This means that $\psi \equiv F$, according to the uniqueness part in [12]. Therefore $\psi_{\ell_j} \longrightarrow \psi$ pointwise in $\mathbb{R}^n$, in contradiction to (28), and the proof is complete.

$\square$

3   A Preliminary Weak Bound Using the Maximum Principle

In this section we prove a rather weak form of Theorem 1, which will be needed for the proof of the theorem later on in Sect. 5. Throughout this section, μ is a log-concave probability measure on $\mathbb{R}^n$ with barycenter at the origin, supported on a convex body $K \subset \mathbb{R}^n$, with density $e^{-\rho}$ satisfying the regularity conditions (2). Also, $\psi : \mathbb{R}^n \to \mathbb{R}$ is the smooth, convex function whose moment measure is μ, which is uniquely defined up to translation, and $\varphi = \psi^*$ is its Legendre transform. In this section we make the following strict-convexity assumptions:

($\star$) The convex body K has a smooth boundary and its Gauss curvature is positive everywhere. Additionally, there exists $\varepsilon_0 > 0$ with

$$\nabla^2 \rho(x) \geq \varepsilon_0 \cdot Id \qquad (x \in K), \qquad (29)$$

in the sense of symmetric matrices.

Denote by $\|A\|$ the operator norm of the matrix A. Our goal in this section is to prove the following:

Proposition 2. *Under the above assumptions,*

$$\sup_{x \in \mathbb{R}^n} \|\nabla^2 \psi(x)\| < +\infty.$$

The argument we present for the demonstration of Proposition 2 closely follows the proof of Caffarelli's contraction theorem [10, Theorem 11]. An alternative approach to Proposition 2 is outlined in Kolesnikov [22, Sect. 6]. We begin the proof of Proposition 2 with the following lemma, which is due to Berman and Berndtsson [5]. Their proof is reproduced here for completeness.

Lemma 3. $\sup_{x \in K} \varphi(x) < +\infty.$

Proof. Since K is bounded, it suffices to show that φ is α-Hölder for some $\alpha > 0$. According to the Sobolev inequality in the convex domain $K \subset \mathbb{R}^n$ (see, e.g., [27, Chap. 1]), it is sufficient to prove that

$$\int_K |\nabla\varphi(x)|^p dx < +\infty, \tag{30}$$

for some $p > n$. Fix $p > n$. The map $x \mapsto \nabla\varphi(x)$ pushes the measure μ forward to $\exp(-\psi(x))dx$. Hence,

$$\int_K |\nabla\varphi|^p d\mu = \int_{\mathbb{R}^n} |x|^p e^{-\psi(x)} dx < +\infty, \tag{31}$$

where we used the fact that $e^{-\psi}$ decays exponentially at infinity (see, e.g., (9) above or [19, Lemma 2.1]). Since ρ is a bounded function on K and $e^{-\rho}$ is the density of μ, then (30) follows from (31). $\qquad\square$

For $x \in \mathbb{R}^n$ denote $h_K(x) = \sup_{y \in K} x \cdot y$, the supporting functional of K. The following lemma is analogous to [10, Lemma 4].

Lemma 4. $\lim\limits_{R \to \infty} \sup\limits_{|x| \geq R} |\nabla\psi(x) - \nabla h_K(x)| = 0.$

Proof. The function $\varphi : K \to \mathbb{R}$ is convex, hence bounded from below by some affine function, which in turn is greater than some constant on the bounded set K. According to Lemma 3, the function φ is also bounded from above. Set $M = \sup_{x \in K} |\varphi(x)|$. By elementary properties of the Legendre transform, for any $x \in \mathbb{R}^n$,

$$\psi(x) = x \cdot \nabla\psi(x) - \varphi(\nabla\psi(x)) \leq x \cdot \nabla\psi(x) + M. \tag{32}$$

Recall that $x/|x|$ is the outer unit normal to K at the boundary point $\nabla h_K(x)$ whenever $0 \neq x \in \mathbb{R}^n$, and that $\sup_{y \in K} x \cdot y = x \cdot \nabla h_K(x)$. Therefore, for any $x \in \mathbb{R}^n$,

$$\psi(x) = \sup_{y \in K} [x \cdot y - \varphi(y)] \geq -M + \sup_{y \in K} x \cdot y = -M + x \cdot \nabla h_K(x). \tag{33}$$

Using (32) and (33),

$$(\nabla h_K(x) - \nabla\psi(x)) \cdot \frac{x}{|x|} \leq \frac{2M}{|x|} \qquad (0 \neq x \in \mathbb{R}^n). \tag{34}$$

Recall that $\nabla\psi(x) \in K$ for any $x \in \mathbb{R}^n$. Since ∂K is smooth with positive Gauss curvature, inequality (34) implies that there exist $R_K, \alpha_K > 0$, depending only on K, with

$$|\nabla h_K(x) - \nabla\psi(x)| \leq \alpha_K \sqrt{\frac{2M}{|x|}} \qquad \text{for } |x| \geq R_K. \tag{35}$$

The lemma follows from (35). $\qquad\square$

For $\varepsilon > 0, \theta \in \mathbb{R}^n$ and a function $f : \mathbb{R}^n \to \mathbb{R}$ denote

$$\delta_{\theta\theta}^{(\varepsilon)} f(x) = f(x + \varepsilon\theta) + f(x - \varepsilon\theta) - 2f(x) \qquad (x \in \mathbb{R}^n).$$

For a smooth f and a small ε, the quantity $\delta_{\theta\theta}^{(\varepsilon)} f(x)/\varepsilon^2$ approximates the pure second derivative $f_{\theta\theta}(x)$. We would like to use the maximum principle for the function $\psi_{\theta\theta}(x)$, but we do not know whether or not it attains its supremum. This is the reason for using the approximate second derivative $\delta_{\theta\theta}^{(\varepsilon)} \psi(x)$ as a substitute.

Corollary 2. *Fix $0 < \varepsilon < 1$. Then the supremum of $\delta_{\theta\theta}^{(\varepsilon)} \psi(x)$ over all $x \in \mathbb{R}^n$ and $\theta \in S^{n-1}$ is attained.*

Proof. According to Lemma 4 and the continuity and 0-homogeneity of $\nabla h_K(x)$,

$$\lim_{R \to \infty} \sup_{\substack{|x| \geq R \\ x_1, x_2 \in B(x,1)}} |\nabla\psi(x_1) - \nabla\psi(x_2)| = \lim_{R \to \infty} \sup_{\substack{|x| \geq R \\ x_1, x_2 \in B(x,1)}} |\nabla h_K(x_1) - \nabla h_K(x_2)|$$

$$= \lim_{R \to \infty} \sup_{\substack{|x|=1 \\ x_1, x_2 \in B(x,1/R)}} |\nabla h_K(x_1) - \nabla h_K(x_2)| = 0, \tag{36}$$

where $B(x, r) = \{y \in \mathbb{R}^n; |x - y| < r\}$. From Lagrange's mean value theorem,

$$\delta_{\theta\theta}^{(\varepsilon)} \psi(x) = (\psi(x + \varepsilon\theta) - \psi(x)) - (\psi(x) - \psi(x - \varepsilon\theta))$$

$$\leq \varepsilon \sup_{x_1, x_2 \in B(x,\varepsilon)} |\nabla\psi(x_1) - \nabla\psi(x_2)|. \tag{37}$$

According to (36) and (37),

$$\lim_{R \to \infty} \sup_{\substack{|x| \geq R \\ \theta \in S^{n-1}}} \delta_{\theta\theta}^{(\varepsilon)} \psi(x) \leq \varepsilon \lim_{R \to \infty} \sup_{\substack{|x| \geq R \\ x_1, x_2 \in B(x,\varepsilon)}} |\nabla\psi(x_1) - \nabla\psi(x_2)| = 0. \tag{38}$$

Since ψ is convex and smooth, then the function $\delta_{\theta\theta}^{(\varepsilon)} \psi$ is non-negative and continuous in $(x, \theta) \in \mathbb{R}^n \times S^{n-1}$. It thus follows from (38) that its supremum is attained. $\qquad \square$

We shall apply the well-known matrix inequality, which states that when A and B are symmetric, positive-definite $n \times n$ matrices, then

$$\log \det B \leq \log \det A + Tr\left[A^{-1}(B - A)\right] = \log \det A + Tr\left[A^{-1}B\right] - n, \tag{39}$$

where $Tr(A)$ stands for the trace of the matrix A. Recall that the transport equation (4) is valid, hence,

$$\log \det \nabla^2 \psi(x) = -\psi(x) + (\rho \circ \nabla\psi)(x) \qquad (x \in \mathbb{R}^n). \tag{40}$$

In particular, $\nabla^2\psi(x)$ is always an invertible matrix which is in fact positive-definite. We denote its inverse by $\left(\nabla^2\psi(x)\right)^{-1} = (\psi^{ij}(x))_{i,j=1,\ldots,n}$. For a smooth function $u : \mathbb{R}^n \to \mathbb{R}$ denote

$$Au(x) = Tr\left[\left(\nabla^2\psi(x)\right)^{-1}\nabla^2 u(x)\right] = \psi^{ij}(x)u_{ij}(x) \qquad (x \in \mathbb{R}^n), \qquad (41)$$

where we adhere to the Einstein convention: When an index is repeated twice in an expression, once as a subscript and once as a superscript, then we sum over this index from 1 to n. According to (39) for any $\theta \in \mathbb{R}^n$,

$$\log \det \nabla^2\psi(x + \theta) \leq \log \det \nabla^2\psi(x) + \psi^{ij}(x)\psi_{ij}(x + \theta) - n \qquad (x \in \mathbb{R}^n), \tag{42}$$

with an equality for $\theta = 0$.

Proof of Proposition 2. We follow Caffarelli's argument [10, Theorem 11]. Our assumption (29) yields that the function $\rho(x) - \varepsilon_0|x|^2/2$ is convex. Hence, for any x, y such that $x - y, x + y, x \in K$,

$$\rho(x+y)+\rho(x-y)-2\rho(x) \geq \frac{\varepsilon_0}{2}\left(|x + y|^2 + |x - y|^2 - 2|x|^2\right) = \varepsilon_0|y|^2. \tag{43}$$

Fix $0 < \varepsilon < 1$ and abbreviate $\delta_{\theta\theta} f = \delta_{\theta\theta}^{(\varepsilon)} f$. From (40) and (42) as well as some simple algebraic manipulations, for any $\theta \in \mathbb{R}^n$,

$$A(\delta_{\theta\theta}\psi) \geq \delta_{\theta\theta}\left(\log \det \nabla^2\psi\right) = -\delta_{\theta\theta}\psi + \delta_{\theta\theta}(\rho \circ \nabla\psi). \tag{44}$$

According to Corollary 2, the maximum of $(x, \theta) \mapsto \delta_{\theta\theta}\psi(x)$ over $\mathbb{R}^n \times S^{n-1}$ is attained at some $(x_0, e) \in \mathbb{R}^n \times S^{n-1}$. Since ψ is smooth, then at the point x_0,

$$0 = \nabla(\delta_{ee}\psi)(x_0) = \nabla\psi(x_0 + \varepsilon e) + \nabla\psi(x_0 + \varepsilon e) - 2\nabla\psi(x_0).$$

In other words, there exists a vector $u \in \mathbb{R}^n$ such that

$$\nabla\psi(x_0 + \varepsilon e) = \nabla\psi(x_0) + u, \qquad \nabla\psi(x_0 - \varepsilon e) = \nabla\psi(x_0) - u.$$

Setting $v = \nabla\psi(x_0)$ and using (43), we obtain

$$\delta_{ee}(\rho \circ \nabla\psi)(x_0) = \rho(v + u) + \rho(v - u) - 2\rho(v) \geq \varepsilon_0|u|^2. \tag{45}$$

The smooth function $x \mapsto \delta_{ee}\psi(x)$ reaches a maximum at x_0, hence the matrix $\nabla^2\left(\delta_{ee}\psi\right)(x_0)$ is negative semi-definite. Since the matrix $(\nabla^2\psi)^{-1}(x_0)$ is positive-definite, then from the definition (41),

$$0 \geq A(\delta_{ee}\psi)(x_0). \tag{46}$$

Now, (44), (45) and (46) yield

$$\delta_{ee}\psi(x_0) \geq \delta_{ee}(\rho \circ \nabla\psi)(x_0) \geq \varepsilon_0|u|^2. \tag{47}$$

By the convexity of ψ,

$$\psi(x_0 + \varepsilon e) - \psi(x_0) \leq \nabla\psi(x_0 + \varepsilon e) \cdot (\varepsilon e) = (v + u) \cdot (\varepsilon e)$$

and

$$\psi(x_0 - \varepsilon e) - \psi(x_0) \leq \nabla\psi(x_0 - \varepsilon e) \cdot (-\varepsilon e) = (v - u) \cdot (-\varepsilon e).$$

Summing the last two inequalities yields

$$\delta_{ee}\psi(x_0) \leq (v + u) \cdot (\varepsilon e) + (v - u) \cdot (-\varepsilon e) = 2\varepsilon(u \cdot e) \leq 2|u|\varepsilon. \tag{48}$$

The inequalities (47) and (48) imply that $|u| \leq 2\varepsilon/\varepsilon_0$ and hence from (48),

$$\delta_{ee}(\psi)(x_0) \leq 4\varepsilon^2/\varepsilon_0.$$

Consequently, for any $x \in \mathbb{R}^n$ and $\theta \in S^{n-1}$ we have $\delta_{\theta\theta}^{(\varepsilon)}\psi(x) \leq 4\varepsilon^2/\varepsilon_0$, and hence

$$\psi_{\theta\theta}(x) = \lim_{\varepsilon \to 0+} \frac{\delta_{\theta\theta}^{(\varepsilon)}\psi(x)}{\varepsilon^2} \leq \frac{4}{\varepsilon_0}.$$

Therefore $\|\nabla^2\psi(x)\| \leq 4/\varepsilon_0$ for any $x \in \mathbb{R}^n$, and the proof is complete. $\square$

Remark 1. Our proof of Proposition 2 provides the explicit bound

$$\sup_{x \in \mathbb{R}^n} \|\nabla^2\psi(x)\| \leq 4/\varepsilon_0. \tag{49}$$

By arguing as in [11], one may improve the right-hand side of (49) to just $1/\varepsilon_0$. We omit the straightforward details.

4 Diffusion Processes and Stochastic Completeness

In this section we consider a diffusion process associated with transportation of measure. Our point of view owes much to the article by Kolesnikov [23], and we make an effort to maintain a discussion as general as the one in Kolesnikov's work.

Let μ be a probability measure supported on an open set $K \subseteq \mathbb{R}^n$, with density $e^{-\rho}$ where $\rho : K \to \mathbb{R}$ is a smooth function. Let $\psi : \mathbb{R}^n \to \mathbb{R}$ be a smooth, convex function with

$$\lim_{R \to \infty} \left(\inf_{|x| \geq R} \psi(x) \right) = +\infty. \tag{50}$$

Condition (50) holds automatically when $\int e^{-\psi} < \infty$, see (9) above. Rather than requiring that the transport equation (4) hold true, in this section we make the more general assumption that

$$e^{-\rho(\nabla \psi(x))} \det \nabla^2 \psi(x) = e^{-V(x)} \qquad (x \in \mathbb{R}^n) \tag{51}$$

for a certain smooth function $V : \mathbb{R}^n \to \mathbb{R}$. Clearly, when μ is the moment measure of ψ, Eq. (51) holds true with $V = \psi$ and condition (50) holds as well. The transport equation (51) means that the map $x \mapsto \nabla \psi(x)$ pushes the probability measure $e^{-V(x)} dx$ forward to μ. In this section we explain and prove the following:

Proposition 3. *Let $K \subseteq \mathbb{R}^n$ be an open set, and let $V, \psi : \mathbb{R}^n \to \mathbb{R}$ and $\rho : K \to \mathbb{R}$ be smooth functions with ψ being convex. Assume (50) and (51), and furthermore, that*

$$\inf_{x \in K} \nabla \rho(x) \cdot x > -\infty. \tag{52}$$

Then the weighted Riemannian manifold $M = \left(\mathbb{R}^n, \nabla^2 \psi, e^{-V(x)} dx \right)$ is stochastically complete.

Remark 2. Note that in the most interesting case where $V = \psi$, the weighted Riemannian manifold M from Proposition 3 coincides with M_μ^* as defined in (5) and (6) above. Additionally, in the case where μ is log-concave with barycenter at the origin, condition (52) does hold true: In this case, according to Fradelizi [14], we know that $\rho(0) \leq n + \inf_{x \in K} \rho(x)$. By convexity,

$$\nabla \rho(x) \cdot x \geq \rho(x) - \rho(0) \geq -n \qquad (x \in K),$$

and (52) follows. Thus Proposition 3 implies the stochastic completeness of M_μ^* when μ is a log-concave probability measure with barycenter at the origin, which satisfies the regularity conditions (2).

We now turn to a detailed explanation of *stochastic completeness* of a weighted Riemannian manifold. See, e.g., Grigor'yan [15] for more information. The *Dirichlet form* associated with the weighted Riemannian manifold $M = (\Omega, g, \nu)$ is defined as

$$\Gamma(u, v) = \int_\Omega g \left(\nabla_g u, \nabla_g v \right) d\nu, \tag{53}$$

where $u, v : \Omega \to \mathbb{R}$ are smooth functions for which the integral in (53) exists. Here, $\nabla_g u$ stands for the Riemannian gradient of u. The *Laplacian* associated with M is the unique operator L, acting on smooth functions $u : \Omega \to \mathbb{R}$, for which

$$\int_\Omega (Lu)v \, dv = -\Gamma(u, v) \tag{54}$$

for any compactly-supported, smooth function $v : \Omega \to \mathbb{R}$. In the case of the weighted manifold $M = (\mathbb{R}^n, \nabla^2 \psi, e^{-V(x)} dx)$ from Proposition 3, we may express the Dirichlet form as follows:

$$\Gamma(u, v) = \int_{\mathbb{R}^n} \left(\psi^{ij} u_i v_j \right) e^{-V} \tag{55}$$

where $\nabla^2 \psi(x)^{-1} = (\psi^{ij}(x))_{i,j=1,\dots,n}$ and $u_i = \partial u / \partial x^i$. Note that the matrix $\nabla^2 \psi(x)$ is invertible, thanks to (51). As in Sect. 3 above, we use the Einstein summation convention; thus in (55) we sum over i, j from 1 to n. We will also make use of abbreviations such as $\psi_{ijk} = \partial^3 \psi / (\partial x^i \partial x^j \partial x^k)$, and also $\psi^i_{j\ell} = \psi^{ik} \psi_{jk\ell}$ and $\psi^{ij}_k = \psi^{i\ell} \psi^{jm} \psi_{\ell m k}$. Therefore, for example,

$$(\psi^{ij})_k = \frac{\partial \psi^{ij}(x)}{\partial x^k} = -\psi^{i\ell} \psi^{jm} \psi_{\ell m k} = -\psi^{ij}_k.$$

We may now express the Laplacian L associated with $M = (\mathbb{R}^n, \nabla^2 \psi, e^{-V(x)} dx)$ by

$$Lu = \psi^{ij} u_{ij} - (\psi^{ij}_j + \psi^{ij} V_j) u_i \tag{56}$$

as may be directly verified from (55) by integration by parts.

Lemma 5. *For any smooth function $u : \mathbb{R}^n \to \mathbb{R}$,*

$$Lu = \psi^{ij} u_{ij} - \sum_{j=1}^n \rho_j(\nabla \psi(x)) u_j. \tag{57}$$

Proof. We take the logarithmic derivative of (51) and obtain that for $\ell = 1, \dots, n$,

$$\psi^i_{i\ell}(x) = -V_\ell(x) + \sum_{i=1}^n \rho_i(\nabla \psi(x)) \psi_{i\ell}(x) \qquad (x \in \mathbb{R}^n). \tag{58}$$

Multiplying (58) by $\psi^{j\ell}$ and summing over ℓ we see that for $j = 1, \dots, n$,

$$\psi^{ij}_i(x) = -\psi^{j\ell}(x) V_\ell(x) + \rho_j(\nabla \psi(x)) \qquad (x \in \mathbb{R}^n). \tag{59}$$

Now (57) follows from (56) and (59).   □

Lemma 6. *Under the assumptions of Proposition 3, there exists $A \geq 0$ such that for all $x \in \mathbb{R}^n$,*

$$(L\psi)(x) \leq A.$$

Proof. Set $A = \max\{0, n - \inf_{y \in K} \nabla\rho(y) \cdot y\}$, which is a finite number according to our assumption (52). From Lemma 5,

$$L\psi(x) = \psi^{ij}\psi_{ij} - \sum_{j=1}^{n} \rho_j(\nabla\psi(x))\psi_j(x) = n - \sum_{j=1}^{n} \rho_j(\nabla\psi(x))\psi_j(x).$$

It remains to prove that $n - \sum_j \rho_j(\nabla\psi(x))\psi_j(x) \leq A$, or equivalently, we need to show that

$$\nabla\rho(y) \cdot y \geq n - A \qquad \text{for all } y \in K. \tag{60}$$

However, (60) holds true in view of the definition of A above. Therefore $L\psi \leq A$ pointwise in $\mathbb{R}^n$. □

The Laplacian L associated with a weighted Riemannian manifold M is a second-order, elliptic operator with smooth coefficients. We say that M is *stochastically complete* if the Itô diffusion process whose generator is L is well-defined at all times $t \in [0, \infty)$. In the particular case of Proposition 3, this means the following: Let $(B_t)_{t \geq 0}$ be the standard n-dimensional Brownian motion. The diffusion equation with generator L as in (57) is the stochastic differential equation:

$$dY_t = \sqrt{2}\left(\nabla^2\psi(Y_t)\right)^{-1/2} dB_t - \nabla\rho(\nabla\psi(Y_t))dt, \tag{61}$$

where $(\nabla^2\psi(x))^{-1/2}$ is the positive-definite square root of $(\nabla^2\psi(x))^{-1}$. For background on stochastic calculus, the reader may consult sources such as Kallenberg [16] or Øksendal [24]. The *stochastic completeness* of M is equivalent to the existence of a solution $(Y_t)_{t \geq 0}$ to Eq. (61), with an initial condition $Y_0 = z$ for a fixed $z \in \mathbb{R}^n$, that does not explode in finite time. Proposition 3 therefore follows from the next proposition:

Proposition 4. *Let ψ, V and ρ be as in Proposition 3. Fix $z \in \mathbb{R}^n$. Then there exists a unique stochastic process $(Y_t)_{t \geq 0}$, adapted to the filtration induced by the Brownian motion, such that for all $t \geq 0$,*

$$Y_t = z + \int_0^t \sqrt{2}\left(\nabla^2\psi(Y_t)\right)^{-1/2} dB_t - \int_0^t \nabla\rho(\nabla\psi(Y_t))dt, \tag{62}$$

and such that almost-surely, the map $t \mapsto Y_t$ $(t \geq 0)$ is continuous in $[0, +\infty)$.

Proof. Since $\psi(x)$ tends to $+\infty$ when $x \to \infty$, then the convex set $\{\psi \leq R\} = \{x \in \mathbb{R}^n; \psi(x) \leq R\}$ is compact for any $R \in \mathbb{R}$. We use Theorem 21.3 in Kallenberg [16] and the remark following it. We deduce that there exists a unique continuous stochastic process $(Y_t)_{t \geq 0}$ and stopping times $T_k = \inf\{t \geq 0; \psi(Y_t) \geq k\}$ such that for any $k > \psi(z), t \geq 0$,

$$Y_{\min\{t,T_k\}} = z + \int_0^{\min\{t,T_k\}} \sqrt{2} \left(\nabla^2 \psi(Y_t)\right)^{-1/2} dB_t - \int_0^{\min\{t,T_k\}} \nabla\rho(\nabla\psi(Y_t))dt. \tag{63}$$

Denote $T = \sup_k T_k$. We would like to prove that $T = +\infty$ almost-surely. According to Dynkin's formula and Lemma 6, for any $k > \psi(z)$ and $t \geq 0$,

$$\mathbb{E}\psi(Y_{\min\{t,T_k\}}) = \psi(z) + \mathbb{E} \int_0^{\min\{t,T_k\}} (L\psi)(Y_t)dt \leq \psi(z) + 2At,$$

where A is the parameter from Lemma 6. Set $\alpha = -\inf_{x \in \mathbb{R}^n} \psi(x)$, a finite number in view of (50). Then $\psi(x) + \alpha$ is non-negative. By Markov-Chebyshev's inequality, for any $t \geq 0$ and $k > \psi(z)$,

$$\mathbb{P}(T_k \leq t) = \mathbb{P}\left(\psi(Y_{\min\{t,T_k\}}) \geq k\right) \leq \frac{\mathbb{E}\psi(Y_{\min\{t,T_k\}}) + \alpha}{k + \alpha} \leq \frac{2At + \psi(z) + \alpha}{k + \alpha}.$$

Hence, for any $t \geq 0$,

$$\mathbb{P}(T \leq t) \leq \inf_k \mathbb{P}(T_k \leq t) \leq \liminf_{k \to \infty} \frac{2At + \psi(z) + \alpha}{k + \alpha} = 0.$$

Therefore $T = +\infty$ almost surely. We may let k tend to infinity in (63) and deduce (62). The uniqueness of the continuous stochastic process $(Y_t)_{t \geq 0}$ that satisfies (62) follows from the uniqueness of the solution to (63). □

For $z \in \mathbb{R}^n$ write $(Y_t^{(z)})_{t \geq 0}$ for the stochastic process from Proposition 4 with $Y_0 = z$. Denote by ν the probability measure on $\mathbb{R}^n$ whose density is $e^{-V(x)}dx$. The lemma below is certainly part of the standard theory of diffusion processes. We were not able to find a precise reference, hence we provide a proof which relies on the existence of the heat kernel.

Lemma 7. *There exists a smooth function $p_t(x, y)$ $(x, y \in \mathbb{R}^n, t > 0)$ which is symmetric in x and y, such that for any $y \in \mathbb{R}^n$ and $t > 0$, the random vector*

$$Y_t^{(y)}$$

has density $x \mapsto p_t(x, y)$ with respect to ν.

Proof. We appeal to Theorems 7.13 and 7.20 in Grigor'yan [15], which deal with heat kernels on weighted Riemannian manifolds. According to these theorems, there exists a heat kernel, that is, a non-negative function $p_t(x, y)$ $(x, y \in \mathbb{R}^n, t > 0)$ symmetric in x and y and smooth jointly in (t, x, y), that satisfies the following two properties:

(i) For any $y \in \mathbb{R}^n$, the function $u(t, x) = p_t(x, y)$ satisfies

$$\frac{\partial u(t, x)}{\partial t} = L_x u(t, x) \qquad (x \in \mathbb{R}^n, t > 0)$$

where by $L_x u(t, x)$ we mean that the operator L is acting on the x-variables.

(ii) For any smooth, compactly-supported function $f : \mathbb{R}^n \to \mathbb{R}$ and $x \in \mathbb{R}^n$,

$$\int_{\mathbb{R}^n} p_t(x, y) f(y) d\nu(y) \xrightarrow{t \to 0^+} f(x), \tag{64}$$

and the convergence in (64) is locally uniform in $x \in \mathbb{R}^n$.

Theorem 7.13 in Grigor'yan [15] also guarantees that $\int p_t(x, y) d\nu(x) \leq 1$ for any y. It remains to prove that the random vector $Y_t^{(y)}$ has density $x \mapsto p_t(x, y)$ with respect to ν. Equivalently, we need to show that for any smooth, compactly-supported function $f : \mathbb{R}^n \to \mathbb{R}$ and $y \in \mathbb{R}^n, t > 0$,

$$\mathbb{E} f\left(Y_t^{(y)}\right) = \int_{\mathbb{R}^n} f(x) p_t(x, y) d\nu(x). \tag{65}$$

Denote by $v(t, y)$ $(t > 0, y \in \mathbb{R}^n)$ the right-hand side of (65), a smooth, bounded function. We also set $v(0, y) = f(y)$ $(y \in \mathbb{R}^n)$ by continuity, according to (ii). Then the function $v(t, y)$ is continuous and bounded in $(t, y) \in [0, +\infty) \times \mathbb{R}^n$. Since f is compactly-supported then we may safely differentiate under the integral sign with respect to y and t, and obtain

$$\frac{\partial v(t, y)}{\partial t} = \int_{\mathbb{R}^n} f(x) \frac{\partial p_t(x, y)}{\partial t} d\nu(y), \quad L_y v(t, y) = \int_{\mathbb{R}^n} f(x)\left(L_y p_t(x, y)\right) d\nu(y).$$

From (i) we learn that

$$\frac{\partial v(t, y)}{\partial t} = L_y v(t, y) \qquad (y \in \mathbb{R}^n, t > 0). \tag{66}$$

Fix $t_0 > 0$ and $y \in \mathbb{R}^n$. Denote $Z_t = v\left(t_0 - t, Y_t^{(y)}\right)$ for $0 \leq t \leq t_0$. Then $(Z_t)_{0 \leq t \leq t_0}$ is a continuous stochastic process. From Itô's formula and (66), for $0 \leq t \leq t_0$,

$$Z_t = Z_0 + R_t + \int_0^t \left[L_y v\left(t_0 - t, Y_t^{(y)}\right) - \frac{\partial v}{\partial t}\left(t_0 - t, Y_t^{(y)}\right)\right] dt = Z_0 + R_t$$

where $(R_t)_{0 \le t \le t_0}$ is a local martingale with $R_0 = 0$. Since v is bounded, then $(Z_t)_{0 \le t \le t_0}$ is a bounded process, and $(R_t)_{0 \le t \le t_0}$ is in fact a martingale. In particular $\mathbb{E} R_{t_0} = \mathbb{E} R_0 = 0$. Thus,

$$\mathbb{E} f \left(Y_{t_0}^{(y)} \right) = \mathbb{E} Z_{t_0} = \mathbb{E} Z_0 = v(t_0, y) = \int_{\mathbb{R}^n} f(x) p_{t_0}(x, y) dv(x),$$

and (65) is proven.   □

Corollary 3. *Suppose that Z is a random vector in $\mathbb{R}^n$, distributed according to v, independent of the Brownian motion $(B_t)_{t \ge 0}$ used for the construction of $(Y_t^{(z)})_{t \ge 0, z \in \mathbb{R}^n}$.*

Then, for any $t \ge 0$, the random vector $Y_t^{(Z)}$ is also distributed according to v.

Proof. According to Lemma 7, for any measurable set $A \subset \mathbb{R}^n$,

$$\mathbb{P} \left(Y_t^{(Z)} \in A \right) = \int_{\mathbb{R}^n} \mathbb{P} \left(Y_t^{(z)} \in A \right) dv(z) = \int_{\mathbb{R}^n} \left(\int_A p_t(z, x) dv(x) \right) dv(z)$$

$$= \int_A \left(\int_{\mathbb{R}^n} p_t(x, z) dv(z) \right) dv(x) = v(A).$$

                                                                                                                                                      □

Remark 3. Our choice to use stochastic processes in this paper is just a matter of personal taste. All of the arguments here can be easily rephrased in analytic terminology. For instance, the proof of Proposition 4 relies on the fact that $L\psi$ is bounded from above, similarly to the analytic approach in Grigor'yan [15, Sect. 8.4]. Another example is the use of local martingales towards the end of Lemma 7, which may be replaced by analytic arguments as in [15, Sect. 7.4].

5 Bakry-Émery Technique

In this section we prove Theorem 1. While the viewpoint and ideas of Bakry and Émery [4] are certainly the main source of inspiration for our analysis, we are not sure whether the abstract framework in [3, 4] entirely encompasses the subtlety of our specific weighted Riemannian manifold. For instance, Lemma 9 below seems related to the positivity of the *carré du champ* Γ_2 and to property (ii) in Sect. 1 above. In the case $\varepsilon \ge 1/2$, Lemma 9 actually follows from an application of [3, Lemma 2.4] with $f(x) = x^1$ and $\rho = 1/2$. Yet, in general, it appears to us advantageous to proceed by analyzing our model for itself, rather than viewing it as an abstract diffusion semigroup satisfying a curvature-dimension bound.

Let μ be a log-concave probability measure on $\mathbb{R}^n$ satisfying the regularity assumptions (2), whose barycenter lies at the origin. Let $\psi : \mathbb{R}^n \to \mathbb{R}$ be convex and smooth, such that the transport equation (4) holds true. In Sect. 4 we proved that M_μ^* is stochastically complete. Since M_{μ^*} is isomorphic to M_μ, then M_μ is also stochastically complete.

Let us describe in greater detail the diffusion process associated with $M_\mu = (K, \nabla^2 \varphi, \mu)$. Recall that the Legendre transform $\varphi = \psi^*$ is smooth and convex on K, and that

$$\varphi(x) + \psi(\nabla\varphi(x)) = x \cdot \nabla\varphi(x) \qquad (x \in K).$$

We may rephrase (4) in terms of $\varphi = \psi^*$, and using $(\nabla^2\varphi(x))^{-1} = \nabla^2\psi(\nabla\varphi(x))$, we arrive at the equation

$$\det \nabla^2\varphi(x) = e^{x \cdot \nabla\varphi(x) - \varphi(x) - \rho(x)} \qquad (x \in K). \qquad (67)$$

The Hessian matrix $\nabla^2\varphi$ is invertible everywhere, so we write $(\nabla^2\varphi(x))^{-1} = (\varphi^{ij}(x))_{i,j=1,\ldots,n}$, and as before we use abbreviations such as $\varphi_i^{jk} = \varphi^{j\ell}\varphi^{km}\varphi_{i\ell m}$. In this section, for a smooth function $u : K \to \mathbb{R}$, denote

$$Lu(x) = \varphi^{ij} u_{ij} - x^i u_i \qquad \text{for } x = (x^1, \ldots, x^n) \in K. \qquad (68)$$

The following lemma is "dual" to Lemma 5.

Lemma 8. *The operator L from (68) is the Laplacian associated with the weighted Riemannian manifold M_μ.*

Proof. By taking the logarithmic derivative of (67) and arguing as in the proof of Lemma 5, we obtain that for any $x \in K, i = 1, \ldots, n$,

$$\varphi_j^{ij} = x^i - \varphi^{ij}\rho_j. \qquad (69)$$

Integrating by parts and using (69), we see that for any two smooth functions $u, v : K \to \mathbb{R}$ with one of them compactly-supported,

$$\int_K \varphi^{ij} u_i v_j \, d\mu = -\int_K v(\varphi^{ij} u_{ij} - (\varphi_j^{ij} + \varphi^{ij}\rho_j)u_i)e^{-\rho} = -\int_K v(Lu)d\mu.$$

$\square$

Lemma 9. *Fix $\varepsilon > 0$. For $x \in K$ set $f(x) = \varphi^{11}(x)$. Then, for the function $f^\varepsilon(x) = f(x)^\varepsilon$ we have*

$$L(f^\varepsilon) + \varepsilon f^\varepsilon \geq 0.$$

Proof. For $i, j = 1, \ldots, n$,

$$f_i = (\varphi^{11})_i = -\varphi^{1k}\varphi^{1\ell}\varphi_{ik\ell}, \qquad f_{ij} = -\varphi_{ij}^{11} + 2\varphi_j^{1k}\varphi_{ik}^1.$$

Therefore,

$$Lf = \varphi^{ij} f_{ij} - x^i f_i = -\varphi_j^{11j} + 2\varphi_i^{1j} \varphi_j^{1i} + x^j \varphi_j^{11}. \tag{70}$$

Taking the logarithm of (67) and differentiating with respect to x^i and x^ℓ, we see that

$$\varphi_{ji\ell}^j - \varphi_i^{jk} \varphi_{jk\ell} = -\rho_{i\ell} + \varphi_{i\ell} + x^j \varphi_{i\ell j} \qquad (i, \ell = 1, \dots, n).$$

Multiplying by $\varphi^{1i} \varphi^{1\ell}$ and summing yields

$$\varphi_j^{j11} - \varphi_k^{1j} \varphi_j^{1k} = -\varphi^{1i} \varphi^{1\ell} \rho_{i\ell} + \varphi^{11} + x^j \varphi_j^{11}. \tag{71}$$

Since ρ is convex then its Hessian matrix is non-negative definite and $\rho_{i\ell} \varphi^{1i} \varphi^{1\ell} \geq 0$. From (70) and (71),

$$Lf = \varphi_k^{1j} \varphi_j^{1k} - \varphi^{11} + \rho_{i\ell} \varphi^{1i} \varphi^{1\ell} \geq \varphi_k^{1j} \varphi_j^{1k} - \varphi^{11} = \varphi_k^{1j} \varphi_j^{1k} - f. \tag{72}$$

The chain rule of the Laplacian is $L(\lambda(f)) = \lambda'(f)Lf + \lambda''(f)\varphi^{ij} f_i f_j$, as may be verified directly. Using the chain rule with $\lambda(t) = t^\varepsilon$ we see that (72) leads to

$$L(f^\varepsilon) = \varepsilon f^{\varepsilon-1} Lf + \varepsilon(\varepsilon - 1) f^{\varepsilon-2} \varphi^{11j} \varphi_j^{11}$$
$$\geq \varepsilon f^{\varepsilon-1} \varphi_k^{1j} \varphi_j^{1k} - \varepsilon f^\varepsilon + \varepsilon(\varepsilon - 1) f^{\varepsilon-2} \varphi^{11j} \varphi_j^{11}.$$

That is,

$$L(f^\varepsilon) + \varepsilon f^\varepsilon \geq \varepsilon f^{\varepsilon-1} \left[\varphi_k^{1j} \varphi_j^{1k} + (\varepsilon - 1) \frac{\varphi^{11j} \varphi_j^{11}}{\varphi^{11}} \right]$$
$$\geq \varepsilon f^{\varepsilon-1} \left[\varphi_k^{1j} \varphi_j^{1k} - \frac{\varphi^{11j} \varphi_j^{11}}{\varphi^{11}} \right], \tag{73}$$

where we used the fact that $\varphi^{11j} \varphi_j^{11} \geq 0$ in the last passage (or more generally, $\varphi^{ij} h_i h_j \geq 0$ for any smooth function h). It remains to show that the right-hand side of (73) is non-negative. Denote $A = (\varphi_k^{1j})_{j,k=1,\dots,n}$. The matrix $B = (\varphi^{1jk})_{j,k=1,\dots,n}$ is a symmetric matrix, since $\varphi^{1jk} = \varphi^{1\ell} \varphi^{jm} \varphi^{kr} \varphi_{\ell m r}$. We have $A = (\nabla^2 \varphi) B$, and hence

$$\varphi_k^{1j} \varphi_j^{1k} = Tr(A^2) = Tr\left[\left((\nabla^2 \varphi)^{1/2} B (\nabla^2 \varphi)^{1/2} \right)^2 \right]$$
$$= \left\| (\nabla^2 \varphi)^{1/2} B (\nabla^2 \varphi)^{1/2} \right\|_{HS}^2,$$

since the matrix $(\nabla^2\varphi)^{1/2}B(\nabla^2\varphi)^{1/2}$ is symmetric, where $\|T\|_{HS}$ stands for the Hilbert-Schmidt norm of the matrix T. We will use the fact that the Hilbert-Schmidt norm is at least as large as the operator norm, that is, $\|T\|^2_{HS} \geq |Tx|^2/|x|^2$ for any $0 \neq x \in \mathbb{R}^n$. Setting $e_1 = (1,0,\ldots,0)$, we conclude that

$$\varphi^{1j}_k\varphi^{1k}_j \geq \frac{\left|(\nabla^2\varphi)^{1/2}B(\nabla^2\varphi)^{1/2}(\nabla^2\varphi)^{-1/2}e_1\right|^2}{\left|(\nabla^2\varphi)^{-1/2}e_1\right|^2} = \frac{\varphi^{11i}\varphi_{ij}\varphi^{11j}}{\varphi^{11}} = \frac{\varphi^{11}_j\varphi^{11j}}{\varphi^{11}}.$$

Therefore the right-hand side of (73) is non-negative, and the lemma follows.   □

Let $(B_t)_{t\geq 0}$ be the standard n-dimensional Brownian motion. From the results of Sect. 4, the diffusion process whose generator is L from (68) is well-defined. That is, there exists a unique stochastic process $(X^{(z)}_t)_{t\geq 0, z\in K}$, continuous in t and adapted to the filtration induced by the Brownian motion, such that for all $t \geq 0$,

$$X^{(z)}_t = z + \int_0^t \sqrt{2}\left(\nabla^2\varphi\left(X^{(z)}_t\right)\right)^{-1/2} dB_t - \int_0^t X^{(z)}_t dt. \tag{74}$$

Our proof of Theorem 1 relies on a few lemmas in which the main technical obstacle is to prove the integrability of certain local martingales.

Lemma 10. *Fix $z \in K$ and set $X_t = X^{(z)}_t$ ($t \geq 0$). Then for any $t \geq 0$,*

$$\mathbb{E}X_t = e^{-t}z, \tag{75}$$

and for any $\theta \in S^{n-1}$,

$$e^{2t}\mathbb{E}(X_t \cdot \theta)^2 \geq (z \cdot \theta)^2 + 2\int_0^t e^{2s}\mathbb{E}\left[(\nabla^2\varphi)^{-1}(X_s)\theta \cdot \theta\right] ds. \tag{76}$$

Proof. From Itô's formula and (74),

$$d(e^t X_t) = e^t dX_t + e^t X_t dt = \sqrt{2}e^t\left(\nabla^2\varphi(X_t)\right)^{-1/2} dB_t.$$

Therefore $(e^t X_t)_{0\leq t\leq T}$ is a local martingale, for any fixed number $T > 0$. However, $e^t X_t \in e^T K$ for $0 \leq t \leq T$, and $K \subset \mathbb{R}^n$ is a bounded set. Therefore $(e^t X_t)_{0\leq t\leq T}$ is a bounded process, and hence it is a martingale. We conclude that

$$\mathbb{E}e^t X_t = \mathbb{E}e^0 X_0 = z \qquad (t \geq 0),$$

and (75) is proven. It remains to prove (76). Without loss of generality we may assume that $\theta = e_1 = (1,0,\ldots,0)$. Denoting $Y_t = X_t \cdot e_1$, we obtain from (74) that

$$dY_t = \sqrt{2}\left(\nabla^2\varphi(X_t)\right)^{-1/2}e_1 \cdot dB_t - Y_t dt.$$

Set $Z_t = e^{2t} Y_t^2 = e^{2t} (X_t \cdot e_1)^2$. According to Itô's formula,

$$dZ_t = 2e^{2t} Y_t^2 dt + 2e^{2t} Y_t dY_t + \frac{1}{2} \cdot (2e^{2t}) \cdot 2\varphi^{11}(X_t) dt = 2e^{2t} \varphi^{11}(X_t) dt + dM_t$$

where $(M_t)_{t \geq 0}$ is a local martingale with $M_0 = 0$. This implies that for any $t \geq 0$,

$$Z_t = (z \cdot e_1)^2 + M_t + \int_0^t \left(2e^{2s} \varphi^{11}(X_s) \right) ds. \tag{77}$$

Since φ^{11} is positive, then for any $t \geq 0$,

$$Z_t - (z \cdot e_1)^2 \geq M_t. \tag{78}$$

The convex body K is bounded, and hence $(Z_t)_{0 \leq t \leq T}$ is a bounded process for any number $T > 0$. According to (78), the local martingale $(M_t)_{0 \leq t \leq T}$ is bounded from above, and by Fatou's lemma it is a sub-martingale. In particular $\mathbb{E} M_t \geq \mathbb{E} M_0 = 0$ for any t. From (77),

$$\mathbb{E} Z_t \geq (z \cdot e_1)^2 + 2\mathbb{E} \int_0^t e^{2s} \varphi^{11}(X_s) ds \qquad (t \geq 0).$$

Since $\mathbb{E} Z_t < +\infty$ and φ^{11} is positive, we may use Fubini's theorem to conclude that for any $t \geq 0$,

$$\mathbb{E} Z_t \geq (z \cdot e_1)^2 + 2\int_0^t e^{2s} \mathbb{E} \varphi^{11}(X_s) ds.$$

$\square$

Remark 4. Once Theorem 1 is established, we can prove that equality holds in (76). Indeed, it follows from Theorem 1 and (77) that $(M_t)_{0 \leq t \leq T}$ is a bounded process and hence a martingale.

Lemma 11. *Assume that the convex body K has a smooth boundary and that its Gauss curvature is positive everywhere. Assume also that there exists $\varepsilon_0 > 0$ with*

$$\nabla^2 \rho(x) \geq \varepsilon_0 \cdot Id \qquad (x \in K) \tag{79}$$

in the sense of symmetric matrices. Fix $z \in K$ and set $X_t = X_t^{(z)}$ ($t \geq 0$). Denote $f(x) = \varphi^{11}(x)$ for $x \in K$. Then, for any $t, \varepsilon > 0$,

$$f(z) \leq e^t \left(\mathbb{E} f^\varepsilon(X_t) \right)^{1/\varepsilon}. \tag{80}$$

Proof. Our assumptions enable the application of Proposition 2. According to the conclusion of Proposition 2, there exists $M > 0$ such that

$$\nabla^2 \psi(y) \leq M \cdot Id \qquad (y \in \mathbb{R}^n).$$

Since $(\nabla^2 \varphi)^{-1}(x) = \nabla^2 \psi(\nabla \varphi(x))$, then,

$$f(x) = \varphi^{11}(x) \leq M \qquad (x \in K). \qquad (81)$$

From Itô's formula and (74),

$$e^{\varepsilon t} f^{\varepsilon}(X_t) = f^{\varepsilon}(z) + M_t + \int_0^t e^{\varepsilon s} \left[(Lf^{\varepsilon})(X_s) + \varepsilon f^{\varepsilon}(X_s) \right] ds, \qquad (82)$$

where M_t is a local martingale with $M_0 = 0$. According to (82) and Lemma 9, for any $t \geq 0$,

$$e^{\varepsilon t} f^{\varepsilon}(X_t) \geq f^{\varepsilon}(z) + M_t. \qquad (83)$$

We may now use (81) and (83) in order to conclude that the local martingale $(M_t)_{0 \leq t \leq T}$ is bounded from above, for any number $T > 0$. Hence it is a sub-martingale, and $\mathbb{E} M_t \geq \mathbb{E} M_0 = 0$ for any $t \geq 0$. Now (80) follows by taking the expectation of (83).   □

Remark 5. We will only use (80) for $\varepsilon = 1$, even though the statement for a small ε is much stronger. In the limit where ε tends to zero, it is not too difficult to prove that the right-hand side of (80) approaches $\exp(t + \mathbb{E} \log f(X_t))$.

The covariance matrix of a square-integrable random vector $Z = (Z_1, \ldots, Z_n) \in \mathbb{R}^n$ is defined to be

$$Cov(Z) = \left(\mathbb{E} Z_i Z_j - \mathbb{E} Z_i \cdot \mathbb{E} Z_j \right)_{i,j=1,\ldots,n}.$$

Corollary 4. *Assume that the convex body K has a smooth boundary and that its Gauss curvature is positive everywhere. Assume also that there exists $\varepsilon_0 > 0$ with*

$$\nabla^2 \rho(x) \geq \varepsilon_0 \cdot Id \qquad (x \in K). \qquad (84)$$

Then for any $z \in K$ and $t > 0$,

$$(\nabla^2 \varphi)^{-1}(z) \leq \frac{e^{2t}}{2(e^t - 1)} \cdot Cov\left(X_t^{(z)} \right)$$

in the sense of symmetric matrices.

Proof. Fix $z \in K, t > 0$ and $\theta \in S^{n-1}$. We need to prove that

$$\left(\nabla^2 \varphi(z) \right)^{-1} \theta \cdot \theta \leq \frac{e^{2t}}{2(e^t - 1)} Var(X_t^{(z)} \cdot \theta). \qquad (85)$$

Without loss of generality we may assume that $\theta = e_1 = (1, 0, \ldots, 0)$. We use Lemma 10 and also Lemma 11 with $\varepsilon = 1$, and obtain

$$e^{2t} \mathbb{E}(X_t^{(z)} \cdot e_1)^2 \geq (z \cdot e_1)^2 + 2 \int_0^t e^{2s} \mathbb{E}\varphi^{11}(X_s^{(z)}) ds \geq (z \cdot e_1)^2 + 2\varphi^{11}(z) \int_0^t e^s ds.$$

Recall that $\mathbb{E}X_t^{(z)} = e^{-t}z$, according to Lemma 10. Consequently,

$$\varphi^{11}(z) \leq \frac{e^{2t}}{2(e^t - 1)} \left(\mathbb{E}(X_t^{(z)} \cdot e_1)^2 - (e^{-t}z \cdot e_1)^2 \right) = \frac{e^{2t}}{2(e^t - 1)} Var(X_t^{(z)} \cdot e_1),$$

and (85) is proven for $\theta = e_1$.   □

Proof of Theorem 1. Assume first that the convex body K has a smooth boundary, that its Gauss curvature is positive everywhere, and that there exists ε_0 for which (84) holds true. We apply Corollary 4 with $t = \log 2$, and conclude that for any $z \in K$,

$$Tr\left[(\nabla^2\varphi)^{-1}(z)\right] \leq 2Tr\left[Cov(X_t^{(z)})\right] \leq 2\mathbb{E}\left|X_t^{(z)}\right|^2 \leq 2R^2(K)$$

as $X_t^{(z)} \in K$ almost surely. Therefore, for any $x \in \mathbb{R}^n$, setting $z = \nabla\psi(x)$ we have

$$\Delta\psi(x) = Tr\left[\nabla^2\psi(x)\right] = Tr\left[(\nabla^2\varphi)^{-1}(z)\right] \leq 2R^2(K). \tag{86}$$

It still remains to eliminate the extra strict-convexity assumptions. To that end, we select a sequence of smooth convex bodies $K_\ell \subset \mathbb{R}^n$, each with a positive Gauss curvature, that converge in the Hausdorff metric to K. We then consider a sequence of log-concave probability measures μ_ℓ with barycenter at the origin that converge weakly to μ, such that μ_ℓ is supported on K_ℓ and such that the smooth density of μ_ℓ satisfies (84) with, say, $\varepsilon_0 = 1/\ell$. We also assume that μ_ℓ and K_ℓ satisfy the regularity conditions (2).

It is not very difficult to construct the μ_ℓ's: For instance, convolve μ with a tiny Gaussian (this preserves log-concavity), multiply the density by $\exp(-|x|^2/\ell)$, truncate with K_ℓ and translate a little so that the barycenter would lie at the origin. This way we obtain a sequence of smooth, convex functions $\psi_\ell : \mathbb{R}^n \to \mathbb{R}$ such that μ_ℓ is the moment measure of ψ_ℓ. We may translate, and assume that ψ and each of the $\psi_\ell's$ are *centered*, in the terminology of Sect. 2. According to (86), we know that

$$\Delta\psi_\ell(x) \leq 2R^2(K_\ell) \qquad (x \in \mathbb{R}^n, \ell \geq 1). \tag{87}$$

Furthermore, $\mu_\ell \longrightarrow \mu$ weakly, and by Proposition 1, also $\psi_\ell \longrightarrow \psi$ pointwise in $\mathbb{R}^n$. Since ψ_ℓ and ψ are smooth, then [26, Theorem 24.5] implies that

$$\nabla\psi_\ell(x) \xrightarrow{\ell \to \infty} \nabla\psi(x) \qquad (x \in \mathbb{R}^n).$$

The function ψ_ℓ is $R(K_\ell)$-Lipschitz, and $R(K_\ell) \longrightarrow R(K)$. Hence $\sup_{\ell,x} |\nabla \psi_\ell(x)|$ is finite. By the bounded convergence theorem, for any $x_0 \in \mathbb{R}^n$ and $\varepsilon > 0$,

$$\int_{B(x_0,\varepsilon)} \Delta \psi_\ell = \int_{\partial B(x_0,\varepsilon)} \nabla \psi_\ell \cdot N \overset{\ell \to \infty}{\longrightarrow} \int_{\partial B(x_0,\varepsilon)} \nabla \psi \cdot N = \int_{B(x_0,\varepsilon)} \Delta \psi, \qquad (88)$$

where N is the outer unit normal. From (87) and (88) we conclude that for any $x_0 \in \mathbb{R}^n$ and $\varepsilon > 0$,

$$\int_{B(x_0,\varepsilon)} \Delta \psi \leq Vol_n(B(x_0,\varepsilon)) \cdot \limsup_{\ell \to \infty} 2R^2(K_\ell) = 2Vol_n(B(x_0,\varepsilon))R^2(K),$$

where Vol_n is the Lebesgue measure in $\mathbb{R}^n$. Since ψ is smooth, then we may let ε tend to zero and conclude that $\Delta \psi(x_0) \leq 2R^2(K)$, for any $x_0 \in \mathbb{R}^n$. $\qquad \square$

Posteriori, we may strengthen Corollary 4 and eliminate the strict-convexity assumptions. These assumptions were used only in the proof of Lemma 11, to deduce the existence of some number $M > 0$ for which $\nabla^2 \psi(x) \leq M \cdot Id$, for all $x \in \mathbb{R}^n$. Theorem 1 provides such a number $M = 2R^2(K)$, without any strict-convexity assumptions on ρ or K. We may therefore upgrade Corollary 4, and conclude that

Corollary 5. *Suppose that μ is a log-concave probability measure in $\mathbb{R}^n$ with barycenter at the origin, satisfying the regularity conditions (2). Let $(X_t^{(z)})_{t \geq 0, z \in K}$ be the stochastic process given by (74). Then this process is well-defined and bounded, and for any $z \in K$ and $t > 0$,*

$$(\nabla^2 \varphi)^{-1}(z) \leq \frac{e^{2t}}{2(e^t - 1)} \cdot Cov\left(X_t^{(z)}\right)$$

in the sense of symmetric matrices.

6   The Brascamp-Lieb Inequality as a Poincaré Inequality

We retain the assumptions and notation of the previous section. That is, μ is a log-concave probability measure on $\mathbb{R}^n$, with barycenter at the origin, that satisfies the regularity assumptions (2). The measure μ is the moment-measure of the smooth and convex function $\psi : \mathbb{R}^n \to \mathbb{R}$. Equation (4) holds true, and we denote $\varphi = \psi^*$. According to the Brascamp-Lieb inequality [8], for any smooth function $u : \mathbb{R}^n \to \mathbb{R}$ such that $ue^{-\psi}$ is integrable,

$$\int_{\mathbb{R}^n} ue^{-\psi} = 0 \quad \Longrightarrow \quad \int_{\mathbb{R}^n} u^2 e^{-\psi} \leq \int_{\mathbb{R}^n} \left[(\nabla^2 \psi)^{-1} \nabla u \cdot \nabla u\right] e^{-\psi}. \qquad (89)$$

Equality in (89) holds when $u(x) = \nabla\psi(x) \cdot \theta$ for some $\theta \in \mathbb{R}^n$. Note that (89) is precisely the Poincaré inequality with the best constant of the weighted Riemannian manifold M_μ^*. By using the isomorphism between M_μ and M_μ^*, we translate (89) as follows: For any smooth function $f : K \to \mathbb{R}$ which is μ-integrable,

$$Var_\mu(f) \leq \int_K \left(\varphi^{ij} f_i f_j\right) d\mu, \tag{90}$$

where $Var_\mu(f) = \int f^2 d\mu - (\int f d\mu)^2$. Equality in (90) holds when $f(x) = A + x \cdot \theta$ for some $\theta \in \mathbb{R}^n$ and $A \in \mathbb{R}$. This is in accordance with the fact that linear functions are eigenfunctions, i.e.,

$$Lx^i = -x^i \qquad (i = 1, \ldots, n)$$

where $Lu = \varphi^{ij} u_{ij} - x^i u_i$ is the Laplacian of the weighted Riemannian manifold M_μ. In fact, (90) means that the spectrum of the (Friedrich extension of the) operator L cannot intersect the interval $(-1, 0)$, and that the restriction of $-L$ to the subspace of mean-zero functions is at least the identity operator, in the sense of symmetric operators.

Theorem 1 states that $\Delta\psi(x) \leq 2R^2(K)$ everywhere in $\mathbb{R}^n$. A weak conclusion is that $\nabla^2\psi(x) \leq 2R^2(K) \cdot Id$, or rather, that $(\nabla^2\varphi(x))^{-1} \leq 2R^2(K) \cdot Id$. By substituting this information into (90), we see that for any smooth function $f \in L^1(\mu)$,

$$Var_\mu(f) \leq 2R^2(K) \int_K |\nabla f|^2 d\mu. \tag{91}$$

This completes the proof of Corollary 1. See [20, 21] for more Poincaré-type inequalities that are obtained by imposing a Riemannian structure on the convex body K. The Kannan-Lovász-Simonovits conjecture speculates that $R^2(K)$ in (91) may be replaced by a universal constant times $\|Cov(\mu)\|$, where $Cov(\mu)$ is the covariance matrix of the random vector that is distributed according to μ, and $\|\cdot\|$ is the operator norm.

A potential way to make progress towards the Kannan-Lovász-Simonovits conjecture is to try to bound the matrices $(\nabla^2\varphi)^{-1}(x)$ $(x \in K)$ in terms of $Cov(\mu)$. The following proposition provides a modest step in this direction:

Proposition 5. *Fix $\theta \in S^{n-1}$ and denote*

$$V = \int_{\mathbb{R}^n} (x \cdot \theta)^2 d\mu(x).$$

Then, for any $p \geq 1$,

$$\left(\int_K \left|\frac{(\nabla^2\varphi)^{-1}\theta \cdot \theta}{V}\right|^p d\mu\right)^{1/p} \leq 4p^2.$$

Proof. Without loss of generality, assume that $\theta = e_1 = (1, 0, \ldots, 0)$. According to Corollary 5, for any $z \in K$ and $t > 0$,

$$\varphi^{11}(z) \leq \frac{e^{2t}}{2(e^t - 1)} Var\left(X_t^{(z)} \cdot e_1\right) \leq \frac{e^{2t}}{2(e^t - 1)} \mathbb{E}\left(X_t^{(z)} \cdot e_1\right)^2. \tag{92}$$

Let Z be a random vector that is distributed according to μ, independent of the Brownian motion used in the construction of the process $(X_t^{(z)})_{t \geq 0, z \in K}$. It follows from Corollary 3 that for any fixed $t \geq 0$ the random vector $X_t^{(Z)}$ is also distributed according to μ. By setting $t = \log 2$ in (92) and applying Hölder's inequality, we see that for any $p \geq 1$,

$$\mathbb{E}\left|\varphi^{11}(Z)\right|^p \leq 2^p \mathbb{E}\left|X_t^{(Z)} \cdot e_1\right|^{2p} = 2^p \mathbb{E}|Z \cdot e_1|^{2p}. \tag{93}$$

The random vector Z has a log-concave density. According to the Berwald inequality [6, 7],

$$\left(\mathbb{E}|Z \cdot e_1|^{2p}\right)^{1/(2p)} \leq \frac{\Gamma(2p+1)^{1/(2p)}}{\Gamma(3)^{1/2}}\sqrt{\mathbb{E}|Z \cdot e_1|^2} \leq \frac{2p}{\sqrt{2}}\sqrt{V}. \tag{94}$$

(The Berwald inequality is formulated in [6, 7] for the uniform measure on a convex body, but it is well-known that is applies for all log-concave probability measures. For instance, one may deduce the log-concave version from the convex-body version by using a marginal argument as in [18]). The proposition follows from (93) and (94). □

Acknowledgements The author would like to thank Bo Berndtsson, Dario Cordero-Erausquin, Ronen Eldan, Alexander Kolesnikov, Eveline Legendre, Emanuel Milman, Ron Peled, Yanir Rubinstein and Boris Tsirelson for interesting discussions related to this work. Supported by a grant from the European Research Council (ERC).

References

1. M. Abreu, Kähler geometry of toric manifolds in symplectic coordinates, in *Symplectic and Contact Topology: Interactions and Perspectives*. Fields Institute Communications, vol. 35 (American Mathematical Society, Providence, 2003), pp. 1–24
2. S. Alesker, S. Dar, V. Milman, A remarkable measure preserving diffeomorphism between two convex bodies in $\mathbb{R}^n$. Geom. Dedicata **74**(2), 201–212 (1999)
3. D. Bakry, On Sobolev and logarithmic Sobolev inequalities for Markov semigroups, in *New trends in stochastic analysis (Charingworth, 1994)* (World Scientific, River Edge, 1997), pp. 43–75
4. D. Bakry, M. Émery, Diffusions hypercontractives, in *Séminaire de Probabilités, XIX, 1983/84*. Springer Lecture Notes in Mathematics, vol. 1123 (1985), pp. 177–206

5. R.J. Berman, B. Berndtsson, Real Monge-Ampère equations and Kähler-Ricci solitons on toric log Fano varieties. Ann. Fac. Sci. Toulouse Math. (6) **22**(4), 649–711 (2013)
6. L. Berwald, Verallgemeinerung eines Mittelwertsatzes von J. Favard, für positive konkave Funktionen. Acta Math. **79**, 17–37 (1947)
7. C. Borell, Complements of Lyapunov's inequality. Math. Ann. **205**, 323–331 (1973)
8. H.J. Brascamp, E.H. Lieb, On extensions of the Brunn-Minkowski and Prékopa-Leindler theorems, including inequalities for log concave functions, and with an application to the diffusion equation. J. Funct. Anal. **22**(4), 366–389 (1976)
9. L. Caffarelli, The regularity of mappings with a convex potential. J. Am. Math. Soc. **5**, 99–104 (1992)
10. L. Caffarelli, Monotonicity properties of optimal transportation and the FKG and related inequalities. Commun. Math. Phys. **214**(3), 547–563 (2000)
11. L. Caffarelli, Erratum: "Monotonicity of optimal transportation and the FKG and related inequalities". Commun. Math. Phys. **225**(2), 449–450 (2002)
12. D. Cordero-Erausquin, B. Klartag, Moment measures. Preprint. arXiv:1304.0630
13. S.K. Donaldson, Kähler geometry on toric manifolds, and some other manifolds with large symmetry, in *Handbook of Geometric Analysis*. Advanced Lectures in Mathematics (ALM), vol. 7, No. 1 (International Press, Somerville, 2008), pp. 29–75
14. M. Fradelizi, Sections of convex bodies through their centroid. Arch. Math. **69**(6), 515–522 (1997)
15. A. Grigor'yan, *Heat Kernel and Analysis on Manifolds*. AMS/IP Studies in Advanced Mathematics, vol. 47 (American Mathematical Society/International Press, Providence/Boston, 2009)
16. O. Kallenberg, *Foundation of Modern Probability*, 2nd edn. Probability and Its Applications (New York) (Springer, New York, 2002)
17. R. Kannan, L. Lovász, M. Simonovits, Isoperimetric problems for convex bodies and a localization lemma. Discrete Comput. Geom. **13**(3–4), 541–59 (1995)
18. B. Klartag, Marginals of geometric inequalities, in *Geometric Aspects of Functional Analysis*. Springer Lecture Notes in Mathematics, vol. 1910 (2007), pp. 133–166
19. B. Klartag, Uniform almost sub-gaussian estimates for linear functionals on convex sets. Algebra i Analiz (St. Petersburg Math. J.) **19**(1), 109–148 (2007)
20. B. Klartag, *Poincaré inequalities and moment maps*. Ann. Fac. Sci. Toulouse Math., **22**(1), 1–41 (2013)
21. B. Klartag, A.V. Kolesnikov, Eigenvalue distribution of optimal transportation. Preprint arXiv:1402.2636
22. A.V. Kolesnikov, On Sobolev regularity of mass transport and transportation inequalities. Theory Probab. Appl. **57**(2), 243–264 (2013)
23. A.V. Kolesnikov, Hessian metrics, $CD(K, N)$-spaces, and optimal transportation of log-concave measures. Discrete Contin. Dyn. Syst. **34**(4), 1511–1532 (2014)
24. B. Øksendal, *Stochastic Differential Equations. An Introduction with Applications*, 6th edn. Universitext (Springer, Berlin, 2003)
25. L.E. Payne, H.F. Weinberger, An optimal Poincaré inequality for convex domains. Arch. Ration. Mech. Anal. **5**, 286–292 (1960)
26. R.T. Rockafellar, *Convex Analysis*. Princeton Mathematical Series, No. 28 (Princeton University Press, Princeton, 1970)
27. L. Saloff-Coste, *Aspects of Sobolev-Type Inequalities*. London Mathematical Society Lecture Note Series, vol. 289 (Cambridge University Press, Cambridge, 2002)
28. J. Urbas, On the second boundary value problem for equations of Monge-Ampère type. J. Reine Angew. Math. **487**, 115–124 (1997)
29. X.-J. Wang, X. Zhu, Kähler-Ricci solitons on toric manifolds with positive first Chern class. Adv. Math. **188**, 87–103 (2004)

Estimates for Measures of Sections of Convex Bodies

Alexander Koldobsky

Abstract A $\sqrt{n}$ estimate in the hyperplane problem with arbitrary measures has recently been proved in [12]. In this note we present analogs of this result for sections of lower dimensions and in the complex case. We deduce these inequalities from stability in comparison problems for different generalizations of intersection bodies.

1 Introduction

The following inequality has recently been proved in [12]. Let K be an origin symmetric convex body in $\mathbb{R}^n$, and let μ be a measure on K with even continuous non-negative density f so that $\mu(B) = \int_B f$ for every Borel subset of K. Then

$$\mu(K) \leq \sqrt{n}\frac{n}{n-1}c_n \max_{\xi \in S^{n-1}} \mu(K \cap \xi^\perp)\,|K|^{1/n}\,, \qquad (1)$$

where $c_n = \left|B_2^n\right|^{\frac{n-1}{n}} / \left|B_2^{n-1}\right| < 1$, B_2^n is the unit Euclidean ball in $\mathbb{R}^n$, and $|K|$ stands for volume of proper dimension. Note that $c_n < 1$ for every n.

In the case of volume, when $f = 1$ everywhere on K, inequality (1) was proved in [17, p. 96]. Another argument follows from [6, Theorem 8.2.13]; in [6] this argument is attributed to Rolf Schneider. Also, in the case of volume the constant $\sqrt{n}$ can be improved to $Cn^{1/4}$, where C is an absolute constant, as shown by Klartag [9] who improved an earlier estimate of Bourgain [4]. These results are much more involved. The question of whether $n^{1/4}$ can also be removed in the case of volume is the matter of the hyperplane conjecture [1–3, 17]; see the book [5] for the current state of the problem.

In this note we prove analogs of inequality (1) for sections of lower dimensions and in the complex case; see Theorems 2 and 4, respectively. As in [12], the proofs are based on certain stability results for generalizations of intersection bodies.

A. Koldobsky (✉)
Department of Mathematics, University of Missouri, Columbia, MO 65211, USA
e-mail: koldobskiya@missouri.edu

© Springer International Publishing Switzerland 2014
B. Klartag, E. Milman (eds.), *Geometric Aspects of Functional Analysis*,
Lecture Notes in Mathematics 2116, DOI 10.1007/978-3-319-09477-9__17

2 Lower Dimensional Sections

We need several definitions and facts. A closed bounded set K in $\mathbb{R}^n$ is called a *star body* if every straight line passing through the origin crosses the boundary of K at exactly two points different from the origin, the origin is an interior point of K, and the *Minkowski functional* of K defined by

$$\|x\|_K = \min\{a \geq 0 : x \in aK\}$$

is a continuous function on $\mathbb{R}^n$.

The *radial function* of a star body K is defined by

$$\rho_K(x) = \|x\|_K^{-1}, \qquad x \in \mathbb{R}^n.$$

If $x \in S^{n-1}$ then $\rho_K(x)$ is the radius of K in the direction of x.

If μ is a measure on K with even continuous density f, then

$$\mu(K) = \int_K f(x)\, dx = \int_{S^{n-1}} \left(\int_0^{\|\theta\|_K^{-1}} r^{n-1} f(r\theta)\, dr \right) d\theta. \tag{2}$$

Putting $f = 1$, one gets

$$|K| = \frac{1}{n} \int_{S^{n-1}} \rho_K^n(\theta)\, d\theta = \frac{1}{n} \int_{S^{n-1}} \|\theta\|_K^{-n}\, d\theta. \tag{3}$$

For $1 \leq k \leq n - 1$, denote by Gr_{n-k} the Grassmanian of $(n - k)$-dimensional subspaces of $\mathbb{R}^n$. The $(n - k)$-*dimensional spherical Radon transform* $R_{n-k} : C(S^{n-1}) \mapsto C(Gr_{n-k})$ is a linear operator defined by

$$R_{n-k} g(H) = \int_{S^{n-1} \cap H} g(x)\, dx, \qquad \forall H \in Gr_{n-k}$$

for every function $g \in C(S^{n-1})$.

The polar formulas (2) and (3), applied to sections of K, express volume in terms of the spherical Radon transform:

$$\mu(K \cap H) = \int_{K \cap H} f = \int_{S^{n-1} \cap H} \left(\int_0^{\|\theta\|_K^{-1}} r^{n-k-1} f(r\theta)\, dr \right) d\theta$$

$$= R_{n-k} \left(\int_0^{\|\cdot\|_K^{-1}} r^{n-k-1} f(r\, \cdot)\, dr \right)(H). \tag{4}$$

and

$$|K \cap H| = \frac{1}{n-k} \int_{S^{n-1} \cap \xi^\perp} \|\theta\|_K^{-n+k} d\theta = \frac{1}{n-k} R_{n-k}(\|\cdot\|_K^{-n+k})(H). \quad (5)$$

The class of intersection bodies was introduced by Lutwak [15] and played a crucial role in the solution of the Busemann-Petty problem; see [6, 10] for definition and properties. A more general class of bodies was introduced by Zhang [18] in connection with the lower dimensional Busemann-Petty problem. Denote

$$R_{n-k}\left(C(S^{n-1})\right) = X \subset C(Gr_{n-k}).$$

Let $M^+(X)$ be the space of linear positive continuous functionals on X, i.e. for every $\nu \in M^+(X)$ and non-negative function $f \in X$, we have $\nu(f) \geq 0$.

An origin symmetric star body K in $\mathbb{R}^n$ is called a *generalized k-intersection body* if there exists a functional $\nu \in M^+(X)$ so that for every $g \in C(S^{n-1})$,

$$\int_{S^{n-1}} \|x\|_K^{-k} g(x)\, dx = \nu(R_{n-k}g). \quad (6)$$

When $k = 1$ we get the class of intersection bodies. It was proved by Grinberg and Zhang [7, Lemma 6.1] that every intersection body in $\mathbb{R}^n$ is a generalized k-intersection body for every $k < n$. More generally, as proved later by Milman [16], if m divides k, then every generalized m-intersection body is a generalized k-intersection body.

We need the following stability result for generalized k-intersection bodies.

Theorem 1. *Suppose that $1 \leq k \leq n-1$, K is a generalized k-intersection body in $\mathbb{R}^n$, f is an even continuous function on K, $f \geq 1$ everywhere on K, and $\varepsilon > 0$. If*

$$\int_{K \cap H} f \leq |K \cap H| + \varepsilon, \qquad \forall H \in Gr_{n-k}, \quad (7)$$

then

$$\int_K f \leq |K| + \frac{n}{n-k} c_{n,k} |K|^{k/n} \varepsilon, \quad (8)$$

where $c_{n,k} = |B_2^n|^{\frac{n-k}{n}}/|B_2^{n-k}| < 1$.

Proof. Use polar formulas (4) and (5) to write the condition (7) in terms of the $(n-k)$-dimensional spherical Radon transform: for all $H \in Gr_{n-k}$

$$R_{n-k}\left(\int_0^{\|\cdot\|_K^{-1}} r^{n-k-1} f(r \cdot)\, dr\right)(H) \leq \frac{1}{n-k} R_{n-k}\left(\|\cdot\|_K^{-n+k}\right)(H) + \varepsilon.$$

Let v be the functional corresponding to K by (6), apply v to both sides of the latter inequality (the direction of the inequality is preserved because v is a positive functional) and use (6). We get

$$\int_{S^{n-1}} \|\theta\|_K^{-k} \left(\int_0^{\|\theta\|_K^{-1}} r^{n-k-1} f(r\theta) \, dr \right) d\theta$$

$$\leq \frac{1}{n-k} \int_{S^{n-1}} \|\theta\|_K^{-n} \, d\theta + \varepsilon v(1). \tag{9}$$

Split the integral in the left-hand side into two integrals and then use $f \geq 1$ as follows:

$$\int_{S^{n-1}} \left(\int_0^{\|\theta\|_K^{-1}} r^{n-1} f(r\theta) \, dr \right) d\theta$$

$$+ \int_{S^{n-1}} \left(\int_0^{\|\theta\|_K^{-1}} (\|\theta\|_K^{-k} - r^k) r^{n-k-1} f(r\theta) \, dr \right) d\theta$$

$$\geq \int_K f + \int_{S^{n-1}} \left(\int_0^{\|\theta\|_K^{-1}} (\|\theta\|_K^{-k} - r^k) r^{n-k-1} \, dr \right) d\theta$$

$$= \int_K f + \frac{k}{n-k} |K|. \tag{10}$$

Now estimate $v(1)$ by first writing $1 = R_{n-k} 1/|S^{n-k-1}|$ and then using definition (6), Hölder's inequality and $|S^{n-1}| = n|B_2^n|$:

$$v(1) = \frac{1}{|S^{n-k-1}|} v(R_{n-k}1) = \frac{1}{|S^{n-k-1}|} \int_{S^{n-1}} \|\theta\|_K^{-k} \, d\theta$$

$$\leq \frac{1}{|S^{n-k-1}|} |S^{n-1}|^{\frac{n-k}{n}} \left(\int_{S^{n-1}} \|\theta\|_K^{-n} \, d\theta \right)^{\frac{k}{n}}$$

$$= \frac{1}{|S^{n-k-1}|} |S^{n-1}|^{\frac{n-k}{n}} n^{k/n} |K|^{k/n} = \frac{n}{n-k} c_{n,k} |K|^{k/n}. \tag{11}$$

Combining (9)–(11) we get

$$\int_K f + \frac{k}{n-k} |K| \leq \frac{n}{n-k} |K| + \frac{n}{n-k} c_{n,k} |K|^{k/n} \varepsilon.$$

$\square$

It was proved in [13] (generalizing the result for $k = 1$ from [11]) that if L is a generalized k-intersection body and μ is a measure with even continuous density, then

$$\mu(L) \leq \frac{n}{n-k} c_{n,k} \max_{H \in Gr_{n-k}} \mu(L \cap H) |L|^{k/n}.$$

We show now that it is possible to extend this inequality to arbitrary origin symmetric convex bodies in $\mathbb{R}^n$ at the expense of an extra constant $n^{k/2}$.

Theorem 2. *Suppose that L is an origin symmetric convex body in $\mathbb{R}^n$, and μ is a measure with even continuous non-negative density g on L. Then*

$$\mu(L) \leq n^{k/2} \frac{n}{n-k} c_{n,k} \max_{H \in Gr_{n-k}} \mu(L \cap H) |L|^{k/n}. \tag{12}$$

Proof. By John's theorem [8], there exists an origin symmetric ellipsoid K such that

$$\frac{1}{\sqrt{n}} K \subset L \subset K.$$

The ellipsoid K is an intersection body [6, Corollary 8.1.7], and every intersection body is a generalized k-intersection body for every k [7, Lemma 6.1]. Let $f = \chi_K + g\chi_L$, where χ_K, χ_L are the indicator functions of K and L, then $f \geq 1$ everywhere on K. Put

$$\varepsilon = \max_{H \in Gr_{n-k}} \left(\int_{K \cap H} f - |K \cap H| \right) = \max_{H \in Gr_{n-k}} \int_{L \cap H} g.$$

Now we can apply Theorem 1 to f, K, ε (the function f is not necessarily continuous on K, but the result holds by a simple approximation argument). We get

$$\mu(L) = \int_L g = \int_K f - |K|$$

$$\leq \frac{n}{n-k} c_{n,k} |K|^{k/n} \max_{H \in Gr_{n-k}} \int_{L \cap H} g$$

$$\leq n^{k/2} \frac{n}{n-k} c_{n,k} |L|^{k/n} \max_{H \in Gr_{n-k}} \mu(L \cap H),$$

because $K \subset \sqrt{n}L$, so $|K| \leq n^{n/2}|L|$. $\square$

3   The Complex Case

Origin symmetric convex bodies in $\mathbb{C}^n$ are the unit balls of norms on $\mathbb{C}^n$. We denote by $\| \cdot \|_K$ the norm corresponding to the body K:

$$K = \{z \in \mathbb{C}^n : \|z\|_K \leq 1\}.$$

In order to define volume, we identify $\mathbb{C}^n$ with $\mathbb{R}^{2n}$ using the standard mapping

$$\xi = (\xi_1, \ldots, \xi_n) = (\xi_{11} + i\xi_{12}, \ldots, \xi_{n1} + i\xi_{n2}) \mapsto (\xi_{11}, \xi_{12}, \ldots, \xi_{n1}, \xi_{n2}).$$

Since norms on $\mathbb{C}^n$ satisfy the equality

$$\|\lambda z\| = |\lambda| \|z\|, \quad \forall z \in \mathbb{C}^n, \ \forall \lambda \in \mathbb{C},$$

origin symmetric complex convex bodies correspond to those origin symmetric convex bodies K in $\mathbb{R}^{2n}$ that are invariant with respect to any coordinate-wise two-dimensional rotation, namely for each $\theta \in [0, 2\pi]$ and each $\xi = (\xi_{11}, \xi_{12}, \ldots, \xi_{n1}, \xi_{n2}) \in \mathbb{R}^{2n}$

$$\|\xi\|_K = \|R_\theta(\xi_{11}, \xi_{12}), \ldots, R_\theta(\xi_{n1}, \xi_{n2})\|_K, \tag{13}$$

where R_θ stands for the counterclockwise rotation of $\mathbb{R}^2$ by the angle θ with respect to the origin. We shall say that K is a *complex convex body in* $\mathbb{R}^{2n}$ if K is a convex body and satisfies Eq. (13). Similarly, complex star bodies are R_θ-invariant star bodies in $\mathbb{R}^{2n}$.

For $\xi \in \mathbb{C}^n, |\xi| = 1$, denote by

$$H_\xi = \{z \in \mathbb{C}^n : (z, \xi) = \sum_{k=1}^n z_k \overline{\xi_k} = 0\}$$

the complex hyperplane through the origin, perpendicular to ξ. Under the standard mapping from $\mathbb{C}^n$ to $\mathbb{R}^{2n}$ the hyperplane H_ξ turns into a $(2n - 2)$-dimensional subspace of $\mathbb{R}^{2n}$.

Denote by $C_c(S^{2n-1})$ the space of R_θ-invariant continuous functions, i.e. continuous real-valued functions f on the unit sphere S^{2n-1} in $\mathbb{R}^{2n}$ satisfying $f(\xi) = f(R_\theta(\xi))$ for all $\xi \in S^{2n-1}$ and all $\theta \in [0, 2\pi]$. The *complex spherical Radon transform* is an operator $\mathcal{R}_c : C_c(S^{2n-1}) \to C_c(S^{2n-1})$ defined by

$$\mathcal{R}_c f(\xi) = \int_{S^{2n-1} \cap H_\xi} f(x) dx.$$

We say that a finite Borel measure μ on S^{2n-1} is R_θ-invariant if for any continuous function f on S^{2n-1} and any $\theta \in [0, 2\pi]$,

$$\int_{S^{2n-1}} f(x) d\mu(x) = \int_{S^{2n-1}} f(R_\theta x) d\mu(x).$$

The complex spherical Radon transform of an R_θ-invariant measure μ is defined as a functional $\mathcal{R}_c \mu$ on the space $C_c(S^{2n-1})$ acting by

$$(\mathcal{R}_c \mu, f) = \int_{S^{2n-1}} \mathcal{R}_c f(x) d\mu(x).$$

Complex intersection bodies were introduced and studied in [14]. An origin symmetric complex star body K in $\mathbb{R}^{2n}$ is called a *complex intersection body* if there exists a finite Borel R_θ-invariant measure μ on S^{2n-1} so that $\| \cdot \|_K^{-2}$ and $\mathcal{R}_c \mu$ are equal as functionals on $C_c(S^{2n-1})$, i.e. for any $f \in C_c(S^{2n-1})$

$$\int_{S^{2n-1}} \|x\|_K^{-2} f(x) \, dx = \int_{S^{2n-1}} \mathcal{R}_c f(\theta) d\mu(\theta). \tag{14}$$

Theorem 3. *Suppose that K is a complex intersection body in $\mathbb{R}^{2n}$, f is an even continuous R_θ-invariant function on K, $f \geq 1$ everywhere on K, and $\varepsilon > 0$. If*

$$\int_{K \cap H_\xi} f \leq |K \cap H_\xi| + \varepsilon, \qquad \forall \xi \in S^{2n-1}, \tag{15}$$

then

$$\int_K f \leq |K| + \frac{n}{n-1} d_n |K|^{1/n} \varepsilon, \tag{16}$$

where $d_n = |B_2^{2n}|^{\frac{n-1}{n}} / |B_2^{2n-2}| < 1.$

Proof. Use the polar formulas (4) and (5) to write the condition (15) in terms of the complex spherical Radon transform: for all $\xi \in S^{2n-1}$

$$\mathcal{R}_c \left(\int_0^{\|\cdot\|_K^{-1}} r^{2n-3} f(r \cdot) \, dr \right)(\xi) \leq \frac{1}{2n-2} \mathcal{R}_c \left(\| \cdot \|_K^{-2n+2} \right)(\xi) + \varepsilon.$$

Let μ be the measure on S^{2n-1} corresponding to K by (14). Integrate the latter inequality over S^{2n-1} with the measure μ and use (14):

$$\int_{S^{2n-1}} \|\theta\|_K^{-2} \left(\int_0^{\|\theta\|_K^{-1}} r^{2n-3} f(r\theta) \, dr \right) d\theta$$

$$\leq \frac{1}{2n-2} \int_{S^{2n-1}} \|\theta\|_K^{-2n} \, d\theta + \varepsilon \int_{S^{2n-1}} d\mu(\xi)$$

$$= \frac{n}{n-1} |K| + \varepsilon \int_{S^{2n-1}} d\mu(\xi). \tag{17}$$

Recall (2), (3) and the assumption that $f \geq 1$. We estimate the integral in the left-hand side of (17) as follows:

$$\int_{S^{2n-1}} \|\theta\|_K^{-2} \left(\int_0^{\|\theta\|_K^{-1}} r^{2n-3} f(r\theta) \, dr \right) d\theta$$

$$= \int_{S^{2n-1}} \left(\int_0^{\|\theta\|_K^{-1}} r^{2n-1} f(r\theta) \, dr \right) d\theta$$

$$+ \int_{S^{2n-1}} \left(\int_0^{\|\theta\|_K^{-1}} (\|\theta\|_K^{-2} - r^2) r^{2n-3} f(r\theta) \, dr \right) d\theta$$

$$\geq \int_K f + \int_{S^{2n-1}} \left(\int_0^{\|\theta\|_K^{-1}} (\|\theta\|_K^{-2} - r^2) r^{2n-3} \, dr \right) d\theta$$

$$= \int_K f + \frac{1}{2(n-1)n} \int_{S^{2n-1}} \|\theta\|_K^{-2n} \, d\theta = \int_K f + \frac{1}{n-1} |K|. \quad (18)$$

Let us estimate the second term in the right-hand side of (17) by adding the complex spherical Radon transform of the unit constant function under the integral ($\mathcal{R}_c 1(\xi) = |S^{2n-3}|$ for every $\xi \in S^{2n-1}$), using again (14) and then applying Hölder's inequality:

$$\varepsilon \int_{S^{2n-1}} d\mu(\xi) = \frac{\varepsilon}{|S^{2n-3}|} \int_{S^{2n-1}} \mathcal{R}_c 1(\xi) \, d\mu(\xi)$$

$$= \frac{\varepsilon}{|S^{2n-3}|} \int_{S^{2n-1}} \|\theta\|_K^{-2} \, d\theta$$

$$\leq \frac{\varepsilon}{|S^{2n-3}|} |S^{2n-1}|^{\frac{n-1}{n}} \left(\int_{S^{2n-1}} \|\theta\|_K^{-2n} \, d\theta \right)^{\frac{1}{n}}$$

$$= \frac{\varepsilon}{|S^{2n-3}|} |S^{2n-1}|^{\frac{n-1}{n}} (2n)^{1/n} |K|^{1/n} = \frac{n}{n-1} d_n |K|^{1/n} \varepsilon. \quad (19)$$

In the last step we used $|S^{2n-1}| = 2n|B_2^{2n}|$. Combining (17)–(19) we get

$$\int_K f + \frac{1}{n-1} |K| \leq \frac{n}{n-1} |K| + \frac{n}{n-1} d_n |K|^{1/n} \varepsilon.$$

$\square$

It was proved in [14] that if K is a complex intersection body in $\mathbb{R}^{2n}$ and γ is an arbitrary measure on $\mathbb{R}^{2n}$ with even continuous density, then

$$\gamma(K) \leq \frac{n}{n-1} d_n \max_{\xi \in S^{2n-1}} \gamma(K \cap H_\xi) |K|^{\frac{1}{n}}.$$

In Theorem 4 below, we remove the condition that K is a complex intersection body at the expense of an extra constant. We use a result from [14, Theorem 4.1] that a complex star body is a complex intersection body if and only if $\| \cdot \|_K^{-2}$ is a positive definite distribution, i.e. its Fourier transform in the sense of distributions assumes non-negative values on non-negative test functions. We refer the reader to [10, 14] for details.

Theorem 4. *Suppose that L is an origin symmetric complex convex body in $\mathbb{R}^{2n}$ and γ is an arbitrary measure on $\mathbb{R}^{2n}$ with even continuous density g, then*

$$\gamma(L) \leq 2n \frac{n}{n-1} d_n \max_{\xi \in S^{2n-1}} \gamma(L \cap H_\xi) \, |L|^{\frac{1}{n}}.$$

Proof. By John's theorem [8], there exists an origin symmetric ellipsoid K such that

$$\frac{1}{\sqrt{2n}} K \subset L \subset K.$$

Construct a new body K_c by

$$\|x\|_{K_c}^{-2} = \frac{1}{2\pi} \int_0^{2\pi} \|R_\theta x\|_K^{-2} d\theta.$$

Clearly, K_c is R_θ-invariant, so it is a complex star body. For every $\theta \in [0, 2\pi]$ the distribution $\|R_\theta x\|_K^{-2}$ is positive definite, because this is a linear transformation of the Euclidean norm. So $\|x\|_{K_c}^{-2}$ is also a positive definite distribution, and, by Koldobsky et al. [14, Theorem 4.1], K_c is a complex intersection body. Since $\frac{1}{\sqrt{2n}} K \subset L \subset K$ and L is R_θ-invariant as a complex convex body, we have

$$\frac{1}{\sqrt{2n}} R_\theta K \subset L \subset R_\theta K, \quad \forall \theta \in [0, 2\pi],$$

so

$$\frac{1}{\sqrt{2n}} K_c \subset L \subset K_c.$$

Let $f = \chi_{K_c} + g\chi_L$, where χ_{K_c}, χ_L are the indicator functions of K_c and L. Clearly, f is R_θ-invariant and $f \geq 1$ everywhere on K. Put

$$\varepsilon = \max_{\xi \in S^{2n-1}} \left(\int_{K_c \cap H_\xi} f - |K_c \cap H_\xi| \right) = \max_{\xi \in S^{2n-1}} \int_{L \cap H_\xi} g.$$

and apply Theorem 3 to f, K_c, ε (the function f is not necessarily continuous on K_c, but the result holds by a simple approximation argument). We get

$$
\begin{aligned}
\mu(L) = \int_L g = \int_{K_c} f - |K_c| \\
\le \frac{n}{n-1} d_n |K_c|^{1/n} \max_{\xi \in S^{2n-1}} \int_{L \cap H_\xi} g \\
\le 2n \, \frac{n}{n-1} d_n |L|^{1/n} \max_{\xi \in S^{2n-1}} \mu(L \cap H_\xi),
\end{aligned}
$$

because $|K_c|^{1/n} \le 2n \, |L|^{1/n}$.   □

Theorem 4 shows that if bodies have additional symmetries then maximum in the slicing inequality can be taken over a rather small set of subspaces.

Acknowledgements I wish to thank the US National Science Foundation for support through grant DMS-1265155.

References

1. K. Ball, Logarithmically concave functions and sections of convex sets in R^n. Studia Math. **88**, 69–84 (1988)
2. J. Bourgain, On high-dimensional maximal functions associated to convex bodies. Am. J. Math. **108**, 1467–1476 (1986)
3. J. Bourgain, Geometry of Banach spaces and harmonic analysis, in *Proceedings of the International Congress of Mathematicians (Berkeley, 1986)* (American Mathematical Society, Providence, 1987), pp. 871–878
4. J. Bourgain, On the distribution of polynomials on high-dimensional convex sets, in *Geometric Aspects of Functional Analysis, Israel Seminar (1989–90)*. Lecture Notes in Mathematics, vol. 1469 (Springer, Berlin, 1991), pp. 127–137
5. S. Brazitikos, A. Giannopoulos, P. Valettas, B. Vritsiou, *Geometry of Isotropic Log-Concave Measures* (American Mathematical Society, Providence, 2014)
6. R.J. Gardner, *Geometric Tomography*, 2nd edn. (Cambridge University Press, Cambridge, 2006)
7. E. Grinberg, G. Zhang, Convolutions, transforms and convex bodies. Proc. Lond. Math. Soc. **78**, 77–115 (1999)
8. F. John, *Extremum Problems with Inequalities as Subsidiary Conditions*. Courant Anniversary Volume (Interscience, New York, 1948), pp. 187–204
9. B. Klartag, On convex perturbations with a bounded isotropic constant. Geom. Funct. Anal. **16**, 1274–1290 (2006)
10. A. Koldobsky, *Fourier Analysis in Convex Geometry* (American Mathematical Society, Providence, 2005)
11. A. Koldobsky, A hyperplane inequality for measures of convex bodies in $\mathbb{R}^n, n \le 4$. Dicrete Comput. Geom. **47**, 538–547 (2012)
12. A. Koldobsky, A $\sqrt{n}$ estimate for measures of hyperplane sections of convex bodies. Adv. Math. **254**, 33–40 (2014)

13. A. Koldobsky, D. Ma, Stability and slicing inequalities for intersection bodies. Geom. Dedicata **162**, 325–335 (2013)
14. A. Koldobsky, G. Paouris, M. Zymonopoulou, Complex intersection bodies. J. Lond. Math. Soc. **88**(2), 538–562 (2013)
15. E. Lutwak, Intersection bodies and dual mixed volumes. Adv. Math. **71**, 232–261 (1988)
16. E. Milman, Generalized intersection bodies. J. Funct. Anal. **240**(2), 530–567 (2006)
17. V. Milman, A. Pajor, Isotropic position and inertia ellipsoids and zonoids of the unit ball of a normed n-dimensional space, in *Geometric Aspects of Functional Analysis*, ed. by J. Lindenstrauss, V. Milman. Lecture Notes in Mathematics, vol. 1376 (Springer, Heidelberg, 1989), pp. 64–104
18. G. Zhang, Section of convex bodies. Am. J. Math. **118**, 319–340 (1996)

Remarks on the KLS Conjecture and Hardy-Type Inequalities

Alexander V. Kolesnikov and Emanuel Milman

Abstract We generalize the classical Hardy and Faber-Krahn inequalities to arbitrary functions on a convex body $\Omega \subset \mathbb{R}^n$, not necessarily vanishing on the boundary $\partial\Omega$. This reduces the study of the Neumann Poincaré constant on Ω to that of the cone and Lebesgue measures on $\partial\Omega$; these may be bounded via the curvature of $\partial\Omega$. A second reduction is obtained to the class of harmonic functions on Ω. We also study the relation between the Poincaré constant of a log-concave measure μ and its associated K. Ball body K_μ. In particular, we obtain a simple proof of a conjecture of Kannan–Lovász–Simonovits for unit-balls of ℓ_p^n, originally due to Sodin and Latała–Wojtaszczyk.

1 Introduction

Given a compact connected set Ω with non-empty interior in Euclidean space $(\mathbb{R}^n, |\cdot|)$ ($n \geq 2$) and a smooth function f on Ω vanishing on $\partial\Omega$, a version of the classical Hardy inequality (e.g. [15]) states that:

$$\int_\Omega f^2 dx \leq \frac{4}{n^2} \inf_{x_0 \in \mathbb{R}^n} \int_\Omega |x - x_0|^2 |\nabla f|^2 dx. \qquad (1)$$

The classical Faber–Krahn inequality (e.g. [4]) states that under the same conditions:

$$\int_\Omega f^2 dx \leq P_{\Omega*}^D \int_\Omega |\nabla f|^2 dx, \qquad (2)$$

where $P_{\Omega*}^D$ is the best constant in the above inequality under the same conditions with $\Omega = \Omega^*$, the Euclidean Ball having the same volume as Ω. P^D is called the

A.V. Kolesnikov
Faculty of Mathematics, National Research University Higher School of Economics, Moscow, Russia
e-mail: sascha77@mail.ru

E. Milman (✉)
Department of Mathematics, Technion - Israel Institute of Technology, Haifa 32000, Israel
e-mail: emilman@tx.technion.ac.il

© Springer International Publishing Switzerland 2014

273

B. Klartag, E. Milman (eds.), *Geometric Aspects of Functional Analysis*,
Lecture Notes in Mathematics 2116, DOI 10.1007/978-3-319-09477-9_18

Poincaré constant with *zero Dirichlet boundary conditions*; it is elementary to verify that $P_{\Omega^*}^D \simeq \frac{1}{n} |\Omega^*|^{2/n}$ (see Remark 3 for more precise information).

In this note we explore what may be said when f does not necessarily vanish on the boundary, and develop applications for estimating the Poincaré constant with Neumann boundary conditions. Here and elsewhere, we use $|M|$ to denote the k-dimensional Hausdorff measure $\mathscr{H}^k$ of the k-dimensional manifold M, and $A \simeq B$ to denote that $c \leq A/B \leq C$, for some universal numeric constants $c, C > 0$. All constants c, c', C, C', C_1, C_2, etc. appearing in this work are positive and universal, i.e. do not depend on Ω, n or any other parameter, and their value may change from one occurrence to the next.

Let λ_Ω denote the uniform (Lebesgue) probability measure on Ω, and let P_Ω^N denote the Poincaré constant of Ω, i.e. the best constant satisfying:

$$\mathbb{V}ar_{\lambda_\Omega} f \leq P_\Omega^N \int_\Omega |\nabla f|^2 d\lambda_\Omega \quad \forall \text{ smooth } f : \Omega \to \mathbb{R}, \tag{3}$$

without assuming any boundary conditions on f. Here and throughout $\mathbb{V}ar_\mu(f) := \int f^2 d\mu - (\int f d\mu)^2$ for any probability measure μ. It is well-known that when $\partial\Omega$ is smooth, $1/P_\Omega^N$ coincides with the first non-zero eigenvalue ("spectral-gap") of the Laplacian on Ω with zero *Neumann* boundary conditions, explaining the superscript N in our notation for P_Ω^N. The classical Szegö–Weinberger inequality (e.g. [4]) states that $P_\Omega^N \geq P_{\Omega^*}^N \simeq Vol(\Omega^*)^{2/n}$. By inspecting domains with very narrow bottlenecks, or even convex domains which are very narrow and elongated in a certain direction, it is clear that without some additional information on Ω, P_Ω^N is not bounded from above. However, a conjecture of Kannan–Lovász–Simonovits [17] asserts that on a *convex* domain Ω, the Poincaré inequality (3) will be saturated by linear functions f, up to a universal constant $C > 0$ independent of n and Ω, i.e.:

$$P_\Omega^N \leq C P_\Omega^{Lin}, \quad P_\Omega^{Lin} := \sup_{\theta \in S^{n-1}} \mathbb{V}ar_{\lambda_\Omega} \langle \cdot, \theta \rangle$$

It is easy to reduce the KLS conjecture to the case that Ω is *isotropic*, meaning that its barycenter is at the origin and the variance of all unit linear functionals is 1, i.e.:

$$\int x_i d\lambda_\Omega = 0, \quad \int x_i x_j d\lambda_\Omega = \delta_{ij} \quad \forall i, j = 1, \ldots, n.$$

The conjecture then asserts that $P_\Omega^N \leq C$ for any convex isotropic domain Ω in $\mathbb{R}^n$.

Given a Borel probability measure μ on $\mathbb{R}^n$ (not necessarily absolutely continuous), we denote by P_μ^∞ the best constant in the following weak L^2-L^∞ Poincaré inequality:

$$\mathbb{V}ar_\mu f \leq P_\mu^\infty \||\nabla f(x)|\|_{L^\infty(\mu)}^2 \quad \forall \text{ smooth } f : \mathbb{R}^n \to \mathbb{R}. \tag{4}$$

Set $P_\Omega^\infty := P_{\lambda_\Omega}^\infty$; clearly $P_\Omega^\infty \leq P_\Omega^N$. In [26], the second-named author showed that when Ω is convex, the latter inequality may be reversed:

$$P_\Omega^N \le C P_\Omega^\infty \,, \tag{5}$$

where $C > 1$ is a universal numeric constant. This reduces the KLS conjecture to the class of 1-Lipschitz functions f (satisfying $\||\nabla f|\|_{L^\infty} \le 1$). Another remarkable reduction was obtained by R. Eldan, who showed [11] that it is essentially enough (up to logarithmic factors in n) to establish the conjecture for the Euclidean norm function $f(x) = |x|$, but simultaneously for *all* isotropic convex domains in $\mathbb{R}^n$. Employing an estimate on the variance of $|x|$ due to O. Guédon and the second-named author [16], it follows from Eldan's reduction that for a general convex body in $\mathbb{R}^n$, $P_\Omega^N \le C n^{2/3} \log(1 + n) P_\Omega^{Lin}$.

In this work, we obtain several additional reductions of the KLS conjecture. First, we obtain a sufficient condition by reducing to the study of $P_{\sigma_{\partial\Omega}}^\infty$ and $P_{\lambda_{\partial\Omega}}^\infty$, the cone and Lebesgue measures on $\partial\Omega$, respectively. In particular, it suffices to bound the variance of homogeneous functions which are 1-Lipschitz on the boundary. This is achieved by obtaining Neumann versions of the Hardy and Faber-Krahn inequalities (1) and (2) for general functions (not necessarily vanishing on the boundary). The parameters $P_{\sigma_{\partial\Omega}}^\infty$ or $P_{\lambda_{\partial\Omega}}^\infty$ may then be bounded using a result from our previous work [21], by averaging certain curvatures on $\partial\Omega$ (see Theorem 11).

Second, we reduce the KLS conjecture to the class of harmonic functions. Thirdly, we consider the Poincaré constant of an unconditional convex body bounded by the principle hyperplanes, when a certain mixed Dirichlet–Neumann boundary condition is imposed. It is interesting to check which of the boundary conditions will dominate this Poincaré constant, and we determine that it is the Dirichlet ones, resulting in a Faber–Krahn / Hardy-type upper bound.

Lastly, we reveal a general relation between the Poincaré constant of a log-concave measure μ and its associated K. Ball body K_μ, assuming that the latter has finite-volume-ratio. In particular, we obtain a quick proof of the KLS conjecture for unit-balls of ℓ_p^n, $p \in [1, 2]$ (first established by S. Sodin [32]), which avoids using the concentration estimates of Schechtman and Zinn [31]. This is also extended to arbitrary $p \ge 2$ bounded away from ∞.

Our proofs follow classical arguments for establishing the Hardy inequality, which can be viewed as a Lyapunov function or vector-field method, in which one is searching for a vector-field whose magnitude is bounded from above on one hand, and whose divergence is bounded from below on the other. For more applications of Lyapunov functions to the study of Sobolev-type inequalities, see [9].

2 Hardy-Type Inequalities

Let Ω denote a compact connected set in $\mathbb{R}^n$ with smooth boundary and having the origin in its interior. We denote by ν the unit exterior normal-field to $\partial\Omega$. We denote by $\lambda_{\partial\Omega}$ the uniform probability measure on $\partial\Omega$ induced by the Lebesgue measure, i.e. $\mathcal{H}^{n-1}|_{\partial\Omega} / |\partial\Omega|$.

Our basic starting point is the following integration-by-parts formula. Let g denote a smooth function and ξ a smooth vector field on Ω. Then:

$$\int_\Omega div(\xi)g\,dx = -\int_\Omega \langle \xi, \nabla g \rangle\,dx + \int_{\partial\Omega} \langle \xi, v \rangle\,g\,d\mathcal{H}^{n-1}. \tag{6}$$

Applying this to $g = f^2$ and using the Cauchy-Schwartz inequality (in additive form), we obtain for any positive function λ on Ω:

$$\int_\Omega div(\xi)f^2dx \le \int_\Omega \lambda f^2 dx + \int_\Omega \frac{1}{\lambda}|\langle \xi, \nabla f \rangle|^2\,dx + \int_{\partial\Omega} \langle \xi, v \rangle\,f^2\,d\mathcal{H}^{n-1},$$

or equivalently:

$$\int_\Omega (div(\xi) - \lambda)\,f^2dx \le \int_\Omega \frac{1}{\lambda}|\langle \xi, \nabla f \rangle|^2\,dx + \int_{\partial\Omega} \langle \xi, v \rangle\,f^2\,d\mathcal{H}^{n-1}. \tag{7}$$

Let us apply this to several different vector fields ξ.

2.1   Radial Vector Field

In this subsection, assume in addition that Ω is star-shaped, meaning that $\Omega = \{x ; \|x\| \le 1\}$, where $\|x\| := \inf\{\lambda > 0; x \in \lambda\Omega\}$ denotes its associated gauge function. We denote by $\sigma_{\partial\Omega}$ the induced cone probability measure on $\partial\Omega$, i.e. the push-forward of λ_Ω via the map $x \mapsto \frac{x}{\|x\|}$. It is well-known and immediate to check that:

$$\sigma_{\partial\Omega} = \frac{1}{|\Omega|}\frac{\langle x, v \rangle}{n} \cdot \mathcal{H}^{n-1}|_{\partial\Omega}.$$

Theorem 1 (Hardy with Boundary). *Let f denote a smooth function on Ω. Then:*

$$\mathbb{V}ar_{\lambda_\Omega} f \le \frac{4}{n^2}\int_\Omega \langle x, \nabla f \rangle^2 d\lambda_\Omega + 2\mathbb{V}ar_{\sigma_{\partial\Omega}} f. \tag{8}$$

Proof. Apply (7) with $\xi(x) = x$, so that $div(\xi) = n$, and $\lambda \equiv n/2$. We obtain:

$$\int_\Omega f^2dx \le \frac{4}{n^2}\int_\Omega \langle x, \nabla f \rangle^2 dx + \frac{2}{n}\int_{\partial\Omega} \langle x, v \rangle f^2 d\mathcal{H}^{n-1}(x). \tag{9}$$

In particular, we see that (8) immediately follows when f vanishes on $\partial\Omega$. For general functions, we divide (9) by $Vol(\Omega)$ and apply the resulting inequality to $f - a$ with $a := \int_{\partial\Omega} f d\sigma_{\partial\Omega}$:

$$\mathbb{V}ar_{\lambda_\Omega} f \le \int (f - a)^2 d\lambda_\Omega \le \frac{4}{n^2} \int_\Omega \langle x, \nabla f \rangle^2 d\lambda_\Omega + 2\mathbb{V}ar_{\sigma_{\partial\Omega}} f|_{\partial\Omega}.$$

This is the desired assertion.   □

2.2 Optimal Transport to Euclidean Ball

A remarkable theorem of Y. Brenier [5] asserts that between any two absolutely continuous probability measures μ, η on $\mathbb{R}^n$ (say having second moments), there exists a unique (μ a.e.) map T which minimizes the transport-cost $\int |T(x) - x|^2 d\mu(x)$, among all maps pushing forward μ onto η; moreover, this optimal transport map T is characterized as being the gradient of a convex function φ. See also [25] for refinements and extensions. The regularity properties of T have been studied by Caffarelli [6–8], who discovered that a necessary condition for T to be smooth is that η have convex support; in particular, Caffarelli's results imply that when $\mu = \lambda_\Omega$, $\eta = \lambda_{B_2^n}$ and $\partial\Omega$ is smooth, then so is the Brenier map $T_0 := \nabla\varphi_0$ pushing forward μ onto ν, on the entire closed Ω (i.e. all the way up to the boundary). By the change-of-variables formula, we obviously have:

$$\text{Jac } T_0 = \det dT_0 = \frac{|B_2^n|}{|\Omega|}.$$

Theorem 2 (Faber–Krahn with Boundary). *Let f denote a smooth function on Ω. Then:*

$$\mathbb{V}ar_{\lambda_\Omega} f \le \frac{4 |\Omega|^{2/n}}{n^2 |B_2^n|^{2/n}} \int_\Omega |\nabla f|^2 d\lambda_\Omega + \frac{2 |\partial\Omega|}{n |B_2^n|^{1/n} |\Omega|^{(n-1)/n}} \mathbb{V}ar_{\lambda_{\partial\Omega}} f|_{\partial\Omega}. \tag{10}$$

Proof. Identifying $\mathbb{R}^n$ with its tangent spaces, we set $\xi = \nabla\varphi_0$ (where φ_0 was defined above). Note that since φ_0 is convex, hence Hess φ_0 is positive-definite, we may apply the arithmetic-geometric means inequality:

$$div(\xi) = \Delta\varphi_0 = tr(Hess\varphi_0) \ge n(\det Hess\varphi_0)^{1/n} = n(\det dT_0)^{1/n} = n \frac{|B_2^n|^{1/n}}{|\Omega|^{1/n}} =: \alpha.$$

Applying (7) with $\lambda \equiv \alpha/2$, and using that $\xi = \nabla\varphi_0 \in B_2^n$, we obtain:

$$\int_\Omega f^2 dx \le \frac{4 |\Omega|^{2/n}}{n^2 |B_2^n|^{2/n}} \int_\Omega |\nabla f|^2 dx + \frac{2 |\Omega|^{1/n}}{n |B_2^n|^{1/n}} \int_{\partial\Omega} f^2 d\mathcal{H}^{n-1}.$$

In particular, when f vanishes on $\partial\Omega$, we deduce (2) with a slightly inferior constant; however, this constant is asymptotically (as $n \to \infty$) best possible, see Remark 3 below. Dividing by $|\Omega|$ and applying the resulting inequality to $f - a$ with $a := \int_{\partial\Omega} f d\lambda_{\partial\Omega}$, the assertion follows.   □

Remark 3. It is known (e.g. [13, p. 139]) that $1/P_{B_2^n}^D$ is equal to the square of the first positive zero of the Bessel function of order $(n - 2)/2$. According to [33, p. 516], the first zero of the Bessel function of order β is $\beta + c_0\beta^{1/3} + O(1)$, for a constant $c_0 \simeq 1.855$, and so consequently $P_{B_2^n}^D = \frac{4}{n^2}(1 + o(1))$. By homogeneity, it follows that $P_{\Omega*}^D = 4|\Omega*|^{2/n}/(n^2|B_2^n|^{2/n})(1+o(1))$, confirming that the constant in Theorem 2 is asymptotically best possible.

Remark 4. Note that if we start from (6) and avoid employing the Cauchy-Schwartz inequality used to derive (7), the above proof (using $\xi = \nabla\varphi_0$ and $g \equiv 1$) yields the isoperimetric inequality with sharp constant for smooth bounded domains:

$$|\partial\Omega| \geq n|B_2^n|^{1/n}|\Omega|^{(n-1)/n}. \quad (11)$$

This proof was first noted by McCann [24], extending an analogous proof by Knothe and subsequently Gromov of the Brunn-Minkowski inequality [28] using the Knothe map [20]. See [12] for rigorous extensions of such an approach to non-smooth domains.

2.3 Normal Vector Field

In this subsection, we assume in addition that Ω is strictly convex. We employ the vector field:

$$\xi(x) = \nu(x/\|x\|),$$

the exterior unit normal-field to the convex set $\Omega_x := \|x\|\Omega$. Note that this field is not well defined (and in particular not continuous) at the origin, so strictly speaking we cannot appeal to (7). However, this is not an issue, since $div(\xi)$ is homogeneous of degree -1, and so the Jacobian term in polar coordinates r^{n-1} will absorb the blow-up of the divergence near the origin (recall $n \geq 2$). To make this rigorous, we simply repeat the derivation of (7) by integrating by parts on $\Omega \setminus \epsilon B_2^n$, and note that we may take the limit as $\epsilon \to 0$, since the contribution of the additional boundary $\partial\epsilon B_2^n$ goes to zero as ξ and f are bounded.

Now, observe that:

$$div(\xi)(x) = H_{\partial\Omega_x}(x) = \frac{1}{\|x\|}H_{\partial\Omega}(x/\|x\|),$$

where $H_S(y)$ denotes the mean-curvature (trace of the second fundamental form II_S) of a smooth oriented hypersurface S at x. Indeed, by definition $\nabla \xi|_{\xi^\perp} = II_{\partial \Omega_x}$, and $2\nabla_\xi \xi = \nabla \langle \xi, \xi \rangle = 0$, and so $div(\xi) = tr(\nabla \xi) = H_{\partial \Omega_x}$.

Theorem 5 (Mean-Curvature Weighted Hardy). *For any strictly convex Ω and smooth function f defined on it:*

$$\int_\Omega \frac{H_{\partial \Omega}(x/\|x\|)}{\|x\|} f^2(x) dx \leq 4 \int_\Omega \frac{\|x\|}{H_{\partial \Omega}(x/\|x\|)} \langle \nabla f(x), \nu(x/\|x\|) \rangle^2 dx$$

$$+ 2 \int_{\partial \Omega} f^2 d\mathcal{H}^{n-1}.$$

Proof. Immediate after appealing to (7) with $\lambda(x) = \frac{1}{2} \frac{H_{\partial \Omega}(x/\|x\|)}{\|x\|}$. $\quad\square$

Remark 6. We note for future reference that by (6) with $g \equiv 1$ we have:

$$\int_\Omega \frac{H_{\partial \Omega}(x/\|x\|)}{\|x\|} d\lambda_\Omega = \frac{|\partial \Omega|}{|\Omega|}.$$

Also, integration in polar coordinates immediately verifies:

$$\int_\Omega \frac{H_{\partial \Omega}(x/\|x\|)}{\|x\|} d\lambda_\Omega(x) = \frac{n}{n-1} \int_{\partial \Omega} H_{\partial \Omega} d\sigma_{\partial \Omega},$$

$$\int_\Omega \frac{\|x\|}{H_{\partial \Omega}(x/\|x\|)} d\lambda_\Omega(x) = \frac{n}{n+1} \int_{\partial \Omega} \frac{d\sigma_{\partial \Omega}}{H_{\partial \Omega}}.$$

2.4 Unconditional Sets

Finally, we consider one additional vector-field for the Lyapunov method, which is useful when Ω is the intersection of an unconditional convex set with the first orthant $Q := [0, \infty)^n$ under a certain mixed Dirichlet–Neumann boundary condition. Let $int(Q)$ denote the interior of Q.

Theorem 7. *Let $\Omega \subset Q$ denote a set having smooth boundary, such that every outer normal ν to $\partial \Omega \cap int(Q)$ has only non-negative coordinates. Let f denote a smooth function vanishing on ∂Q. Then:*

$$\int_\Omega \frac{f^2}{|x|^2} dx \leq \frac{4}{n^2} \int_\Omega |\nabla f|^2 dx.$$

Proof. Consider the vector field:

$$\xi = - \left(\frac{1}{x_1}, \cdots, \frac{1}{x_n} \right).$$

Since $\langle \xi, \nu \rangle \leq 0$ in $int(Q) \cap \partial \Omega$ and $f|_{\partial \Omega \cap \partial \Omega} = 0$, we have:

$$\int_\Omega \sum_{i=1}^n \frac{1}{x_i^2} f^2 dx = \int_\Omega div(\xi) f^2 dx = -2 \int_\Omega f \langle \nabla f, \xi \rangle dx + \int_{\partial \Omega} \langle \xi, \nu \rangle f^2 d\mathcal{H}^{n-1}$$

$$\leq -2 \int_\Omega f \langle \nabla f, \xi \rangle dx \leq 2 \sqrt{\int_\Omega \sum_{i=1}^n \frac{1}{x_i^2} f^2 dx} \sqrt{\int_\Omega |\nabla f|^2 dx}.$$

Finally, by the arithmetic-harmonic means inequality, we obtain:

$$n^2 \int_\Omega \frac{f^2}{|x|^2} dx \leq \int_\Omega \sum_{i=1}^n \frac{1}{x_i^2} f^2 dx \leq 4 \int_\Omega |\nabla f|^2 dx.$$

$\square$

We stress that this result is very similar to the following variant of the Hardy inequality:

$$\int_\Omega \frac{f^2}{|x|^2} dx \leq \left(\frac{2}{n-2} \right)^2 \int_\Omega |\nabla f|^2 dx,$$

which holds for any smooth f vanishing on $\partial \Omega$ (see [15]).

3   Reduction of KLS Conjecture to Subclasses of Functions

3.1   Reduction to the Boundary

Let us now see how the Hardy-type inequalities of the previous section may be used to reduce the KLS conjecture to the behaviour of 1-Lipschitz functions on the boundary $\partial \Omega$. We remark that we use here the term "reduction" in a rather loose sense—we obtain a sufficient condition for the KLS conjecture to hold, but we were unable to show that this is also a necessary one.

Together with (5), Theorem 1 immediately yields:

Corollary 8. *For any smooth convex domain Ω with barycenter at the origin:*

$$P_\Omega^N \leq CP_\Omega^\infty \leq C \left(\frac{4}{n} P_\Omega^{Lin} + 2P_{\sigma \partial \Omega}^\infty \right). \tag{12}$$

where $C > 0$ is a universal constant.

Proof. Apply Theorem 1 to an arbitrary 1-Lipschitz function f, and note that $\int |x|^2 d\lambda_\Omega = \sum_{i=1}^n Var_{\lambda_\Omega}(x_i) \leq nP_\Omega^{Lin}$. $\square$

Consequently, a sufficient criterion for verifying the KLS conjecture is to establish that $P^\infty_{\sigma_{\partial\Omega}} \leq C'$ for any isotropic convex Ω—a "weak KLS conjecture for cone measures". This suggests that the most difficult part of the conjecture concerns the behavior of 1-Lipschitz functions on the boundary.

It may be more desirable to work with the Lebesgue measure $\lambda_{\partial\Omega}$ instead of the cone measure $\sigma_{\partial\Omega}$. Since:

$$P^{Lin}_\Omega \geq P^{Lin}_{\Omega^*} \simeq |\Omega^*|^{2/n} \; , \; |B^n_2|^{1/n} \simeq \frac{1}{\sqrt{n}}, \tag{13}$$

(see e.g. [27]), Theorem 2 together with (5) immediately yields:

Corollary 9. *For any smooth convex domain Ω with barycenter at the origin:*

$$P^N_\Omega \leq C_1 P^\infty_\Omega \leq C_2 \left(\frac{4}{n} P^{Lin}_\Omega + 2I P^\infty_{\lambda_{\partial\Omega}} \right) \; , \; I := \frac{|\partial\Omega|}{n \, |B^n_2|^{1/n} \, |\Omega|^{(n-1)/n}}. \tag{14}$$

Note that for an isotropic convex body, the isoperimetric ratio term I satisfies:

$$1 \leq I \leq C' \sqrt{n} \, |\Omega|^{1/n} . \tag{15}$$

The left-hand side in fact holds for any arbitrary set Ω by the sharp isoperimetric inequality (11). The right-hand side follows since when Ω is convex and isotropic, it is known that $\Omega \supset \frac{1}{C} B^n_2$ (e.g. [27]). Consequently (see e.g. [2]):

$$|\partial\Omega| = \lim_{\epsilon\to 0} \frac{|\Omega + \epsilon B^n_2| - |\Omega|}{\epsilon} \leq \lim_{\epsilon\to 0} \frac{|\Omega + \epsilon C\Omega| - |\Omega|}{\epsilon} = nC \, |\Omega| ,$$

and so (15) immediately follows. Up to the value of C', the right-hand side is also sharp, as witnessed by the n-dimensional cube. Note that by (13), $|\Omega|^{1/n} \simeq P^{Lin}_{\Omega^*} \leq P^{Lin}_\Omega = 1$ for any isotropic convex body Ω, and so in fact $I \leq C'' \sqrt{n}$.

To avoid the isoperimetric ratio term I which may be too large, we can instead invoke Theorem 5:

Corollary 10. *For any strictly convex smooth domain Ω:*

$$P^N_\Omega \leq C_2 \left(A^2 + A \frac{|\partial\Omega|}{|\Omega|} P^\infty_{\lambda_{\partial\Omega}} \right) \; , \; A := \int_{\partial\Omega} \frac{d\sigma_{\partial K}}{H_{\partial\Omega}}. \tag{16}$$

Note that by Jensen's inequality and Remark 6:

$$A \frac{|\partial\Omega|}{|\Omega|} \geq \frac{1}{\int_{\partial\Omega} H_{\partial\Omega} d\sigma_{\partial K}} \frac{|\partial\Omega|}{|\Omega|} = \frac{n}{n-1},$$

but perhaps the term $A \frac{|\partial\Omega|}{|\Omega|}$ is nevertheless still more favorable than I.

For the proof, we require the following variant of the notion of P_Ω^∞:

$$P_\Omega^{1,\infty} := \sup\left\{\left(\int |f - med_{\lambda_\Omega} f|\, d\lambda_\Omega\right)^2 ; \; \|\nabla f\|_{L^\infty(\lambda_\Omega)} \le 1\right\}.$$

It follows from the results of [26] that for any convex Ω:

$$P_\Omega^N \le C_1 P_\Omega^{1,\infty} \le C_2 P_\Omega^\infty \le C_2 P_\Omega^N. \tag{17}$$

Proof of Corollary 10. By Cauchy-Schwartz:

$$\left(\int_\Omega |f|\, d\lambda_\Omega\right)^2 \le \int_\Omega \frac{\|x\|}{H_{\partial\Omega}(x/\|x\|)} d\lambda_\Omega(x) \int_\Omega \frac{H_{\partial\Omega}(x/\|x\|)}{\|x\|} f^2(x)\, d\lambda_\Omega(x)$$

Assuming that f is 1-Lipschitz and invoking Theorem 5, it follows that:

$$\left(\int_\Omega |f|\, d\lambda_\Omega\right)^2 \le B\left(4B + 2\frac{|\partial\Omega|}{|\Omega|}\int_{\partial\Omega} f^2 d\lambda_{\partial\Omega}\right), \; B := \int_\Omega \frac{\|x\|}{H_{\partial\Omega}(x/\|x\|)} d\lambda_\Omega(x).$$

Applying this to $f - a$ where $a := \int_{\partial\Omega} f d\lambda_{\partial\Omega}$, we obtain:

$$P_\Omega^{1,\infty} \le B\left(4B + 2\frac{|\partial\Omega|}{|\Omega|} P_{\lambda_{\partial\Omega}}^\infty\right).$$

But $B = \frac{n}{n+1} A$ by Remark 6, and so the assertion follows from (17). $\qquad\square$

3.2 A Concrete Bound

To control the variance of 1-Lipschitz functions on the boundary $\partial\Omega$, we recall an argument from our previous work [21], where a generalization of the following inequality of A. Colesanti [10] was obtained:

$$\int_{\partial\Omega} H f^2 d\mathcal{H}^{n-1} - \frac{n-1}{n}\frac{\left(\int_{\partial\Omega} f d\mathcal{H}^{n-1}\right)^2}{Vol(\Omega)} \le \int_{\partial\Omega} \langle \mathrm{II}_{\partial\Omega}^{-1} \nabla_{\partial\Omega} f, \nabla_{\partial\Omega} f\rangle d\mathcal{H}^{n-1}, \tag{18}$$

for any strictly convex Ω with smooth boundary and smooth function f on $\partial\Omega$. Applying the Cauchy-Schwartz inequality, we obtain for any 1-Lipschitz function f with $\int_{\partial\Omega} f d\lambda_{\partial\Omega} = 0$:

$$\left(\int_{\partial\Omega} |f - med_{\lambda_{\partial\Omega}} f|\, d\lambda_{\partial\Omega}\right)^2 \le \left(\int_{\partial\Omega} |f|\, d\lambda_{\partial\Omega}\right)^2 \le \int_{\partial\Omega} \frac{d\lambda_{\partial\Omega}}{H_{\partial\Omega}} \int_{\partial\Omega} \frac{d\lambda_{\partial\Omega}}{\kappa_{\partial\Omega}},$$

where $\kappa_{\partial\Omega}(x)$ denotes the (positive) minimal principle curvature of $\partial\Omega$ at x, so that $\mathrm{II}_{\partial\Omega} \geq \kappa Id$. Consequently, the right-hand-side is an upper bound on $P^{1,\infty}_{\lambda_{\partial\Omega}}$. Using the equivalence (17) in a more general Riemannian setting, we were able to deduce in [21] that:

$$(P^{\infty}_{\lambda_{\partial\Omega}} \leq)\ P^{N}_{\lambda_{\partial\Omega}} \leq CP^{1,\infty}_{\lambda_{\partial\Omega}} \leq C \int_{\partial\Omega} \frac{d\lambda_{\partial\Omega}}{H_{\partial\Omega}} \int_{\partial\Omega} \frac{d\lambda_{\partial\Omega}}{\kappa_{\partial\Omega}}. \tag{19}$$

Plugging this estimate into the estimates of the previous subsection, we obtain:

Theorem 11. *For n larger than a universal constant and any isotropic strictly convex body Ω with smooth boundary in $\mathbb{R}^n$:*

$$P^{N}_{\Omega} \leq C_2 \frac{|\partial\Omega|}{\sqrt{n}\,|\Omega|^{\frac{n-1}{n}}} \int_{\partial\Omega} \frac{d\lambda_{\partial\Omega}}{H_{\partial\Omega}} \int_{\partial\Omega} \frac{d\lambda_{\partial\Omega}}{\kappa_{\partial\Omega}}.$$

Proof. The easiest option is to invoke Corollary 9, but note that Corollary 8 or 10 would also work after an appropriate application of Cauchy-Schwartz. Coupled with (19), it follows that:

$$P^{N}_{\Omega} \leq C_1 \left(\frac{4}{n} P^{Lin}_{\Omega} + \frac{|\partial\Omega|}{\sqrt{n}\,|\Omega|^{\frac{n-1}{n}}} \int_{\partial\Omega} \frac{d\lambda_{\partial\Omega}}{H_{\partial\Omega}} \int_{\partial\Omega} \frac{d\lambda_{\partial\Omega}}{\kappa_{\partial\Omega}} \right).$$

But since $P^{Lin}_{\Omega} \leq P^{N}_{\Omega}$, the assertion follows for e.g. $n \geq 8C_1$. □

Note that this estimate yields the correct result, up to constants, for the Euclidean ball. A concrete class of isotropic convex bodies for which the first term above $\frac{|\partial\Omega|}{\sqrt{n}}$ is upper bounded by a constant, is the class of quadratically uniform convex bodies Ω, since in isotropic position $\Omega \supset c\sqrt{n}B^n_2$ and $|\Omega|^{1/n} \simeq 1$ (see e.g. [19]). It is not hard to show that when in addition $\Omega \subset C_1\sqrt{n}B^n_2$—i.e. Ω is an isotropic quadratically uniform convex body which is isomorphic to a Euclidean ball—then $P^{N}_{\Omega} \leq C_2$. It would be very interesting to see if the additional assumption $\Omega \subset C_1\sqrt{n}B^n_2$ could be removed by employing the estimate given by Theorem 11.

3.3 Reduction to Harmonic Functions

We conclude this section by providing another different reduction of the KLS conjecture:

Theorem 12 (Reduction to Harmonic Functions). *There exists a universal constant $C > 1$ so that:*

$$P^{N}_{\Omega} \leq CP^{H}_{\Omega}\ ,\ \ P^{H}_{\Omega} := \sup_{h \in H} \frac{\mathbb{V}ar_{\lambda_{\Omega}}h}{\int |\nabla h|^2 \, d\lambda_{\Omega}},$$

where H denotes the class of harmonic functions h on Ω. In fact, for large enough n, one can use $C = 2$.

Proof. Fix an arbitrary smooth function f on Ω, and solve the Poisson equation $\Delta h = 0$, $h|_{\partial\Omega} = f|_{\partial\Omega}$. One has:

$$\mathbb{V}ar_{\lambda_\Omega} f \leq 2(\mathbb{V}ar_{\lambda_\Omega}(f - h) + \mathbb{V}ar_{\lambda_\Omega} h).$$

Since $f - h$ vanishes on $\partial\Omega$, the Faber-Krahn inequalities (2) or (10) imply:

$$\mathbb{V}ar_{\lambda_\Omega}(f - h) \leq \frac{4\,|\Omega|^{2/n}}{n^2\,|B_2^n|^{2/n}} \int |\nabla f - \nabla h|^2\, d\lambda_\Omega.$$

It follows that:

$$\mathbb{V}ar_{\lambda_\Omega} f \leq \max\left(\frac{C_1}{n}\,|\Omega|^{2/n}, 2P_\Omega^H\right)\left(\int |\nabla f - \nabla h|^2\, d\lambda_\Omega + \int |\nabla h|^2\, d\lambda_\Omega\right).$$

But since h is harmonic and $(f - h)|_{\partial\Omega} = 0$ we have $\int \langle \nabla f - \nabla h, \nabla h\rangle d\lambda_\Omega = 0$, and consequently:

$$\int \left(|\nabla f - \nabla h|^2 + |\nabla h|^2\right) d\lambda_\Omega = \int |\nabla f|^2 d\lambda_\Omega.$$

It remains to note that since linear functions are harmonic, $P_\Omega^H \geq P_\Omega^{Lin} \geq P_{\Omega^*}^{Lin} \simeq |\Omega^*|^{2/n}$, concluding the proof. $\qquad\square$

Remark 13. It is not clear to us if it enough to only control the variance of harmonic functions h, so that the restriction $h|_{\partial\Omega}$ is 1-Lipschitz. The reason is that we do not know whether h has bounded Lipschitz constant on the entire Ω, and so we cannot apply (5). We believe that the latter would be an interesting property of convex domains which is worth investigating. A small observation in this direction is that $|\nabla h|^2$ is subharmonic and hence satisfies the maximum principle, but we do not know how to control the derivative in the normal direction to $\partial\Omega$.

4   Transferring Poincaré Inequalities from μ to K_μ

Given an absolutely continuous probability measure μ on $\mathbb{R}^n$ having upper-semi-continuous density f, the following set was considered by K. Ball [1]:

$$K_\mu := \left\{x \in \mathbb{R}^n \;;\; \|x\|_{K_\mu} \leq 1\right\}, \quad \text{where} \quad \frac{1}{\|x\|_{K_\mu}} = \left(n \int_0^\infty r^{n-1} f(rx) dr\right)^{1/n}.$$

Integration in polar coordinates immediately verifies that $|K_\mu| = \|\mu\| = 1$. A remarkable observation of Ball is that when f is log-concave (i.e. $\log f : \mathbb{R}^n \to \mathbb{R} \cup \{-\infty\}$ is concave), then K_μ is a compact convex set (see [18] for the case that f is non-even). If in addition the origin is in the interior of the support of μ, then it will also be in the interior of K_μ—we will say in that case that K_μ is a convex body.

Given a convex body K, consider the map $T(x) = \frac{x}{\|x\|_K}$ where $\|\cdot\|_K$ denotes the gauge function of K (when K is origin-symmetric, this function defines a norm). It is an elementary exercise to show that $T_*\mu = \sigma_{\partial K}$ if and only if $K = cK_\mu$ for some $c > 0$ (see [29, Proposition 3.1]).

Proposition 14. *Let $\mu = f(x)dx$ denote a probability measure with log-concave density on $\mathbb{R}^n$ ($n \geq 3$) and barycenter at the origin, and set $T(x) = \frac{x}{\|x\|_{K_\mu}}$. Assume that $K_\mu \supset RB_2^n$. Then:*

$$\int_{\mathbb{R}^n} \|dT^*(x)\|_{op}^2 \, d\mu(x) \leq C \frac{f(0)^{2/n}}{R^2} \int_{K_\mu} |x|^2 \, dx,$$

where $\|\cdot\|_{op}$ denotes the operator norm, and $dT^*(x)$ is the dual operator to the differential $dT(x) : T_x\mathbb{R}^n \to T_{T(x)}\partial K_\mu$.

For the proof, we first require:

Lemma 15. *If $T(x) = \frac{x}{\|x\|_K}$ and ∂K is smooth then:*

$$\|dT^*(x)\|_{op} = \frac{|x| \, |\nabla \|x\|_K|}{\|x\|_K^2} = \frac{1}{\|x\|_K \langle x/|x|, \nu_{\partial K}(T(x))\rangle} = \frac{|x|}{\|x\|_K^2 \, h_K(\nu_{\partial K}(T(x)))},$$

where $h_K(\theta) = \sup\{\langle x, \theta\rangle ; x \in K\}$ denotes the support function of K.

Proof. Since $\nabla \|x\|_K$ is parallel to $\nu = \nu_{\partial K}(T(x))$, taking the partial derivative in the direction of x verifies that:

$$\nabla \|x\|_K = \frac{\|x\|_K}{\langle x, \nu\rangle}\nu.$$

Consequently $dT(x) = \frac{1}{\|x\|}(Id - \frac{x \otimes \nu}{\langle x, \nu\rangle})$. Now observe that:

$$\|dT(x)^*\|_{op}^2 = \sup\left\{\langle dT(x)dT(x)^*v, v\rangle ; v \in T_{T(x)}^*\partial K , |v| \leq 1\right\}.$$

But $dT(x)dT(x)^* = \frac{1}{\|x\|^2}(Id + u \otimes u)$, where $u = v - \frac{x}{\langle x, \nu\rangle}$. Consequently, its top eigenvalue is:

$$\frac{1}{\|x\|^2}(1 + |u|^2) = \frac{1}{\|x\|_K^2} \frac{|x|^2}{\langle x, \nu\rangle^2}.$$

It remains to note that when $x \in \partial K$ then $\langle x, \nu_{\partial K}(x) \rangle$ is precisely the support function of K in the direction of the latter normal. Consequently, $\langle x, \nu \rangle = \|x\|_K h_K(\nu)$, and the assertion follows.   □

Proof of Proposition 14. It is easy to see that if the density f of μ is smooth, then so is ∂K_μ, and so by approximation we may assume that this is indeed the case. Consequently, if $K_\mu \supset RB_2^n$, we have by Lemma 15:

$$\int_{\mathbb{R}^n} \|dT^*(x)\|_{op}^2 d\mu(x) = \int_{\mathbb{R}^n} \frac{|x|^2}{\|x\|_{K_\mu}^4 h_{K_\mu}^2(\nu_{\partial K_\mu}(T(x)))} d\mu(x) \le \frac{1}{R^2} \int_{\mathbb{R}^n} \frac{|x|^2}{\|x\|_{K_\mu}^4} d\mu(x). \tag{20}$$

Integrating in polar coordinates, we have:

$$\int_{\mathbb{R}^n} \frac{|x|^2}{\|x\|_{K_\mu}^4} d\mu(x) = \int_{S^{n-1}} \frac{1}{\|\theta\|_{K_\mu}^4} \int_0^\infty r^{n-3} f(r\theta) dr d\theta. \tag{21}$$

Denoting $k_p(\theta) := (p \int_0^\infty r^{p-1} f(r\theta) dr)^{1/p}$, we use that for any non-negative function f on $[0, \infty)$:

$$0 < p_1 \le p_2 \Rightarrow \frac{k_{p_1}(\theta)}{M_\theta^{1/p_1}} \le \frac{k_{p_2}(\theta)}{M_\theta^{1/p_2}},$$

where $M_\theta = \sup_{r \in [0,\infty)} f(r\theta)$. See [1, 3, 27] for case that f is even and [18, Lemmas 2.5 and 2.6] or [30, Lemma 3.2 and (3.12)] for the general case. Applying this to (21) with $p_1 = n - 2$ and $p_2 = n$, denoting $M = \max_{x \in \mathbb{R}^n} f(x)$, and using polar integration again, it follows that:

$$\int_{\mathbb{R}^n} \frac{|x|^2}{\|x\|_{K_\mu}^4} d\mu(x) \le M^{2/n} \frac{1}{n-2} \int_{S^{n-1}} \frac{1}{\|\theta\|_{K_\mu}^{n+2}} d\theta = M^{2/n} \frac{n+2}{n-2} \int_{K_\mu} |x|^2 dx.$$

It remains to apply a result of M. Fradelizi [14] stating that for a log-concave measure $\mu = f(x)dx$ with barycenter at the origin:

$$M \le e^n f(0).$$

Plugging all of these estimates into (20), the assertion is proved.   □

We can now obtain:

Theorem 16. *Let $\mu = f(x)dx$ denote a log-concave probability measure on $\mathbb{R}^n$ having barycenter at the origin. Assume that $K_\mu \supset RB_2^n$. Then for large-enough n:*

$$P_{K_\mu}^N \le C \frac{\int |x|^2 d\lambda_{K_\mu}(x)}{R^2} f(0)^{2/n} P_\mu^N.$$

In particular, if μ satisfies the KLS conjecture then so does λ_{K_μ}, as soon as $\frac{\int |x|^2 d\lambda_{K_\mu}}{R^2}$ is bounded above by a constant.

Remark 17. This result was already noticed by Bo'az Klartag and the second-named author using a more elaborate computation which was never published. The idea is to control the average Lipschitz constant of the radial map from [29] pushing forward μ onto λ_{K_μ} instead of $\sigma_{\partial K_\mu}$.

Proof. We employ Corollary 8 and Proposition 14. When n is large-enough, $C\frac{4}{n}P_{K_\mu}^{Lin} \leq \frac{1}{2}P_{K_\mu}^{Lin} \leq \frac{1}{2}P_{K_\mu}^N$, and hence by Corollary 8:

$$P_{K_\mu}^N \leq C' P_{\sigma_{\partial K_\mu}}^\infty.$$

Denoting $T(x) = \frac{x}{\|x\|_{K_\mu}}$, we see by Proposition 14 that for any 1-Lipschitz function f on ∂K_μ:

$$\mathbb{V}ar_{\sigma_{\partial K_\mu}}(f) = \mathbb{V}ar_\mu(f \circ T) \leq P_\mu^N \int |\nabla(f \circ T)|^2 \, d\mu$$

$$\leq P_\mu^N \int |\nabla_{\partial K_\mu} f|^2 (T(x)) \|dT^*(x)\|_{op}^2 \, d\mu(x)$$

$$\leq P_\mu^N \int \|dT^*(x)\|_{op}^2 \, d\mu(x) \leq C \frac{f(0)^{2/n}}{R^2} \int_{K_\mu} |x|^2 \, dx \, P_\mu^N.$$

This implies the first part of the assertion.

The second part follows since, as shown by Ball [1] (see [18] for the non-even case):

$$P_{K_\mu}^{Lin} \simeq f(0)^{2/n} P_\mu^{Lin}. \tag{22}$$

Consequently:

$$P_\mu^N \leq A P_\mu^{Lin} \quad \Rightarrow \quad P_{K_\mu}^N \leq C \frac{\int |x|^2 \, d\lambda_{K_\mu}}{R^2} A \, P_{K_\mu}^{Lin}.$$

$\square$

We thus obtain a simple recipe for obtaining good spectral-gap estimates on certain convex bodies K having in-radius R so that $\int |x|^2 d\lambda_K(x)/R^2$ is bounded above by a constant: if we can find a log-concave measure μ having good spectral-gap so that $K_\mu = K$, Theorem 16 will imply that K also has good spectral-gap.

Remark 18. An inspection of the proofs of Proposition 14 and Theorem 16 shows that we may replaces $\frac{1}{R^2}$ in all of the occurrences above, with the more refined expression $\int_{\partial K_\mu} \frac{d\sigma_{\partial K_\mu}}{h_{K_\mu}^2(\nu_{\partial K_\mu})}$. However, we do not know how to effectively control the latter quantity.

4.1 An Example: Unit-Balls of ℓ_p^n, $p \in [1, 2]$

We illustrate this for unit-balls B_p^n of ℓ_p^n, $p \in [1, 2]$. It was first shown by S. Sodin [32] that these convex bodies satisfy the KLS conjecture. An alternative derivation was obtain in [26] by using the weaker P^∞ parameter and the equivalence (5). Both approaches relied on the Schechtman–Zinn concentration estimates for these bodies [31].

Using Theorem 16, we avoid passing through the Schechtman–Zinn concentration results. Indeed, let μ_p denote the one-dimensional probability measure $\frac{1}{2\Gamma(1/p+1)} \exp(-|t|^p)dt$. The n-fold product measure $\mu_p^n := \mu_p^{\otimes n}$ has density $f_p^n(x)$ where:

$$f_p^n(x) = \frac{1}{2^n \Gamma(1/p + 1)^n} \exp(-\sum_{i=1}^{n} |x_i|^p).$$

By the tensorization property of the Poincaré inequality [23], $P_{\mu_p^n}^N = P_{\mu_p}^N$, and since any one-dimensional log-concave measure satisfies the KLS conjecture, then so does any log-concave product measure. Now, since all level sets of f_p^n are homothetic copies of B_p^n, it is immediate to see that $K_{\mu_p^n}$ must be (the necessarily volume one) homothetic copy $\tilde{B}_p^n$ of B_p^n. In the range $p \in [1, 2]$, it is known (e.g. [28]) and easy to check that $\tilde{B}_p^n$ are finite volume-ratio bodies, meaning that $\tilde{B}_p^n \supset c\sqrt{n} B_2^n$. On the other hand, by (22):

$$\int |x|^2 \lambda_{\tilde{B}_p^n} \simeq f_p^n(0)^{2/n} \int |x|^2 d\mu_p^n(x) = \frac{1}{2^2 \Gamma(1/p+1)^2} n \int_{-\infty}^{\infty} |t|^2 d\mu_p(t) \leq Cn,$$

uniformly in $p \in [1, 2]$. Consequently, Theorem 16 implies that $\tilde{B}_p^n$ (and hence B_p^n) satisfy the KLS conjecture, uniformly in n and $p \in [1, 2]$. Similar versions may easily be obtained for convex functions more general than $|t|^p$; we leave this to the interested reader.

4.2 Another Example: Unit-Balls of ℓ_p^n, $p \in (2, \infty)$

To conclude, we use the unit-balls of ℓ_p^n for $p \in (2, \infty)$ to further illustrate the advantage and disadvantage of the method we propose in this section. Note that $\tilde{B}_p^n$ are not finite volume-ratio bodies when $p \in (2, \infty]$, and so Theorem 16 does not directly apply. However, by inspecting its proof and avoiding using the wasteful bound (20), we can still deduce the KLS conjecture for these bodies when p is bounded away from ∞. It was first shown by R. Latala and J.O. Wojtaszczyk [22] that in the entire range $p \in [2, \infty]$, there exists a globally Lipschitz map pushing forward μ_p^n onto $\lambda_{\tilde{B}_p^n}$, different from the radial map we have considered in this section. It is interesting to note that the radial-map is nevertheless Lipschitz on-average, at least when $p < \infty$.

Indeed, by inspecting the proof of Theorem 16 and employing Lemma 15, we see that we just need to control:

$$\int_{\mathbb{R}^n} \|dT^*(x)\|_{op}^2 \, d\mu_p^n(x) = \int_{\mathbb{R}^n} \frac{|x|^2 \, |\nabla \|x\|_{\tilde{B}_p^n}|^2}{\|x\|_{\tilde{B}_p^n}^4} d\mu_p^n(x)$$

$$= c_{p,n}^2 \int_{\mathbb{R}^n} \frac{|x|^2 \, |\nabla \|x\|_p|^2}{\|x\|_p^4} d\mu_p^n(x),$$

· where $\tilde{B}_p^n = c_{p,n} B_p^n$. It is well-known and easy to calculate that $c_{p,n} \simeq n^{1/p}$. Using that $|x|^2 \le n^{1-2/p} \|x\|_p^2$ (since $p \ge 2$), that:

$$|\nabla \|x\|_p|^2 = \frac{\sum_{i=1}^n |x_i|^{2p-2}}{\|x\|_p^{2p-2}},$$

and the invariance under permutation of coordinates, we conclude that:

$$\int_{\mathbb{R}^n} \|dT^*(x)\|_{op}^2 \, d\mu_p^n(x) \simeq n^2 \int \frac{|x_1|^{2p-2}}{\|x\|_p^{2p}} d\mu_p^n(x).$$

Integrating by parts, we have:

$$\int_{\mathbb{R}^n} \frac{\exp(-\|x\|_p^p)}{\|x\|_p^{2p}} |x_1|^{2p-2} \, dx = \int_{\mathbb{R}^n} \int_{\|x\|_p^p}^\infty \exp(-t) \left(\frac{1}{t^2} + \frac{2}{t^3} \right) dt \, |x_1|^{2p-2} \, dx$$

$$= \int_0^\infty \exp(-t) \left(\frac{1}{t^2} + \frac{2}{t^3} \right) \int_{t^{1/p} B_p^n} |x_1|^{2p-2} \, dx \, dt$$

$$= \int_0^\infty \exp(-t) \left(\frac{1}{t^2} + \frac{2}{t^3} \right) t^{\frac{n+2p-2}{p}} \, dt \int_{B_p^n} |x_1|^{2p-2} \, dx,$$

and so by a similar computation we conclude:

$$\int \frac{|x_1|^{2p-2}}{\|x\|_p^{2p}} d\mu_p^n(x) = AB,$$

$$A := \frac{\int_0^\infty \exp(-t) \left(\frac{1}{t^2} + \frac{2}{t^3} \right) t^{\frac{n+2p-2}{p}} \, dt}{\int_0^\infty \exp(-t) t^{\frac{n+2p-2}{p}} \, dt}, \quad B := \int |x_1|^{2p-2} \, d\mu_p^n(x).$$

Now:

$$B = \int_{-\infty}^{\infty} |t|^{2p-2} d\mu_p(t) = \frac{\Gamma(-1/p)}{p\Gamma(1+1/p)} \le \frac{C_1}{p},$$

uniformly in $p \in [2, \infty]$, whereas it is elementary to verify that in that range:

$$A \le C_2 \min\left(\frac{p^2}{n^2}, \frac{p}{n}\right).$$

Putting everything together, we see that:

$$\int_{\mathbb{R}^n} \|dT^*(x)\|_{op}^2 d\mu_p^n(x) \le C \min(p, n). \tag{23}$$

Consequently, the same argument as in the previous subsection shows that $\tilde{B}_p^n$ verify the KLS conjecture uniformly in n, as long as p is bounded above.

It is natural to wonder whether the only inequality we have used to derive the above estimate, namely $|x|^2 \le n^{1-2/p} \|x\|_p^2$, was perhaps too crude. However, this is not the case, and unfortunately it is the method of working with the map $T(x) = x/\|x\|_{K_\mu}$ which is too crude. Indeed, when $p = \infty$, so that μ_∞^n is the uniform measure on $[-1, 1]^n$ and $K = K_{\mu_\infty^n} = [-1/2, 1/2]^n$, we see by Lemma 15 that:

$$\|dT^*(x)\|_{op} = \frac{|x| |\nabla \|x\|_K|}{\|x\|_K^2} = \frac{|x|}{4 \|x\|_\infty^2},$$

and consequently:

$$\int_{\mathbb{R}^n} \|dT^*(x)\|_{op}^2 d\mu_p^n(x) \simeq n,$$

confirming that our estimate (23) is tight. This example suggests that perhaps it is better to work with the radial map from [29] pushing forward μ onto λ_{K_μ} instead of our map T which pushes μ onto $\sigma_{\partial K_\mu}$.

Acknowledgements We would like to thank Bo'az Klartag for his interest and fruitfull discussions. The first-named author was supported by RFBR project 12-01-33009 and the DFG project CRC 701. This study (research grant No 14-01-0056) was supported by The National Research University–Higher School of Economics' Academic Fund Program in 2014/2015. The second-named author was supported by ISF (grant no. 900/10), BSF (grant no. 2010288), Marie-Curie Actions (grant no. PCIG10-GA-2011-304066) and the E. and J. Bishop Research Fund.

References

1. K. Ball, Logarithmically concave functions and sections of convex sets in $\mathbb{R}^n$. Studia Math. **88**(1), 69–84 (1988)
2. K. Ball, Volume ratios and a reverse isoperimetric inequality. J. Lond. Math. Soc. **44**(2), 351–359 (1991)
3. R.E. Barlow, A.W. Marshall, F. Proschan, Properties of probability distributions with monotone hazard rate. Ann. Math. Stat. **34**, 375–389 (1963)
4. R.D. Benguria, Isoperimetric inequalities for eigenvalues of the Laplacian, in *Entropy and the Quantum II*. Contemporary Mathematics, vol. 552 (American Mathematical Society, Providence, 2011), pp. 21–60
5. Y. Brenier, Polar factorization and monotone rearrangement of vector-valued functions. Commun. Pure Appl. Math. **44**(4), 375–417 (1991)
6. L.A. Caffarelli, Interior $W^{2,p}$ estimates for solutions of the Monge-Ampère equation. Ann. Math. (2) **131**(1), 135–150 (1990)
7. L.A. Caffarelli, Boundary regularity of maps with convex potentials. Commun. Pure Appl. Math. **45**(9), 1141–1151 (1992)
8. L.A. Caffarelli, The regularity of mappings with a convex potential. J. Am. Math. Soc. **5**(1), 99–104 (1992)
9. P. Cattiaux, A. Guillin, Functional inequalities via Lyapunov conditions, in *Proceedings of the Summer School on Optimal Transport*, Grenoble (2009). arXiv:1001.1822
10. A. Colesanti, From the Brunn-Minkowski inequality to a class of Poincaré-type inequalities. Commun. Contemp. Math. **10**(5), 765–772 (2008)
11. R. Eldan, Thin shell implies spectral gap up to polylog via a stochastic localization scheme. Geom. Funct. Anal. **23**(2), 532–569 (2013)
12. A. Figalli, Quantitative isoperimetric inequalities, with applications to the stability of liquid drops and crystals, in *Concentration, Functional Inequalities and Isoperimetry*. Contemporary Mathematics, vol. 545 (American Mathematical Society, Providence, 2011), pp. 77–87
13. G.B. Folland, *Introduction to Partial Differential Equations*, 2nd edn. (Princeton University Press, Princeton, 1995)
14. M. Fradelizi, Sections of convex bodies through their centroid. Arch. Math. (Basel) **69**(6), 515–522 (1997)
15. N. Ghoussoub, A. Moradifam, *Functional Inequalities: New Perspectives and New Applications*. Mathematical Surveys and Monographs, vol. 187 (American Mathematical Society, Providence, 2013)
16. O. Guédon, E. Milman, Interpolating thin-shell and sharp large-deviation estimates for isotropic log-concave measures. Geom. Funct. Anal. **21**(5), 1043–1068 (2011)
17. R. Kannan, L. Lovász, M. Simonovits, Isoperimetric problems for convex bodies and a localization lemma. Discrete Comput. Geom. **13**(3–4), 541–559 (1995)
18. B. Klartag, On convex perturbations with a bounded isotropic constant. Geom. Funct. Anal. **16**(6), 1274–1290 (2006)
19. B. Klartag, E. Milman, On volume distribution in 2-convex bodies. Isr. J. Math. **164**, 221–249 (2008)
20. H. Knothe, Contributions to the theory of convex bodies. Mich. Math. J. **4**, 39–52 (1957)
21. A.V. Kolesnikov, E. Milman, Poincaré and Brunn-Minkowski inequalities on weighted manifolds with boundary. Submitted (2014). arxiv.org/abs/1310.2526
22. R. Latała, J.O. Wojtaszczyk, On the infimum convolution inequality. Studia Math. **189**(2), 147–187 (2008)
23. M. Ledoux, *The Concentration of Measure Phenomenon*. Mathematical Surveys and Monographs, vol. 89 (American Mathematical Society, Providence, 2001)
24. R.J. McCann, A convexity principle for interacting gases. Adv. Math. **128**(1), 153–179 (1997)
25. R.J. McCann, Polar factorization of maps on Riemannian manifolds. Geom. Funct. Anal. **11**(3), 589–608 (2001)

26. E. Milman, On the role of convexity in isoperimetry, spectral-gap and concentration. Invent. Math. **177**(1), 1–43 (2009)
27. V.D. Milman, A. Pajor, Isotropic position and interia ellipsoids and zonoids of the unit ball of a normed n-dimensional space, in *Geometric Aspects of Functional Analysis*. Lecture Notes in Mathematics, vol. 1376 (Springer, Berlin, 1987–1988), pp. 64–104
28. V.D. Milman, G. Schechtman, *Asymptotic Theory of Finite-Dimensional Normed Spaces*. Lecture Notes in Mathematics, vol. 1200 (Springer, Berlin, 1986). With an appendix by M. Gromov
29. E. Milman, S. Sodin, An isoperimetric inequality for uniformly log-concave measures and uniformly convex bodies. J. Funct. Anal. **254**(5), 1235–1268 (2008)
30. G. Paouris, Small ball probability estimates for log-concave measures. Trans. Am. Math. Soc. **364**(1), 287–308 (2012)
31. G. Schechtman, J. Zinn, Concentration on the l_p^n ball, in *Geometric Aspects of Functional Analysis*. Lecture Notes in Mathematics, vol. 1745 (Springer, Berlin, 2000), pp. 245–256
32. S. Sodin, An isoperimetric inequality on the ℓ_p balls. Ann. Inst. H. Poincaré Probab. Stat. **44**(2), 362–373 (2008)
33. G.N. Watson, *A Treatise on the Theory of Bessel Functions*. Cambridge Mathematical Library (Cambridge University Press, Cambridge, 1995). Reprint of the second (1944) edition

Modified Paouris Inequality

Rafał Latała

Abstract The Paouris inequality gives the large deviation estimate for Euclidean norms of log-concave vectors. We present a modified version of it and show how the new inequality may be applied to derive tail estimates of l_r-norms and suprema of norms of coordinate projections of isotropic log-concave vectors.

1 Introduction and Main Results

A random vector X is called *log-concave* if it has a logarithmically concave distribution, i.e. $\mathbb{P}(X \in \lambda K + (1 - \lambda)L) \geq \mathbb{P}(X \in K)^{\lambda}\mathbb{P}(X \in L)^{1-\lambda}$ for all nonempty compact sets K, L and $\lambda \in [0, 1]$. The result of Borell [4] states that a random vector with the full dimensional support is log-concave iff it has a logconcave density, i.e. a density of the form $e^{-h(x)}$, where h is a convex function with values in $(-\infty, \infty]$. A typical example of a log-concave vector is a vector uniformly distributed over a convex body. In recent years the study of log-concave vectors attracted attention of many researchers, cf. the monograph [5].

The fundamental result of Paouris [8] gives the large deviation estimate for Euclidean norms of log-concave vectors. It may be stated, c.f. [1], in the form

$$(\mathbb{E}|X|^{p})^{1/p} \leq C_1(\mathbb{E}|X| + \sigma_X(p)) \quad \text{for any } p \geq 1,$$

and any log-concave vector X, where here and in the sequel C_i denote universal constants, $|x|$ is the canonical Euclidean norm on $\mathbb{R}^n$ and

$$\sigma_X(p) := \sup_{|t|=1}(\mathbb{E}|\langle t, X \rangle|^{p})^{1/p}, \quad p \geq 1.$$

In particular if X is additionally *isotropic*, i.e. it is centered and has identity covariance matrix then

$$(\mathbb{E}|X|^{p})^{1/p} \leq C_1(\sqrt{n} + \sigma_X(p)) \quad \text{for } p \geq 1. \tag{1}$$

R. Latała (✉)
Institute of Mathematics, University of Warsaw, Banacha 2, 02-097 Warszawa, Poland
e-mail: rlatala@mimuw.edu.pl

© Springer International Publishing Switzerland 2014
B. Klartag, E. Milman (eds.), *Geometric Aspects of Functional Analysis*,
Lecture Notes in Mathematics 2116, DOI 10.1007/978-3-319-09477-9_19

Together with Chebyshev's inequality this implies

$$\mathbb{P}(|X| \geq 2eC_1 t \sqrt{n}) \leq \exp(-\sigma_X^{-1}(t\sqrt{n})) \quad \text{for } t \geq 1. \tag{2}$$

In this note we show the following modification of the Paouris inequality.

Theorem 1. *For any isotropic log-concave n-dimensional random vector X and $p \geq 1$,*

$$\mathbb{E}\left(\sum_{i=1}^{n} X_i^2 \mathbf{1}_{\{|X_i| \geq t\}}\right)^p \leq (C_2 \sigma_X(p))^{2p} \quad \text{for } t \geq C_2 \log\left(\frac{n}{\sigma_X(p)^2}\right). \tag{3}$$

Obviously $\sum_{i=1}^{n} X_i^2 \mathbf{1}_{\{|X_i| \geq t\}} \geq t^2 N_X(t)$, where

$$N_X(t) := \sum_{i=1}^{n} \mathbf{1}_{\{|X_i| \geq t\}}, \quad t > 0,$$

thus (3) generalizes the estimate derived in [2]:

$$\mathbb{E}(t^2 N_X(t))^p \leq (C\sigma_X(p))^{2p} \quad \text{for } t \geq C \log\left(\frac{n}{\sigma_X(p)^2}\right).$$

It is also not hard to see that Theorem 1 implies Paouris' inequality (1). To see this let $p' := \inf\{q \geq p: \sigma_X(q) \geq \sqrt{n}\}$. Then

$$(\mathbb{E}|X|^p)^{1/p} \leq (\mathbb{E}|X|^{2p'})^{1/2p'} \leq C_2 \sigma_X(p') \leq C_2(\sqrt{n} + \sigma_X(p)),$$

where the second inequality follows by (3) aplied with $p = p'$ and $t = 0$.

In fact we may extend estimate (1) replacing the Euclidean norm by the l_r-norm, $\|x\|_r := (\sum_i |x_i|^r)^{1/r}$, $r \geq 2$.

Theorem 2. *For any $r \geq 2$ and any isotropic log-concave n-dimensional random vector X,*

$$(\mathbb{E}\|X\|_r^p)^{1/p} \leq C_3(rn^{1/r} + \sigma_X(p)) \quad \text{for } p \geq 1. \tag{4}$$

Theorem 2 gives better bounds than presented in [6], since the constant does not explode for $r \to 2+$ and the parameter p is replaced by the smaller quantity $\sigma_X(p)$. Estimate (4) and Chebyshev's inequality imply for $t \geq 1$,

$$\mathbb{P}(\|X\|_r \geq 2eC_3 trn^{1/r}) \leq \exp(-\sigma_X^{-1}(trn^{1/r})).$$

In general (4) is sharp up to a multiplicative constant, since for a random vector X with i.i.d. symmetric exponential coordinates with variance 1 we have $\sigma_X(p) \leq p\sigma_X(2) = p$ and

$$(\mathbb{E}\|X\|_r^p)^{1/p} \geq \max\{\mathbb{E}\|X\|_r, (\mathbb{E}|X_1|^p)^{1/p}\} \geq \frac{1}{C}\max\{rn^{1/r}, p\}.$$

However there are reasons to believe that the following stronger estimate may hold for log-concave vectors (c.f. [7])

$$(\mathbb{E}\|X\|_r^p)^{1/p} \leq C\left(\mathbb{E}\|X\|_r + \sup_{\|t\|_{r'}\leq 1}(\mathbb{E}|\langle t, X\rangle|^p)^{1/p}\right).$$

Another consequence of Theorem 3 is the uniform version of the Paouris inequality. For $I \subset \{1,\ldots,n\}$ by P_I we denote the coordinate projection from $\mathbb{R}^n$ into $\mathbb{R}^I$.

Theorem 3. *For any isotropic log-concave n-dimensional random vector X and $1 \leq m \leq n$ we have*

$$\left(\mathbb{E}\max_{|I|=m}|P_IX|^p\right)^{1/p} \leq C_4\left(\sqrt{m}\log\left(\frac{en}{m}\right) + \sigma_X(p)\right) \quad \text{for } p \geq 1. \tag{5}$$

Again the example of a vector with the product isotropic exponential distribution shows that in general estimate (5) is sharp. Theorem 3 and Chebyshev's inequality yield for $t \geq 1$,

$$\mathbb{P}\left(\max_{|I|=m}|P_IX| \geq 2eC_4t\sqrt{m}\log\left(\frac{en}{m}\right)\right) \leq \exp\left(-\sigma_X^{-1}\left(t\sqrt{m}\log\left(\frac{en}{m}\right)\right)\right),$$

which removes an exponential factor from Theorem 3.4 in [2].

The paper is organised as follows. In Sect. 2 we recall basic facts about log-concave vectors and prove Theorem 1. In Sect. 3 we show how to use (3) to get estimates for the joint distribution of order statistics of X and derive Theorems 2 and 3.

Notation. For a r.v. Y and $p > 0$ we set $\|Y\|_p := (\mathbb{E}|Y|^p)^{1/p}$. We write $|I|$ for the cardinality of a set I. By a letter C we denote absolute constants, value of C may differ at each occurrence. Whenever we want to fix a value of an absolute constant we use letters $C_1, C_2, \ldots$.

2 Proof of Theorem 1

The result of Barlow et al. [3] imply that for symmetric log-concave random variables Y, and $p \geq q > 0$, $\|Y\|_p \leq \Gamma(p+1)^{1/p}/\Gamma(q+1)^{1/q}\|Y\|_q$. If Y

is centered and log-concave and Y' is an independent copy of Y then $Y - Y'$ is symmetric and log-concave, hence for $p \geq q \geq 2$,

$$\|Y\|_p \leq \|Y - Y'\|_p \leq \frac{\Gamma(p+1)^{1/p}}{\Gamma(q+1)^{1/q}} \|Y - Y'\|_q \leq 2\frac{\Gamma(p+1)^{1/p}}{\Gamma(q+1)^{1/q}} \|Y\|_q \leq 2\frac{p}{q}\|Y\|_q.$$

Thus for isotropic log-concave vectors X,

$$\sigma_X(\lambda p) \leq 2\lambda\sigma_X(p) \quad \text{and} \quad \sigma_X^{-1}(\lambda t) \geq \frac{\lambda}{2}\sigma_X^{-1}(t) \quad \text{for } p \geq 2, \, t, \lambda \geq 1.$$

In particular $\sigma_X(p) \leq p$ for $p \geq 2$.

If Y is a log-concave r.v. (not necessarily centered) then for $p \geq 2$, $\|Y\|_p \leq |\mathbb{E}Y| + \|Y - Y'\|_p \leq (p+1)\|Y\|_2$ and Chebyshev's inequality yields $\mathbb{P}(|Y| \geq e(p+1)\|Y\|_2) \leq e^{-p}$. Thus we obtain a Ψ_1-estimate for log-concave r.v's

$$\mathbb{P}(|Y| \geq t) \leq \exp\left(2 - \frac{t}{2e\|Y\|_2}\right) \quad \text{for } t \geq 0. \tag{6}$$

We start with a variant of Proposition 7.1 from [2].

Proposition 4. *There exists an absolute positive constant C_5 such that the following holds. Let X be an isotropic log-concave n-dimensional random vector, $A = \{X \in K\}$, where K is a convex set in $\mathbb{R}^n$ satisfying $0 < \mathbb{P}(A) \leq 1/e$. Then for every $t \geq 1$,*

$$\sum_{i=1}^{n} \mathbb{E}X_i^2 \mathbf{1}_{A \cap \{X_i \geq t\}} \leq C_5\mathbb{P}(A)\left(\sigma_X^2(-\log(\mathbb{P}(A))) + nt^2 e^{-t/C_5}\right) \tag{7}$$

and for every $t > 0$, $u \geq 1$,

$$\sum_{k=0}^{\infty} 4^k |\{i \leq n: \mathbb{P}(A \cap \{X_i \geq 2^k t\}) \geq e^{-u}\mathbb{P}(A)\}|$$

$$\leq \frac{C_5 u^2}{t^2}\left(\sigma_X^2(-\log(\mathbb{P}(A))) + n\mathbf{1}_{\{t \leq uC_5\}}\right). \tag{8}$$

Proof. Let Y be a random vector defined by

$$\mathbb{P}(Y \in B) = \frac{\mathbb{P}(A \cap \{X \in B\})}{\mathbb{P}(A)} = \frac{\mathbb{P}(X \in B \cap K)}{\mathbb{P}(X \in K)},$$

i.e. Y is distributed as X conditioned on A. Clearly, for every measurable set B one has $\mathbb{P}(X \in B) \geq \mathbb{P}(A)\mathbb{P}(Y \in B)$. It is easy to see that Y is log-concave, but not necessarily isotropic.

The Paouris inequality (2) (applied for the isotropic vector $P_I X$) implies that for any $\emptyset \neq I \subset \{1, \ldots, n\}$ and $t \geq (2eC_1)^2 |I|$,

$$\mathbb{P}\left(\sum_{i \in I} X_i^2 \geq t\right) = \mathbb{P}(|P_I X| \geq \sqrt{t}) \leq \exp\left(-\sigma_X^{-1}\left(\frac{1}{2eC_1}\sqrt{t}\right)\right). \tag{9}$$

Let

$$I := \{i \leq n : \mathbb{E}Y_i^2 \geq 2(2eC_1)^2\}.$$

Log-concavity of Y (and as a consequence also of $P_I Y$) yields $\mathbb{E}|P_I Y|^4 \leq C(\mathbb{E}|P_I Y|^2)^2$. The Paley-Zygmund inequality implies

$$\mathbb{P}\left(\sum_{i \in I} Y_i^2 \geq \frac{1}{2} \sum_{i \in I} \mathbb{E}Y_i^2\right) \geq \frac{1}{4} \frac{(\mathbb{E}\sum_{i \in I} Y_i^2)^2}{\mathbb{E}(\sum_{i \in I} Y_i^2)^2} \geq \frac{1}{C_6}.$$

Therefore

$$\mathbb{P}\left(\sum_{i \in I} X_i^2 \geq \frac{1}{2} \sum_{i \in I} \mathbb{E}Y_i^2\right) \geq \mathbb{P}(A)\,\mathbb{P}\left(\sum_{i \in I} Y_i^2 \geq \frac{1}{2} \sum_{i \in I} \mathbb{E}Y_i^2\right) \geq \frac{1}{C_6}\mathbb{P}(A).$$

Together with (9) this gives

$$\frac{1}{C_6}\mathbb{P}(A) \leq \exp\left(-\sigma_X^{-1}\left(\frac{1}{2eC_1}\sqrt{\frac{1}{2}\sum_{i \in I} \mathbb{E}Y_i^2}\right)\right),$$

hence

$$\sum_{i \in I} \mathbb{E}Y_i^2 \leq C\sigma_X^2(-\log \mathbb{P}(A)).$$

Moreover if $i \notin I$, i.e. $\mathbb{E}Y_i^2 \leq 2(2eC_1)^2$ then (6) yields $\mathbb{E}Y_i^2 \mathbf{1}_{\{|Y_i| \geq t\}} \leq Ct^2 e^{-t/C}$ for $t \geq 1$. Therefore

$$\sum_{i=1}^n \mathbb{E}X_i^2 \mathbf{1}_{A \cap \{|X_i| \geq t\}} = \mathbb{P}(A) \sum_{i=1}^n \mathbb{E}Y_i^2 \mathbf{1}_{\{|Y_i| \geq t\}}$$

$$\leq \mathbb{P}(A)\left(\sum_{i \in I} \mathbb{E}Y_i^2 + nCt^2 e^{-t/C}\right)$$

$$\leq C\mathbb{P}(A)\left(\sigma_X^2(-\log(\mathbb{P}(A))) + nt^2 e^{-t/C}\right).$$

To show (8) note first that for every i the random variable Y_i is log-concave, hence for $s \geq 0$,

$$\frac{\mathbb{P}(A \cap \{X_i \geq s\})}{\mathbb{P}(A)} = \mathbb{P}(Y_i \geq s) \leq \exp\left(2 - \frac{t}{2e\|Y_i\|_2}\right).$$

Thus, if $\mathbb{P}(A \cap \{X_i \geq 2^k t\}) \geq e^{-u}\mathbb{P}(A)$ and $u \geq 1$ then $\|Y_i\|_2 \geq 2^k t/(2e(u+2)) \geq 2^k t/(6eu)$. In particular it cannot happen if $i \notin I$, $k \geq 0$ and $u \leq t/C_5$ with C_5 large enough.

Therefore

$$\sum_{k=0}^{\infty} 4^k |\{i \leq n: \mathbb{P}(A \cap \{X_i \geq 2^k t\}) \geq e^{-u}\mathbb{P}(A)\}|$$

$$\leq \left(\sum_{i \in I} + 1_{\{t \leq uC_5\}} \sum_{i \notin I}\right) \sum_{k=0}^{\infty} 4^k 1_{\{(\mathbb{E}Y_i^2)^{1/2} \geq 2^k t/(6eu)\}}$$

$$\leq \frac{2(6eu)^2}{t^2} \left(\sum_{i \in I} + 1_{\{t \leq uC_5\}} \sum_{i \notin I}\right) \mathbb{E}Y_i^2$$

$$\leq \frac{Cu^2}{t^2} \left(\sigma_X^2(-\log(\mathbb{P}(A))) + n1_{\{t \leq uC_5\}}\right).$$

$\square$

We will also use the following simple combinatorial lemma (Lemma 11 in [6]).

Lemma 5. *Let* $\ell_0 \geq \ell_1 \geq \ldots \geq \ell_s$ *be a fixed sequence of positive integers and*

$$\mathcal{F} := \{f : \{1, 2, \ldots, \ell_0\} \to \{0, 1, 2, \ldots, s\}: \forall_{1 \leq i \leq s} |\{r: f(r) \geq i\}| \leq \ell_i\}.$$

Then

$$|\mathcal{F}| \leq \prod_{i=1}^{s} \left(\frac{e\ell_{i-1}}{\ell_i}\right)^{\ell_i}.$$

Proof of Theorem 1. We have by the Paouris estimate (1)

$$\mathbb{E}\left(\sum_{i=1}^{n} X_i^2 1_{\{|X_i| \geq t\}}\right)^p \leq \mathbb{E}|X|^{2p} \leq (C_1(\sqrt{n} + \sigma_X(2p)))^{2p},$$

so the estimate (3) is obvious if $\sigma_X(p) \geq \frac{1}{8}\sqrt{n}$, we will thus assume that $\sigma_X(p) \leq \frac{1}{8}\sqrt{n}$.

Observe that for $l = 1, 2, \ldots,$

$$\mathbb{E}\left(\sum_{i=1}^{n} X_i^2 \mathbf{1}_{\{X_i \geq t\}}\right)^l \leq \mathbb{E}\left(\sum_{i=1}^{n} \sum_{k=0}^{\infty} 4^{k+1} t^2 \mathbf{1}_{\{X_i \geq 2^k t\}}\right)^l$$

$$= (2t)^{2l} \sum_{i_1,\ldots,i_l=1}^{n} \sum_{k_1,\ldots,k_l=0}^{\infty} 4^{k_1+\ldots+k_l} \mathbb{P}(B_{i_1,k_1\ldots,i_l,k_l}),$$

where

$$B_{i_1,k_1\ldots,i_l,k_l} := \{X_{i_1} \geq 2^{k_1} t, \ldots, X_{i_l} \geq 2^{k_l} t_l\}.$$

Define a positive integer l by

$$p < l \leq 2p \quad \text{and} \quad l = 2^m \text{ for some positive integer } m.$$

Then $\sigma_X(p) \leq \sigma_X(l) \leq \sigma_X(2p) \leq 4\sigma_X(p)$. Since $-X$ is also isotropic log-concave and for any nonnegative r.v. Y, $(\mathbb{E}Y^p)^{1/p} \leq (\mathbb{E}Y^l)^{1/l}$, it is enough to show that

$$m(l) := \sum_{k_1,\ldots,k_l=0}^{\infty} \sum_{i_1,\ldots,i_l=1}^{n} 4^{k_1+\ldots+k_l} \mathbb{P}(B_{i_1,k_1\ldots,i_l,k_l}) \leq \left(\frac{C\sigma_X(l)}{t}\right)^{2l} \tag{10}$$

provided that $t \geq C_2 \log(\frac{n}{\sigma_X(l)^2})$. Since $\sigma_X(l) \leq 4\sigma_X(p) \leq \frac{1}{2}\sqrt{n}$ this in particular implies that $t \geq C_2$.

We divide the sum in $m(l)$ into several parts. Define sets

$$I_0 := \left\{(i_1, k_1, \ldots, i_l, k_l): \mathbb{P}(B_{i_1,k_1,\ldots,i_l,k_l}) > e^{-l}\right\},$$

and for $j = 1, 2, \ldots,$

$$I_j := \left\{(i_1, k_1, \ldots, i_l, k_l): \mathbb{P}(B_{i_1,k_1,\ldots,i_l,k_l}) \in (e^{-2^j l}, e^{-2^{j-1} l}]\right\}.$$

Then $m(l) = \sum_{j \geq 0} m_j(l)$, where

$$m_j(l) := \sum_{(i_1,k_1,\ldots,i_l,k_l) \in I_j} 4^{k_1+\ldots+k_l} \mathbb{P}(B_{i_1,k_1\ldots,i_l,k_l}).$$

To estimate $m_0(l)$ define for $1 \leq s \leq l,$

$$P_s I_0 := \{(i_1, k_1, \ldots, i_s, k_s): (i_1, k_1, \ldots, i_l, k_l) \in I_0 \text{ for some } i_{s+1}, \ldots, k_l\}.$$

We have by (6) (if C_2 is large enough)

$$\mathbb{P}(B_{i_1,k_1\ldots,i_s,k_s}) \leq \mathbb{P}(B_{i_1,k_1}) \leq \exp(2 - 2^{k_1-1}t/e) \leq e^{-1}.$$

Thus for $s = 1,\ldots,l-1$,

$$t^2 \sum_{(i_1,\ldots,k_{s+1})\in P_{s+1}I_0} 4^{k_1+\ldots+k_{s+1}}\mathbb{P}(B_{i_1,\ldots,k_{s+1}})$$

$$\leq \sum_{(i_1,\ldots,k_s)\in P_s I_0} 4^{k_1+\ldots+k_s} \sum_{i_{s+1}=1}^{n} \sum_{k_{s+1}=0}^{\infty} 4^{k_{s+1}} t^2 \mathbb{P}(B_{i_1,\ldots,k_s} \cap \{X_{i_{s+1}} \geq 2^{k_{s+1}}t\})$$

$$\leq \sum_{(i_1,\ldots,k_s)\in P_s I_0} 4^{k_1+\ldots+k_s} \sum_{i_{s+1}=1}^{n} \mathbb{E}2X^2_{i_{s+1}} \mathbf{1}_{B_{i_1,\ldots,k_s} \cap \{X_{i_{s+1}}\geq t\}}$$

$$\leq 2C_5 \sum_{(i_1,\ldots,k_s)\in P_s I_0} 4^{k_1+\ldots+k_s}\mathbb{P}(B_{i_1,\ldots,k_s})(\sigma_X^2(-\log\mathbb{P}(B_{i_1,\ldots,k_s})) + nt^2 e^{-t/C_5}),$$

where the last inequality follows by (7). Note that for $(i_1,\ldots,k_s) \in P_s I_0$ we have $\mathbb{P}(B_{i_1,\ldots,k_s}) \geq e^{-l}$ and, by our assumptions on t (if C_2 is sufficiently large) $nt^2 e^{-t/C_5} \leq ne^{-t/(2C_5)} \leq \sigma_X^2(l)$. Therefore

$$\sum_{(i_1,\ldots,k_{s+1})\in P_{s+1}I_0} 4^{k_1+\ldots+k_{s+1}}\mathbb{P}(B_{i_1,\ldots,k_{s+1}})$$

$$\leq 4C_5 t^{-2}\sigma_X^2(l) \sum_{(i_1,\ldots,k_s)\in P_s I_0} 4^{k_1+\ldots+k_s}\mathbb{P}(B_{i_1,\ldots,k_s}).$$

By induction we get

$$m_0(l) = \sum_{(i_1,\ldots,k_l)\in I_0} 4^{k_1+\ldots+k_l}\mathbb{P}(B_{i_1,\ldots,k_l})$$

$$\leq (4C_5 t^{-2}\sigma_X^2(l))^{l-1} \sum_{(i_1,k_1)\in P_1 I_0} 4^{k_1}\mathbb{P}(B_{i_1,k_1})$$

$$\leq (4C_5 t^{-2}\sigma_X^2(l))^{l-1} t^{-2} \sum_{i=1}^{n} 2\mathbb{E}X_i^2 \mathbf{1}_{\{X_i\geq t\}}$$

$$\leq (4C_5 t^{-2}\sigma_X^2(l))^{l-1} nCe^{-t/C} \leq \left(\frac{C\sigma_X(l)}{t}\right)^{2l},$$

where the last inequality follows from the assumptions on t.

Now we estimate $m_j(l)$ for $j > 0$. Fix $j > 0$ and define a positive integer r_1 by

$$2^{r_1-1} < \frac{t}{C_5} \leq 2^{r_1}.$$

For all $(i_1, k_1, \ldots, i_l, k_l) \in I_j$ define a function $f_{i_1,k_1,\ldots,i_l,k_l} \colon \{1,\ldots,\ell\} \to \{0,1,2,\ldots\}$ by

$$f_{i_1,k_1,\ldots,i_l,k_l}(s) := \begin{cases} 0 \text{ if } \frac{\mathbb{P}(B_{i_1,k_1,\ldots,i_s,k_s})}{\mathbb{P}(B_{i_1,k_1,\ldots,i_s-1,k_s-1})} > e^{-1}, \\ r \text{ if } e^{-2^r} < \frac{\mathbb{P}(B_{i_1,k_1,\ldots,i_s,k_s})}{\mathbb{P}(B_{i_1,k_1,\ldots,i_s-1,k_s-1})} \leq e^{-2^{r-1}}, \ r \geq 1. \end{cases}$$

Note that for every $(i_1, k_1, \ldots, i_l, k_l) \in I_j$ one has

$$1 = \mathbb{P}(B_\emptyset) \geq \mathbb{P}(B_{i_1,k_1}) \geq \mathbb{P}(B_{i_1,k_1,i_2,k_2}) \geq \ldots \geq \mathbb{P}(B_{i_1,k_1\ldots i_l,k_l}) > \exp(-2^j l).$$

Denote

$$\mathcal{F}_j := \{f_{i_1,k_1,\ldots,i_l,k_l} \colon (i_1, k_1, \ldots, i_l, k_l) \in I_j\}.$$

Then for $f = f_{i_1,k_1,\ldots,i_l,k_l} \in \mathcal{F}_j$ and $r \geq 1$ one has

$$\exp(-2^j l) < \mathbb{P}(B_{i_1,k_1,\ldots,i_l,k_l}) = \prod_{s=1}^{\ell} \frac{\mathbb{P}(B_{i_1,k_1\ldots,i_s,k_s})}{\mathbb{P}(B_{i_1,k_1,\ldots,i_s-1,k_s-1})}$$

$$\leq \exp(-2^{r-1}|\{s \colon f(s) \geq r\}|).$$

Hence for every $r \geq 1$ one has

$$|\{s \colon f(s) \geq r\}| \leq \min\{2^{j+1-r}l, l\} =: l_r. \tag{11}$$

In particular f takes values in $\{0, 1, \ldots, j + 1 + \lfloor \log_2 l \rfloor\}$. Clearly, $\sum_{r \geq 1} l_r = (j+2)l$ and $l_{r-1}/l_r \leq 2$, so by Lemma 5

$$|\mathcal{F}_j| \leq \prod_{r=1}^{j+1+\lfloor \log_2 l \rfloor} \left(\frac{e l_{r-1}}{l_r}\right)^{l_r} \leq e^{2(j+2)l}.$$

Now fix $f \in \mathcal{F}_j$ and define

$$I_j(f) := \{(i_1, k_1, \ldots, i_l, k_l) \colon f_{i_1,k_1,\ldots,i_l,k_l} = f\}$$

and for $s \leq l$,

$$I_{j,s}(f) := \{(i_1, k_1, \ldots, i_s, k_s) \colon f_{i_1,k_1,\ldots,i_l,k_l} = f \text{ for some } i_{s+1}, k_{s+1} \ldots, i_l, k_l\}.$$

Recall that for $s \geq 1$, $\mathbb{P}(B_{i_1,k_1,\ldots,i_s,k_s}) \leq e^{-1}$, moreover for $s \leq l$,

$$\sigma_X(-\log \mathbb{P}(B_{i_1,k_1,\ldots,i_s,k_s})) \leq \sigma_X(-\log \mathbb{P}(B_{i_1,k_1,\ldots,i_l,k_l})) \leq \sigma_X(2^j l)$$
$$\leq 2^{j+1}\sigma_X(l).$$

Hence estimate (8) applied with $u = 2^{f(s+1)}$ implies for $1 \leq s \leq l - 1$,

$$\sum_{(i_1,k_1,\ldots,i_{s+1},k_{s+1})\in I_{j,s+1}(f)} 4^{k_1+\ldots+k_{s+1}}\mathbb{P}(B_{i_1,k_1,\ldots,i_{s+1},k_{s+1}})$$

$$\leq g(f(s+1)) \sum_{(i_1,k_1,\ldots,i_s,k_s)\in I_{j,s}(f)} 4^{k_1+\ldots+k_s}\mathbb{P}(B_{i_1,k_1,\ldots,i_s,k_s}),$$

where

$$g(r) := \begin{cases} C_5 t^{-2}4^{j+1}\sigma_X^2(l) & \text{for } r = 0, \\ C_5 t^{-2}4^{r+j+1}\sigma_X^2(l)\exp(-2^{r-1}) & \text{for } 1 \leq r < r_1, \\ C_5 t^{-2}4^r(4^{j+1}\sigma_X^2(l) + n)\exp(-2^{r-1}) & \text{for } r \geq r_1. \end{cases}$$

Suppose that $(i_1, k_1) \in I_1(f)$ and $f(1) = r$ then

$$\exp(-2^r) \leq \mathbb{P}(X_{i_1} \geq 2^{k_1}t) \leq \exp(2 - 2^{k_1-1}t/e),$$

hence $2^{k_1}t \leq e2^{r+2}$. W.l.o.g. $C_5 > 4e$, therefore $r \geq r_1$. Moreover, $4^{k_1} \leq 16e^24^r t^{-2}$, hence

$$\sum_{(i_1,k_1)\in I_{j,1}(f)} 4^{k_1}\mathbb{P}(B_{i_1,k_1}) \leq n32e^2t^{-2}4^r\exp(-2^{r-1}) \leq g(r) = g(f(1)),$$

since we may assume that $C_5 \geq 32e^2$. Thus the easy induction shows that

$$m_j(f) := \sum_{(i_1,\ldots,k_l)\in I_j(f)} 4^{k_1+\ldots+k_l}\mathbb{P}(B_{i_1,k_1,\ldots,i_l,k_l}) \leq \prod_{s=1}^{l} g(f(s)) = \prod_{r=0}^{\infty} g(r)^{n_r},$$

where $n_r := |f^{-1}(r)|$.

Observe that

$$e^{-2^{j-1}l} \geq \mathbb{P}(B_{i_1,k_1,\ldots,i_l,k_l}) = \prod_{s=1}^{l} \frac{\mathbb{P}(B_{i_1,k_1,\ldots,i_s,k_s})}{\mathbb{P}(B_{i_1,k_1,\ldots,i_{s-1},k_{s-1}})} \geq e^{-l}\prod_{s:f(s)\geq 1} e^{-2^{f(s)}},$$

therefore

$$\sum_{r=1}^{\infty} n_r 2^{r-1} = \frac{1}{2} \sum_{s: f(s) \geq 1} 2^{f(s)} \geq \frac{1}{2} l(2^{j-1} - 1).$$

Moreover $4^{j+1} \sigma_X^2(l) + n \leq 2 \cdot 4^{j+1} n$ and

$$\sum_{r \geq 1} r n_r \leq (j+1)l + \sum_{r \geq j+2} r l_r = (2j+4)l.$$

Hence

$$\prod_{r=0}^{\infty} g(r)^{n_r} \leq \left(\frac{C_5 4^{j+1} \sigma_X^2(l)}{t^2} \right)^l 4^{(2j+4)l} \left(\frac{2n}{\sigma_X^2(l)} \right)^m \exp\left(-\frac{l}{2}(2^{j-1} - 1) \right).$$

where $m = \sum_{r \geq r_1} n_r \leq l_{r_1} \leq 2^{j+1-r_1} l$. By the assumption on l we have $(2n/\sigma_X^2(l)) \leq 2\exp(t/C_2) \leq \exp(2^{r_1-4})$ if C_2 is large enough with respect to C_5. Hence

$$m_j(l) \leq |\mathcal{F}_j| \left(\frac{\sqrt{e} C_5 4^{3j+5} \sigma_X^2(l)}{t^2} \right)^l \exp(-l2^{j-3})$$

and we get

$$m(l) = \sum_{j=0}^{\infty} m_j(l) \leq \left(\frac{C \sigma_X(l)}{t} \right)^{2l} + \sum_{j \geq 1} \left(\frac{C_5 e^{2j+5} 4^{3j+5} \sigma_X^2(l)}{t^2} \right)^l \exp(-l2^{j-3})$$

and (10) easily follows. □

3 Estimates for Joint Distribution of Order Statistics

For a random vector $X = (X_1, \ldots, X_n)$ by $X_1^* \geq X_2^* \geq \ldots \geq X_n^*$ we denote the nonincreasing rearrangement of $|X_1|, \ldots, |X_n|$, in particular $X_1^* = \max\{|X_1|, \ldots, |X_n|\}$ and $X_n^* = \min\{|X_1|, \ldots, |X_n|\}$. The following consequence of Theorem 1 generalizes Theorem 3.3 from [2].

Theorem 6. *Let X be an isotropic log-concave vector, $0 = l_0 < l_1 < l_2 < \ldots < l_k \leq n$ and $t_1, \ldots, t_k \geq 0$ be such that*

$$t_r \geq C_7 \log\left(\frac{C_7^2 n}{\sum_{j=1}^{s} t_j^2(l_j - l_{j-1})} \right) \quad \text{for } 1 \leq r \leq k.$$

Then

$$\mathbb{P}\left(X_{l_1}^* \geq t_1, \ldots, X_{l_k}^* \geq t_k\right) \leq \exp\left(-\sigma_X^{-1}\left(\frac{1}{C_7}\sqrt{\sum_{j=1}^{k} t_j^2(l_j - l_{j-1})}\right)\right).$$

Proof. Let $t := \min\{t_1, \ldots, t_k\}$, $u := (\sum_{j=1}^{k} t_j^2(l_j - l_{j-1}))^{1/2}$ and $p := \sigma_X^{-1}(e^{-1/2}u/C_2)$. It is not hard to see that if C_7 is large enough then $u \geq \sqrt{e}C_2$, so $p \geq 2$. Assumptions imply (if C_7 is large enough) that $C_2 \log(n/\sigma_X^2(p)) = C_2 \log(enC_2^2/u^2) \leq t$. Therefore Chebyshev's inequality and Theorem 1 yield

$$\mathbb{P}\left(X_{l_1}^* \geq t_1, \ldots, X_{l_k}^* \geq t_k\right) \leq \mathbb{P}\left(\sum_{i=1}^{n} X_i^2 \mathbf{1}_{\{|X_i| \geq t\}} \geq u^2\right)$$

$$\leq u^{-2p}\mathbb{E}\left(\sum_{i=1}^{n} X_i^2 \mathbf{1}_{\{|X_i| \geq t\}}\right)^p \leq \left(\frac{C_2\sigma_X(p)}{u}\right)^{2p} \leq e^{-p}.$$

$\square$

Corollary 7. *Let X be an isotropic log-concave vector and*

$$Y_j := \left(X_{2^{j-1}}^* - C_7 \log(4n2^{-j})\right)_+, \quad 1 \leq j \leq 1 + \log_2 n.$$

Then for any $1 \leq s \leq 1 + \log_2 n$ and $u_1, \ldots, u_s \geq 0$ with $\sum_{j=1}^{s} u_s > 0$ we have

$$\mathbb{P}(Y_1 \geq u_1, \ldots, Y_s \geq u_s) \leq \exp\left(-\sigma_X^{-1}\left(\frac{1}{2C_7}\sqrt{\sum_{j=1}^{s} 2^j u_j^2}\right)\right).$$

Proof. Let

$$I = \{j \geq 0: u_j > 0\} = \{i_1 < \ldots < i_k\}.$$

By our assumptions $I \neq \emptyset$, hence $k \geq 1$. Let $l_0 = 0$, $l_j = 2^{i_j-1}$, $t_j := C_7 \log(4n2^{-i_j}) + u_{i_j}$ for $1 \leq j \leq k$ and $u := (\sum_{j=1}^{k}(l_j - l_{j-1})t_j^2)^{1/2}$. Then for $1 \leq j \leq k$, $u^2 \geq C_7^2 2^{i_j-2}$ therefore $t_j \geq C_7 \log(C_7^2 n/u^2)$ for all j and we may apply Theorem 6 and get

$$\mathbb{P}(Y_1 \geq u_1, \ldots, Y_s \geq u_s) = \mathbb{P}(X_{l_1}^* \geq t_1, \ldots, X_{l_k}^* \geq t_k) \leq \exp\left(-\sigma_X^{-1}\left(\frac{1}{C_7}u\right)\right)$$

$$\leq \exp\left(-\sigma_X^{-1}\left(\frac{1}{2C_7}\sqrt{\sum_{j=1}^{s} 2^j u_j^2}\right)\right).$$

$\square$

Lemma 8. *For nonnegative r.v.'s* $Y_1, \ldots, Y_s$ *and* $u > 0$ *we have*

$$\mathbb{P}\left(\sum_{i=1}^{s} Y_i \geq u\right) \leq \sum_{(k_1,\ldots,k_s)\in I_s} \mathbb{P}\left(Y_1 \geq \frac{k_1 u}{2s}, \ldots, Y_s \geq \frac{k_s u}{2s}\right),$$

where

$$I_s := \{k_1, \ldots, k_s \in \{0, 1, \ldots, s\}^s \colon k_1 + \ldots + k_s = s\}.$$

Proof. It is enough to observe that if $y_1 + \ldots + y_s \geq u$ and we set $l_i := \lfloor 2sy_i/u \rfloor$ then $y_i \geq l_i u/(2s)$ and $\sum_{i=1}^{s} l_i \geq \sum_{i=1}^{s}(2sy_i/u - 1) \geq s$. □

Proof of Theorem 2. Let $s := 1 + \lfloor \log_2 n \rfloor$ and $Y_j, 1 \leq j \leq s$ be as in Corollary 7. We have

$$\|X\|_r^r = \sum_{i=1}^{n} |X_i^*|^r \leq \sum_{j=1}^{s} 2^{j-1}|X_{2^{j-1}}^*|^r \leq \sum_{j=1}^{s} 2^{r+j-1}(Y_j^r + C_7^r \log^r(4n2^{-j}))$$

$$\leq (C_8 r)^r n + \sum_{j=1}^{s} 2^{r+j-1} Y_j^r.$$

By Lemma 8

$$\mathbb{P}\left(\sum_{j=1}^{s} 2^{r+j-1} Y_j^r \geq u^r\right) \leq \sum_{(k_1,\ldots,k_s)\in I_s} \mathbb{P}\left(2Y_1^r \geq \frac{k_1 u^r}{s2^r}, \ldots, 2^s Y_s^r \geq \frac{k_s u^r}{s2^r}\right).$$

Moreover for any $(k_1, \ldots, k_s) \in I_s$,

$$\sum_{j=1}^{s} 2^{j-2j/r}\left(\frac{k_j}{s}\right)^{2/r} \geq \sum_{j=1}^{s}\left(\frac{k_j}{s}\right)^{2/r} \geq \left(\sum_{j=1}^{s}\frac{k_j}{s}\right)^{2/r} = 1.$$

Therefore Corollary 7 yields

$$\mathbb{P}\left(\sum_{j=1}^{s} 2^{r+j-1} Y_j^r \geq u^r\right) \leq |I_s| \exp\left(-\sigma_X^{-1}\left(\frac{u}{4C_7}\right)\right).$$

However $|I_s| = \binom{2s-1}{s-1} \leq 2^{2s-2} \leq n^2$, so we obtain for $u \geq 2C_8 rn^{1/r}$,

$$\mathbb{P}(\|X\|_r \geq u) \leq n^2 \exp\left(-\sigma_X^{-1}\left(\frac{u}{8C_7}\right)\right).$$

Since $rn^{1/r} \geq e \log n$ and for $\lambda, s \geq 1$, $\sigma_X^{-1}(2\lambda s) \geq \lambda \sigma_X^{-1}(s)$ and $\sigma_X^{-1}(s) \geq s$ we get

$$\mathbb{P}(\|X\|_r \geq Ct) \leq \exp(-\sigma_X^{-1}(t)) \quad \text{for } t \geq rn^{1/r}.$$

Integration by parts easily yields (4). □

Proof of Theorem 3. Let $s := 1 + \lfloor \log_2 m \rfloor$ and Y_j, $1 \leq j \leq s$ be as in Corollary 7. Then

$$\sup_{|I|=m} |P_I X|^2 = \sum_{i=1}^{m} |X_i^*|^2 \leq \sum_{j=1}^{s} 2^{j-1} |X_{2^{j-1}}^*|^2 \leq \sum_{j=1}^{s} 2^j \left(C_7^2 \log^2(4n2^{-j}) + Y_j^2 \right)$$

$$\leq C_9 m \log^2(en/m) + \sum_{j=1}^{s} 2^j Y_j^2.$$

Moreover,

$$\mathbb{P}\left(\sum_{j=1}^{s} 2^j Y_j^2 \geq u^2 \right) \leq \sum_{(k_1,\ldots,k_s) \in I_s} \mathbb{P}\left(2Y_1^2 \geq \frac{k_1 u^2}{2s}, \ldots, 2^s Y_s^2 \geq \frac{k_s u^2}{2s} \right)$$

$$\leq |I_s| \exp\left(-\sigma_X^{-1}\left(\frac{u}{2\sqrt{2}C_7} \right) \right),$$

where the first inequality follows by Lemma 8 and the second one by Corollary 7. Observe that $|I_s| = \binom{2s-1}{s-1} \leq 2^{2s-2} \leq m^2$, thus we showed that for $u \geq \sqrt{2C_9} m \log(en/m)$

$$\mathbb{P}\left(\max_{|I|=m} |P_I X| \geq u \right) \leq m^2 \exp\left(-\sigma_X^{-1}\left(\frac{u}{4C_7} \right) \right).$$

Since for $\lambda, s \geq 1$, $\sigma_X^{-1}(2\lambda s) \geq \lambda \sigma_X^{-1}(s)$ and $\sigma_X^{-1}(s) \geq s$ we easily get for $t \geq 1$,

$$\mathbb{P}\left(\max_{|I|=m} |P_I X| \geq Ct\sqrt{m} \log(en/m) \right) \leq \exp\left(-\sigma_X^{-1}\left(t\sqrt{m} \log(en/m) \right) \right).$$

Theorem 3 follows by integration by parts. □

Acknowledgements Research supported by NCN grant 2012/05/B/ST1/00412.

References

1. R. Adamczak, R. Latała, A.E. Litvak, K. Oleszkiewicz, A. Pajor N. Tomczak-Jaegermann, A short proof of Paouris' inequality. Can. Math. Bull. **57**, 3–8 (2014)
2. R. Adamczak, R. Latała, A.E. Litvak, A. Pajor, N. Tomczak-Jaegermann, Tail estimates for norms of sums of log-concave random vectors. Proc. Lond. Math. Soc. **108**, 600–637 (2014)
3. R.E. Barlow, A.W. Marshall, F. Proschan, Properties of probability distributions with monotone hazard rate. Ann. Math. Stat. **34**, 375–389 (1963)
4. C. Borell, Convex measures on locally convex spaces. Ark. Math. **12**, 239–252 (1974)
5. S. Brazitikos, A. Giannopoulos, P. Valettas, B.H. Vritsiou, *Geometry of Isotropic Convex Bodies*. Mathematical Surveys and Monographs vol. 196 (American Mathematical Society, Providence, 2014)
6. R. Latała, Order statistics and concentration of l_r norms for log-concave vectors. J. Funct. Anal. **261**, 681–696 (2011)
7. R. Latała, Weak and strong moments of random vectors, in: *Marcinkiewicz Centenary Volume*, Banach Center Publications vol. 95 (Polish Academy of Sciences, Warsaw, 2011), pp. 115–121
8. G. Paouris, Concentration of mass on convex bodies. Geom. Funct. Anal. **16**, 1021–1049 (2006)

Remarks on Gaussian Noise Stability, Brascamp-Lieb and Slepian Inequalities

Michel Ledoux

Abstract E. Mossel and J. Neeman recently provided a heat flow monotonicity proof of Borell's noise stability theorem. In this note, we develop the argument to include in a common framework noise stability, Brascamp-Lieb inequalities (including hypercontractivity), and even a weak form of Slepian inequalities. The scheme applies furthermore to families of measures with are more log-concave than the Gaussian measure.

1 Introduction

Borell's noise stability theorem [12] expresses that if γ is the standard Gaussian measure $d\gamma(x) = d\gamma^n(x) = e^{-|x|^2/2}\frac{dx}{(2\pi)^{n/2}}$ on $\mathbb{R}^n$, and if A, B are Borel measurable sets in $\mathbb{R}^n$ and H, K parallel half-spaces

$$H = \{(x_1, \ldots, x_n) \in \mathbb{R}^n; x_1 \leq a\}, \quad K = \{(x_1, \ldots, x_n) \in \mathbb{R}^n; x_1 \leq b\}$$

with respectively the same Gaussian measures $\gamma(H) = \gamma(A)$, $\gamma(K) = \gamma(B)$, then, for every $t \geq 0$,

$$\int_{\mathbb{R}^n} 1_A \, Q_t(1_B) d\gamma \leq \int_{\mathbb{R}^n} 1_H \, Q_t(1_K) d\gamma. \tag{1}$$

Here $(Q_t)_{t\geq 0} = (Q_t^n)_{t\geq 0}$ is the Ornstein-Uhlenbeck semigroup defined, on suitable functions $f : \mathbb{R}^n \to \mathbb{R}$, by

$$Q_t f(x) = \int_{\mathbb{R}^n} f\left(e^{-t}x + \sqrt{1 - e^{-2t}}\, y\right) d\gamma(y), \quad t \geq 0, \ x \in \mathbb{R}^n. \tag{2}$$

M. Ledoux (✉)
Institut de Mathématiques de Toulouse, Université de Toulouse, 31062 Toulouse, France

Institut Universitaire de France, France
e-mail: ledoux@math.univ-toulouse.fr

© Springer International Publishing Switzerland 2014
B. Klartag, E. Milman (eds.), *Geometric Aspects of Functional Analysis*,
Lecture Notes in Mathematics 2116, DOI 10.1007/978-3-319-09477-9_20

According to this representation, setting $\rho = e^{-t}$, if $X = X^n$ and $Y = Y^n$ are independent with distribution $\gamma = \gamma^n$,

$$\int_{\mathbb{R}^n} 1_A \, Q_t(1_B) d\gamma = \mathbb{P}\big(X \in A, \rho X + \sqrt{1 - \rho^2}\, Y \in B\big)$$

so that the conclusion (1) equivalently reads as

$$\mathbb{P}\big(X \in A, \rho X + \sqrt{1 - \rho^2}\, Y \in B\big) \leq \mathbb{P}\big(X \in H, \rho X + \sqrt{1 - \rho^2}\, Y \in K\big). \qquad (3)$$

The result then extends to any $\rho \in [-1, +1]$, with however the inequality in (3) reversed when $\rho \in [-1, 0]$. For simplicity in the exposition, we mostly only consider $\rho \in [0, 1]$ below (actually $\rho \in (0, 1)$ since the cases $\rho = 0$ and $\rho = 1$ are straightforward). The content of (3) is that the (Gaussian) noise stability (of a set A)

$$\mathbb{P}\big(X \in A, \rho X + \sqrt{1 - \rho^2}\, Y \in A\big)$$

is maximal for half-spaces.

Towards the proof of (1), C. Borell [12] developed symmetrization arguments with respect to the Gaussian measure introduced by A. Ehrhard in [20] (see also [10,15]). Recently, E. Mossel and J. Neeman [28] proposed an alternative semigroup proof. The purpose of this note is to somewhat broaden their argument to cover in the same mould various related inequalities such as hypercontractivity, Brascamp-Lieb and Slepian inequalities. Heat flow arguments towards Brascamp-Lieb inequalities [13] have been investigated in the recent years by E. Carlen et al. [16] and J. Bennett et al. [8] (see also [5, 7, 17]). Section 2 describes the main theorem of [28] as an equivalent concavity property covering at the same time hypercontractivity and Borell's noise stability theorem. In Sect. 3, we consider multidimensional versions which were recently emphasized in [30], and discuss their applications to various families of concave functions towards Brascamp-Lieb and (a weak form of) Slepian-type inequalities. In the next section, we address extensions from the Gaussian model to families of measures $d\mu = e^{-V} dx$ with a lower bound on the Hessian of V following the basic semigroup interpolation argument. Section 5 comments on some analogous issues on the discrete cube which raise questions on a family of concave functions in connection with the recent discrete proof by A. De et al. [18] of the "Majority is Stablest" theorem of [29].

2   Hypercontractivity and Gaussian Noise Stability

The main result of E. Mossel and J. Neeman [28] expresses an integral concavity property for correlated Gaussian vectors for a specific family of functions on $\mathbb{R}^2$. Say that a C^2 function J on $\mathbb{R}^2$, or some open rectangle $\mathcal{R} = I_1 \times I_2 \subset \mathbb{R}^2$, where I_1 and I_2 are open intervals, is ρ-concave for some $\rho \in \mathbb{R}$ if the matrix

$$\begin{pmatrix} \partial_{11} J & \rho\, \partial_{12} J \\ \rho\, \partial_{12} J & \partial_{22} J \end{pmatrix}$$

is (uniformly) semi-negative definite. $\rho = 1$ amounts to standard concavity while $\rho = 0$ amounts to concavity along each coordinate.

Theorem 1. *Let $\rho \in (0,1)$ and let J on $\mathcal{R} = I_1 \times I_2 \subset \mathbb{R}^2$ be of class C^2. Then,*

$$\int_{\mathbb{R}^n} \int_{\mathbb{R}^n} J\big(f(x), g(\rho x + \sqrt{1-\rho^2}\, y)\big) d\gamma(x) d\gamma(y)$$

$$\leq J\left(\int_{\mathbb{R}^n} f\, d\gamma, \int_{\mathbb{R}^n} g\, d\gamma \right) \tag{4}$$

for every suitably integrable functions $f : \mathbb{R}^n \to I_1$, $g : \mathbb{R}^n \to I_2$ if and only if J is ρ-concave.

Let us sketch at this stage the heat flow proof of Theorem 1 following [28], the detailed argument being developed in the more general context of Sect. 4. Consider, for $t\ (> 0)$ fixed and (smooth) functions $f : \mathbb{R}^n \to I_1$, $g : \mathbb{R}^n \to I_2$,

$$\psi(s) = \int_{\mathbb{R}^n} \int_{\mathbb{R}^n} J\big(Q_s f(x), Q_s g(\rho x + \sqrt{1-\rho^2}\, y)\big) d\gamma(x) d\gamma(y), \quad s \geq 0.$$

By ergodicity, $Q_s f \to \int_{\mathbb{R}^n} f d\gamma$ and $Q_s g \to \int_{\mathbb{R}^n} g d\gamma$ as $s \to \infty$ so that it is enough to show that ψ is non-decreasing in order that $\psi(0) \leq \psi(\infty)$. Differentiating ψ and integrating by parts with respect to the infinitesimal generator $L = \Delta - x \cdot \nabla$ of the Ornstein-Uhlenbeck semigroup $(Q_s)_{s \geq 0}$ yields

$$\psi'(s) = \int_{\mathbb{R}^n} \int_{\mathbb{R}^n} [\partial_1 J\, L Q_s f + \partial_2 J\, L Q_s g] d\gamma\, d\gamma$$

$$= -\int_{\mathbb{R}^n} \int_{\mathbb{R}^n} \big[\partial_{11} J\, |\nabla Q_s f|^2 + \rho \partial_{12} J\, \nabla Q_s f \cdot \nabla Q_s g\big] d\gamma\, d\gamma$$

$$+ \int_{\mathbb{R}^n} \int_{\mathbb{R}^n} \partial_2 J\, L Q_s g\, d\gamma\, d\gamma.$$

By Gaussian rotational invariance, setting $(x, y)_\rho = \rho x + \sqrt{1-\rho^2}\, y$,

$$\int_{\mathbb{R}^n} \int_{\mathbb{R}^n} \partial_2 J\, L Q_s g\, d\gamma\, d\gamma$$

$$= \int_{\mathbb{R}^n} \int_{\mathbb{R}^n} \partial_2 J\big(Q_s f(x), Q_s g((x,y)_\rho)\big) L Q_s g((x,y)_\rho) d\gamma(x) d\gamma(y)$$

$$= \int_{\mathbb{R}^n} \int_{\mathbb{R}^n} \partial_2 J\big(Q_s f((x,-y)_\rho), Q_s g(x)\big) L Q_s g(x) d\gamma(x) d\gamma(y)$$

$$= -\int_{\mathbb{R}^n}\int_{\mathbb{R}^n}\Big[\rho\partial_{12}J\big(Q_sf((x,-y)_\rho),Q_sg(x)\big)\nabla Q_sf((x,-y)_\rho)\cdot\nabla Q_sg(x)$$

$$+ \partial_{22}J\big(Q_sf((x,-y)_\rho),Q_sg(x)\big)\big|\nabla Q_sg(x)\big|^2\Big]d\gamma(x)d\gamma(y)$$

$$= -\int_{\mathbb{R}^n}\int_{\mathbb{R}^n}\Big[\rho\partial_{12}J\big(Q_sf(x),Q_sg((x,y)_\rho)\big)\nabla Q_sf(x)\cdot\nabla Q_sg((x,y)_\rho)$$

$$+ \partial_{22}J\big(Q_sf(x),Q_sg((x,y)_\rho)\big)\big|\nabla Q_sg((x,y)_\rho)\big|^2\Big]d\gamma\,d\gamma.$$

Finally

$$\psi'(s) = \int_{\mathbb{R}^n}\int_{\mathbb{R}^n}\Big[(-\partial_{11}J)|\nabla Q_sf|^2 + (-\partial_{22}J)|\nabla Q_sg|^2$$

$$-2\rho\,\partial_{12}J\,\nabla Q_sf\cdot\nabla Q_sg\Big]d\gamma\,d\gamma$$

From the hypothesis of ρ-concavity on J, it follows that $\psi' \geq 0$ which is the result.

The converse was observed by R. O'Donnell, and communicated to us by J. Neeman. Indeed, applying (4) to $f(x) = a+\varepsilon x$ and $g(y) = b+\varepsilon y$ (in dimension one) and letting $\varepsilon \to 0$ shows that J is ρ-concave.

It may be mentioned that due to the product structure of the Gaussian measure γ^n, the inequality of Theorem 1 immediately tensorizes so that it is actually enough to establish it in dimension one.

Let us now illustrate the application of Theorem 1 to two main examples of ρ-concave function J, covering hypercontractivity and noise stability at the same time.

Let first

$$J^{\mathrm{H}}(u,v) = u^\alpha v^\beta, \quad (u,v) \in [0,\infty)^2.$$

Since

$$\partial_{11}J^{\mathrm{H}} = \alpha(\alpha-1)u^{\alpha-2}v^\beta, \quad \partial_{22}J^{\mathrm{H}} = \beta(\beta-1)u^\alpha v^{\beta-2}, \quad \partial_{12}J^{\mathrm{H}} = \alpha\beta u^{\alpha-1}v^{\beta-1},$$

J^{H} is ρ-concave on $(0,\infty)^2$ as soon as $\alpha, \beta \in [0,1]$ and

$$\rho^2\alpha\beta \leq (\alpha-1)(\beta-1). \tag{5}$$

The function J^{H} will be called the hypercontractive function in this context.

Indeed, let $1 < p < q < \infty$ and $\rho = e^{-t} \in (0,1)$ be such that

$$\frac{1}{\rho^2} = \frac{q-1}{p-1}.$$

Denote by q' the conjugate of q, $\frac{1}{q} + \frac{1}{q'} = 1$. Then, according to (5), the function J^{H} with $\alpha = \frac{1}{q'}$ and $\beta = \frac{1}{p}$ is ρ-concave on $(0,\infty)^2$. By Theorem 1, for strictly positive functions $f, g : \mathbb{R}^n \to \mathbb{R}$,

$$\int_{\mathbb{R}^n} \int_{\mathbb{R}^n} f^{1/q'}(x) g^{1/p} \left(e^{-t} x + \sqrt{1 - e^{-2t}} \, y \right) d\gamma(x) d\gamma(y)$$

$$\leq \left(\int_{\mathbb{R}^n} f \, d\gamma \right)^{1/q'} \left(\int_{\mathbb{R}^n} g \, d\gamma \right)^{1/p}.$$

In other words, changing f into $f^{q'}$ and g into g^p,

$$\int_{\mathbb{R}^n} f \, Q_t g \, d\gamma \leq \|f\|_{q'} \|g\|_p.$$

By duality

$$\|Q_t g\|_q \leq \|g\|_p$$

which amounts to hypercontractivity of the Ornstein-Uhlenbeck semigroup [23, 31] (cf. e.g. [3]). Clearly, the conclusion of Theorem 1 for J^H is actually equivalent to hypercontractivity. Note that a prior to the proof of hypercontractivity along these lines may be found in [24].

The second example involves a new function introduced in [28] defined for $(u, v) \in [0, 1]^2$ by

$$J^B(u, v) = J^B_\rho(u, v) = \mathbb{P}\left(X^1 \leq \Phi^{-1}(u), \rho X^1 + \sqrt{1 - \rho^2} \, Y^1 \leq \Phi^{-1}(v) \right)$$

where $\Phi(a) = \gamma^1((-\infty, a])$, $a \in \mathbb{R}$, is the distribution of the standard normal on $\mathbb{R}$ and $\rho \in [-1, +1]$. For the connection with Borell's theorem, observe that if H and K are the (parallel) half-spaces $H = \{x_1 \leq a\}$ and $K = \{x_1 \leq b\}$ for some $a, b \in \mathbb{R}$, with $\rho = e^{-t}$ and the integral representation (2) of Q_t,

$$J^B\left(\gamma(H), \gamma(K) \right) = \mathbb{P}\left(X^1 \leq a, \rho X^1 + \sqrt{1 - \rho^2} \, Y^1 \leq b \right)$$
$$= \int_{\mathbb{R}^n} 1_H \, Q_t(1_K) d\gamma. \tag{6}$$

Apply now Theorem 1 to the function J^B. Since $J^B(u, 0) = J^B(0, v) = 0$ and $J^B(1, 1) = 1$, for f and g approaching 1_A and 1_B respectively,

$$\int_{\mathbb{R}^n} \int_{\mathbb{R}^n} J\left(f(x), g\left(\rho x + \sqrt{1 - \rho^2} \, y \right) \right) d\gamma(x) d\gamma(y) = \int_{\mathbb{R}^n} 1_A \, Q_t(1_B) d\gamma.$$

We then recover Borell's noise stability theorem (1) since by (6),

$$J^B\left(\int_{\mathbb{R}^n} f \, d\gamma, \int_{\mathbb{R}^n} g \, d\gamma \right) = J^B\left(\gamma(A), \gamma(B) \right) = J^B\left(\gamma(H), \gamma(K) \right)$$

$$= \int_{\mathbb{R}^n} 1_H \, Q_t(1_K) d\gamma$$

for parallel half-spaces H and K such that respectively $\gamma(A) = \gamma(H)$ and $\gamma(B) = \gamma(K)$. When $\rho \in [-1, 0]$, observe that

$$J_\rho^B(u, v) = u - J_{-\rho}^B(u, 1 - v)$$

so that J^B is ρ-convex in this case, and the conclusion of Theorem 1 for the function J_ρ^B is thus reversed. As pointed out in [28], (1) on sets may actually be turned to Theorem 1 (for J^B) through epigraphs of functions on $\mathbb{R}^{n-1}$.

It remains to check the ρ-concavity of J^B. To this task, it is convenient to recall that $Q_t f(x)$ may be given alternatively by the Mehler kernel

$$Q_t f(x) = \int_{\mathbb{R}^n} f(y) q_t(x, y) d\gamma(y) \tag{7}$$

where, for $t > 0$, $(x, y) \in \mathbb{R}^n \times \mathbb{R}^n$,

$$
\begin{aligned}
q_t(x, y) &= q_t^n(x, y) \\
&= \frac{1}{\sqrt{1 - e^{-2t}}} \exp\left(-\frac{e^{-2t}}{2(1 - e^{-2t})} \left[|x|^2 + |y|^2 - 2 e^t x \cdot y \right] \right).
\end{aligned}
\tag{8}
$$

In particular, if $\rho = e^{-t}$,

$$J^B(u, v) = \int_{-\infty}^{\Phi^{-1}(u)} \int_{-\infty}^{\Phi^{-1}(v)} q_t^1(x, y) d\gamma^1(x) d\gamma^1(y).$$

Observe also that $(\Phi^{-1})' = \frac{1}{\varphi \circ \Phi^{-1}}$ where $\varphi = \Phi'$ is the density of γ^1 on $\mathbb{R}$. Hence,

$$\partial_1 J^B(u, v) = \int_{-\infty}^{\Phi^{-1}(v)} q_t^1(\Phi^{-1}(u), y) d\gamma^1(y)$$

and

$$\partial_{12} J^B(u, v) = q_t^1(\Phi^{-1}(u), \Phi^{-1}(v)).$$

On the other hand, by the integral representations (2) and (7), for h smooth enough,

$$\partial_x \int_{\mathbb{R}} h(y) q_t^1(x, y) d\gamma^1(y) = \partial_x Q_t^1 h(x) = \rho \, Q_t^1 h'(x) = \rho \int_{\mathbb{R}} h'(y) q_t^1(x, y) d\gamma^1(y).$$

With h a smooth approximation of $1_{(-\infty, b]}$,

$$\partial_x \int_{-\infty}^b q_t^1(x, y) d\gamma^1(y) = -\rho \, q_t^1(x, b) \varphi(b).$$

Therefore,

$$\partial_{11} J^B(u, v) = -\rho q_t^1(\Phi^{-1}(u), \Phi^{-1}(v)) \frac{\varphi \circ \Phi^{-1}(v)}{\varphi \circ \Phi^{-1}(u)}.$$

Similarly,

$$\partial_{22} J^B(u, v) = -\rho q_t^1(\Phi^{-1}(u), \Phi^{-1}(v)) \frac{\varphi \circ \Phi^{-1}(u)}{\varphi \circ \Phi^{-1}(v)}.$$

Hence, on $(0, 1)^2$,

$$\partial_{11} J^B \partial_{22} J^B - \rho^2 (\partial_{12} J^B)^2 = 0$$

and $\partial_{11} J^B \leq 0$, $\partial_{22} J^B \leq 0$ so that J^B is indeed ρ-concave.

It would be of interest to find other relevant examples of function J. It is also of interest to directly compare the conclusion of Theorem 1 for the hypercontractive function J^H and for the Borell noise stability function J^B, and namely to show that noise stability is a stronger statement implying hypercontractivity. One way towards this end, however along a rather long detour, is to observe, as emphasized in [26], that Borell's noise stability theorem may be used to reach the Gaussian isoperimetric inequality. Now, the latter implies in turn the standard logarithmic Sobolev inequality for the Gaussian measure, equivalent to hypercontractivity (cf. [3, 26]).

There is an alternative direct argument towards this relationship, applying Theorem 1 for J^B to εf and δg and letting $\varepsilon, \delta \to 0$. To this task, it is necessary to investigate the asymptotics of $J^B(\varepsilon u, \delta v)$ as $\varepsilon, \delta \to 0$. Similar asymptotics are investigated in [19].

Set $\rho = e^{-t} \in (0, 1)$ and fix $0 < u, v < 1$. Let furthermore $0 < \varepsilon < 1, \delta = \varepsilon^{\kappa^2}$ where $\rho < \kappa < \frac{1}{\rho}$, and

$$Z = \sqrt{2 \log \frac{1}{\varepsilon}}, \quad U = \log \frac{1}{u}, \quad V = \log \frac{1}{v}.$$

In this notation, after a change of variables,

$$J^B(\varepsilon u, \delta v) = \frac{UV}{\kappa Z^2} \int_c^\infty \int_d^\infty \tilde{q}_t^1 \left(-Z - \frac{Ux}{Z}, -\kappa Z - \frac{Vy}{\kappa Z} \right) dx dy$$

where

$$c = -\frac{Z}{U} [Z + \Phi^{-1}(\varepsilon u)] \quad \text{and} \quad d = -\frac{Z}{\kappa V} [\kappa Z + \Phi^{-1}(\delta v)],$$

and

$$\tilde{q}_t^1(x, y) = (2\pi)^{-1} q_t^1(x, y) e^{-(x^2+y^2)/2}, \quad (x, y) \in \mathbb{R} \times \mathbb{R}.$$

After some algebra,

$$J^B(\varepsilon u, \delta v) = \frac{UV e^{\sigma Z^2}}{2\pi \sqrt{1 - \rho^2 \kappa Z^2}} \int_c^\infty \int_d^\infty e^{-\alpha Ux - \beta Vy - R(x,y)} dx dy$$

where

$$\sigma = -\frac{1 - 2\kappa\rho + \kappa^2}{2(1 - \rho^2)}, \quad \alpha = \frac{1 - \kappa\rho}{1 - \rho^2}, \quad \beta = \frac{1 - \kappa^{-1}\rho}{1 - \rho^2}$$

and

$$R(x, y) = -\frac{1}{2(1 - \rho^2)} \left(\frac{U^2 x^2}{Z^2} + \frac{V^2 x^2}{\kappa^2 Z^2} - 2\rho \frac{UVxy}{\kappa Z^2} \right).$$

It is classical that

$$\Phi^{-1}(\varepsilon) = -\sqrt{2 \log \frac{1}{\varepsilon}} + o\left(\sqrt{2 \log \frac{1}{\varepsilon}} \right)$$

as $\varepsilon \to 0$, so that

$$\Phi^{-1}(\varepsilon u) = -Z - \frac{U}{Z} + o(Z)$$

as $Z \to \infty$. Moreover, $o(Z)$ can be made uniform over $\eta \leq u \leq 1 - \eta$ for $\eta > 0$ fixed. As a consequence, as $\varepsilon \to 0$, $c, d \to 1$ and

$$2\pi \sqrt{1 - \rho^2} \kappa Z^2 e^{-\sigma Z^2} J^B(\varepsilon u, \delta v) \to UV \int_1^\infty \int_1^\infty e^{-\alpha Ux - \beta Vy} dx dy = \frac{1}{\alpha\beta} e^{-\alpha U - \beta V}.$$

By definition of U and V, the right-hand side is $\frac{1}{\alpha\beta} u^\alpha v^\beta$.

Let now f, g on $\mathbb{R}^n$ such that $\eta \leq f, g \leq 1 - \eta$ for some fixed $\eta > 0$. Translating the preceding asymptotics into the inequality

$$\int_{\mathbb{R}^n} \int_{\mathbb{R}^n} J^B\left(\varepsilon f(x), \delta g\left(\rho x + \sqrt{1 - \rho^2}\, y\right)\right) d\gamma(x) d\gamma(y)$$

$$\leq J^B\left(\varepsilon \int_{\mathbb{R}^n} f \, d\gamma, \delta \int_{\mathbb{R}^n} g \, d\gamma \right)$$

yields

$$\int_{\mathbb{R}^n} \int_{\mathbb{R}^n} f^\alpha(x) g^\beta(\rho x + \sqrt{1-\rho^2}\, y) d\gamma(x) d\gamma(y) \leq \left(\int_{\mathbb{R}^n} f\, d\gamma\right)^\alpha \left(\int_{\mathbb{R}^n} g\, d\gamma\right)^\beta.$$

This inequality extends to all positive measurable functions $f, g : \mathbb{R}^n \to \mathbb{R}$ by homogeneity. Now, as is immediately checked, for the values of α, β defined above,

$$(\alpha - 1)(\beta - 1) = \rho^2 \alpha \beta,$$

that is condition (5) of hypercontractivity holds. Given therefore any $\alpha, \beta \in (0, 1)$ satisfying this relation, one may choose $\rho < \kappa < \frac{1}{\rho}$ such that $\alpha = \frac{1-\kappa\rho}{1-\rho^2}$ and $\beta = \frac{1-\kappa^{-1}\rho}{1-\rho^2}$ as above. The announced claim follows.

It would be worthwhile to examine similarly noise stability for the Lebesgue measure λ with respect to the standard heat kernel expressing that for Borel sets A, B in $\mathbb{R}^n$ with finite volume,

$$\int_{\mathbb{R}^n} 1_A\, H_t(1_B) dx \leq \int_{\mathbb{R}^n} 1_C\, H_t(1_D) dx$$

where

$$H_t f(x) = \int_{\mathbb{R}^n} f(y)\, e^{-|x-y|^2/4t} \frac{dy}{(4\pi t)^{n/2}}, \quad t > 0, \ x \in \mathbb{R}^n,$$

and C and D are centered balls in $\mathbb{R}^n$ such that $\lambda(A) = \lambda(C)$ and $\lambda(B) = \lambda(D)$. This classical result is going back to the Riesz rearrangement inequality [33] (see also [14,27,34]), and one might wonder for a heat flow proof. A similar question may be formulated on the sphere (cf. [1, 10, 15]).

3 Multidimensional Extensions

On the basis of the heat flow proof of Theorem 1, we address in this section multidimensional extensions and develop connections to Brascamp-Lieb and Slepian-type inequalities. The multidimensional version of noise stability was already put forward by J. Neeman in [30]. The Brascamp-Lieb applications are essentially contained with the same approach in [8, 16] (see also [17]). At the same time, the investigation provides a somewhat different analytical treatment of the conclusions of Sect. 2.

Let J be a (smooth) real-valued function on some rectangle subset $\mathcal{R}$ of $\mathbb{R}^m$. It will implicitly be assumed below that a composition like $J \circ f$ is meant for functions f with values in $\mathcal{R}$.

Let $f_1, \ldots, f_m$ be (smooth) functions on $\mathbb{R}^n$ and consider, for $f = (f_1, \ldots, f_m)$,

$$\psi(s) = \int_{\mathbb{R}^n} J \circ Q_s f \, d\gamma, \quad s \geq 0,$$

where $(Q_s)_{s \geq 0}$ is the Ornstein-Uhlenbeck semigroup on $\mathbb{R}^n$ (extended to functions with values in $\mathbb{R}^m$). Arguing as in Sect. 2, by integration by parts with respect to the Ornstein-Uhlenbeck generator,

$$\begin{aligned}
\psi'(s) &= \sum_{k=1}^{n} \int_{\mathbb{R}^n} \partial_k J \circ Q_s f \, L Q_s f_k \, d\gamma \\
&= - \sum_{k,\ell=1}^{m} \int_{\mathbb{R}^n} \partial_{k\ell} J \circ Q_s f \, \nabla Q_s f_k \cdot \nabla Q_s f_\ell \, d\gamma.
\end{aligned} \tag{9}$$

Definition. Given a smooth function J on an open subset of $\mathbb{R}^m$ and $\Gamma = (\Gamma_{k\ell})_{1 \leq k, \ell \leq m}$ where $\Gamma_{k\ell}$ are $p \times p$ matrices ($p \geq 1$), say that J is Γ-concave if

$$\sum_{k,\ell=1}^{m} \partial_{k\ell} J \, \Gamma_{k\ell} v_k \cdot v_\ell \leq 0 \tag{10}$$

for all vectors v_k, $k = 1, \ldots, m$, in $\mathbb{R}^p$. If $p = 1$, the meaning of this condition is that the point-wise (Hadamard) product $\mathrm{Hess}(J) \circ \Gamma$ of the Hessian of J and of Γ is (semi-) negative definite.

When $m = 2$, $p = n$ and Γ is the $2n \times 2n$ matrix

$$\begin{pmatrix} \mathrm{Id}_n & \rho \, \mathrm{Id}_n \\ \rho \, \mathrm{Id}_n & \mathrm{Id}_n \end{pmatrix} \tag{11}$$

where $\rho \in \mathbb{R}$, the Γ-concavity of J (on $\mathbb{R}^2$) amounts to its ρ-concavity.

In (9), replace now n by qn, $q \geq 1$ integer, and assume that for every $k = 1, \ldots, m$,

$$f_k = g_k \circ A_k$$

where $g_k : \mathbb{R}^p \to \mathbb{R}$ and A_k is a (constant) $p \times qn$ matrix such that $A_k {}^t A_k$ is the identity matrix (of $\mathbb{R}^p$). By the integral representation (2) of Q_s,

$$\nabla Q_s f_k = \mathrm{e}^{-s} \, {}^t A_k (\nabla Q_s g_k) \circ A_k$$

where on the left-hand side the semigroup Q_s is acting on $\mathbb{R}^{qn}$ and on the right-hand side, it is acting on $\mathbb{R}^p$. Hence

$$\psi'(s) = -\mathrm{e}^{-2s} \sum_{k,\ell=1}^{m} \int_{\mathbb{R}^{qn}} \partial_{k\ell} J \circ Q_s f \, \Gamma_{k\ell} (\nabla Q_s g_k) \circ A_k \cdot (\nabla Q_s g_\ell) \circ A_\ell \, d\gamma$$

where $\Gamma_{k\ell} = A_\ell{}^t A_k$ (which is a $p \times p$ matrix).

With this choice of $\Gamma = (\Gamma_{k\ell})_{1 \leq k, \ell \leq m}$, the following proposition summarizes the conclusion at this level of generality.

Proposition 2. *In the preceding setting, assume that J is Γ-concave. Then* $\int_{\mathbb{R}^{qn}} J \circ f \, d\gamma \leq J(\int_{\mathbb{R}^{qn}} f \, d\gamma)$, *that is*

$$\int_{\mathbb{R}^{qn}} J(g_1 \circ A_1, \ldots, g_m \circ A_m) d\gamma \leq J\left(\int_{\mathbb{R}^{qn}} g_1 \circ A_1 d\gamma, \ldots, \int_{\mathbb{R}^{qn}} g_m \circ A_m d\gamma \right).$$

To connect with Sect. 2, take for example $p = n$ and $q = m = 2$ and let A_1 and A_2 be the $n \times 2n$ matrices $A_1 = (\text{Id}_n; 0_n)$ and $A_2 = (\rho \, \text{Id}_n; \sqrt{1 - \rho^2} \, \text{Id}_n)$ so that

$$f_1(x, y) = g_1(x) \quad \text{and} \quad f_2(x, y) = g_2(\rho x + \sqrt{1 - \rho^2}\, y), \quad (x, y) \in \mathbb{R}^n \times \mathbb{R}^n.$$

Since Γ is given by (11), the monotonicity property follows from the ρ-concavity of J.

We next systematically investigate illustrations of Proposition 2 for some main examples of interest. For simplicity, we only consider $p = q = 1$, the multidimensional cases being often obtained by tensor products with the identity matrix (as in the preceding example). The Γ-concavity thus amounts to $\text{Hess}(J) \circ \Gamma \leq 0$ in the following.

(i) The first illustration examines Brascamp-Lieb inequalities under geometric conditions. Consider unit vectors $A_1, \ldots, A_m$ which decompose the identity in $\mathbb{R}^n$ in the sense that for $0 \leq c_k \leq 1, k = 1, \ldots, m$,

$$\sum_{k=1}^{m} c_k A_k \otimes A_k = \text{Id}_n. \tag{12}$$

Then, for

$$J(u_1, \ldots, u_m) = u_1^{c_1} \cdots u_m^{c_m}$$

on $(0, \infty)^m$ and $f_k(x) = g_k(A_k \cdot x)$, $g_k : \mathbb{R} \to \mathbb{R}$, $k = 1, \ldots, m$, the Γ-concavity with respect to $\Gamma_{k\ell} = A_k \cdot A_\ell$, $k, \ell = 1, \ldots, m$, is expressed by

$$\sum_{k,\ell=1}^{m} c_k c_\ell A_k \cdot A_\ell v_k v_\ell \leq \sum_{k=1}^{m} c_k v_k^2 \tag{13}$$

for all $v_1, \ldots, v_m \in \mathbb{R}$. Now, if $x = \sum_{k=1}^{m} c_k A_k v_k$,

$$|x|^2 = \sum_{k=1}^{m} c_k A_k v_k \cdot x \leq \left(\sum_{k=1}^{m} c_k v_k^2 \right)^{1/2} \left(\sum_{k=1}^{m} c_k (A_k \cdot x)^2 \right)^{1/2}$$

Since, by the decomposition (12), $|x|^2 = \sum_{k=1}^{m} c_k (A_k \cdot x)^2$, it follows that

$$|x|^2 = \left| \sum_{k=1}^{m} c_k A_k v_k \right|^2 \le \sum_{k=1}^{m} c_k v_k^2$$

which is precisely the requested condition (13). We therefore conclude to the following result.

Corollary 3. *Under the decomposition (12), for non-negative functions g_k on $\mathbb{R}$,* $k = 1, \ldots, m,$

$$\int_{\mathbb{R}^n} \prod_{k=1}^{m} g_k^{c_k} (A_k \cdot x) d\gamma \le \prod_{k=1}^{m} \left(\int_{\mathbb{R}} g_k d\gamma \right)^{c_k}.$$

This inequality is part of the Brascamp-Lieb inequalities under the geometric Ball condition (12) [4] (cf. e.g. [7, 8]). It is more classically stated with respect to the Lebesgue measure as

$$\int_{\mathbb{R}^n} \prod_{k=1}^{m} f_k^{c_k} (A_k \cdot x) dx \le \prod_{k=1}^{m} \left(\int_{\mathbb{R}} f_k dx \right)^{c_k}$$

which is immediately obtained after the change $f_k(x) = g_k(x) e^{-x^2/2}$ (using that $\sum_{k=1}^{m} c_k = n$).

The heat flow proof of Corollary 3 is thus going back to [16] and [8] in which more general statements are considered and achieved in this way. One of the motivations of [16] was actually to investigate similar inequalities for coordinates on the sphere. Let $\mathbb{S}^{n-1}$ be the standard n-sphere in $\mathbb{R}^n$ and denote by σ the uniform (normalized) measure on it. In this framework, one result then reads as follows. If $g_k, k = 1, \ldots, n$, are, say bounded measurable, functions on $\mathbb{R}$, then

$$\int_{\mathbb{S}^{n-1}} J\big(g_1(x_1), \ldots, g_n(x_n)\big) d\sigma \le J\left(\int_{\mathbb{S}^{n-1}} g_1(x_1) d\sigma, \ldots, \int_{\mathbb{S}^{n-1}} g_n(x_n) d\sigma \right)$$

as soon as J on $\mathbb{R}^n$, or some open (convex) set in $\mathbb{R}^n$, is separately concave in any two variables. The proof proceeds as the one of Proposition 2 along now the heat flow of the Laplace operator

$$\Delta = \frac{1}{2} \sum_{k,\ell=1}^{n} (x_k \partial_\ell - x_\ell \partial_k)^2$$

on $\mathbb{S}^{n-1}$. The monotonicity condition on J then takes the form

$$\sum_{k,\ell=1}^{n} \partial_{k\ell} J \, (\delta_{k\ell} - x_k x_\ell) v_k v_\ell \le 0$$

which is easily seen to be satisfied under concavity of J in any two variables. The case considered in [16] simply corresponds to

$$J(u_1, \ldots, u_n) = (u_1 \cdots u_n)^{1/2}$$

on $\mathbb{R}^n_+$. More general forms under decompositions (12) of the identity have been considered in [6, 7].

In the further illustrations, consider $X = (X_1, \ldots, X_m)$ a centered Gaussian vector on $\mathbb{R}^m$ with covariance matrix $\Gamma = A^t A$ such that $\Gamma_{kk} = 1$ for every $k = 1, \ldots, m$. The vector X has the distribution of Ax, $x \in \mathbb{R}^n$, under the standard normal distribution γ on $\mathbb{R}^n$. Applying the general Proposition 2 to the unit vectors $(1 \times n$ matrices) A_k, $k = 1, \ldots, m$, which are the lines of the matrix A, and to $f_k(x) = g_k(A_k \cdot x)$, $x \in \mathbb{R}^n$, where $g_k : \mathbb{R} \to \mathbb{R}$, $k = 1, \ldots, m$, with respect to γ, yields that whenever $\mathrm{Hess}(J) \circ \Gamma \leq 0$, under suitable integrability properties on the g_k's,

$$\mathbb{E}\Big(J\big(g_1(X_1), \ldots, g_m(X_m)\big) \Big) \leq J\Big(\mathbb{E}\big(g_1(X_1)\big), \ldots, \mathbb{E}\big(g_m(X_m)\big) \Big). \tag{14}$$

Note that, as in Sect. 1, the condition $\mathrm{Hess}(J) \circ \Gamma \leq 0$ is actually necessary and sufficient for (14) to hold.

(ii) This illustration deals with a correlation inequality for Gaussian vectors which covers in particular the classical hypercontractivity property. For a Gaussian vector X as above, let as in the first illustration,

$$J(u_1, \ldots, u_m) = u_1^{c_1} \cdots u_m^{c_m}$$

on $(0, \infty)^m$, with $0 \leq c_k \leq 1$, $k = 1, \ldots, m$. This function J is the suitable multidimensional analogue of the hypercontractive function J^H. For this choice of J, the condition $\mathrm{Hess}(J) \circ \Gamma \leq 0$ (where Γ is the covariance matrix of X) amounts to

$$\sum_{k,\ell=1}^{m} c_k c_\ell \, \Gamma_{k\ell} \, v_k v_\ell \leq \sum_{k=1}^{m} c_k v_k^2 \tag{15}$$

for all $v_k \in \mathbb{R}$, $k = 1, \ldots, m$. Note that this condition expresses equivalently that $\Gamma \leq \Delta_c$ in the sense of symmetric matrices where Δ_c is the diagonal matrix $\big(\frac{1}{c_k}\big)_{1 \leq k \leq m}$. While the next Corollary 4 is somewhat part of the folklore (implicit for example in [7]), it has been emphasized recently in [17] together with reverse and multidimensional versions (in particular, if $\Gamma \geq \Delta_c$, the conclusion is reversed in (16)).

Corollary 4. *Under (15), for all non-negative functions $g_k : \mathbb{R} \to \mathbb{R}$, $k = 1, \ldots, m$,*

$$\mathbb{E}\left(\prod_{k=1}^{m} g_k^{c_k}(X_k) \right) \leq \prod_{k=1}^{m} \mathbb{E}(g_k(X_k))^{c_k}. \tag{16}$$

One application concerns the Ornstein-Uhlenbeck process $Z = (Z_t)_{t \geq 0}$ (in dimension one) with stationary measure $\gamma = \gamma^1$ and associated Markov semigroup $(Q_t)_{t \geq 0} = (Q_t^1)_{t \geq 0}$. If X is the vector $(Z_{t_1}, \ldots, Z_{t_m})$ with $0 \leq t_1 \leq \cdots \leq t_m$, the covariance matrix Γ has entries $\Gamma_{k\ell} = \mathrm{e}^{-|t_k - t_\ell|}$, $k, \ell = 1, \ldots, m$. In particular, for $t_1 = 0$ and $t_2 = t > 0$, (15) reads

$$2\,\mathrm{e}^{-t} c_1 c_2 v_1 v_2 \leq c_1 (1 - c_1) v_1^2 + c_2 (1 - c_2) v_2^2$$

for all $v_1, v_2 \in \mathbb{R}$ which amounts to (5)

$$\mathrm{e}^{-2t} c_1 c_2 \leq (c_1 - 1)(c_2 - 1)$$

and the conclusion of Corollary 4 leads to hypercontractivity. The condition

$$\sum_{k,\ell=1}^{m} c_k c_\ell \, \mathrm{e}^{-|t_k - t_\ell|} v_k v_\ell \leq \sum_{k=1}^{m} c_k v_k^2$$

yields a multidimensional form of hypercontractivity

$$\mathbb{E}\left(\prod_{k=1}^{m} g_k^{c_k}(Z_{s_k}) \right) \leq \prod_{k=1}^{m} \mathbb{E}(g_k(Z_{s_k}))^{c_k}.$$

In terms of the Mehler kernel (8),

$$\int_{\mathbb{R}} \cdots \int_{\mathbb{R}} \prod_{k=1}^{m} g_k^{c_k}(x_k)\, q_{t_2 - t_1}(x_1, x_2) \cdots q_{t_m - t_{m-1}}(x_{m-1}, x_m) d\gamma(x_1) \cdots d\gamma(x_m)$$

$$\leq \prod_{k=1}^{m} \left(\int_{\mathbb{R}} g_k d\gamma \right)^{c_k}.$$

(iii) We next turn to the multidimensional versions of Gaussian noise stability following [30]. As above, let $X = (X_1, \ldots, X_m)$ be a centered Gaussian vector on $\mathbb{R}^m$ with (non-degenerate) covariance matrix Γ. Define, for $u_1, \ldots, u_m$ in $(0, 1)$,

$$J(u_1, \ldots, u_m) = \mathbb{P}(X_1 \leq \alpha_1(u_1), \ldots, X_m \leq \alpha_m(u_m)) \tag{17}$$

where $\alpha_1, \ldots, \alpha_m$ are smooth functions on $(0, 1)$. For specific choices of α_k, this function will turn as the multidimensional analogue of the noise stability function J^{B}. Denoting by p the density of the distribution of X with respect to the Lebesgue measure, elementary (although a bit tedious, see [30]) differential calculus leads to

$$\partial_{k\ell} J = \alpha_k'(u_k)\alpha_\ell'(u_\ell) \int_{-\infty}^{\alpha_1(u_1)} \cdots \int_{-\infty}^{\alpha_m(u_m)} p_{k\ell}\, dx$$

for $k \neq \ell$ and

$$\partial_{kk} J = \left(\alpha_k''(u_k) - \frac{\alpha_k(u_k)\alpha_k'(u_k)^2}{\Gamma_{kk}} \right) \int_{-\infty}^{\alpha_1(u_1)} \cdots \int_{-\infty}^{\alpha_m(u_m)} p_k\, dx$$

$$- \alpha_k'(u_k)^2 \sum_{\ell \neq k} \frac{\Gamma_{k\ell}}{\Gamma_{kk}} \int_{-\infty}^{\alpha_1(u_1)} \cdots \int_{-\infty}^{\alpha_m(u_m)} p_{k\ell}\, dx$$

where

$$p_k = p\big(x_1, \ldots, \alpha_k(u_k), \ldots, x_m\big),$$

$$p_{k\ell} = p\big(x_1, \ldots, \alpha_k(u_k), \ldots, \alpha_\ell(u_\ell), \ldots, x_m\big).$$

Choose now $\alpha_k = \Phi^{-1}$, $k = 1, \ldots, m$, where we recall the distribution function Φ of the standard normal, and φ its derivative. Since

$$\alpha_k' = \frac{1}{\varphi \circ \Phi^{-1}} \quad \text{and} \quad \alpha_k'' = \frac{\Phi^{-1}}{(\varphi \circ \Phi^{-1})^2},$$

in order for the condition $\mathrm{Hess}(J) \circ \Gamma \leq 0$ to hold it is thus sufficient that $\Gamma_{kk} = 1$ for every $k = 1, \ldots, m$ and

$$\sum_{k=1}^{m} \sum_{\ell \neq k} \Gamma_{k\ell}\, p_{k\ell}\, v_k^2 - \sum_{k \neq \ell} p_{k\ell}\, \Gamma_{k\ell}\, v_k v_\ell \geq 0$$

for all $v_1, \ldots, v_m \in \mathbb{R}$. This holds as soon as $\Gamma_{k\ell} \geq 0$ for all k, ℓ.

For the application to the following corollary, recall that for the choice of $\alpha_k = \Phi^{-1}$, the function J of (17) is equal to 0 if one of the u_k's is (approaches) 0, and is equal to (approach) 1 if all the u_k's are equal to 1. The following corollary, thus due to J. Neeman [30], is then a consequence of (14) applied to $g_k = 1_{B_k}$, $k = 1, \ldots, m$. The restriction $\Gamma_{kk} = 1$, $k = 1, \ldots, m$, is lifted after a simple scaling of the Gaussian vector and the Borel sets.

Corollary 5. *Let* $X = (X_1, \ldots, X_m)$ *be a centered Gaussian vector in* $\mathbb{R}^m$ *with (non-degenerate) covariance matrix* Γ *such that* $\Gamma_{k\ell} \geq 0$ *for all* $k, \ell = 1, \ldots, m$. *Then, for any Borel sets* $B_1, \ldots, B_m$ *in* $\mathbb{R}$,

$$\mathbb{P}(X_1 \in B_1, \ldots, X_m \in B_m) \leq \mathbb{P}(X_1 \leq b_1, \ldots, X_m \leq b_m)$$

where $\mathbb{P}(X_k \in B_k) = \Phi(b_k/\sigma_k)$, $\sigma_k = \sqrt{\Gamma_{kk}}$, $k = 1, \ldots, m$.

When $\Gamma_{k\ell} \leq 0$ whenever $k \neq \ell$, the inequality in the conclusion of Corollary 5 is reversed. As developed in [30], the result applies similarly to Gaussian vectors $X_1, \ldots, X_m$ with covariance identity matrix. A related work by M. Isaksson and E. Mossel [25] establishes the conclusion of Corollary 5 under the (stronger) hypothesis that the off-diagonal elements of the inverse of Γ are non-positive. Their approach relies on a rearrangement inequality for kernels on the sphere. Corollary 5 (as well as actually, after some work, the result of [25]—see [30]) covers the example of the Ornstein-Uhlenbeck process, and thus of C. Borell's result [12] in the form of the following corollary.

Corollary 6. *Let* $(Z_t)_{t \geq 0}$ *be the Ornstein-Uhlenbeck process on the line, and let* $0 \leq t_1 \leq \cdots \leq t_m$. *For any Borel sets* $B_1, \ldots, B_m$ *in* $\mathbb{R}$,

$$\mathbb{P}(Z_{t_1} \in B_1, \ldots, Z_{t_m} \in B_m) \leq \mathbb{P}(Z_{t_1} \leq b_1, \ldots, Z_{t_m} \leq b_m)$$

where $\mathbb{P}(Z_{t_k} \in B_k) = \gamma(B_k) = \Phi(b_k)$, $k = 1, \ldots, m$.

(iv) This illustration is a variation on the previous multidimensional noise stability result which actually leads to a weak form of the classical Slepian inequalities. Let as above $X = (X_1, \ldots, X_m)$ be a centered Gaussian vector on $\mathbb{R}^m$ with covariance matrix $\Gamma = \Gamma^X$ such that $\Gamma_{kk}^X = 1$ for every $k = 1, \ldots, m$. Consider furthermore $Y = (Y_1, \ldots, Y_m)$ a centered Gaussian vector on $\mathbb{R}^m$ with covariance matrix Γ^Y also such that $\Gamma_{kk}^Y = 1$ for every $k = 1, \ldots, m$, yielding a J function (17)

$$J(u_1, \ldots, u_m) = \mathbb{P}\big(Y_1 \leq \alpha_1(u_1), \ldots, Y_m \leq \alpha_m(u_m)\big), \quad u_1, \ldots, u_m \in (0, 1).$$

Choose now again $\alpha_k = \Phi^{-1}$. Arguing as in *(iii)* towards $\mathrm{Hess}(J) \circ \Gamma^X \leq 0$, the condition is now that

$$\sum_{k=1}^{m} \sum_{\ell \neq k} \Gamma_{k\ell}^Y \, p_{k\ell} \, v_k^2 - \sum_{k \neq \ell} p_{k\ell} \, \Gamma_{k\ell}^X \, v_k v_\ell \geq 0$$

for all $v_1, \ldots, v_m \in \mathbb{R}$ (where p is here the density of the law of Y). This holds as soon as $\Gamma_{k\ell}^Y \geq 0$ and

$$\big(\Gamma_{k\ell}^X\big)^2 \leq \big(\Gamma_{k\ell}^Y\big)^2$$

for all $k \neq \ell$. As a conclusion

Corollary 7. *Let $X = (X_1, \ldots, X_m)$ and $Y = (Y_1, \ldots, Y_m)$ be centered Gaussian vectors on $\mathbb{R}^m$ with respective (non-degenerate) covariance matrices Γ^X and Γ^Y. Assume that $\Gamma^X_{kk} = \Gamma^Y_{kk} = 1$ and*

$$\left| \Gamma^X_{k\ell} \right| \leq \Gamma^Y_{k\ell}$$

for all $k, \ell = 1, \ldots, m$. Then, for any Borel sets $B_1, \ldots, B_m$ in $\mathbb{R}$,

$$\mathbb{P}(X_1 \in B_1, \ldots, X_m \in B_m) \leq \mathbb{P}(Y_1 \leq b_1, \ldots, Y_m \leq b_m)$$

where $\mathbb{P}(X_k \in B_k) = \Phi(b_k)$, $k = 1, \ldots, m$. In particular, for every $r_1, \ldots, r_m$ in $\mathbb{R}$,

$$\mathbb{P}(X_1 \leq r_1, \ldots, X_m \leq r_m) \leq \mathbb{P}(Y_1 \leq r_1, \ldots, Y_m \leq r_m).$$

This result is of course a weak form, in particular due to the constraint $\Gamma^Y_{k\ell} \geq 0$ (with however a somewhat stronger conclusion), of the classical Slepian lemma which indicates that for Gaussian vectors X and Y in $\mathbb{R}^m$, the conclusion of Corollary 7 holds whenever $\Gamma^X_{kk} = \Gamma^Y_{kk}$ and $\Gamma^X_{k\ell} \leq \Gamma^Y_{k\ell}$ for all $k, \ell = 1, \ldots, m$. Note that the traditional proof of Slepian's lemma [21, 22, 32, 36] is an interpolation between the covariances Γ^X and Γ^Y which is not exactly the same as the one at the root of Corollary 7.

4 Log-Concave Measures

In this section, we develop the heat flow proof of Theorem 1 of E. Mossel and J. Neeman in the somewhat extended context of probability measures $d\mu = e^{-V} dx$ on $\mathbb{R}^n$ such that V is a smooth potential with a uniform lower bound on its Hessian. The typical application actually concerns potentials V which are more convex than the quadratic one, corresponding to Gaussian measures. The argument may be extended to the more general context of Markov diffusion semigroups and the Γ-calculus as exposed in [3] although for the simplicity of this note, we stay in the familiar Euclidean setting.

Consider therefore a probability measure $d\mu = e^{-V} dx$ on the Borel sets of $\mathbb{R}^n$, invariant and symmetric measure of the second order differential operator $\mathrm{L} = \Delta - \nabla V \cdot \nabla$ where V is a smooth potential on $\mathbb{R}^n$. The (symmetric) semigroup $(P_t)_{t \geq 0}$ with generator L may be represented by (smooth) probability kernels

$$P_t h(x) = \int_{\mathbb{R}^n} h(y) \, p_t(x, dy). \tag{18}$$

It will be assumed that $V - c \frac{|x|^2}{2}$ is convex for some $c \in \mathbb{R}$, in other words the Hessian of V is bounded from below by $c \, \mathrm{Id}_n$ as symmetric matrices. It is by now classical (cf. [3]) that this convexity assumption ensures that for all (smooth) $h : \mathbb{R}^n \to \mathbb{R}$,

$$|\nabla P_t h| \leq e^{-ct} P_t(|\nabla h|). \tag{19}$$

The Gaussian example of the Ornstein-Uhlenbeck semigroup $(Q_t)_{t \geq 0}$ with invariant measure γ is included with $c = 1$. In this case, due to the representation (2), the gradient bound (19) actually turns into the identity $\nabla Q_t h = e^{-t} Q_t(\nabla h)$.

We start with the analogue of Theorem 1 in this context following therefore the argument of [28].

Theorem 8. *Let J be ρ-concave, $\rho > 0$, on $\mathcal{R} = I_1 \times I_2 \subset \mathbb{R}^2$ where I_1 and I_2 are open intervals. Then, for every $f : \mathbb{R}^n \to I_1$, $g : \mathbb{R}^n \to I_2$ suitably integrable, and with $\rho = e^{-ct}$, $t > 0$,*

$$\int_{\mathbb{R}^n} \int_{\mathbb{R}^n} J(f(x), g(y)) \, p_t(x, dy) d\mu(x) \leq J\left(\int_{\mathbb{R}^n} f \, d\mu, \int_{\mathbb{R}^n} g \, d\mu\right).$$

Proof. It is enough to assume that f and g are taking values in respective compact sub-intervals of I_1 and I_2. Set

$$\psi(s) = \int_{\mathbb{R}^n} \int_{\mathbb{R}^n} J(P_s f(x), P_s g(y)) \, p_t(x, dy) d\mu(x), \quad s \geq 0.$$

The task is to show that ψ is non-decreasing. Taking derivative in time s,

$$\psi'(s) = \int_{\mathbb{R}^n} \int_{\mathbb{R}^n} \partial_1 J(P_s f(x), P_s g(y)) L P_s f(x) \, p_t(x, dy) d\mu(x)$$

$$+ \int_{\mathbb{R}^n} \int_{\mathbb{R}^n} \partial_2 J(P_s f(x), P_s g(y)) L P_s g(y) \, p_t(x, dy) d\mu(x).$$

By integration by parts in space with respect to the operator L, expressed (for smooth functions $\xi, \zeta : \mathbb{R}^n \to \mathbb{R}$) by

$$\int_{\mathbb{R}^n} \xi(-L\zeta) d\mu = \int_{\mathbb{R}^n} \nabla \xi \cdot \nabla \zeta \, d\mu,$$

it holds

$$\int_{\mathbb{R}^n} \int_{\mathbb{R}^n} \partial_1 J(P_s f(x), P_s g(y)) L P_s f(x) \, p_t(x, dy) d\mu(x)$$

$$= -\int_{\mathbb{R}^n} \int_{\mathbb{R}^n} \partial_{11} J(P_s f(x), P_s g(y)) |\nabla P_s f(x)|^2 p_t(x, dy) d\mu(x)$$

$$- \int_{\mathbb{R}^n \times \mathbb{R}^n} \partial_1 J(P_s f(x), P_s g(y)) \nabla P_s f(x) \cdot \nabla_x p_t(x, dy) d\mu(x).$$

For $x \in \mathbb{R}^n$ fixed, consider $h(y) = \partial_1 J(P_s f(x), P_s g(y))$, $y \in \mathbb{R}^n$. Since

$$\nabla P_t h(z) = \int_{\mathbb{R}^n} h(y) \nabla_z p_t(z, dy), \quad z \in \mathbb{R}^n,$$

at $z = x$,

$$\int_{\mathbb{R}^n} \partial_1 J\big(P_s f(x), P_s g(y)\big) \nabla P_s f(x) \cdot \nabla_x p_t(x, dy) = \nabla P_t h(x) \cdot \nabla P_s f(x).$$

Now, by (19),

$$\big|\nabla P_t h(x)\big| \leq e^{-ct} P_t\big(|\nabla h|\big)(x) = e^{-ct} \int_{\mathbb{R}^n} \big|\nabla h(y)\big| p_t(x, dy).$$

Since

$$\nabla h(y) = \partial_{12} J\big(P_s f(x), P_s g(y)\big) \nabla P_s g(y),$$

it follows that

$$\int_{\mathbb{R}^n} \int_{\mathbb{R}^n} \partial_1 J\big(P_s f(x), P_s g(y)\big) \nabla_x p_t(x, dy) \cdot \nabla P_s f(x) d\mu(x)$$

$$\leq e^{-ct} \int_{\mathbb{R}^n} \int_{\mathbb{R}^n} |\partial_{12} J|\big(P_s f(x), P_s g(y)\big) \big|\nabla P_s g(y)\big| \big|\nabla P_s f(x)\big| p_t(x, dy) d\mu(x).$$

Summarizing, and by the symmetric conclusion in the y variable, $\psi'(s)$ is bounded from below by

$$\int_{\mathbb{R}^n} \int_{\mathbb{R}^n} \Big[(-\partial_{11} J)|\nabla P_s f|^2 + (-\partial_{22} J)|\nabla P_s g|^2 - 2 e^{-cs} |\partial_{12} J| |\nabla P_s f| |\nabla P_s g|\Big]$$

$$\times p_t(x, dy) d\mu(x).$$

From the hypothesis on the Hessian of J, it follows that $\psi' \geq 0$ which is the result.
$\square$

As in the Gaussian case, the examples of illustration of Theorem 8 cover both hypercontractivity and noise stability for the choices of $J = J^H$ or $J = J^B$. Under $c > 0$, the choice of J^H yields hypercontractivity of the semigroup associated to this family of invariant measures, and thus the equivalent logarithmic Sobolev inequality for μ (cf. [3]). On the other hand, the noise stability part actually turns into a comparison theorem.

Corollary 9. *Let $(P_t)_{t \geq 0}$ be the Markov semigroup with invariant reversible measure $d\mu = e^{-V} dx$ where V is a smooth potential on $\mathbb{R}^n$ such that $\mathrm{Hess}(V) \geq c\, \mathrm{Id}_n$*

with $c > 0$. Then, whenever A, B are Borel sets in $\mathbb{R}^n$ and H, K are respective parallel half-spaces such that $\mu(A) = \gamma(H)$, $\mu(B) = \gamma(K)$, then

$$\int_{\mathbb{R}^n} 1_A P_t(1_B) d\mu \leq \int_{\mathbb{R}^n} 1_H \, Q_{ct}(1_K) d\gamma.$$

Again, as in the Gaussian setting (cf. [26]), the comparison property of Corollary 9 may be shown to imply the isoperimetric comparison theorem of [2] (see [3]) comparing the isoperimetric profile of measures $d\mu = e^{-V} dx$ to the Gaussian one.

Next, we turn to the multidimensional version of the preceding result, with therefore in the following $c > 0$. Let $X = (X_t)_{t \geq 0}$ be the Markov process with generator $L = \Delta - \nabla V \cdot \nabla$ and initial invariant distribution $d\mu = e^{-V} dx$. We are interested in the distribution of $(X_{t_1}, \ldots, X_{t_m})$ where $0 \leq t_1 \leq \cdots \leq t_m$. Consider the covariance matrix Γ the Ornstein-Uhlenbeck process at speed ct, that is $\Gamma_{k\ell} = e^{-c|t_k - t_\ell|}$, $k, \ell = 1, \ldots, m$. In the Gaussian case, this extension (for thus the Ornstein-Uhlenbeck process) was achieved by the study of general Gaussian vectors. In the present case, we deal with the kernels as given by (18), for simplicity one-dimensional.

Theorem 10. *In the preceding notation, assume that the Hadamard product of $(|\partial_{k\ell} J|)_{1 \leq k, \ell \leq m}$ and Γ is (semi-) negative-definite. Then, for every $f_i : \mathbb{R} \to I_i$, $i = 1, \ldots, m$, suitably integrable,*

$$\int_{\mathbb{R}} \cdots \int_{\mathbb{R}} J\big(f_1(x_1), \ldots, f_m(x_m)\big) p_{t_m - t_{m-1}}(x_{m-1}, dx_m) \cdots p_{t_2 - t_1}(x_1, dx_2) d\mu(x_1)$$

$$\leq J\left(\int_{\mathbb{R}} f_1 \, d\mu, \ldots, \int_{\mathbb{R}} f_m \, d\mu\right).$$

We outline the argument when $m = 3$. Consider

$$\psi(s) = \int_{\mathbb{R}} \int_{\mathbb{R}} \int_{\mathbb{R}} J\big(P_s f(x), P_s g(y), P_s h(z)\big) p_{t-u}(y, dz) p_t(x, dy) d\mu(x), \quad s \geq 0,$$

for $t > u > 0$ and three functions f, g, h. Differentiating ψ and integrating by parts in space leads to consider expressions such as

$$\int_{\mathbb{R}} \int_{\mathbb{R}} \int_{\mathbb{R}} \partial_1 J \, p_{t-u}(y, dz) \partial_x p_u(x, dy) \partial_x P_s f \, d\mu(x).$$

Arguing as in the proof of Theorem 8, this expression is equal to

$$\int_{\mathbb{R}} \partial_x P_s k \, \partial_x P_s f \, d\mu(x)$$

where $k = k(y) = \int_{\mathbb{R}} \partial_1 J \, p_{t-u}(y, dz)$. Now by (19)

$$|\partial_x P_s k| \le e^{-cs} P_s(|\partial_y k|).$$

Since

$$\partial_y k = \int_{\mathbb{R}} \partial_{12} J \, p_{t-u}(y, dz) + \int_{\mathbb{R}} \partial_1 J \, \partial_y \, p_{t-u}(y, dz),$$

similarly

$$|\partial_y k| \le \int_{\mathbb{R}} |\partial_{12} J| \, p_{t-u}(y, dz) + e^{-c(t-u)} \int_{\mathbb{R}} |\partial_{13} J| \, |\partial_y \, p_{t-u}(y, dz).$$

The proof is then completed in the same way.

With the J function (17) associated to a finite-dimensional distribution of the Ornstein-Uhlenbeck process, the following consequence holds true.

Corollary 11. *Let $c > 0$ and $0 \le t_1 \le \cdots \le t_m$. For any Borel sets $B_1, \ldots, B_m$ in $\mathbb{R}$,*

$$\mathbb{P}(X_{t_1} \in B_1, \ldots, X_{t_m} \in B_m) \le \mathbb{P}(Z_{ct_1} \le b_1, \ldots, Z_{ct_m} \le b_m)$$

where $\mathbb{P}(X_{t_k} \in B_k) = \mu(B_k) = \Phi(b_k)$, $k = 1, \ldots, m$ and where $(Z_{ct})_{t \ge 0}$ is the Ornstein-Uhlenbeck process with speed ct.

As suggested by J. Neeman following his arguments developed in [30], Corollary 11 may be used towards a comparison property between hitting times. For a Borel set B in $\mathbb{R}$, let $e_B^X = \inf\{t \ge 0 ; X_t \notin B\}$ be the exit time of the Markov process $X = (X_t)_{t \ge 0}$ from the set B.

Corollary 12. *Under the preceding notation, for any $s \ge 0$,*

$$\mathbb{P}(e_B^X \ge s) \le \mathbb{P}(e_H^Z \ge s)$$

where H is a half-line in $\mathbb{R}$ such that $\gamma(H) = \mu(B)$ and $Z = (Z_{ct})_{t \ge 0}$ the Ornstein-Uhlenbeck process at speed ct.

5 The Discrete Cube

To conclude this note, we briefly address in this last section the corresponding noise stability issue on the discrete cube, and collect a few remarks in connection with the recent development [18] on the "Majority is Stablest" theorem.

By Theorem 1, a function J is ρ-concave in the sense that

$$\begin{pmatrix} \partial_{11} J & \rho\,\partial_{12} J \\ \rho\,\partial_{12} J & \partial_{22} J \end{pmatrix} \leq 0$$

if and only if for all suitable functions f and g on $\mathbb{R}^n$,

$$\int_{\mathbb{R}^n} \int_{\mathbb{R}^n} J\big(f(x), g(\rho x + \sqrt{1-\rho^2}\, y)\big) d\gamma(x) d\gamma(y)$$
$$\leq J\Big(\int_{\mathbb{R}^n} f\, d\gamma, \int_{\mathbb{R}^n} g\, d\gamma\Big). \tag{20}$$

The analogue of the Gaussian couple $(X, \rho X + \sqrt{1-\rho^2}\, Y)$ with correlation $\rho\,\mathrm{Id}_n$, $\rho \in [-1, 1]$, on the discrete cube $\Sigma^n = \{-1, +1\}^n$, with $n = 1$ to start with, leads to consider a couple with distribution

$$(1 + \rho x y) d\mu(x) d\mu(y)$$

on Σ^2, where μ is the uniform probability measure on $\Sigma = \{-1, +1\}$. The latter inequality (20) on the two-point space $\Sigma = \{-1, +1\}$ therefore amounts to

$$\int_{\Sigma} \int_{\Sigma} J\big(f(x), g(y)\big) K_\rho(x, y) d\mu(x) d\mu(y) \leq J\Big(\int_{\Sigma} f\, d\mu, \int_{\Sigma} g\, d\mu\Big) \tag{21}$$

for every functions $f, g : \Sigma \to \mathbb{R}$, where $K_\rho(x, y) = 1 + \rho x y$. This inequality (21) is stable under product. On $\Sigma^n = \{-1, +1\}^n$ equipped with the uniform product measure μ^n, let for $\rho \in \mathbb{R}$ and $x = (x_1, \ldots, x_n) \in \Sigma^n$, $y = (y_1, \ldots, y_n) \in \Sigma^n$,

$$K_\rho(x, y) = \prod_{i=1}^{n} (1 + \rho\, x_i y_i).$$

If (21) holds, for every f, g on Σ^n,

$$\int_{\Sigma^n} \int_{\Sigma^n} J\big(f(x), g(y)\big) K_\rho(x, y) d\mu^n(x) d\mu^n(y) \leq J\Big(\int_{\Sigma^n} f d\mu^n, \int_{\Sigma^n} g d\mu^n\Big).$$

One may of course wonder for the equivalence of (21) with the ρ-concavity of J. Actually, (21) expresses equivalently a 4-point inequality similar to the standard characterization of concavity. Say namely that a function J on some open convex set $\mathcal{O}$ of $\mathbb{R}^2$ is strongly ρ-concave for some $\rho \in \mathbb{R}$ if for all $(u, v) \in \mathcal{O}$, $(u', v') \in \mathcal{O}$,

$$\tfrac{1+\rho}{4} J(u, v) + \tfrac{1-\rho}{4} J(u', v) + \tfrac{1-\rho}{4} J(u, v') + \tfrac{1+\rho}{4} J(u', v')$$
$$\leq J\Big(\tfrac{u+u'}{2}, \tfrac{v+v'}{2}\Big). \tag{22}$$

Lemma 13. *Strong ρ-concavity implies ρ-concavity (for smooth functions).*

Proof. By a Taylor expansion, at any $(a,b) \in \mathcal{O}$, $(h,k) \in \mathbb{R}^2$, such that $(a \pm h, b \pm k) \in \mathcal{O}$,

$$(1+\rho)\big[J(a+h,b+k) + J(a-h,b-k) - 2J(a,b)\big]$$
$$+ (1-\rho)\big[J(a+h,b-k) + J(a-h,b+k) - 2J(a,b)\big]$$
$$= 2h^2 \partial_{11} J(a,b) + 4\rho h k \partial_{12} J(a,b) + 2k^2 \partial_{22} J(a,b) + o(h^2 + k^2).$$

With $u = a + h$, $v = b + k$, $u' = a - h$, $v' = b - k$, (22) implies the ρ-concavity of J as $h, k \to 0$.   □

It is a main result, namely the Bonami-Beckner hypercontractivity theorem [9, 11], that the hypercontractive function J^H is strongly ρ-concave under (5) (along the equivalence between hypercontractivity and Theorem 1 described in Sect. 2 for the Ornstein-Uhlenbeck semigroup). However, we could not establish directly the strong ρ-concavity of J^H in this case. Such a proof could give a better understanding of the strong ρ-concavity property.

On the other hand, it is not true in general that ρ-concavity implies back strong ρ-concavity and one example, taken from [18], is simply Borell's noise stability function J^B (with parameter $\rho \in (0,1)$). Indeed, for $u = v = 1$ and $u' = v' = 0$ (in the Boolean analysis terminology, this choice corresponds to the dictator functions $f(x) = \frac{1}{2} + \frac{x}{2}$, $g(y) = \frac{1}{2} + \frac{y}{2}$), (22) would imply that

$$1 + \rho \leq 4 J^B \left(\frac{1}{2}, \frac{1}{2} \right) \tag{23}$$

since $J^B(1,1) = 1$ and $J^B(1,0) = J^B(0,1) = J^B(0,0) = 0$. But

$$J^B \left(\frac{1}{2}, \frac{1}{2} \right) = \int_{-\infty}^{0} \int_{-\infty}^{0} q_t^1(x,y) d\gamma^1(x) d\gamma^1(y) = \int_0^{\infty} \Phi(\alpha x) d\gamma^1(x)$$

where $\alpha = \frac{\rho}{\sqrt{1-\rho^2}}$ and $\rho = e^{-t}$. Taking the derivative in α easily shows that

$$J^B \left(\frac{1}{2}, \frac{1}{2} \right) = \frac{1}{4} + \frac{1}{2\pi} \arctan(\alpha) = \frac{1}{2} - \frac{1}{2\pi} \arccos(\rho)$$

so that (23) indeed fails as $\rho \to 0$. This value of $J^B\left(\frac{1}{2}, \frac{1}{2}\right)$ appears in Sheppard's formula put forward in [35] as early as 1899 as the asymptotic noise stability of the Majority function (see [18, 29]).

It would be of interest to understand which additional property to ρ-concavity ensures strong ρ-concavity. A. De et al. [18] recently observed by a suitable Taylor expansion that there exists, for any $\rho \in (0,1)$, $C(\rho) > 0$ such that

$$\left| \frac{\partial^3 J_\rho^B(u,v)}{\partial^i u \, \partial^j v} \right| \leq C(\rho) \big[uv(1-u)(1-v) \big]^{-C(\rho)}$$

for all $i, j \geq 0$ with $i + j = 3$. This property then implies the approximate validity of (22) in the sense that for every $u, u', v, v' \in [\varepsilon, 1 - \varepsilon]$ for some $\varepsilon > 0$,

$$
\begin{aligned}
&\tfrac{1+\rho}{4} J_\rho^{\mathrm{B}}(u, v) + \tfrac{1-\rho}{4} J_\rho^{\mathrm{B}}(u', v) + \tfrac{1-\rho}{4} J_\rho^{\mathrm{B}}(u, v') + \tfrac{1+\rho}{4} J_\rho^{\mathrm{B}}(u', v') \\
&\leq J_\rho^{\mathrm{B}}\left(\tfrac{u+u'}{2}, \tfrac{v+v'}{2}\right) + C'(\rho)\, \varepsilon^{-C'(\rho)}\left(|u - u'|^3 + |v - v'|^3\right).
\end{aligned}
\tag{24}
$$

As a main achievement, the authors of [18] develop from this conclusion and tensorization a fully discrete proof of the "Majority is Stablest" theorem of [29] (with Sheppard's constant $J^{\mathrm{B}}\left(\tfrac{1}{2}, \tfrac{1}{2}\right)$ as stability value) by suitably controlling the error term via the influences of the Boolean functions under investigation.

One further observation of [18] is that the preceding two-point inequality (24) is still good enough to reach, after tensorization and the central limit theorem, Borell's noise stability theorem for the Ornstein-Uhlenbeck semigroup.

Acknowledgements This note grew up out of discussions and exchanges with J. Neeman and E. Mossel around their works [28, 30] and [18] . In particular, Sect. 2, the first part of Sects. 3 and 5 are directly following their contributions. I sincerely thank them for their interest and comments. I also thank K. Oleszkiewicz for exchanges on ρ-concave functions, R. Bouyrie for several comments and corrections, F. Barthe for pointing out the reference [17], and the referee for helpful comments in improving the exposition.

References

1. A. Baernstein II, B.A. Taylor, Spherical rearrangements, subharmonic functions and *-functions in n-space. Duke Math. J. **43**, 245–268 (1976)
2. D. Bakry, M. Ledoux, Lévy-Gromov's isoperimetric inequality for an infinite-dimensional diffusion generator. Invent. Math. **123**, 259–281 (1996)
3. D. Bakry, I. Gentil, M. Ledoux, *Analysis and Geometry of Markov Diffusion Operators.* Grundlehren der Mathematischen Wissenschaften, vol. 348 (Springer, Berlin, 2014)
4. K. Ball, Volumes of sections of cubes and related problems, in *Geometric Aspects of Functional Analysis (1987–88).* Lecture Notes in Mathematics, vol. 1376 (Springer, Berlin, 1989), pp. 251–260
5. F. Barthe, D. Cordero-Erausquin, Inverse Brascamp-Lieb inequalities along the heat equation, in *Geometric Aspects of Functional Analysis.* Lecture Notes in Mathematics, vol. 1850 (Springer, Berlin, 2004), pp. 65–71
6. F. Barthe, D. Cordero-Erausquin, B. Maurey, Entropy of spherical marginals and related inequalities. J. Math. Pures Appl. **86**, 89–99 (2006)
7. F. Barthe, D. Cordero-Erausquin, M. Ledoux, B. Maurey, Correlation and Brascamp-Lieb inequalities for Markov semigroups. Int. Math. Res. Not. **10**, 2177–2216 (2011)
8. J. Bennett, A. Carbery, M. Christ, T. Tao, The Brascamp-Lieb inequalities: finiteness, structure and extremals. Geom. Funct. Anal. **17**, 1343–1415 (2008)
9. W. Beckner, Inequalities in Fourier analysis. Ann. Math. **102**, 159–182 (1975)
10. W. Beckner, Sobolev inequalities, the Poisson semigroup, and analysis on the sphere $\mathbb{S}^n$. Proc. Natl. Acad. Sci. USA **89**, 4816–4819 (1992)
11. A. Bonami, Étude des coefficients de Fourier des fonctions de $L^p(G)$. Ann. Inst. Fourier **20**, 335–402 (1970)

12. C. Borell, Geometric bounds on the Ornstein-Uhlenbeck velocity process. Z. Wahrsch. Verw. Gebiete **70**, 1–13 (1985)
13. H. Brascamp, E. Lieb, Best constants in Young's inequality, its converse, and its generalization to more than three functions. Adv. Math. **20**, 151–173 (1976)
14. H. Brascamp, E. Lieb, J.M. Luttinger, A general rearrangement inequality for multiple integrals. J. Funct. Anal. **17**, 227–237 (1974)
15. E. Carlen, M. Loss, Extremals of functionals with competing symmetries. J. Funct. Anal. **88**, 437–456 (1990)
16. E. Carlen, E. Lieb, M. Loss, A sharp analog of Young's inequality on s^N and related entropy inequalities. J. Geom. Anal. **14**, 487–520 (2004)
17. W.-K. Chen, N. Dafnis, G. Paouris, Improved Hölder and reverse Hölder inequalities for correlated Gaussian random vectors (2013)
18. A. De, E. Mossel, J. Neeman, Majority is Stablest: Discrete and SoS, in *Proceedings of the 45th Annual ACM Symposium on Theory of Computing*, (2013).
19. E. De Klerk, D. Pasechnik, J. Warners, On approximate graph colouring and MAX-k-CUT algorithms based on the #-function. J. Comb. Optim. **8**, 267–294 (2004)
20. A. Ehrhard, Symétrisation dans l'espace de Gauss. Math. Scand. **53**, 281–301 (1983)
21. X. Fernique, Régularité des trajectoires des fonctions aléatoires gaussiennes, in *École d'Été de Probabilités de St-Flour 1974*. Lecture Notes in Mathematics, vol. 480 (Springer, Berlin, 1975), pp. 1–96
22. Y. Gordon, Some inequalities for Gaussian processes and applications. Israel J. Math. **50**, 265–289 (1985)
23. L. Gross, Logarithmic Sobolev inequalities. Am. J. Math. **97**, 1061–1083 (1975)
24. Y. Hu, A unified approach to several inequalities for Gaussian and diffusion measures, in *Séminaire de Probabilités XXXIV*. Lecture Notes in Mathematics, vol. 1729 (Springer, Berlin, 2000), pp. 329–335
25. M. Isaksson, E. Mossel, New maximally stable Gaussian partitions with discrete applications. Israel J. Math. **189**, 347–396 (2012)
26. M. Ledoux, Isoperimetry and Gaussian analysis, in *Ecole d'Été de Probabilités de St-Flour 1994*. Lecture Notes in Mathematics, vol. 1648 (Springer, Berlin, 1996), pp. 165–294
27. E. Lieb, M. Loss, *Analysis*, 2nd edn. Graduate Studies in Mathematics, vol. 14 (American Mathematical Society, Providence, 2001)
28. E. Mossel, J. Neeman, Robust optimality of Gaussian noise stability. J. Eur. Math Soc. (2012, to appear)
29. E. Mossel, R. O'Donnell, K. Oleszkiewicz, Noise stability of functions with low influences: invariance and optimality. Ann. Math. **171**, 295–341 (2010)
30. J. Neeman, A multidimensional version of noise stability (2014)
31. E. Nelson, The free Markoff field. J. Funct. Anal. **12**, 211–227 (1973)
32. R. Plackett, A reduction formula for normal multivariate integrals. Biometrika **41**, 351–360 (1954)
33. F. Riesz, Sur une inégalité intégrale. J. Lond. Math. Soc. **5**, 162–168 (1930)
34. C. A. Rogers, A single integral inequality. J. Lond. Math. Soc. **32**, 102–108 (1957)
35. W. Sheppard, On the application of the theory of error to cases of normal distribution and normal correlation. Philos. Trans. R. Soc. Lond. **192**, 101–168 (1899)
36. D. Slepian, The one-sided barrier problem for Gaussian noise. Bell. Syst. Tech. J. **41**, 463–501 (1962)

Quantitative Version of a Silverstein's Result

Alexander E. Litvak and Susanna Spektor

Abstract We prove a quantitative version of a Silverstein's Theorem on the 4-th moment condition for convergence in probability of the norm of a random matrix. More precisely, we show that for a random matrix with i.i.d. entries, satisfying certain natural conditions, its norm cannot be small.

Let w be a real random variable with $\mathbb{E}w = 0$ and $\mathbb{E}w^2 = 1$, and let w_{ij}, $i, j \geq 1$ be its i.i.d. copies. For integers n and $p = p(n)$ consider the $p \times n$ matrix $W_n = \{w_{ij}\}_{i \leq p, j \leq n}$, and consider its sample covariance matrix $\Gamma_n := \frac{1}{n} W_n W_n^T$. We also denote by $X_j = (w_{j1}, \ldots, w_{jn})$, $j \leq p$, the rows of W_n.

The questions on behavior of eigenvalues are of great importance in random matrix theory. We refer to [4, 6, 15] for the relevant results, history and references.

In this note we study lower bounds on $\max_{i \leq p} |X_i|$, where $|\cdot|$ denotes the Euclidean norm of a vector, and on the operator (spectral) norms of matrices W_n and Γ_n. Note, as Γ_n is symmetric, its largest singular value $\lambda_{\max}$ is equal to the norm and that in general we have

$$\lambda_{\max}(\Gamma_n) = \|\Gamma_n\| = \frac{1}{n} \|W_n\|^2 \geq \frac{1}{n} \max_{i \leq p} |X_i|^2. \tag{1}$$

Assume that $p(n)/n \to \beta > 0$ as $n \to \infty$. In [20] it was proved that then $\|\Gamma_n\| \to (1 + \sqrt{\beta})^2$ a.s., while in [7] it was shown that $\limsup_{n \to \infty} \|\Gamma_n\| = \infty$ a.s. if $\mathbb{E}w^4 = \infty$.

In [17] Silverstein studied the weak behavior of $\|\Gamma_n\|$. In particular, he proved that assuming $p(n)/n \to \beta > 0$ as $n \to \infty$, $\|\Gamma_n\|$ converges to a non-random quantity (which must be $(1 + \sqrt{\beta})^2$) in probability if and only if $n^4 \mathbb{P}(|w| \geq n) = o(1)$.

The purpose of this note is to provide the quantitative counterpart of Silverstein's result. More precisely, we want to show an estimate of the type $\mathbb{P}(\|\Gamma_n\| \geq K) \geq$

A.E. Litvak (✉) • S. Spektor
Department of Mathematical and Statistical Sciences, University of Alberta, Edmonton, AB, Canada T6G 2G1
e-mail: aelitvak@gmail.com; sanaspek@gmail.com

© Springer International Publishing Switzerland 2014   335
B. Klartag, E. Milman (eds.), *Geometric Aspects of Functional Analysis*,
Lecture Notes in Mathematics 2116, DOI 10.1007/978-3-319-09477-9_21

$\delta = \delta(K)$ for an arbitrary large K, provided that w has heavy tails (in particular, provided that w does not have 4-th moment). Our proof essentially follows ideas of [17]. It gives a lower bound on $\max_{i \leq p} |X_i|$ as well.

Theorem 1. *Let $\alpha \geq 2$, $c_0 > 0$. Let w be a random variable satisfying $\mathbb{E}w = 0$, $\mathbb{E}w^2 = 1$ and*

$$\forall t \geq 1 \qquad \mathbb{P}(|w| \geq t) \geq \frac{c_0}{t^{\alpha}}. \tag{2}$$

Let $W_n = \{w_{ij}\}_{i \leq p, j \leq n}$ be a $p \times n$ matrix whose entries are i.i.d. copies of w and let $X_i, i \leq p$, be the rows of W_n. Then, for every $K \geq 1$,

$$\mathbb{P}\left(\max_{i \leq p} |X_i| \geq \sqrt{Kn}\right) \geq \min\left\{\frac{c_0 p}{4n^{(\alpha-2)/2}K^{\alpha/2}}, \frac{1}{2}\right\}. \tag{3}$$

In particular, $\Gamma_n = \frac{1}{n}W_n W_n^T$ satisfies for every $K \geq 1$,

$$\mathbb{P}(\|\Gamma_n\| \geq K) \geq \min\left\{\frac{c_0 p}{4n^{(\alpha-2)/2}K^{\alpha/2}}, \frac{1}{2}\right\}.$$

Remark 2. Taking $K = (c_0 pn/2)^{2/\alpha}/n$ we observe

$$\mathbb{P}\left(\|W_n\| \geq (c_0 pn/2)^{1/\alpha}\right) \geq \mathbb{P}\left(\max_{i \leq p} |X_i| \geq (c_0 pn/2)^{1/\alpha}\right) \geq \frac{1}{2}.$$

This estimate seems to be sharp in view of the following result (see Corollary 2 in [5]). Let $0 < \alpha < 4$ and let w be defined by

$$\mathbb{P}(|w| > t) = \min\{1, t^{-\alpha}\} \quad \text{for } t > 0.$$

Let W_n and X_i's be as in Theorem 1. Assume that $p/n \to \beta > 0$ as $n \to \infty$. Then

$$\lim_{n \to \infty} \mathbb{P}\left(\|W_n\| \leq (pn)^{1/\alpha} t\right) = \exp(-t^{-\alpha}).$$

Remark 3. If p is proportional to n, say $p = \beta n$, the theorem gives

$$\mathbb{P}(\|\Gamma_n\| \geq K) \geq \mathbb{P}\left(\max_{i \leq p} |X_i| \geq \sqrt{Kn}\right) \geq \min\left\{\frac{c_0 \beta}{4n^{(\alpha-4)/2}K^{\alpha/2}}, \frac{1}{2}\right\},$$

in particular, taking $K = (c_0\beta/2)^{2/\alpha} n^{4/\alpha-1}$, we observe

$$\mathbb{P}\left(\|W_n\| \geq (c_0\beta/2)^{1/\alpha} n^{2/\alpha}\right) \geq \mathbb{P}\left(\max_{i \leq p} |X_i| \geq (c_0\beta/2)^{1/\alpha} n^{2/\alpha}\right) \geq \frac{1}{2}.$$

Remark 4. Note that by Chebychev's inequality one has $\mathbb{P}(|w| \geq t) \leq t^{-2}$. Note also that we use condition (2) in the proof only once, with $t = \sqrt{Kn}$.

Remark 5. If $p \geq (2/c_0)K^{\alpha/2}n^{(\alpha-2)/2}$, then, by condition (2), we have

$$\frac{n}{2}\mathbb{P}(w^2 \geq Kn) \geq \frac{nc_0}{2(Kn)^{\alpha/2}} = \frac{c_0}{2K^{\alpha/2}n^{(\alpha-2)/2}} \geq \frac{1}{p}.$$

Therefore in this case the proof below gives

$$\mathbb{P}(\|\Gamma_n\| \geq K) \geq \mathbb{P}\left(\max_{i \leq p}|X_i| \geq \sqrt{Kn}\right) \geq \frac{1}{2}.$$

In particular, if $\alpha = 4$ and $p \geq (2K^2/c_0)n$ then $\|\Gamma_n\| \geq K$ with probability at least $1/2$.

Before we prove the theorem we would like to mention that last decade many works appeared on non-limit behavior of the norms of random matrices with random entries. In most of them $\max_{i \leq p}|X_i|$ appears naturally (or $\sqrt{n}$, when X_i is with high probability bounded by $\sqrt{n}$). For earlier works on Gaussian matrices we refer to [10, 11, 19] and references therein. For the general case of centered i.i.d. $w_{i,j}$ (as in our setting) Seginer [16] proved that

$$\mathbb{E}\|W_n\| \leq C\left(\mathbb{E}\max_{i \leq p}|X_i| + \mathbb{E}\max_{j \leq n}|Y_j|\right),$$

where Y_j, $j \leq n$, are the columns of W_n. Later Latała [12] was able to remove the condition that $w_{i,j}$ are identically distributed (his formula involves 4-th moments). Moreover, Mendelson and Paouris [13] have recently proved that for centered i.i.d. $w_{i,j}$ of variance one satisfying $\mathbb{E}|w_{1,1}|^q \leq L$ for some $q > 4$ and $L > 0$ with high probability one has

$$\mathbb{E}\|W_n\| \leq \max\{\sqrt{p}, \sqrt{n}\} + C(q, L)\min\{\sqrt{p}, \sqrt{n}\}.$$

In [1, 3, 13, 14, 18] matrices with independent columns (which can have dependent coordinates) were investigated. In particular, in [1] (see Theorem 3.13 there) it was shown that if columns of $p \times n$ matrix A satisfy

$$\sup_{q \geq 1}\sup_{i \leq p}\sup_{y \in S^{n-1}}\frac{1}{q}\left(\mathbb{E}|\langle X_i, y\rangle|^q\right)^{1/q} \leq \psi$$

then with probability at least $1 - \exp(-c\sqrt{p})$ one has

$$\|A\| \leq 6\max_{i \leq p}|X_i| + C\psi\sqrt{p} \tag{4}$$

(using Theorem 5.1 in [2] the factor 6 can be substituted by $(1 + \varepsilon)$ in which case constants C and c will be substituted with $C \ln(2/\varepsilon)$ and $c \ln(2/\varepsilon)$ correspondingly). Moreover, very recently (4) was extended to the case of matrices whose (independent) columns satisfy

$$\sup_{i \leq p} \sup_{y \in S^{n-1}} (\mathbb{E}|\langle X_i, y \rangle|^q)^{1/q} \leq \psi$$

for some $q > 4$ with the constant C depending on q ([8,9]).

Proof of the Theorem. By (1) the "In particular" part of the Theorem follows immediately from (3). Thus, it is enough to prove (3).

Since $X_1, \ldots, X_p$ are i.i.d. random vectors and since $|X_1|^2$ is distributed as $\sum_{j=1}^{n} w_{1,j}^2$, we observe for every $K \geq 1$,

$$\mathbb{P}\left(\max_{i \leq p} |X_i| \geq \sqrt{Kn}\right) = 1 - \mathbb{P}\left(\max_{i \leq p} |X_i| < \sqrt{Kn}\right) = 1 - \mathbb{P}\left(\left\{\forall i \ : \ |X_i| < \sqrt{Kn}\right\}\right)$$

$$(5)$$

$$= 1 - \left(\mathbb{P}(|X_1| < \sqrt{Kn})\right)^p = 1 - \left(\mathbb{P}\left(\sum_{j=1}^{n} w_{1,j}^2 < Kn\right)\right)^p.$$

For $j \leq n$ consider the events $A_j := \{w_{1,j}^2 \geq nK\}$. Clearly,

$$A := \left\{\sum_{j=1}^{n} w_{1,j}^2 \geq nK\right\} \supset \bigcup_{j=1}^{n} A_j.$$

By the inclusion-exclusion principle, we have

$$\mathbb{P}(A) \geq \mathbb{P}\left\{\bigcup_{j=1}^{n} A_j\right\} \geq \sum_{j=1}^{n} \mathbb{P}(A_j) - \sum_{j \neq k} \mathbb{P}(A_j \cap A_k)$$

$$= \sum_{j=1}^{n} \mathbb{P}(w^2 \geq nK) - \sum_{j \neq k} (\mathbb{P}(w^2 \geq nK))^2$$

$$= n\mathbb{P}(w^2 \geq nK) - \frac{n^2 - n}{2}(\mathbb{P}(w^2 \geq nK))^2$$

$$= \frac{n}{2}\mathbb{P}(w^2 \geq nK)(2 - (n-1)\mathbb{P}(w^2 \geq nK)).$$

By Chebychev's inequality we have $\mathbb{P}(w^2 \geq nK) \leq \dfrac{1}{nK}$, hence,

$$2 - (n-1)\,\mathbb{P}(w^2 \geq nK) \geq 1.$$

Thus, by (5),

$$\mathbb{P}\left(\max_{i \leq p} |X_i| \geq \sqrt{Kn}\right) \geq 1 - \left(1 - \mathbb{P}\left(\frac{1}{n}\sum_{j=1}^{n} w_{1,j}^2 \geq K\right)\right)^p \geq 1 - \left(1 - \frac{n}{2}\mathbb{P}\left(w^2 \geq nK\right)\right)^p.$$

If $\dfrac{n}{2}\mathbb{P}(w^2 \geq Kn) \geq \dfrac{1}{p}$, then

$$\mathbb{P}\left(\max_{i \leq p} |X_i| \geq \sqrt{Kn}\right) \geq 1 - \left(1 - \frac{1}{p}\right)^p \geq 1 - \frac{1}{e} \geq \frac{1}{2}.$$

Finally assume that

$$\frac{n}{2}\mathbb{P}(w^2 \geq Kn) \leq \frac{1}{p}. \tag{6}$$

Using that $(1-x)^p \leq (1 + px)^{-1}$ on $[0, 1]$, we get

$$\mathbb{P}\left(\max_{i \leq p} |X_i| \geq \sqrt{Kn}\right) \geq 1 - \frac{1}{(np/2)\mathbb{P}(w^2 \geq Kn) + 1}.$$

Applying condition (2) with $t = \sqrt{Kn}$ and using (6) again, we observe

$$1 \geq \frac{np}{2}\mathbb{P}(w^2 \geq Kn) \geq \frac{np}{2}\frac{c_0}{(Kn)^{\alpha/2}}.$$

Thus,

$$\mathbb{P}\left(\max_{i \leq p} |X_i| \geq \sqrt{Kn}\right) \geq \frac{c_0 p}{4n^{(\alpha-2)/2}K^{\alpha/2}},$$

which completes the proof. □

Acknowledgements We are grateful to A. Pajor for useful comments and to S. Sodin for bringing reference [5] to our attention. Research partially supported by the E.W.R. Steacie Memorial Fellowship.

References

1. R. Adamczak, A.E. Litvak, A. Pajor, N. Tomczak-Jaegermann, Quantitative estimates of the convergence of the empirical covariance matrix in log-concave ensembles. J. Am. Math. Soc. **23**, 535–561 (2010)
2. R. Adamczak, A.E. Litvak, A. Pajor, N. Tomczak-Jaegermann, Restricted isometry property of matrices with independent columns and neighborly polytopes by random sampling. Constr. Approx. **34**, 61–88 (2011)
3. R. Adamczak, A.E. Litvak, A. Pajor, N. Tomczak-Jaegermann, Sharp bounds on the rate of convergence of empirical covariance matrix. C.R. Math. Acad. Sci. Paris **349**, 195–200 (2011)
4. G.W. Anderson, A. Guionnet, O. Zeitouni, *An Introduction to Random Matrices*. Cambridge Studies in Advanced Mathematics, vol. 118 (Cambridge University Press, Cambridge, 2010)
5. A. Auffinger, G. Ben Arous, S. Péché, Sandrine Poisson convergence for the largest eigenvalues of heavy tailed random matrices. Ann. Inst. Henri Poincaré Probab. Stat. **45**, 589–610 (2009)
6. Z.D. Bai, J.W. Silverstein, *Spectral Analysis of Large Dimensional Random Matrices*, 2nd edn. Springer Series in Statistics (Springer, New York, 2010)
7. Z.D. Bai, J. Silverstein, Y.Q. Yin, A note on the largest eigenvalue of a large dimensional sample covariance matrix. J. Multivar. Anal. **26**(2), 166–168 (1988)
8. O. Guédon, A.E. Litvak, A. Pajor, N. Tomczak-Jaegermann, Restricted isometry property for random matrices with heavy tailed columns. C.R. Math. Acad. Sci. Paris **352**, 431–434 (2014)
9. O. Guédon, A.E. Litvak, A. Pajor, N. Tomczak-Jaegermann, On the interval of fluctuation of the singular values of random matrices (submitted)
10. Y. Gordon, On Dvoretzky's theorem and extensions of Slepian's lemma, in *Israel Seminar on Geometrical Aspects of Functional Analysis (1983/84), II* (Tel Aviv University, Tel Aviv, 1984)
11. Y. Gordon, Some inequalities for Gaussian processes and applications. Isr. J. Math. **50**, 265–289 (1985)
12. R. Latala, Some estimates of norms of random matrices. Proc. Am. Math. Soc. **133**, 1273–1282 (2005)
13. S. Mendelson, G. Paouris, On generic chaining and the smallest singular values of random matrices with heavy tails. J. Funct. Anal. **262**, 3775–3811 (2012)
14. S. Mendelson, G. Paouris, On the singular values of random matrices. J. Eur. Math. Soc. **16**, 823–834 (2014)
15. L. Pastur, M. Shcherbina, *Eigenvalue Distribution of Large Random Matrices*. Mathematical Surveys and Monographs, vol. 171 (American Mathematical Society, Providence, 2011)
16. Y. Seginer, The expected norm of random matrices. Combin. Probab. Comput. **9**, 149–166 (2000)
17. J. Silverstein, On the weak limit of the largest eigenvalue of a large dimensional sample covariance matrix. J. Multivar. Anal. **30**(2), 307–311 (1989)
18. N. Srivastava, R. Vershynin, Covariance estimation for distributions with 2+epsilon moments. Ann. Probab. **41**, 3081–3111 (2013)
19. S.J. Szarek, Condition numbers of random matrices. J. Complex. **7**(2), 131–149 (1991)
20. Y.Q. Yin, Z.D. Bai, P.R. Krishnaiah, On the limit of the largest eigenvalue of the large dimensional sample covariance matrix. Probab. Theory Relat. Fields **78**, 509–527 (1988)

The (B) Conjecture for Uniform Measures in the Plane

Amir Livne Bar-on

Abstract We prove that for any two centrally-symmetric convex shapes $K, L \subset \mathbb{R}^2$, the function $t \mapsto |e^t K \cap L|$ is log-concave. This extends a result of Cordero-Erausquin, Fradelizi and Maurey in the two dimensional case. Possible relaxations of the condition of symmetry are discussed.

1 Introduction

It was conjectured by Banaszcyk (see Latała [1]) that for any convex set $K \subset \mathbb{R}^n$ that is centrally-symmetric (i.e., $K = -K$) and for a centered Gaussian measure γ,

$$\gamma(s^{1-\lambda}t^\lambda K) \geq \gamma(sK)^{1-\lambda}\gamma(tK)^\lambda \tag{1}$$

for any $\lambda \in [0, 1]$ and $s, t > 0$.

This conjecture was proven in [2], in the equivalent form that the function $t \mapsto \gamma(e^t K)$ is log-concave. The same paper raises the question whether (1) remains valid when γ is replaced by other log-concave measures. The proof of (1) for *unconditional* sets and log-concave measures was given in [2] as well:

Theorem 1 ([2], Proposition 9). *Let $K \subset \mathbb{R}^d$ be a convex set and let μ be a log-concave measure on $\mathbb{R}^d$, and assume that both are invariant under coordinate reflections. Then $t \mapsto \mu(e^t K)$ is a log-concave function.*

This paper explores the situation in $\mathbb{R}^2$. To distinguish this special case, we call a convex set $K \subset \mathbb{R}^2$, which is compact and has a non-empty interior, a *shape*. The main result is

This paper is a part of an M.Sc. thesis written under the supervision of Bo'az Klartag.

A. Livne Bar-on (✉)
Tel Aviv University, Tel Aviv 69978, Israel
e-mail: livnebaron@mail.tau.ac.il

© Springer International Publishing Switzerland 2014
B. Klartag, E. Milman (eds.), *Geometric Aspects of Functional Analysis*,
Lecture Notes in Mathematics 2116, DOI 10.1007/978-3-319-09477-9_22

Theorem 2. *Let $K, L \subset \mathbb{R}^2$ be centrally-symmetric convex shapes. Then*

$$t \mapsto |e^t K \cap L|$$

is a log-concave function.

Here, $|\cdot|$ is the Lebesgue measure, so Theorem 2 is an analog of (1) for *uniform* measures—with density $d\mu(x) = 1_L(x)dx$. Note that a uniform measure on a set is log-concave if and only if the set is convex.

This theorem also follows from the logarithmic Brunn-Minkowski theory, as shown in [5].

The condition of central symmetry in Theorem 2 can be replaced by dihedral symmetry. For an integer $n \geq 2$, let D_n be the group of symmetries of $\mathbb{R}^2$ that is generated by two reflections, one across the axis $\mathrm{Span}\{(1,0)\}$ and the other across the axis $\mathrm{Span}\{(\cos \frac{\pi}{n}, \sin \frac{\pi}{n})\}$. The dihedral group D_n contains $2n$ transformations. A D_n-symmetric shape $A \subset \mathbb{R}^2$ is one invariant under the action of D_n.

Theorem 3. *Let $n \geq 2$ be an integer, and let $K, L \subset \mathbb{R}^2$ be D_n-symmetric convex shapes. Then $t \mapsto |e^t K \cap L|$ is a log-concave function.*

Examples and Open Questions. For what sets and measures is (1) valid?

The (B)-conjecture or (1) is not necessarily true for measures and sets with just one axis of symmetry in $\mathbb{R}^2$. An example with a log-concave uniform measure is

$$L = \mathrm{conv}\{(-5, -2), (0, 3), (5, -2)\},$$

$$K = [-6, 6] \times [-3, 1].$$

The function $t \mapsto |e^t K \cap L|$ is not log-concave in a neighbourhood of $t = 0$.

Another negative result is for quasi-concave measures. These are measures with density $d\mu(x) = \varphi(x)dx$ satisfying $\varphi((1 - \lambda)x + \lambda y) \geq \max\{\varphi(x), \varphi(y)\}$ for all $0 \leq \lambda \leq 1$. If

$$\mu(A) = |A \cap Q| + |A|, \qquad Q = [-1, 1] \times [-1, 1],$$

then the corresponding function $t \mapsto \mu(e^t Q)$ is not log-concave in a neighbourhood of $t = 1$.

The (B)-conjecture for general centrally-symmetric log-concave measures is not settled yet, even in two dimensions. It is also of interest to generalize the method of this paper to higher dimensions.

Notation. For a convex shape $K \subset \mathbb{R}^2$, its boundary is denoted by ∂K. The support function is denoted $h_K(x) = \sup_{y \in K} \langle x, y \rangle$. The normal map $\nu_K : \partial K \to S^1$ is defined for all smooth points on the boundary, and $\nu_K(p)$ is the unique direction that satisfies $\langle \nu_K(p), x \rangle = h_K(x)$. We denote the unit square by $Q = [-1, 1] \times [-1, 1] = B_\infty^2$. The Hausdorff distance between sets $A, B \subset \mathbb{R}^n$ is defined as $d_H(A, B) = \max\{\sup_{a \in A} d(a, B), \sup_{b \in B} d(b, A)\}$. The radial function $\rho_K : \mathbb{R} \to \mathbb{R}$ of a convex shape $K \subset \mathbb{R}^2$ is $\rho_K(\theta) = \max\{r \in \mathbb{R} : (r \cos \theta, r \sin \theta) \in K\}$, with period 2π.

2 Main Result

This section proves Theorem 2:

Theorem. *Let $K, L \subset \mathbb{R}^2$ be centrally-symmetric convex shapes. Then the function $f_{K,L}(t) = |e^t K \cap L|$ is log-concave.*

Obviously, it suffices to show log-concavity around $t = 0$.

If we consider the space of centrally-symmetric convex shapes in the plane, equipped with the Hausdorff metric d_H, then the operations $K, L \mapsto K \cap L$ and $K \mapsto |K|$ are continuous. This means that the correspondence $K, L \mapsto f_{K,L}$ is continuous as well. Since the condition of log-concavity in the vicinity of a point is a closed condition in the space $C(\mathbb{R})$ of bounded continuous functions, the class of pairs of centrally-symmetric shapes $K, L \subset \mathbb{R}^2$ for which $f_{K,L}(t)$ is log-concave near $t = 0$ is closed with respect to Hausdorff distance. Thus in order to prove Theorem 2 it suffices to prove that $f_{K,L}(t)$ is a log-concave function near $t = 0$ for a dense set in the space of pairs of centrally-symmetric convex shapes.

As a dense subset, we shall pick the class of transversely-intersecting convex polygons. This class will be denoted by $\mathcal{F}$. The elements of $\mathcal{F}$ are pairs (K, L) of shapes $K, L \subset \mathbb{R}^2$ that satisfy:

- The pairs (K, L) are pairs of centrally-symmetric convex polygons in $\mathbb{R}^2$.
- The intersection $\partial K \cap \partial L$ is finite.
- None of the points $x \in \partial K \cap \partial L$ are vertices of K or of L. That is, there is some $\varepsilon > 0$ such that $B(x, \varepsilon) \cap \partial K$ and $B(x, \varepsilon) \cap \partial L$ are line segments.
- For every $x \in \partial K \cap \partial L$, $\nu_K(x) \neq \nu_L(x)$.

Claim. *The class $\mathcal{F}$ is dense in the space of pairs of centrally-symmetric convex shapes (with respect to the Hausdorff metric).*

Hence, in order to prove Theorem 2, it is enough to consider polygons with transversal intersection.

Deriving a Concrete Inequality.

Lemma 4. *If $(K, L) \in \mathcal{F}$, then $f_{K,L}(t)$ is twice differentiable in some neighbourhood of $t = 0$.*

Remark. In this case, log-concavity around $t = 0$ amounts to the inequality

$$\frac{d^2}{dt^2} \log f(t) \bigg|_{t=0} \leq 0$$

$$f(0) \cdot f''(0) \leq f'(0)^2. \tag{2}$$

Proof. The area of the intersection is

$$|aK \cap L| = \int_0^a dr \int_{x \in r\partial K \cap L} h_K(v_K(\tfrac{x}{r}))d\ell,$$

where $d\ell$ is the length element.

Denote

$$g_{K,L}(r) = \int_{x \in r\partial K \cap L} h_K(v_K(\tfrac{x}{r}))d\ell.$$

The transversality of the intersection implies that $g_{K,L}(r)$ is continuous near $r = 1$. Therefore $a \mapsto |aK \cap L|$ is continuously differentiable near $a = 1$.

The contour $r\partial K \cap L$ is a finite union of segments in $\mathbb{R}^2$. Transversality implies that the number of connected components does not change with r in a small neighbourhood of $r = 1$. The beginning and end points of each component are smooth functions of r, also in some neighbourhood of $r = 1$. Therefore $g_{K,L}(r)$ is differentiable as claimed. $\square$

Note that in such a neighbourhood of $r = 1$, the function $g_{K,L}(r)$ only depends on the parts of K and L that are close to $\partial K \cap L$, and is in fact a sum of contributions from each of the connected components.

Writing (2) in terms of $g(r)$, we get the following condition:

Definition. For convex shapes $(K, L) \in \mathcal{F}$, we say that K and L satisfy property B, or that $B(K, L)$, if

$$|K \cap L| \cdot [g_{K,L}(1) + g'_{K,L}(1)] \le g_{K,L}(1)^2. \tag{3}$$

The set $\mathcal{F}$ is open with respect to the Hausdorff metric, and in particular, if $(K, L) \in \mathcal{F}$ then $(K, rL) \in \mathcal{F}$ for every r in some neighbourhood of $r = 1$. If $B(K, rL)$ holds for every r in such a neighbourhood, then $f_{K,L}(t)$ is log-concave in some neighbourhood of $t = 0$, as

$$f_{K,L}(t_0 + t) = e^{2t_0} f_{K,e^{-t_0}L}(t).$$

Therefore verifying (3) for all pairs $(K, L) \in \mathcal{F}$ will prove Theorem 2.

Reduction to Parallelograms. Given two polygons $(K, L) \in \mathcal{F}$, the intersection $\partial K \cap L$ consists of a finite number of connected components. Due to central symmetry, they come in opposite pairs. We denote these components by $S_1, \ldots, S_{2n}$, and $S_{i+n} = \{-x : x \in S_i\}$.

We define a pair of convex shapes $K^{(i)}, L^{(i)}$ for each $1 \le i \le n$ via the following properties.

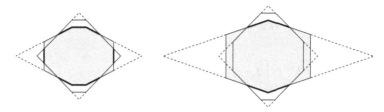

Fig. 1 Two examples of the extension $K, L \implies K^{(1)}, L^{(1)}$. The *shaded shape* in each diagram is K and the *white shape* with a *solid boundary line* is the corresponding L

- The shape $K^{(i)}$ is the largest convex set whose boundary contains $S_i \cup S_{i+n}$. Equivalently, denoting by x_1, x_2 the endpoints of S_i, and by x the solution of the equations

$$
\begin{cases}
\langle \nu_K(x_1), x \rangle = h_K(\nu_K(x_1)) \\
\langle \nu_K(x_2), x \rangle = -h_K(\nu_K(x_2))
\end{cases}
$$

  then $K^{(i)} = \mathrm{conv}\,(S_i \cup S_{i+n} \cup \{x, -x\})$.
- The shape $L^{(i)}$ is the parallelogram defined by the four lines

$$
\langle \nu_L(x_1), x \rangle = \pm h_L(\nu_L(x_1)) \quad , \quad \langle \nu_L(x_2), x \rangle = \pm h_L(\nu_L(x_2))
$$

See Fig. 1 for examples.

If S_i is a segment then $K^{(i)}$ described above is an infinite strip, and if $\nu_L(x_1) = \nu_L(x_2)$ then $L^{(i)}$ is an infinite strip. We would like to work with compact shapes, thus we apply a procedure to modify $K^{(i)}, L^{(i)}$ to become bounded without changing their significant properties. Transversality implies that the intersection $K^{(i)} \cap L^{(i)}$ is bounded, even if both sets are strips. For each $1 \leq i \leq n$ we pick a centrally-symmetric strip $A \subset \mathbb{R}^2$ such that $A \cap K^{(i)}$ and $A \cap L^{(i)}$ are both bounded, and which contains K and L, and whichever of $K^{(i)}, L^{(i)}$ that is bounded. From now on we replace $K^{(i)}$ and $L^{(i)}$ by their intersection with A.

Remark. Note that the sets grow in the process: $K \subset K^{(i)}$ and $L \subset L^{(i)}$ for all $i = 1 \ldots n$. They satisfy $\partial K^{(i)} \cap L^{(i)} = S_i \cup S_{i+n}$. Also note that if K is a parallelogram then so are the $K^{(i)}$, for every i. It is trivial to check that $(K^{(i)}, L^{(i)}) \in \mathcal{F}$ when $(K, L) \in \mathcal{F}$.

Lemma 5. *If* $B(K^{(i)}, L^{(i)})$ *for all* $i = 1 \ldots n$, *then* $B(K, L)$.

Proof. The function $g_{K,L}(r)$ takes non-negative values for $r > 0$. In addition, its value is the sum of contributions from the different connected components of $r \partial K \cap L$. From transversality, these components vary continuously around $r = 1$, hence $g'_{K,L}(1)$ is also a sum of values coming from the different components. Therefore we can write

$$|K \cap L| \cdot [g_{K,L}(1) + g'_{K,L}(1)] = |K \cap L| \cdot \sum_{i=1}^{n} \left[g_{K^{(i)},L^{(i)}}(1) + g'_{K^{(i)},L^{(i)}}(1) \right]$$

$$\leq \sum_{i=1}^{n} |K^{(i)} \cap L^{(i)}| \cdot \left[g_{K^{(i)},L^{(i)}}(1) + g'_{K^{(i)},L^{(i)}}(1) \right]$$

$$\underset{by\, B(K^{(i)},L^{(i)})}{\leq} \sum_{i=1}^{n} g_{K^{(i)},L^{(i)}}(1)^2 \leq \left(\sum_{i=1}^{n} g_{K^{(i)},L^{(i)}}(1) \right)^2 = g_{K,L}(1)^2$$

$\square$

Lemma 6. *If $B(K, L)$ holds for all pairs of parallelograms $(K, L) \in \mathcal{F}$, then Theorem 2 follows.*

Proof. Let $(K, L) \in \mathcal{F}$ be any polygons. Construct the sequence of pairs $K^{(i)}, L^{(i)}$ from K, L. The shape $L^{(i)}$ is a parallelogram for every i. Then construct the pairs $(L^{(i)})^{(j)}, (K^{(i)})^{(j)}$ from $L^{(i)}, K^{(i)}$, for all i. The shapes $(L^{(i)})^{(j)}$ and $(K^{(i)})^{(j)}$ will be parallelograms for every i, j. Under our assumption, we have $B\left((L^{(i)})^{(j)}, (K^{(i)})^{(j)} \right)$. From this and the previous lemma, $B(L^{(i)}, K^{(i)})$ follows.

The property B is symmetric in the shapes. That is, $B(S, T) \iff B(T, S)$ for all $(S, T) \in \mathcal{F}$. This is since $f_{S,T}$ and $f_{T,S}$ differ by a log-linear factor:

$$f_{S,T}(t) = |e^t S \cap T| = e^{2t} f_{T,S}(-t)$$

This means that we have $B(K^{(i)}, L^{(i)})$ as well. Applying the previous lemma again gives $B(K, L)$. $\square$

All that remains in order to deduce Theorem 2 is to analyse the case of centrally-symmetric parallelograms.

If K, L are parallelograms and $K = TQ$ where T is an invertible linear map and $Q = [-1, 1] \times [-1, 1]$,

$$f_{K,L} = \det T \cdot f_{Q,T^{-1}L}.$$

Therefore we can take one of the parallelograms to be a square. In other words, establishing $B(Q, L)$ where Q is the unit square and L is a parallelogram, and $(Q, L) \in \mathcal{F}$, will imply Theorem 2.

In fact, we may place additional geometric constraints on the square and the parallelogram.

If neither Q nor L contains a vertex of the other quadrilateral in its interior, then $\partial Q \cap L$ has four connected components. Applying the reduction above to Q, L gives $Q^{(i)}, L^{(i)}$ with $i = 1, 2$, and the intersection $\partial Q^{(i)} \cap L^{(i)}$ has only two connected components, as remarked above.

Since the shapes are convex, if all the vertices of one shape are contained in the other, we have $Q \subset L$ or $L \subset Q$, and then (3) holds trivially. If L contains vertices of Q but Q does not contain vertices of L, we shall swap them.

These arguments leave two cases to be considered:

1. Q contains two vertices of L, and L does not contain vertices of Q. In this case the intersection $\partial Q \cap L$ is contained in two opposite edges of Q.
   This case is proved in Lemma 7 below.
2. Q contains two vertices of L, and L contains two vertices of Q. In this case the intersection $\partial Q \cap L$ is a subset of the edges around these vertices of Q.
   This case is proved in Lemma 8 below.

Computation of the Special Cases. These cases are defined by four real parameters—the coordinates of the vertices of L. A symbolic expression for $f(t)$ can be derived, and (3) will be a polynomial inequality in these parameters. The geometric conditions given above are also polynomial inequalities in these parameters. Thus each of the two cases can each be expressed by a universally-quantified formula in the language of real closed fields. By Tarski's theorem [3], this first order theory has a decision procedure. This is implemented in the QEPCAD B computer program [4]. Relevant computer files, for generation of the symbolic condition and for running the logic solver, for one of the two cases above, are available at http://www.tau.ac.il/~livnebaron/files/bconj_201311/bconj_corners. mac and http://www.tau.ac.il/~livnebaron/files/bconj_201311/bconj_qelim.txt.

A human-readable proof of both cases is included here as well.

Lemma 7. *If L is a centrally-symmetric parallelogram that satisfies $(Q, L) \in \mathcal{F}$, and if L crosses Q only inside the vertical edges of Q, then $B(Q, L)$.*

Proof. Let α, β, c, d be as in Fig. 2.

The equations for the edges of L are

$$\begin{cases} x \cos\alpha + y \sin\alpha = \pm(c \cos\alpha + d \sin\alpha) \\ x \cos\beta + y \sin\beta = \pm(c \cos\beta + d \sin\beta) \end{cases}$$

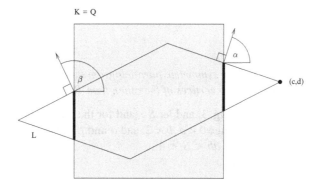

Fig. 2 Geometric setting of Lemma 7

Relevant parameters are computed as follows:

$$\partial Q \cap \partial L = \{\pm(1, (c-1)\cot\alpha + d), \pm(1, (c-1)\cot\beta + d)\}$$

$$g_{Q,L}(1) = 2(c-1)(\cot\alpha - \cot\beta)$$

$$g'_{Q,L}(1) = -2(\cot\alpha - \cot\beta)$$

$$g_{Q,L}(1) + g'_{Q,L}(1) = (2c-4)(\cot\alpha - \cot\beta).$$

The area of L is comprised of $Q \cap L$ and of two triangles. The area of the triangles is $\frac{1}{2}g(1) \cdot (c-1)$ so

$$|Q \cap L| = |L| - (c-1)^2(\cot\alpha - \cot\beta).$$

Note that $0 < \alpha < \frac{\pi}{2} < \beta < \pi$ so $\cot\alpha - \cot\beta$ is a positive quantity, and that if $c < 2$ the value of $g(1) + g'(1)$ is negative so inequality (3) is satisfied immediately. Assume $c > 2$ from now on. What we need to prove is

$$(2c-4)(\cot\alpha - \cot\beta) \cdot \left[|L| - (c-1)^2(\cot\alpha - \cot\beta)\right] \le 4(c-1)^2(\cot\alpha - \cot\beta)^2.$$

Or equivalently

$$(2c-4)|L| \le (c-1)^2(\cot\alpha - \cot\beta) \cdot (4 + 2c - 4),$$

or still

$$|L| \le \left(1 + \frac{2}{c-2}\right) \cdot \frac{1}{2}(c-1)g(1).$$

The amount $\frac{1}{2}(c-1)g(1)$ is the area of the triangles $L \backslash Q$. By convexity the area of L cannot be larger than that times $\left(\frac{c}{c-1}\right)^2$. It remains to verify that for $c > 2$, $\frac{c^2}{(c-1)^2} < 1 + \frac{2}{c-2}$. This is a simple exercise in algebra:

$$\frac{c^2}{(c-1)^2} = 1 + \frac{2c-1}{(c-1)^2} = 1 + \frac{2}{c-2} \cdot \frac{(c-\frac{1}{2})(c-2)}{(c-1)^2} = 1 + \frac{2}{c-2}\left[1 - \frac{c/2}{(c-1)^2}\right] \le 1 + \frac{2}{c-2}$$

$\square$

Lemma 8. *If L is a centrally-symmetric parallelogram that satisfies $(Q, L) \in \mathcal{F}$, and each of Q, L contains two vertices of the other, then $B(Q, L)$.*

Proof. Let a and b be as in Fig. 3, and let S stand for the area $S = |Q \cap L|$. The numbers a and b are in the range $0 < a, b < 2$, and α and β satisfy $\frac{1}{2}\pi < \alpha < \beta < \pi$. The area S is in the range $4 - ab < S < 4$.

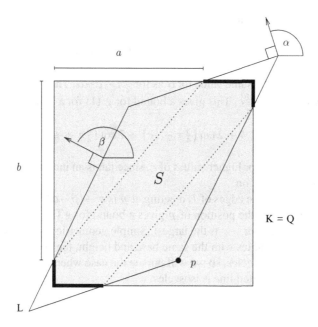

Fig. 3 Geometric setting of Lemma 8

The quantity $g(1)$ is simply $8 - 2a - 2b$, and $g'(1)$ will soon be shown to be bounded by

$$g'(1) \leq -8 \frac{S - (4 - ab)}{(4 - S) + \frac{1}{2}(a - b)^2}.$$

This gives an inequality in the 3 variables a, b, S, which will be proved for values in the prescribed ranges.

The length of each dotted line in Fig. 3 is $(a^2 + b^2)^{1/2}$. Denoting the height of the triangle (the distance between p and the closest dotted line) by h, the area is

$$S = (4 - ab) + 2 \cdot \frac{1}{2} h \cdot (a^2 + b^2)^{1/2},$$

so

$$h = \frac{S - (4 - ab)}{(a^2 + b^2)^{1/2}}.$$

The formula for $g'(1)$ in terms of the angles α, β is

$$g'(1) = 4 + 2 \tan \alpha + 2 \cot \beta.$$

Denote $c = \beta - \alpha$. Holding c fixed, the function

$$\alpha \mapsto g'(1) = 4 + 2\tan\alpha + 2\cot(\alpha + c)$$

is concave and takes the same value for α as for $\frac{3}{2}\pi - c - \alpha$. Therefore its maximum is attained at $\alpha = \frac{3}{4}\pi - \frac{1}{2}c$. This gives a bound for $g'(1)$ for a given $c = \beta - \alpha$:

$$g'(1) \le 4 + 2\tan\left(\tfrac{3}{4}\pi - \tfrac{1}{2}c\right) + 2\cot\left(\tfrac{3}{4}\pi + \tfrac{1}{2}c\right).$$

This bound is stronger for higher values of c, since tan is an increasing function and cot is a decreasing function.

The angle between the edges of L meeting at p is $\pi - (\beta - \alpha) = \pi - c$. When a, b, and h are kept fixed, the position of p gives a bound for $g'(1)$. This bound is the weakest when the angle $\pi - c$ is the largest. Simple geometric considerations show that in a family of triangles with the same base and height, the apex angle is largest when the triangle is isosceles, so we will pursue the case where the triangle formed by p and the nearest dotted line is isosceles.

The value of c in this case is $c = 2\tan^{-1}\frac{\frac{1}{2}(a^2+b^2)^{1/2}}{h}$, and we get

$$g'(1) \le 4 + 2\tan\left(\tfrac{3}{4}\pi - \tfrac{1}{2}\pi + \tan^{-1}\frac{\tfrac{1}{2}(a^2+b^2)^{1/2}}{h}\right)$$

$$+ 2\cot\left(\tfrac{3}{4}\pi + \tfrac{1}{2}\pi - \tan^{-1}\frac{\tfrac{1}{2}(a^2+b^2)^{1/2}}{h}\right)$$

$$= 4 + 4\tan\left(\tfrac{1}{4}\pi + \tan^{-1}\frac{1}{2}\frac{a^2+b^2}{S-(4-ab)}\right)$$

$$= 4 + 4 \cdot \frac{1 + \frac{1}{2}\frac{a^2+b^2}{S-(4-ab)}}{1 - \frac{1}{2}\frac{a^2+b^2}{S-(4-ab)}} = \frac{8}{1 - \frac{1}{2}\frac{a^2+b^2}{S-(4-ab)}} = -8\frac{S-(4-ab)}{(4-S)+\frac{1}{2}(a-b)^2},$$

which proves the forementioned bound for $g'(1)$.

Therefore, to prove (3) it is enough to show

$$S \cdot \left(8 - 2a - 2b - 8\frac{S-(4-ab)}{(4-S)+\frac{1}{2}(a-b)^2}\right) \le (8 - 2a - 2b)^2$$

Rearranging and taking into account that $S < 4$, this is equivalent to

$$\underbrace{(8 - 2a - 2b)(8 - 2a - 2b - S)\left((4-S) + \tfrac{1}{2}(a-b)^2\right) + 8S(S - (4-ab)) \ge 0}_{E}$$

When a and b are held fixed, this is a 2nd degree condition on S. Since $0 < a, b < 2$, the value and the first two derivatives in the point $S = 4 - ab$ are positive:

$$E|_{S=4-ab} = (8 - 2a - 2b)(2 - a)(2 - b) \cdot \tfrac{1}{2}(a^2 + b^2) > 0,$$

$$\left.\frac{\partial E}{\partial S}\right|_{S=4-ab} = (a + b)\left((5 - a - b)^2 - 1\right) + 2(a - b)^2 > 0,$$

$$\left.\frac{\partial^2 E}{\partial S^2}\right|_{S=4-ab} = 18(4 - ab) > 0.$$

This means that the condition stays true for all $S > 4 - ab$, as required.   □

3 Dihedral Symmetry

This section deals with dihedrally symmetric sets. The group D_n is defined in the introduction.

Theorem (3). *Let $n \geq 2$ be an integer, and let $K, L \subset \mathbb{R}^2$ be D_n-symmetric convex shapes. Then $t \mapsto |e^t K \cap L|$ is a log-concave function.*

For $n = 2$ the group D_n is generated by reflections across the standard axes. This corresponds to unconditional sets and functions, and Theorem 1 from [2] solves this case.

The proof for $n \geq 3$ is by reduction to the unconditional case.

A *smooth strongly-convex* shape $K \subset \mathbb{R}^2$ is one whose boundary is a smooth curve with strictly positive curvature everywhere. The radial function ρ_K of a smooth strongly-convex shape $K \subset \mathbb{R}^2$ is a smooth function. The boundary ∂K is the curve

$$\gamma_K(\theta) = (\rho_K(\theta) \cos \theta, \rho_K(\theta) \sin \theta).$$

The convexity of K is reflected in the sign of the curvature of γ_K. Positive curvature can be written as a condition on the radial function:

$$\rho(\theta)^2 + 2\rho'(\theta)^2 - \rho(\theta)\rho''(\theta) > 0. \tag{4}$$

Proof of Theorem 3. For any D_n-symmetric convex shape $K \subset \mathbb{R}^2$ there is a sequence of D_n-symmetric convex shapes whose boundaries are smooth and strongly convex curves, and whose Hausdorff limit is K. By the continuity argument from the previous section, the general case follows from the smooth case. From here on, K and L are smooth D_n-symmetric shapes.

D_n-symmetric shapes correspond to radial functions that are even and have period $\frac{2\pi}{n}$. These shapes are completely determined by their intersection with the sector

$$G_n = \{(r\cos\theta, r\sin\theta) : r \geq 0, \theta \in [0, \tfrac{\pi}{n}]\}.$$

Given two such shapes K, L the area function is

$$f_{K,L}(t) = |e^t K \cap L| = 2nf_{K\cap G_n, L\cap G_n}(t).$$

Let $K \subset \mathbb{R}^2$ be a D_n-symmetric strongly convex shape, and consider the function $\tilde{\rho}(\theta) = \rho_K(\frac{2}{n}\theta)$. This is an even function with period π. The function $\tilde{\rho}(\theta)$ also satisfies (4):

$$\tilde{\rho}(\theta)^2 + \tilde{\rho}'(\theta)^2 - \tilde{\rho}(\theta)\tilde{\rho}''(\theta) =$$

$$\tfrac{4}{n^2}\left(\rho_K(\tfrac{2}{n}\theta)^2 + 2\rho_K'(\tfrac{2}{n}\theta)^2 - \rho_K(\tfrac{2}{n}\theta)\rho_K''(\tfrac{2}{n}\theta)\right) + (1 - \tfrac{4}{n^2})\rho_K(\tfrac{2}{n}\theta)^2 > 0.$$

This means that $\tilde{\rho}(\theta)$ is the radial function of some D_2-symmetric (unconditional) strongly convex shape. We denote this $w(K)$: the unique shape that satisfies $\rho_{w(K)}(\theta) = \rho_K(\frac{2}{n}\theta)$.

The following function, also named w, is defined on G_n:

$$w\begin{pmatrix} r\cos\theta \\ r\sin\theta \end{pmatrix} = \begin{pmatrix} r\cos\frac{n}{2}\theta \\ r\sin\frac{n}{2}\theta \end{pmatrix}. \qquad \left(\text{for } r \geq 0, \ \theta \in [0, \tfrac{\pi}{n}]\right)$$

The point function w is an bijection between G_n and G_2. It relates to the shape function w by the formula

$$\{w(x) : x \in K \cap G_n\} = w(K) \cap G_2.$$

The point function w is differentiable inside G_n, and has a constant Jacobian determinant $\frac{n}{2}$.

Hence

$$f_{K,L}(t) = 2nf_{K\cap G_n, L\cap G_n}(t) = 4f_{w(K)\cap G_2, w(L)\cap G_2}(t) = f_{w(K), w(L)}(t),$$

and the theorem follows from the result in [2]. □

Acknowledgements Supported in part by a grant from the European Research Council.

References

1. R. Latała, On some inequalities for Gaussian measures, in *Proceedings of the International Congress of Mathematicians*, vol. II (2002), pp. 813–822
2. D. Cordero-Erausquin, M. Fradelizi, B. Maurey, The (B) conjecture for the Gaussian measure of dilation of symmetric convex sets and related problems. J. Funct. Anal. **214**(2), 410–427 (2004)
3. A. Tarski, *A Decision Method for Elementary Algebra and Geometry* (University of California Press, Berkeley, 1951)
4. C.W. Brown, QEPCAD B: a program for computing with semi-algebraic sets using CADs. SIGSAM Bull. **37**(4), 97–108 (2003)
5. C. Saroglou, Remarks on the conjectured log-Brunn-Minkowski inequality. Preprint (2013). arXiv:1311.4954

Maximal Surface Area of a Convex Set in $\mathbb{R}^n$ with Respect to Log Concave Rotation Invariant Measures

Galyna Livshyts

Abstract It was shown by K. Ball and F. Nazarov, that the maximal surface area of a convex set in $\mathbb{R}^n$ with respect to the Standard Gaussian measure is of order $n^{\frac{1}{4}}$. In the present paper we establish the analogous result for all rotation invariant log concave probability measures. We show that the maximal surface area with respect to such measures is of order $\frac{\sqrt{n}}{\sqrt[4]{Var|X|}\sqrt{\mathbb{E}|X|}}$, where X is a random vector in $\mathbb{R}^n$ distributed with respect to the measure.

1 Introduction

In this paper we will study geometric properties of the probability measures γ on $\mathbb{R}^n$ with density $C_n e^{-\varphi(|y|)}$, where $\varphi(t)$ is a nonnegative nondecreasing convex function, which may take infinity as a value, and the normalizing constant

$$C_n = \left(\int_{\mathbb{R}^n} e^{-\varphi(|y|)} dy \right)^{-1}.$$

We recall that the Minkowski surface area of a convex set Q with respect to the measure γ is defined to be

$$\gamma(\partial Q) = \liminf_{\varepsilon \to +0} \frac{\gamma((Q + \varepsilon B_2^n)\backslash Q)}{\varepsilon}, \tag{1}$$

where B_2^n denotes the Euclidian unit ball in $\mathbb{R}^n$.

The special case of $\varphi(t) = \frac{t^2}{2}$, which corresponds to the standard Gaussian measure γ_2, has been actively studied. Sudakov and Tsirelson [18] and Borell [5] proved, that among all convex sets of a fixed Gaussian measure, half spaces have the smallest Gaussian surface area. Mushtari and Kwapien asked the reverse version of isoperimetric inequality, i.e. how large the Gaussian surface area of a convex set

G. Livshyts (✉)
Department of Mathematical Sciences, Kent State University, Kent, OH 44242, USA
e-mail: glivshyt@kent.edu

© Springer International Publishing Switzerland 2014
B. Klartag, E. Milman (eds.), *Geometric Aspects of Functional Analysis*,
Lecture Notes in Mathematics 2116, DOI 10.1007/978-3-319-09477-9_23

$A \subset \mathbb{R}^n$ can be. It was shown by Ball [1], that Gaussian surface area of a convex set in $\mathbb{R}^n$ is asymptotically bounded by $Cn^{\frac{1}{4}}$, where C is an absolute constant. Nazarov [17] proved the sharpness of Ball's result and gave the complete solution to this asymptotic problem:

$$0.28n^{\frac{1}{4}} \leq \max_{Q \in \mathcal{K}_n} \gamma_2(\partial Q) \leq 0.64n^{\frac{1}{4}}, \tag{2}$$

where by $\mathcal{K}_n$ we denote the set of all convex sets in $\mathbb{R}^n$. Further estimates for $\gamma_2(\partial Q)$ for the special case of polynomial level set surfaces were provided by D. Kane [10]. He showed that for any polynomial $P(y)$ of degree d, $\gamma_2(P(y) = 0) \leq \frac{d}{\sqrt{2}}$.

Isoperimetric inequalities for a wider class of rotation invariant measures were studied by Sudakov and Tsirelson [18]. Recently, geometric properties for various classes of rotation invariant measures were established by Bobkov [2–4], Bray and Morgan [6], Maurmann and Morgan [15] and others.

The maximal surface area of convex sets for the probability measures γ_p with densities $C_{n,p} e^{-\frac{|x|^p}{p}}$, where $p > 0$, was studied in [14]. It was shown there, that

$$c(p)n^{\frac{3}{4}-\frac{1}{p}} \leq \max_{Q \in \mathcal{K}_n} \gamma_p(\partial Q) \leq C(p)n^{\frac{3}{4}-\frac{1}{p}}, \tag{3}$$

where $c(p)$ and $C(p)$ are constants depending on p only.

In the present paper we obtain a generalization of results due to Ball and Nazarov, and find an expression for the maximal surface area with respect to an arbitrary rotation invariant log concave measure γ. The expression depends on the measure's natural characteristics, i.e. expectation and variance of a random variable, distributed with respect to γ.

We shall use notation $\precsim$ for an asymptotic inequality: we say that $A(n) \precsim B(n)$ if there exists an absolute positive constant C (independent of n), such that $A(n) \leq CB(n)$. Correspondingly, $A(n) \approx B(n)$ means that $B(n) \precsim A(n) \precsim B(n)$.

The following theorem is the main result of the present paper:

Theorem 1. *Fix $n \geq 2$. Let γ be log concave rotation invariant measure on $\mathbb{R}^n$. Consider a random vector X in $\mathbb{R}^n$ distributed with respect to γ. Then*

$$\max_{Q \in \mathcal{K}_n} \gamma(\partial Q) \approx \frac{\sqrt{n}}{\sqrt{\mathbb{E}|X|}\sqrt[4]{Var|X|}},$$

where, as usual, $\mathbb{E}|X|$ and $Var|X|$ denote the expectation and the variance of $|X|$ correspondingly.

Let us note, that the above Theorem implies (2). It also implies (3) in the case $p \geq 1$, and the details of these implications are shown in Sect. 3.

Another classical example of a log concave rotation invariant measure is the normalized Lebesgue measure restricted to the unit ball. In that case $\varphi(t)$ equals to zero for all $t < 1$ and takes infinity as a value for all $t \geq 1$. For that measure $\mathbb{E}|X| \approx 1$ and $Var|X| \approx n^{-2}$, so the maximal surface area is of order $\frac{\sqrt{n}}{\sqrt[4]{n^{-2} \times 1}} = n$. The set with the maximal surface area is the sphere of radius 1, which is also clear by monotonicity of the standard surface area measure.

We outline, that for isotropic measures (see [16] for definitions and details), Theorem 1 together with the result from [12] entails that the maximal perimeter varies between $C_1 n^{\frac{1}{4}}$ and $C_2 n^{\frac{1}{2}}$, where C_1 and C_2 are absolute constants. The standard Gaussian measure is an example of an isotropic measure with the maximal surface area of order $n^{\frac{1}{4}}$, and the Lebesgue measure restricted to a ball of radius $\sqrt{n}$ is an example of an isotropic measure with the maximal surface area of order $n^{\frac{1}{2}}$. If the measures γ_p from (3) are brought to the isotropic position, the maximal surface area with respect to them is of order $n^{\frac{1}{4}}$.

The main definitions, technical lemmas and some preliminary facts are given in Sect. 2. Some connections between the probabilistic and analytic setup are provided in Sect. 3. The upper bound for Theorem 1 is obtained in Sect. 4, and the lower bound is shown in Sect. 5. In Sect. 6 we provide some examples to exhibit the sharpness of Theorem 1.

2 Some Definitions and Lemmas

This section is dedicated to some general properties of spherically invariant log concave measures. We outline some basic facts which are needed for the proof. Some of them have appeared in literature. See [11] for an excellent overview of the properties of log concave measures, in particular the proof of Lemma 4.5, where some portion of the current section appears.

We write all the calculations in $\mathbb{R}^{n+1}$ instead of $\mathbb{R}^n$ for the notational simplicity. We use notation $|\cdot|$ for the norm in Euclidean space $\mathbb{R}^{n+1}$; $|A|$ stands for the Lebesgue measure of a measurable set $A \subset \mathbb{R}^{n+1}$. We will write $B_2^{n+1} = \{x \in \mathbb{R}^{n+1} : |x| \leq 1\}$ for the unit ball in $\mathbb{R}^{n+1}$ and $\mathbb{S}^n = \{x \in \mathbb{R}^{n+1} : |x| = 1\}$ for the unit sphere. We denote $v_{n+1} = |B_2^{n+1}| = \frac{\pi^{\frac{n+1}{2}}}{\Gamma(\frac{n+1}{2}+1)}$. We note that $|\mathbb{S}^n| = (n+1)v_{n+1}$.

We fix a convex nondecreasing function $\varphi(t) : [0, \infty) \to [0, \infty]$. Let γ be a probability measure on $\mathbb{R}^{n+1}$ with density $C_{n+1}e^{-\varphi(|y|)}$. The normalizing constant C_{n+1} equals to $[(n+1)v_{n+1}J_n]^{-1}$, where

$$J_n = \int_0^\infty t^n e^{-\varphi(t)}dt. \tag{4}$$

We introduce the notation $g_n(t) = t^n e^{-\varphi(t)}$. Since we normalize the measure anyway, we may assume that $\varphi(0) = 0$.

Without loss of generality we may assume that $\varphi \in C^2[0, \infty)$. This can be shown by the standard smoothing argument (see, for example, [7]).

We shall use a well known integral formula for $\gamma(\partial Q)$, which holds true, in particular, for the measures with continuous densities:

$$\gamma(\partial Q) = C_{n+1} \int_{\partial Q} e^{-\varphi(|y|)} d\sigma(y), \tag{5}$$

where $d\sigma(y)$ stands for Lebesgue surface measure (see the Appendix for the proof).

The below Lemma shows that the surface area with respect to γ is stable under small perturbations.

Lemma 1. *Fix $n \geq 2$. Let M be a measurable subset of a boundary of a convex set in $\mathbb{R}^{n+1}$. Then*

$$C_{n+1} \int_{\frac{1}{1+\frac{1}{n}}M} e^{-\varphi(|y|)} d\sigma(y) \gtrsim \gamma(M).$$

Proof. We observe, that $\varphi\left(\frac{|y|}{1+\frac{1}{n}}\right) \leq \varphi(|y|)$, since $\varphi(t)$ is nondecreasing. Also,

$$d\sigma\left(\frac{y}{1+\frac{1}{n}}\right) = \left(1 + \frac{1}{n}\right)^{-n} d\sigma(y) \approx d\sigma(y).$$

We conclude:

$$\int_{\frac{1}{1+\frac{1}{n}}M} e^{-\varphi(|y|)} d\sigma(y) = \int_M e^{-\varphi\left(\frac{|y|}{1+\frac{1}{n}}\right)} d\sigma\left(\frac{y}{1+\frac{1}{n}}\right) \gtrsim \gamma(M). \quad \square$$

Remark 1. We observe as well, that the same statement holds for all measures with densities, decreasing along each ray starting at zero.

Definition 2. We define t_0 to be the point of maxima of the function $g_n(t)$, i.e., t_0 is the solution of the equation

$$\varphi'(t)t = n. \tag{6}$$

We note that Eq. (6) has a solution, since $t\varphi'(t)$ is nondecreasing, continuous and $\lim_{t\to+\infty} t\varphi'(t) = +\infty$. This solution is unique, since $t\varphi'(t)$ strictly increases on its support. This definition appears in most of the literature dedicated to rotation invariant log concave measures: see, for example, [13, Sect. 2], [3, Remark 3.4] or [11, Lemma 4.3].

Remark 2. We may define t_n and t_{n-1} by

$$\varphi'(t_{n-1})t_{n-1} = n - 1, \tag{7}$$

$$\varphi'(t_n)t_n = n. \tag{8}$$

We claim that $t_n'' \approx t_{n-1}''$. To see this, we note that function $t\varphi'(t)$ is nondecreasing. Hence $t_n \geq t_{n-1}$. On the other hand, subtracting (7) from (8), we get

$$1 = \varphi'(t_n)t_n - \varphi'(t_{n-1})t_{n-1} \geq \varphi'(t_{n-1})(t_n - t_{n-1}) = \frac{n-1}{t_{n-1}}(t_n - t_{n-1}).$$

The above leads to the following chain of inequalities:

$$1 \leq \frac{t_n}{t_{n-1}} \leq 1 + \frac{1}{n-1}, \tag{9}$$

and therefore $t_n'' \approx t_{n-1}''$.

In a view of the above we introduce the notation "t_0". Everywhere in the paper it is assumed that $t_0 = t_{n+1}$.

We notice in addition, that

$$\varphi(t_0) = \varphi(t_0) - \varphi(0) \leq \varphi'(t_0)t_0 = n, \tag{10}$$

since $\varphi(0) = 0$ by our assumption.

The next lemma provides simple asymptotic bounds for J_n. It was proved in [13], but for the sake of completeness we sketch the proof below.

Lemma 2.

$$\frac{g_n(t_0)t_0}{n+1} \leq J_n \leq \sqrt{2\pi}(1 + o(1))\frac{g_n(t_0)t_0}{\sqrt{n}}.$$

Sketch of the Proof. The integral J_n can be estimated from above by Laplace method, which can be found, for example, in [8]. We rewrite

$$J_n = g_n(t_0) \int_0^\infty e^{n \log \frac{t}{t_0} + \varphi(t_0) - \varphi(t)} \, dt. \tag{11}$$

By the Mean Value theorem, $\varphi(t_0) - \varphi(t) \leq \varphi'(t_0)(t_0 - t) = n(1 - \frac{t}{t_0})$ for any $t \geq 0$. Thus, (11) is less than

$$g_n(t_0)t_0 \int_0^\infty e^{nh(t)} \, dt,$$

where $h(t) = \log t - t + 1$. It is easy to check that $h(t)$ satisfies Laplace's condition (see [8, pp. 85–86] for the formulation), so

$$J_n \leq (1 + o(1))\sqrt{2\pi}\frac{g_n(t_0)t_0}{\sqrt{n}}.$$

On the other hand, since $\varphi(t)$ is nondecreasing and positive,

$$J_n \geq \int_0^{t_0} t^n e^{-\varphi(t)} dt \geq e^{-\varphi(t_0)} \int_0^{t_0} t^n dt = \frac{t_0 g_n(t_0)}{n+1}. \tag{12}$$

$\square$

The next Lemma is a simple fact which we shall apply it to estimate the "tails" of J_n.

Lemma 3. *Let $g(t) = e^{f(t)}$ be a log concave function on $[a,b]$ (where both a and b may be infinite). Assume that $f \in C^2[a,b]$ and that t_0 is the unique point of maxima of $f(t)$. Assume that $t_0 > 0$. Consider $x > 0$ and $\psi > 0$ such that*

$$f(t_0) - f((1+x)t_0) \geq \psi.$$

Then,

$$\int_{(1+x)t_0}^b g(t)dt \leq \frac{xt_0 g(t_0)}{\psi e^{\psi}}.$$

Similarly, if $f(t_0) - f((1-x)t_0) \geq \psi$,

$$\int_a^{(1-x)t_0} g(t)dt \leq \frac{xt_0 g(t_0)}{\psi e^{\psi}}.$$

Proof. We pick any $t > (1+x)t_0$. First, we notice by concavity:

$$\psi \leq f(t_0) - f((1+x)t_0) \leq -f'((1+x)t_0)xt_0.$$

Next, since $f(t)$ is concave,

$$f(t) \leq f'((1+x)t_0)(t - (1+x)t_0) + f((1+x)t_0) \leq$$

$$-\frac{\psi}{xt_0}(t - (1+x)t_0) + f(t_0) - \psi.$$

Thus, for $t > (1+x)t_0$,

$$g(t) \leq g(t_0)e^{-\psi} e^{-\frac{\psi}{xt_0}(t-(1+x)t_0)}. \tag{13}$$

Consequently,

$$\int_{(1+x)t_0}^b g(t)dt \leq g(t_0)e^{-\psi} \int_0^\infty e^{-\frac{\psi}{xt_0}s} ds \leq \frac{xt_0 g(t_0)}{\psi e^{\psi}}.$$

The second part of the Lemma can be obtained similarly. $\square$

We note, that the condition $t_0 > 0$ in the above Lemma is not crucial, and everything can be restated for $t_0 < 0$. For our purposes it is enough to consider $t_0 > 0$.

The function $g_n(t) = t^n e^{-\varphi(t)}$ is log concave on $[0, \infty)$, and we shall apply Lemma 3 with $g(t) = g_n(t)$ and $\psi = 1$.

Definition 3. Define the "outer" λ_o to be a positive number satisfying:

$$\varphi(t_0(1 + \lambda_o)) - \varphi(t_0) - n \log(1 + \lambda_o) = 1. \tag{14}$$

Similarly, define the "inner" λ_i as follows:

$$\varphi(t_0(1 - \lambda_i)) - \varphi(t_0) - n \log(1 - \lambda_i) = 1. \tag{15}$$

We put

$$\lambda := \lambda_i + \lambda_o. \tag{16}$$

We note that (14) is equivalent to

$$g_n(t_0) = e g_n(t_0(1 + \lambda_o)), \tag{17}$$

and (15) is equivalent to

$$g_n(t_0) = e g_n(t_0(1 - \lambda_i)). \tag{18}$$

Parameter λ from (16) has a nice property:

Lemma 4.

$$J_n \approx \lambda t_0 g_n(t_0).$$

Proof. We apply the first part of Lemma 3 with $x = \lambda_o$ and $\psi = 1$. We get

$$\int_{t_0(1+\lambda_o)}^{\infty} g_n(t) dt \le \frac{1}{e} \lambda_o t_0 g_n(t_0). \tag{19}$$

Similarly, the second part of the Lemma applied with $x = \lambda_i$, gives

$$\int_0^{t_0(1-\lambda_i)} g_n(t) dt \le \frac{1}{e} \lambda_i t_0 g_n(t_0). \tag{20}$$

Along with the above, we observe:

$$\int_{t_0(1-\lambda_i)}^{t_0(1+\lambda_o)} g_n(t) dt \le (\lambda_i + \lambda_o) t_0 g_n(t_0) = \lambda t_0 g_n(t_0). \tag{21}$$

From (19), (20) and (21), applied together with the definition of λ, it follows that:

$$J_n \leq \frac{e+1}{e} \lambda t_0 g_n(t_0).$$ (22)

On the other hand,

$$J_n \geq \int_{t_0(1-\lambda_i)}^{t_0(1+\lambda_o)} g_n(t)dt \geq$$

$$\lambda_i t_0 g_n((1+\lambda_i)t_0) + \lambda_o t_0 g_n((1+\lambda_o)t_0) = \frac{1}{e}\lambda t_0 g_n(t_0),$$ (23)

where the last equality is obtained in a view of (18) and (17). □

Remark 3. Let us note, that Lemma 2 together with (22) and (23) leads to the estimates:

$$\frac{e}{e+1}\frac{1}{n+1} \leq \lambda \leq (1+o(1))\sqrt{2\pi e}\frac{1}{\sqrt{n}}.$$ (24)

The above implies also, that both "inner" and "outer" lambdas are asymptotically bounded by $\frac{1}{\sqrt{n}}$. In addition $\lambda_i \gtrsim \frac{1}{n}$. To see this, we write:

$$\int_0^{t_0} t^n e^{-\varphi(t)}dt \geq e^{-\varphi(t_0)}\int_0^{t_0} t^n dt = \frac{t_0 g_n(t_0)}{n+1}.$$ (25)

On the other hand, we estimate:

$$\int_{t_0(1-\lambda_i)}^{t_0} t^n e^{-\varphi(t)}dt \leq \lambda_i t_0 g_n(t_0).$$

Finally, we use (20) and conclude:

$$\int_0^{t_0} t^n e^{-\varphi(t)}dt \leq \int_0^{t_0(1-\lambda_i)} t^n e^{-\varphi(t)}dt + \int_{t_0(1-\lambda_i)}^{t_0} t^n e^{-\varphi(t)}dt$$

$$\leq \frac{1}{e}\lambda_i t_0 g_n(t_0) + \lambda_i t_0 g_n(t_0).$$ (26)

The inequalities (25) and (26) yield the estimate $\lambda_i \geq \frac{e}{e+1}\frac{1}{n+1}$.

However, λ_o can be arbitrarily small: for any $\varepsilon > 0$ there exist a measure with continuous density (close to the one of the normalized Lebesgue measure on the unit ball) so that $\lambda_o < \varepsilon$.

Remark 4. Remark 3 shows that λ_o and λ_i are $o(1)$ when $n \to \infty$. Consequently, for sufficiently large n,

$$\int_{\frac{1}{2}t_0}^{2t_0} g_n(t)dt \approx J_n,$$

The following fact is believed to be well known (see Remark 3.4 from [3] for the best possible estimate).

Lemma 5. *For all $n \geq 2$,*

$$\frac{J_n}{J_{n-1}} \approx t_0.$$

Proof. In a view of Remark 4,

$$J_n = \int_0^\infty t^n e^{-\varphi(t)}dt \approx \int_{\frac{1}{2}t_0}^{2t_0} t^n e^{-\varphi(t)}dt$$

$$\approx t_0 \int_{\frac{1}{2}t_0}^{2t_0} t^{n-1} e^{-\varphi(t)}dt \approx t_0 J_{n-1},$$

which completes the proof of the Lemma. □

Let us consider some computable examples of γ-surface area. The first natural example to look at is the sphere of radius $R > 0$.

$$\gamma(R\mathbb{S}^n) = \frac{1}{(n+1)v_{n+1}J_n} \int_{R\mathbb{S}^n} e^{-\varphi(|y|)}d\sigma(y) = \frac{|R\mathbb{S}^n|e^{-\varphi(R)}}{(n+1)v_{n+1}J_n}$$

$$= \frac{R^n e^{-\varphi(R)}}{J_n} \approx \frac{g_n(R)}{\lambda t_0 g_n(t_0)}.$$

Since t_0 is the maximum point for $g_n(t_0)$, we notice that among all the spheres, $t_0\mathbb{S}^n$ has the maximal γ-surface area, and it is equivalent to $\frac{1}{\lambda t_0}$.

Next, for a unit vector θ we consider the half space $H_\theta = \{y : \langle y, \theta \rangle \leq 0\}$.

$$\gamma(\partial H_\theta) = \frac{1}{(n+1)v_{n+1}J_n} \int_{\mathbb{R}^n} e^{-\varphi(|y|)}dy = \frac{nv_n J_{n-1}}{(n+1)v_{n+1}J_n}. \tag{27}$$

Applying the fact that $\frac{v_n}{v_{n+1}} = \frac{\sqrt{n}}{\sqrt{2\pi}}(1 + o(1))$ together with Lemma 5 and (27), we obtain that $\gamma(H) \approx \frac{\sqrt{n}}{t_0}$.

We shall use a trick from [1] to show a rough upper bound for $\gamma(\partial Q)$.

Lemma 6. $\gamma(\partial Q) \lesssim \frac{n}{t_0}$ *for any convex set Q.*

Proof. We obtain the following integral expression for the density:

$$e^{-\varphi(|y|)} = \int_{|y|}^{\infty} \varphi'(t)e^{-\varphi(t)}dt = \int_0^{\infty} \varphi'(t)e^{-\varphi(t)}\chi_{[0,t]}(|y|)dt,$$

where $\chi_{[0,t]}$ stands for characteristic function of the interval $[0,t]$. Thus

$$\gamma(\partial Q) = \frac{1}{(n+1)v_{n+1}J_n} \int_{\partial Q}\int_0^{\infty} \varphi'(t)e^{-\varphi(t)}\chi_{[0,t]}(|y|)dtd\sigma(y)$$

$$= \frac{1}{(n+1)v_{n+1}J_n} \int_0^{\infty} \varphi'(t)e^{-\varphi(t)}|\partial Q \cap tB_2^{n+1}|dt,$$

which by can be estimated from above by

$$\frac{(n+1)v_{n+1}}{(n+1)v_{n+1}J_n} \int_0^{\infty} t^n \varphi'(t)e^{-\varphi(t)}dt, \tag{28}$$

since $Q \cap tB_2^{n+1} \subset tB_2^{n+1}$, and thus $|\partial Q \cap tB_2^{n+1}| \leq |\partial tB_2^{n+1}|$ by convexity. After integrating (28) by parts and applying Lemma 5, we get

$$\gamma(\partial Q) \leq n\frac{J_{n-1}}{J_n} \approx \frac{n}{t_0}.\ \square$$

The next lemma is an important tool in our proof.

Lemma 7. *Assume that there exists a positive μ such that*

$$\varphi\left(t_0(1+\mu)\right) - \varphi(t_0) - n\log(1+\mu) \geq \log\left(\mu\sqrt{\frac{n}{\lambda}}\right) \geq 1. \tag{29}$$

Define

$$A := (1+\mu)t_0 B_2^{n+1} \setminus \frac{t_0}{e}B_2^{n+1}.$$

Then

$$\gamma(\partial Q \setminus A) \lesssim \frac{\sqrt{n}}{t_0\sqrt{\lambda}}.$$

Proof. First, define the surface $B = \partial Q \cap \frac{t_0}{e}B_2^{n+1}$. Then,

$$\gamma(B) = \frac{1}{(n+1)v_{n+1}J_n} \int_B e^{-\varphi(|y|)} d\sigma(y) \le \frac{|B|}{(n+1)v_{n+1}J_n}$$

$$\le \frac{|\frac{t_0}{e} \mathbb{S}^n|}{(n+1)v_{n+1}J_n} \approx \frac{t_0^n}{e^n \lambda t_0 e^{-\varphi(t_0)} t_0^n} = \frac{1}{\lambda t_0} \frac{e^{\varphi(t_0)}}{e^n}, \tag{30}$$

where the equivalence follows from Lemma 4. Recalling (10), which states that $\varphi(t_0) \le n$, we estimate (30) from above by $\frac{1}{\lambda t_0}$. We recall as well, that $\frac{1}{\lambda t_0} \approx \frac{\sqrt{n}}{t_0 \sqrt{\lambda}}$, since $\lambda \gtrsim \frac{1}{n}$.

Next, let the surface $P = \partial Q \setminus (1+\mu)t_0 B_2^{n+1}$. As in Lemma 6, we make use of the estimate (28) and integrate by parts:

$$\gamma(P) \le \frac{1}{J_n} \int_{(1+\mu)t_0}^{\infty} t^n \varphi'(t) e^{-\varphi(t)} dt \lesssim$$

$$\frac{g_n((1+\mu)t_0) + n \int_{(1+\mu)t_0}^{\infty} g_{n-1}(t) dt}{\lambda t_0 g_n(t_0)}. \tag{31}$$

Lemma 3, applied with $x = \mu$ and $\psi = \log\left(\mu \sqrt{\frac{n}{\lambda}}\right)$, entails that (31) is less than

$$\frac{e^{-\psi}}{\lambda t_0} + \frac{n\mu}{\lambda t_0 \psi e^{\psi}} = \frac{1}{\lambda t_0} \times (1 + \frac{\mu n}{\psi}) e^{-\psi} \lesssim \frac{\sqrt{n}}{t_0 \sqrt{\lambda}},$$

where the last bound follows if we plug $\psi = \log\left(\mu \sqrt{\frac{n}{\lambda}}\right)$ and use the fact that $\psi \ge 1$. We also used Remark 3 which yields the fact that $\frac{1}{\lambda t_0} \lesssim \frac{\sqrt{n}}{t_0 \sqrt{\lambda}}$. $\qquad \square$

The next Lemma shows, that μ in Lemma 7 can be chosen very small.

Lemma 8.

$$\mu = \frac{\log n}{\sqrt{n}}$$

satisfies the condition of Lemma 7 for sufficiently large n.

Proof. First, notice that $\varphi((1+\mu)t_0) - \varphi(t_0) \ge \varphi'(t_0)\mu t_0 = n\mu$. Thus

$$\varphi(t_0(1+\mu)) - \varphi(t_0) - n\log(1+\mu) \ge n(\mu - \log(1+\mu)). \tag{32}$$

Plugging $\mu = \frac{\log n}{\sqrt{n}}$ into (32) and applying the Taylor approximation for logarithm, we get that the right hand side of (32) is approximately equal to

$$\sqrt{n}\log n - n\log\left(1 + \frac{\log n}{\sqrt{n}}\right) = \frac{\log^2 n}{2} + o(1). \tag{33}$$

In order to satisfy (29), we need to estimate $\log\left(\mu\sqrt{\frac{n}{\lambda}}\right)$ from above:

$$\log\left(\mu\sqrt{\frac{n}{\lambda}}\right) = \log\left(\frac{\log n}{\sqrt{n}}\sqrt{\frac{n}{\lambda}}\right) \leq \log(5n\log n), \qquad (34)$$

since $\lambda \geq \frac{e}{e+1}\frac{1}{n}$ (see Remark 3). Observing, that for all $n \geq 12$, $\log(5n\log n) \leq \frac{\log^2 n}{2} + o(1)$, we obtain the Lemma. $\qquad\qquad\square$

3 Connections to Probability

We consider a random vector X in $\mathbb{R}^{n+1}$ distributed with respect to γ. Then $|X|$ is a random variable distributed on $[0, \infty)$ with density $\frac{g_n(t)}{J_n}$. We shall use standard notation for its expectation and variance: $E = \mathbb{E}|X| = \frac{1}{J_n}\int_0^\infty tg_n(t)dt$ and

$$\sigma^2 = Var|X| = \frac{1}{J_n}\int_0^\infty (t - E)^2 g_n(t)dt. \qquad (35)$$

The next two Lemmas give an expression for the expectation and variance of $|X|$ in terms of our parameters λ and t_0, which will be used to restate Theorem 1.

Lemma 9.

$$\mathbb{E}|X| \approx t_0.$$

Proof. We write

$$\mathbb{E}|X| = [(n+1)v_{n+1}J_n]^{-1}\int_{\mathbb{R}^{n+1}} |y|e^{-\varphi(|y|)}dy$$

$$= \frac{1}{J_n}\int_0^\infty t^{n+1}e^{-\varphi(t)}dt = \frac{J_{n+1}}{J_n} \approx t_0,$$

where the last equivalence follows from Lemma 5. $\qquad\qquad\square$

Lemma 10.

$$Var|X| \approx (\lambda t_0)^2.$$

Proof. We notice first that (35) implies:

$$\int_0^\infty g_n(t)\frac{(t - E)^2}{4\sigma^2}dt = \frac{J_n}{4}. \qquad (36)$$

Subtracting (36) from the equation $J_n = \int_0^\infty g_n(t)dt$, we get

$$\int_0^\infty g_n(t)\left(1 - \frac{(t-E)^2}{4\sigma^2}\right) dt = \frac{3}{4}J_n.$$

We observe that $1 - \frac{(t-E)^2}{4\sigma^2}$ is between zero and one whenever $|t - E| \le 2\sigma$, and negative otherwise. Thus

$$\int_{E-2\sigma}^{E+2\sigma} g_n(t)dt \ge \int_{E-2\sigma}^{E+2\sigma} g_n(t)\left(1 - \frac{(t-E)^2}{4\sigma^2}\right) dt \ge \frac{3}{4}J_n. \tag{37}$$

On the other hand,

$$\int_{E-2\sigma}^{E+2\sigma} g_n(t)dt \le 4\sigma \times \max_{t\in[E-2\sigma,E+2\sigma]} g_n(t)$$

$$\le 4\sigma \max_{t\in[0,\infty)} g_n(t) = 4\sigma g_n(t_0). \tag{38}$$

Bringing together Lemma 4, (37) and (38), we get

$$4\sigma g_n(t_0) \ge \frac{3}{4}J_n \approx \lambda t_0 g_n(t_0),$$

and thus $\sigma \gtrsim \lambda t_0$.

Next, we shall obtain the reverse estimate. We note that the expression

$$\int_0^\infty (t - \tau)^2 g_n(t)dt$$

is minimal when $\tau = E$. Thus for $\tau = t_0(1 + \lambda)$ we get:

$$\sigma^2 J_n \le \int_0^\infty (t - t_0(1 + \lambda))^2 g_n(t)dt$$

$$= \int_0^{t_0(1-\lambda)} + \int_{t_0(1-\lambda)}^{t_0(1+\lambda)} + \int_{t_0(1+\lambda)}^\infty (t - t_0(1 + \lambda))^2 g_n(t)dt. \tag{39}$$

The second integral in (39) can be bounded by

$$\max_{t\in[t_0-\lambda t_0,t_0+\lambda t_0]} (t - t_0(1 + \lambda))^2 \int_{t_0(1-\lambda)}^{t_0(1+\lambda)} g_n(t)dt \lesssim (\lambda t_0)^2 J_n. \tag{40}$$

In order to estimate the third integral we apply (13) with $g(t) = g_n(t)$, $\psi = 1$ and $x = \lambda$. It implies that for all $t > t_0(1 + \lambda)$, the following holds:

$$g_n(t) \lesssim g_n(t_0)e^{-\frac{1}{\lambda t_0}(t-t_0(1+\lambda))}.$$

Thus the third integral from (39) can be estimated from above with

$$g_n(t_0) \int_{t_0(1+\lambda)}^{\infty} (t - t_0(1 + \lambda))^2 e^{-\frac{1}{\lambda t_0}(t-t_0(1+\lambda))} dt$$

$$= (\lambda t_0)^3 g_n(t_0) \int_0^{\infty} s^2 e^{-s} ds = 2(\lambda t_0)^2 \lambda t_0 g_n(t_0) \approx (\lambda t_0)^2 J_n,$$

where the last equivalence follows from Lemma 4. The first integral in (39) can be estimated similarly (with the loss of e^{-2}). Adding both of them together with (40), we obtain that

$$\sigma^2 J_n \lesssim (\lambda t_0)^2 J_n,$$

which finishes the proof.   □

Now we are ready to restate Theorem 1:

Theorem 4. *Fix $n \geq 2$. Let t_0 be the solution of $\varphi'(t)t = n - 1$. Define $\widetilde{\lambda} = \frac{\int_0^{\infty} t^{n-1}e^{-\varphi(t)}dt}{t_0^n e^{-\varphi(t_0)}}$. Then*

$$\max_{Q \in \mathscr{K}_n} \gamma(\partial Q) \approx \frac{\sqrt{n}}{\sqrt{\widetilde{\lambda} t_0}}.$$

From now on we will be after proving Theorem 4. Notice, that by Lemma 4, $\widetilde{\lambda}$ is equivalent to λ, defined in the previous section.

Remark 5. The statement of Theorem 4 becomes shorter if the measure is isotropic. We refer to [16] and [11] for the definitions and details. Here we observe only, that $t_0 = \sqrt{n}$ for isotropic measures on $\mathbb{R}^n$, and after making a change of variables $\widetilde{\varphi}(t) = \varphi(\frac{t_0}{\sqrt{n}}t)$, we get a measure $\widetilde{\gamma}$ with density $C(n)e^{-\widetilde{\varphi}(|y|)}$, which has properties similar to γ and for which the statement of Theorem 4 becomes:

$$\max_{Q \in \mathscr{K}_n} \widetilde{\gamma}(\partial Q) \approx \frac{1}{\sqrt{\widetilde{\lambda}}}.$$

Remark 6. For $p \geq 1$ we define γ_p to be a probability measure on $\mathbb{R}^n$ with density $C_{n,p}e^{-\frac{|y|^p}{p}}$ (as in (3)). In this case $\varphi(t) = \frac{t^p}{p}$, and $\varphi'(t)t = t^p$. Thus, for such measures $t_0 = (n - 1)^{\frac{1}{p}}$ (see (6) for the definition of t_0). Also, Laplace method entails, that

$$J_n = c(p)\frac{(n-1)^{\frac{n}{p}}e^{-\frac{n-1}{p}}}{\sqrt{n}} = c(p)\frac{t_0 g_n(t_0)}{\sqrt{n}}.$$

(see [14] for the details.) In a view of Lemma 4 we conclude, that in this case $\lambda \approx \frac{1}{\sqrt{n}}$. So Theorem 4 asserts, that

$$\max_{Q \in \mathcal{K}_n} \gamma_p(\partial Q) \approx C(p) n^{\frac{3}{4}-\frac{1}{p}},$$

which means that the result of [14] for the case $p \geq 1$, the result of [17] for the standard Gaussian measure, and the result from [1] are consequences of the current one.

4 Upper Bound

We will use the approach developed by Nazarov in [17]. We pick a convex set Q. The aim is to estimate $\gamma(\partial Q)$ from above. By log concavity of measure γ, we may assume that Q contains the origin: otherwise we may shift Q towards the origin so that the surface area does not decrease. Indeed, if Q does not contain the origin, let $y_0 \in Q$ be the closest point to the origin. Apply the shift $S(y) = y - y_0$. The body $S(Q)$ contains the origin in it's boundary, and also $|y - y_0| \leq |y|$ for all $y \in Q$. Since φ is increasing, we get $\varphi(|y - y_0|) \leq \varphi(|y|)$, and thus $\gamma(\partial S(Q)) \geq \gamma(\partial Q)$. Moreover, by continuity of $\varphi(t)$ we may assume that the origin is contained not in the boundary, but in the interior of Q.

Let us consider "polar" coordinate system $x = X(y,t)$ in $\mathbb{R}^{n+1}$ with $y \in \partial Q$, $t > 0$. We write

$$C_{n+1} \int_{\mathbb{R}^n} e^{-\varphi(|y|)} d\sigma(y) = C_{n+1} \int_0^\infty \int_{\partial Q} D(y,t) e^{-\varphi(|X(y,t)|)} d\sigma(y) dt,$$

where $D(y,t)$ is the Jacobian of $x \to X(y,t)$. Define

$$\xi(y) = e^{\varphi(|y|)} \int_0^\infty D(y,t) e^{-\varphi(|X(y,t)|)} dt. \tag{41}$$

Then

$$1 = C_{n+1} \int_{\partial Q} e^{-\varphi(|y|)} \xi(y) d\sigma(y),$$

and thus

$$\gamma(\partial Q) = C_{n+1} \int_{\partial Q} e^{-\varphi(|y|)} dy \leq \frac{1}{\min\limits_{y \in \partial Q} \xi(y)}. \tag{42}$$

Following [17], we shall consider two such systems.

4.1 First Coordinate System

We consider "radial" polar coordinate system $X_1(y,t) = yt$. The Jacobian $D_1(y,t) = t^n |y| \alpha$, where

$$\alpha = \alpha(y) = \cos(y, n_y), \tag{43}$$

where n_y stands for a normal vector at y. Without loss of generality we assume that n_y is defined uniquely for every $y \in \partial Q$. Rewriting (41), making a change of variables $\tau = t|y|$ and applying Lemma 5, we get:

$$\xi_1(y) := e^{\varphi(|y|)} \int_0^\infty t^n |y| \alpha e^{-\varphi(|ty|)} dt =$$

$$e^{\varphi(|y|)} \alpha |y|^{-n} J_n \gtrsim t_0 \alpha \lambda \frac{g_n(t_0)}{g_n(|y|)}. \tag{44}$$

We define $x = x(y)$ to satisfy $|y| = (1+x)t_0$ and

$$\psi(x) := \varphi((1+x)t_0) - \varphi(t_0) - n\log(1+x) = \log \frac{g_n(t_0)}{g_n((1+x)t_0)}. \tag{45}$$

Then, by (44),

$$\xi_1(y) \gtrsim t_0 \alpha \lambda e^{\psi(x)}. \tag{46}$$

Remark 7. For the sake of completeness we note, that the above formula might as well be obtained by projecting the set on the unit sphere and passing to new coordinates. Indeed, let $x = \frac{y}{|y|}$. Then the coordinate change writes as $d\sigma(y) = \frac{|y|^n}{\alpha(y)} d\sigma(x)$, and we obtain

$$\gamma(\partial Q) = [(n+1)v_{n+1} J_n]^{-1} \int_{\partial Q} e^{-\varphi(|y|)} d\sigma(y)$$

$$= [(n+1)v_{n+1} J_n]^{-1} \int_{\mathbb{S}^n} e^{-\varphi(|y|)} \frac{|y|^n}{\alpha(y)} d\sigma(x) \leq \max_{y \in \partial Q} \frac{g_n(|y|)}{\alpha(y) J_n},$$

which is equivalent to the bound we obtain from (41) and (46). This observation shows, that no volume argument of the type (42) is needed here. However, we shall need it below.

4.2 Second Coordinate System

We consider "normal" polar coordinate system $X_2(y, t) = y + tn_y$. Then $D_2(y, t) \geq 1$ for all $y \notin Q$. We write

$$\varphi(|X_2(y, t)|) = \varphi(|y + tn_y|) = \varphi\left(\sqrt{|y|^2 + t^2 + 2t|y|\alpha}\right),$$

where $\alpha = \alpha(y)$ was defined by (43). Let $\xi_2(y)$ be $\xi(y)$ from (41), corresponding to $X(y, t) = X_2(y, t)$. Then

$$\xi_2(y) \geq e^{\varphi(|y|)} \int_0^\infty e^{-\varphi\left(\sqrt{|y|^2 + t^2 + 2t|y|\alpha}\right)} dt. \tag{47}$$

Define $t_1 = t_1(y)$ to be the largest number such that:

$$\varphi\left(\sqrt{|y|^2 + t_1^2 + 2t_1|y|\alpha}\right) - \varphi(|y|) = 1.$$

Such number always exists, since the function $\varphi\left(\sqrt{|y|^2 + t^2 + 2t|y|\alpha}\right)$ is a nondecreasing continuous function of t on $[0, \infty)$, and

$$\lim_{t \to +\infty} \varphi\left(\sqrt{|y|^2 + t^2 + 2t|y|\alpha}\right) = +\infty.$$

We shall use an elementary inequality

$$\int f(x)d\mu(x) \geq a \times \mu(f(x) \geq a),$$

which holds for all positive integrable functions f. Notice, that

$$|\{t \geq 0 : e^{-\varphi\left(\sqrt{|y|^2 + t^2 + 2t|y|\alpha}\right)} \geq e^{-\varphi(|y|)-1}\}| = t_1.$$

Thus the right hand side of (47) is asymptotically bounded from below by t_1.

We define $\Lambda(t) : [0, \infty) \to [0, \infty)$ the relation

$$\varphi((1 + \Lambda(t))t) - \varphi(t) = 1. \tag{48}$$

By the definition of $t_1 = t_1(y)$,

$$\sqrt{1 + \frac{t_1^2}{|y|^2} + \frac{2t_1\alpha}{|y|}} = 1 + \Lambda(|y|).$$

We solve the quadratic equation and obtain, that for all $y \in \partial Q$

$$\xi_2(y) \geq \frac{t_1}{e} \gtrsim \frac{|y|\sqrt{\Lambda(|y|) + \Lambda^2(|y|)}}{\frac{\alpha(y)}{\sqrt{\Lambda(|y|)+\Lambda^2(|y|)}} + 1}. \tag{49}$$

4.3 Cases

We shall split the space into several annuli and estimate γ-surface area of ∂Q intersected with each annulus separately. The proof splits into several cases. Below we assume that $y \in \partial Q$.

Case 1: $|y| \leq \frac{1}{2e}t_0$ or $|y| \geq (1 + \frac{\log n}{\sqrt{n}})t_0$.

We define $\partial Q_1 = \{y \in \partial Q : |y| \leq \frac{1}{2e}t_0$ or $|y| \geq (1 + \frac{\log n}{\sqrt{n}})t_0\}$. Direct application of Lemmas 7 and 8 asserts that the desired upper bound holds for $\gamma(\partial Q_1)$ (we remark, that even though the application of Lemma 8 requires $n \geq 12$, we may apply Lemma 6 for $n \leq 12$ and select the proper constant at the end).

Case 2: $\frac{1}{2e}t_0 \leq |y| \leq (1 - \frac{1}{n})t_0$. We define $\partial Q_2 = \{y \in \partial Q : \frac{1}{2e}t_0 \leq |y| \leq (1 - \frac{1}{n})t_0\}$. Pick $y \in \partial Q_2$. We observe:

$$\varphi\left((1 - \frac{1}{n})t_0 \times (1 + \frac{1}{n})\right) - \varphi\left((1 - \frac{1}{n})t_0\right) \leq$$

$$\varphi(t_0) - \varphi\left((1 - \frac{1}{n})t_0\right) \leq \frac{t_0}{n}\varphi'(t_0) = 1. \tag{50}$$

This asserts that $\Lambda((1 - \frac{1}{n})t_0) \geq \frac{1}{n}$. We note, that $\Lambda(t)$ decreases, when t increases. Thus $\Lambda(|y|) \geq \frac{1}{n}$ for any y such that $|y| \leq (1 - \frac{1}{n})t_0$. We rewrite (49) and get the estimate

$$\xi_2(y) \gtrsim \frac{|y|}{\sqrt{n}} \times \frac{1}{\alpha\sqrt{n} + 1}. \tag{51}$$

Since $|y|$ is assumed to be asymptotically equivalent to t_0, (51) rewrites as

$$\xi_2(y) \gtrsim \frac{t_0}{\sqrt{n}} \times \frac{1}{\alpha\sqrt{n}+1}. \tag{52}$$

As for the first system, we apply a rough estimate $\psi(x) \geq 0$ and rewrite (46) as follows:

$$\xi_1(y) \gtrsim t_0 \alpha \lambda. \tag{53}$$

We consider

$$\xi(y) := \xi_1(y) + \xi_2(y) \gtrsim \tag{54}$$

$$t_0 \alpha \lambda + \frac{t_0}{\sqrt{n}} \times \frac{1}{\alpha\sqrt{n}+1}.$$

We minimize the above expression with respect to $\alpha \in [0,1]$. The minimum is attained when $\alpha = \frac{1}{\sqrt{\lambda}n}$, and thus

$$\xi(y) \gtrsim \frac{t_0\sqrt{\lambda}}{\sqrt{n}},$$

which together with (41) and (42) leads to the desired estimate for $\gamma(\partial Q_2)$.

Case 3: $(1 - \frac{1}{n})t_0 \leq |y| \leq t_0$ We define $\partial Q_3 = \{y \in \partial Q : (1 - \frac{1}{n})t_0 \leq |y| \leq t_0\}$. Along the annulus the value of $\varphi(t)$ doesn't change that much. Namely, since $\varphi(t)$ is nondecreasing and by (50),

$$\varphi\left(t_0(1 - \frac{1}{n})\right) \in [\varphi(t_0) - 1, \varphi(t_0)].$$

So for all $y \in \partial Q_3$, $\varphi(|y|) \approx \varphi(t_0)$. Thus we write

$$\gamma(\partial Q_3) = [(n+1)v_{n+1}J_n]^{-1} \int_{\partial Q_3} e^{-\varphi(|y|)} d\sigma(y) \approx$$

$$\frac{e^{-\varphi(t_0)}}{(n+1)v_{n+1}J_n} \int_{\partial Q_3} d\sigma(y) = \frac{e^{-\varphi(t_0)}}{(n+1)v_{n+1}J_n} |\partial Q_3|.$$

Since Q_3 is a convex body contained in $t_0 B_2^{n+1}$, we get $|\partial Q_3| \leq |t_0\mathbb{S}^n|$, so the above is less than

$$\frac{e^{-\varphi(t_0)}|t_0\mathbb{S}^n|}{(n+1)v_{n+1}J_n} = \frac{e^{-\varphi(t_0)}t_0^n}{J_n} \approx \frac{1}{\lambda t_0},$$

where the last equivalence is a direct application of Lemma 4. We conclude that the portion of any convex set in a very thin annulus around the maximal sphere is at least as small as the maximal sphere itself, and, in particular, smaller than our desired upper bound.

Case 4: $t_0 \leq |y| \leq (1 + \frac{\log n}{\sqrt{n}})t_0$.

This case is the hardest one. We face the problem of controlling $\Lambda(y)$: there is no way to get a proper lower bound for it unless we "step inside" the set a little bit. Fortunately, Lemma 1 shows that stepping not too far does not change γ−surface area too much. So we will be estimating $\xi_2\left(\frac{|y|}{(1+\frac{1}{n})^2}\right)$ from below, rather than $\xi_2(y)$. The key estimate in all our computation is the following Proposition.

Proposition 1. *For any y such that $|y| \in [t_0, (1 + \frac{\log n}{\sqrt{n}})t_0]$,*

$$
\Lambda\left(\frac{|y|}{(1 + \frac{1}{n})^2}\right) \gtrsim \frac{1}{n} \times \frac{1}{\psi(x) + 1 + o(1)},
$$

where $\psi(x)$ is defined by (45), $\Lambda(t)$ is defined by (48) and $|y| = (1 + x)t_0$.

Proof. We fix $|y| = (1 + x)t_0$. The parameter x in this case ranges between 0 and $\frac{\log n}{\sqrt{n}}$. Notice that by the Mean Value Theorem,

$$
\Lambda(|y|) \gtrsim \frac{1}{|y|\varphi'((1 + \Lambda(|y|))|y|)}. \tag{55}
$$

For any y such that $|y| \geq t_0$,

$$
\Lambda(|y|) \leq \frac{\varphi((1 + \Lambda(|y|))|y|) - \varphi(|y|)}{|y|\varphi'(|y|)} = \frac{1}{|y|\varphi'(|y|)} \leq \frac{1}{t_0\varphi'(t_0)} = \frac{1}{n}. \tag{56}
$$

Since $\varphi'(t)$ is nondecreasing, (55) is greater than $\frac{1}{|y|\varphi'((1+\frac{1}{n})|y|)}$. We apply (55) with $|y| = \frac{1+x}{(1+\frac{1}{n})^2}t_0$:

$$
\Lambda\left(\frac{1 + x}{(1 + \frac{1}{n})^2}t_0\right) \gtrsim \frac{1}{(1 + x)t_0\varphi'(\frac{(1+x)t_0}{1+\frac{1}{n}})} \approx \frac{1}{t_0\varphi'\left(\frac{(1+x)t_0}{1+\frac{1}{n}}\right)}, \tag{57}
$$

where the last equivalence holds in the current range of x. Next, we write that

$$
\varphi'\left(\frac{(1 + x)t_0}{1 + \frac{1}{n}}\right) \leq \frac{\varphi((1 + x)t_0) - \varphi\left(\frac{(1+x)t_0}{1+\frac{1}{n}}\right)}{(1 + x)t_0 - \frac{(1+x)t_0}{1+\frac{1}{n}}}. \tag{58}
$$

We note, that

$$(1+x)t_0 - \frac{(1+x)t_0}{1+\frac{1}{n}} = \frac{(1+x)t_0}{n+1} \approx \frac{t_0}{n} \tag{59}$$

in the current range of x. We shall invoke the function $\psi(x)$. Applying its definition (45) in the numerator and (59) in the denominator of (58), we get that (58) is equivalent to

$$\frac{\psi(x) + n\log(1+x) + \varphi(t_0) - \varphi\left(\frac{(1+x)t_0}{1+\frac{1}{n}}\right)}{\frac{t_0}{n}}. \tag{60}$$

Notice now, that by the Mean Value Theorem,

$$\varphi\left(\frac{(1+x)t_0}{1+\frac{1}{n}}\right) - \varphi(t_0) \gtrsim \varphi'(t_0)t_0\left(x - \frac{1+o(1)}{n}\right) = nx - 1 + o(1). \tag{61}$$

By (60) and (61),

$$\varphi'\left(\frac{(1+x)t_0}{1+\frac{1}{n}}\right) \gtrsim \frac{n}{t_0} \times (\psi(x) + n\log(1+x) - nx + 1 + o(1)).$$

An elementary inequality $x \geq \log(1+x)$ entails that

$$\varphi'\left(\frac{(1+x)t_0}{1+\frac{1}{n}}\right) \gtrsim \frac{n}{t_0}(\psi(x) + 1 + o(1)). \tag{62}$$

Finally, by (62) and (57) we conclude

$$\Lambda\left(\frac{1+x}{(1+\frac{1}{n})^2}t_0\right) \gtrsim \frac{1}{n} \times \frac{1}{\psi(x) + 1 + o(1)}. \qquad \Box$$

In the next few lines we use notation $\Lambda = \Lambda(\frac{|y|}{(1+\frac{1}{n})^2})$ for clarity of the presentation. We consider

$$\widetilde{\xi(y)} := \xi_1(y) + \xi_2\left(\frac{y}{(1+\frac{1}{n})^2}\right)$$

$$\gtrsim t_0\alpha\lambda e^{\psi(x)} + \frac{t_0\sqrt{\Lambda + \Lambda^2}}{2\frac{\alpha}{\sqrt{\Lambda+\Lambda^2}} + 1}. \tag{63}$$

First, we shall minimize (63) with respect to α. It is minimized whenever

$$\alpha \approx \alpha_{min} := \sqrt{\Lambda + \Lambda^2} \left(\frac{1}{\sqrt{e^{\psi(x)}\lambda}} - 1 \right).$$

Since ψ is increasing on (t_0, ∞), and due to our restrictions of the case 4, we may assume that

$$\psi(x) \leq \psi(t_0(1 + \frac{\log n}{\sqrt{n}})) = \log\left(\sqrt{\frac{n}{\lambda}} x \right)$$

$$\leq \log\left(\frac{\log n}{\sqrt{n}} \sqrt{\frac{n}{\lambda}} \right) = \log\left(\frac{\log n}{\sqrt{\lambda}} \right).$$

Consequently,

$$\sqrt{\lambda e^{\psi(x)}} \leq \sqrt[4]{\lambda} \sqrt{\log n} = o(1),$$

and thus $\alpha_{min} \approx \sqrt{\frac{\Lambda + \Lambda^2}{e^{\psi(x)}\lambda}}$. Plugging it into (63), we obtain:

$$\widetilde{\xi(y)} \gtrsim t_0 \sqrt{\lambda(\Lambda + \Lambda^2)e^{\psi(x)}}. \tag{64}$$

Finally, we apply (64) together with Proposition 1:

$$\widetilde{\xi(y)} \gtrsim \frac{t_0\sqrt{\lambda}}{\sqrt{n}} \sqrt{\frac{e^{\psi(x)}}{\psi(x) + 1 + o(1)}} \gtrsim \frac{t_0\sqrt{\lambda}}{\sqrt{n}}, \tag{65}$$

where the last inequality holds since $\psi(x)$ is positive.

4.4   Balancing for Case 4

We restrict our attention on the part of the boundary which satisfies the condition of the Case 4. Namely, denote $\partial Q_4 := \{y \in \partial Q : t_0 \leq |y| \leq (1 + \frac{\log n}{\sqrt{n}})t_0\}$.

We would like to apply (41) and (42) with $\xi(y) = \xi_1(y) + \xi_2(y)$ and finish the proof, but unfortunately we only have a lower bound for $\widetilde{\xi(y)} = \xi_1(y) + \xi_2(\frac{y}{(1+\frac{1}{n})^2})$. So we have to be a little bit more careful. We define $A = \{y \in \partial Q_4 : \xi_1(y) \geq \xi_2(\frac{y}{(1+\frac{1}{n})^2})\}$ and its compliment $B = \{y \in \partial Q_4 : \xi_1(y) < \xi_2(\frac{y}{(1+\frac{1}{n})^2})\}$. Note, that both A and B are γ−measurable, since ξ_1 and ξ_2 are Borell functions and γ is absolutely continuous with respect to Lebesgue measure. We shall apply (41) and (42) with $\xi(y) = \xi_1(y)$ on the set A and with $\xi(y) = \xi_2(y)$ on the set $\frac{1}{(1+\frac{1}{n})^2}B$.

We write that

$$1 \geq [(n+1)v_{n+1}J_n]^{-1} \int_A \int_0^\infty e^{-\varphi(X_1(y,t))} D_1(y,t) dt d\sigma(y)$$

$$= [(n+1)v_{n+1}J_n]^{-1} \int_A e^{-\varphi(|y|)} \xi_1(y) d\sigma(y) \geq \gamma(A) \min_{y \in A} \xi_1(y).$$

Thus,

$$\gamma(A) \leq \frac{1}{\min_{y \in A} \xi_1(y)}. \tag{66}$$

Similarly, we write

$$1 \geq [(n+1)v_{n+1}J_n]^{-1} \int_{\frac{1}{(1+\frac{1}{n})^2}B} \int_0^\infty e^{-\varphi(X_2(y,t))} D_2(y,t) dt d\sigma(y)$$

$$= [(n+1)v_{n+1}J_n]^{-1} \int_{\frac{1}{(1+\frac{1}{n})^2}B} e^{-\varphi(|y|)} \xi_2(y) d\sigma(y)$$

$$\geq \min_{y \in B} \xi_2 \left(\frac{y}{(1+\frac{1}{n})^2} \right) \gamma \left(\frac{1}{(1+\frac{1}{n})^2}B \right). \tag{67}$$

We apply Lemma 1 for $M = \frac{1}{(1+\frac{1}{n})^2}B$ together with (67), and conclude that

$$\gamma(B) \precsim \frac{1}{\min_{y \in B} \xi_2(\frac{y}{(1+\frac{1}{n})^2})}. \tag{68}$$

From (66) and (68) we obtain the following:

$$\gamma(\partial Q_4) = \gamma(A \cup B) \precsim \frac{1}{\min_{y \in A} \xi_1(y)} + \frac{1}{\min_{y \in B} \xi_2(\frac{y}{(1+\frac{1}{n})^2})}.$$

Invoking the definitions of the sets A and B, we notice, that

$$\min_{y \in A} \xi_1(y) \geq \frac{1}{2} \min_{y \in A} \left(\xi_1(y) + \xi_2 \left(\frac{y}{(1+\frac{1}{n})^2} \right) \right)$$

$$\geq \frac{1}{2} \min_{y \in \partial Q_4} \left(\xi_1(y) + \xi_2 \left(\frac{y}{(1+\frac{1}{n})^2} \right) \right),$$

as well as

$$\min_{y \in B} \xi_2 \left(\frac{y}{(1 + \frac{1}{n})^2} \right) \geq \frac{1}{2} \min_{y \in B} \left(\xi_1(y) + \xi_2 \left(\frac{y}{(1 + \frac{1}{n})^2} \right) \right)$$

$$\geq \frac{1}{2} \min_{y \in \partial Q_4} \left(\xi_1(y) + \xi_2 \left(\frac{y}{(1 + \frac{1}{n})^2} \right) \right),$$

since the minimum over the smaller set is greater than the minimum over the larger set. We conclude, that

$$\gamma(\partial Q_4) \lesssim \frac{1}{\min_{y \in \partial Q_4} \widetilde{\xi}(y)},$$

where $\widetilde{\xi}(y) = \xi_1(y) + \xi_2(\frac{y}{(1 + \frac{1}{n})^2})$. The desired lower bound for this quantity was obtained earlier (65), which finishes the proof of the upper bound part for Theorem 4.

5   Lower Bound

It seems impossible to construct an explicit example of a convex set Q with $\gamma(\partial Q) \approx \frac{\sqrt{n}}{\sqrt{\lambda t_0}}$. So we provide a probabilistic construction similar to the one in [17]. Namely, we shall consider a random polytope circumscribed around a sphere of a certain radius. The radius of the sphere and the number of faces shall be chosen so that most of the time $\alpha(y) = \cos(y, n_y) \approx \alpha_{min}$ which appears in the proof of the upper bound, and so that large enough portion of the polytope falls close to the maximal sphere $t_0 \mathbb{S}^n$. As it was shown in Lemma 4, a lot of the measure is concentrated in the thin annulus around $t_0 \mathbb{S}^n$; more precise results describing the decay outside of the annulus were obtained in [11] (Theorem 1.4) and [12] (Theorem 4.4). For simplicity of the calculations, we only look at the portion of the polytope in that annulus, and it turns out to be enough for the lower bound.

We consider N uniformly distributed random vectors $x_i \in \mathbb{S}^n$. Let ϱ and W be positive parameters, let $r = t_0 + w$, where $w \in [-W, W]$. For the purposes of the calculation we assume from the beginning that $W, \varrho \leq \frac{t_0}{20}$. Consider a random polytope Q in $\mathbb{R}^{n+1}$, defined as follows:

$$Q = \{ x \in \mathbb{R}^{n+1} : \langle x, x_i \rangle \leq \varrho, \ \forall i = 1, \ldots, N \}.$$

Passing to the polar coordinates in $H_i = \{ x : \langle x, x_i \rangle = \varrho \}$, we estimate the surface area of the half space $A_i = \{ x : \langle x, x_i \rangle \leq \varrho \}$:

$$\gamma(\partial A_i) = \frac{1}{(n+1)v_{n+1}J_n} \int_{\mathbb{R}^n} e^{-\varphi(\sqrt{|y|^2+\varrho^2})} dy$$

$$\gtrsim \frac{1}{(n+1)v_{n+1}J_n}(n+1)v_n \int_{t_0-W}^{t_0+W} e^{-\varphi(t)}(t^2-\varrho^2)^{\frac{n-1}{2}} \frac{\sqrt{t^2-\varrho^2}}{t} dt$$

$$\gtrsim \frac{\sqrt{n}}{J_n}\left(1 - \frac{\varrho^2}{(t_0-W)^2}\right)^{\frac{n}{2}} \int_{t_0-W}^{t_0+W} e^{-\varphi(t)}t^{n-1} dt.$$

Thus the expectation $\mathbb{E}\gamma(\partial Q)$ can be estimated from below by

$$N\frac{\sqrt{n}}{J_n}\left(1 - \frac{\varrho^2}{(t_0-W)^2}\right)^{\frac{n}{2}} \int_{t_0-W}^{t_0+W} e^{-\varphi(t)}t^{n-1}(1-p(t))^{N-1}dt, \tag{69}$$

where $p(t)$ is the probability that the fixed point on the sphere of radius t is separated from the origin by the hyperplane H_i.

As in [17], we use the formula for a surface area of a body of revolution to obtain the formula for $p(r)$:

$$p(r) = \left(\int_{-r}^{r}(1 - \frac{t^2}{r^2})^{\frac{n-2}{2}}dt\right)^{-1} \int_{\varrho}^{r}(1 - \frac{t^2}{r^2})^{\frac{n-2}{2}}dt. \tag{70}$$

By Laplace method, the first integral is approximately equal to $\frac{r}{\sqrt{n}}$. Thus, after the change of variables $x = \frac{t}{r}$, we obtain

$$p(r) \approx \frac{\sqrt{n}}{r}r \int_{\frac{\varrho}{r}}^{1}(1 - x^2)^{\frac{n-2}{2}}dx = \sqrt{n} \int_{\frac{\varrho}{r}}^{1}(1 - x^2)^{\frac{n-2}{2}} dx. \tag{71}$$

Notice, that for any $z \in (0, 1)$,

$$\int_{z}^{1}(1 - t^2)^m dt \leq \frac{2}{z}\int_{0}^{1-z^2} s^m ds = \frac{2}{z(m+1)}(1 - z^2)^{m+1}. \tag{72}$$

By (71), (72) applied with $z = \frac{\varrho}{r}$ and $m = \frac{n-2}{2}$, and the fact that $r \approx t_0$,

$$p(r) \lesssim \frac{r}{\sqrt{n}\varrho}\left(1 - \frac{\varrho^2}{r^2}\right)^{\frac{n}{2}} \lesssim \frac{t_0}{\sqrt{n}\varrho}\left(1 - \frac{\varrho^2}{(t_0+W)^2}\right)^{\frac{n}{2}} \tag{73}$$

for all $r \in [t_0 - W, t_0 + W]$. At this point we choose

$$N = \frac{\sqrt{n}\varrho}{t_0}(1 - \frac{\varrho^2}{(t_0+W)^2})^{-\frac{n}{2}}.$$

Observe that $(1 - p(r))^{N-1} \lesssim (1 - \frac{1}{N})^N \leq e^{-1}$. Applying the above together with (69) and (73), we get:

$$\mathbb{E}(\gamma(\partial Q)) \gtrsim \frac{n\varrho}{J_n t_0} \left(\frac{1 - \frac{\varrho^2}{(t_0 - W)^2}}{1 - \frac{\varrho^2}{(t_0 + W)^2}} \right)^{\frac{n}{2}} \int_{t_0 - W}^{t_0 + W} e^{-\varphi(t)} t^{n-1} dt. \tag{74}$$

Let us now plug $W = \lambda t_0$. By Lemmas 4 and 5 and Remark 2, we observe, that $J_{n-1} \approx \int_{t_0 - \lambda t_0}^{t_0 + \lambda t_0} t^{n-1} e^{-\varphi(t)}$. Thus,

$$\mathbb{E}(\gamma(\partial Q)) \gtrsim \frac{n\varrho}{t_0} \times \frac{J_{n-1}}{J_n} \left(\frac{1 - \frac{\varrho^2}{(t_0 - W)^2}}{1 - \frac{\varrho^2}{(t_0 + W)^2}} \right)^{\frac{n}{2}} \approx \frac{n\varrho}{t_0^2} \left(\frac{1 - \frac{\varrho^2}{(t_0 - W)^2}}{1 - \frac{\varrho^2}{(t_0 + W)^2}} \right)^{\frac{n}{2}}.$$

We plug $\varrho = \frac{1}{5\sqrt{\lambda n}} t_0$. Then

$$\frac{1 - \frac{\varrho^2}{(t_0 - W)^2}}{1 - \frac{\varrho^2}{(t_0 + W)^2}} \geq 1 - \frac{1}{n},$$

which implies that

$$\mathbb{E}(\gamma(\partial Q)) \gtrsim \frac{\sqrt{n}}{\sqrt{\lambda} t_0}.$$

This finishes the lower bound part of the Theorem 4. $\square$

6   Final Remarks

As was discussed in Sect. 3, Theorem 4 entails Theorem 1. Its conclusion can be understood for any measure which has at least two bounded moments, so it is interesting to explore sufficiency of our conditions, i.e. spherical invariance and log concavity. We shall consider some examples of non rotation invariant or non log concave measures, for which the conclusion of Theorem 1 does not hold.

Example 1. Consider Lebesgue measure concentrated on the cube $[-\frac{1}{2}, \frac{1}{2}]^n$. Due to convexity, the set of maximal surface area for this measure is the cube $[-\frac{1}{2}, \frac{1}{2}]^n$ itself. Its surface area is $2n$. However, $\mathbb{E}|X| \approx \sqrt{n}$ and $Var|X| \approx 1$ (see [9] for the proof), so if Theorem 1 was true, it would give $n^{\frac{1}{4}}$ as a maximal surface area. Thus there is no hope for Theorem 1 to be true for all log concave measures. The isotropicity assumption would not change anything due to the homogeneity of Theorem 1.

Example 2. Pick $\varepsilon \ll \frac{1}{n}$. We consider a rotation invariant non log concave measure γ_ε. Let its density be

$$f(y) = c_n \begin{cases} 0 & \text{if } |y| \in [0, 1 - \varepsilon] \cup [1, \infty) \\ 1 & \text{if } |y| \in (1 - \varepsilon, 1). \end{cases}$$

The normalizing constant

$$c_n = v_{n+1}(1 - (1 - \varepsilon)^{n+1}) \le (n + 1)\varepsilon v_{n+1}.$$

For a random variable X with density f we compute

$$\mathbb{E}|X| = \frac{1 - \frac{n+1}{2}\varepsilon}{1 - \frac{n}{2}\varepsilon} + o((n + 1)\varepsilon) \approx 1$$

and

$$Var|X| \approx \frac{\varepsilon^2}{4}.$$

Thus if Theorem 1 was true, the maximal surface area would be of order $\frac{\sqrt{n}}{\sqrt{\varepsilon}}$. However,

$$\gamma_\varepsilon(\mathbb{S}^n) \ge \frac{1}{n\varepsilon^2},$$

which is greater than $\frac{\sqrt{n}}{\sqrt{\varepsilon}}$ for $\varepsilon \ll \frac{1}{n}$.

Example 2 shows, that for any dimension n there exist a rotation invariant measure for which the conclusion of Theorem 1 fails, but it is hard to find an example of a density function which would serve all sufficiently large dimensions at once. It suggests the following conjecture.

Conjecture 1. Fix any real-valued function $\varphi(t)$ on the positive semi-axes. Then there exists a positive constant C_φ, depending on the function $\varphi(t)$, such that for all $n \ge C_\varphi$,

$$\max_{Q \in \mathcal{K}_n} \gamma(\partial Q) \approx \frac{\sqrt{n}}{\sqrt{\mathbb{E}|X|} \sqrt[4]{Var|X|}},$$

where X is a random vector on $\mathbb{R}^n$ distributed with respect to the density $e^{-\varphi(|X|)}$.

Acknowledgements I would like to thank Artem Zvavitch and Fedor Nazarov for introducing me to the subject, suggesting me this problem and for extremely helpful and fruitful discussions. I would also like to thank Benjamin Jaye for a number of useful remarks. The author is supported in part by U.S. National Science Foundation grant DMS-1101636.

Appendix

In this Appendix we provide a technical lemma which is believed to be well known to the specialists. See [10] for the proof of the same statement in the case of Standard Gaussian Measure and polynomial level sets.

Lemma 11. *Let γ be a probability measure on $\mathbb{R}^{n+1}$ with a continuous density $f(y)$. Then, for any convex set Q in $\mathbb{R}^{n+1}$,*

$$\int_{\partial Q} f(y)d\sigma(y) = \lim_{\varepsilon \to 0} \frac{\gamma(Q + \varepsilon B_2^{n+1}) - \gamma(Q)}{\varepsilon},$$

where, as before, $d\sigma(y)$ stands for Lebesgue surface measure.

Proof. For a convex set Q in $\mathbb{R}^{n+1}$ and $\varepsilon > 0$, we introduce the notation $A_{Q,\varepsilon} = (Q + \varepsilon B_2^{n+1}) \setminus Q$. We remark, that the normal vector n_y is well defined almost everywhere for $y \in \partial Q$ if Q is convex. So the function $f(y + tn_y)$ is defined almost everywhere on ∂Q. We shall apply the second Nazarov's system (40), which we used in the proof of the main result. By convexity of Q,

$$\gamma(A_{Q,\varepsilon}) \geq \int_{\partial Q} \int_0^\varepsilon f(y + tn_y)dtd\sigma(y), \qquad (75)$$

where the integration is understood in the Lebesgue sense. By Lebesgue Differentiation Theorem, for every $y \in \partial Q$ such that n_y is defined,

$$\lim_{\varepsilon \to 0} \frac{1}{\varepsilon} \int_0^\varepsilon f(y + tn_y)dt = f(y).$$

Consequently,

$$\lim_{\varepsilon \to 0} \frac{1}{\varepsilon} \gamma(A_{Q,\varepsilon}) \geq \int_{\partial Q} f(y)d\sigma(y). \qquad (76)$$

On the other hand, we compare the measure of our annulus to the surface area of $Q + \varepsilon B_2^{n+1}$.

We note that for any $\varepsilon > 0$ and $x \in A_{Q,\varepsilon}$ we may find $y \in \partial \left(Q + \varepsilon B_2^{n+1}\right)$ and $t \in [0, \varepsilon]$ so that $x = y - tn_y$.

To see this, inscribe a ball centred at x into $Q + \varepsilon B_2^{n+1}$ and chose y to be a contact point of the ball and $\partial(Q + \varepsilon B_2^{n+1})$. We see, that $|x - y| \leq \varepsilon$, since

$$dist(x, \partial(Q + \varepsilon B_2^{n+1})) \leq \varepsilon.$$

We write

$$\gamma(A_{Q,\varepsilon}) \leq \int_{\partial(Q+\varepsilon B_2^{n+1})} \int_0^\varepsilon f(y - tn_y) dt d\sigma(y). \tag{77}$$

We observe, that

$$\lim_{\varepsilon \to 0} \frac{1}{\varepsilon} \gamma(A_{Q,\varepsilon}) \leq \int_{\partial Q} f(y) d\sigma(y). \tag{78}$$

Finally, (76) and (78) entail the conclusion of the Lemma.   □

References

1. K. Ball, The reverse isoperimetric problem for the Gaussian measure. Discrete Comput. Geom. **10**, 411–420 (1993)
2. S.G. Bobkov, Spectral gap and concentration for some spherically symmetric probability measures. Lect. Notes Math. **1807**, 37–43 (2003)
3. S.G. Bobkov, Gaussian concentration for a class of spherically invariant measures. J. Math. Sci. **167**(3), 326–339 (2010)
4. S.G. Bobkov, Convex bodies and norms associated to convex measures. Probab. Theory Relat. Fields **147**(3), 303–332 (2010)
5. C. Borell, The Brunn-Minkowski inequality in Gauss spaces. Invent. Math **30**, 207–216 (1975)
6. H. Bray, F. Morgan, An isoperimetric comparison theorem for Schwarzschild space and other manifolds. Proc. Am. Math. Soc. **130**, 1467–1472 (2002)
7. D. Cordero-Erausquin, A.M. Fradelizi, B. Maurey, The (B) conjecture for the Gaussian measure of dilates of symmetric convex sets and related problems. J. Funct. Anal. **214**, 410–427 (2004)
8. N.G. De Bruijn, *Asymptotic Methods in Analysis* (Dover, New York, 2010), 200 pp.
9. A. Giannopolus, *Notes on Isotropic Convex Bodies*, Warsaw (2003)
10. D.M. Kane, The Gaussian surface area and noise sensitivity of degree-D polynomial threshold functions, in *IEEE 25th Annual Conference on Computational Complexity (CCC)* (2010), pp. 205–210
11. B. Klartag, A central limit theorem for convex sets. Invent. Math. **168**, 91–131 (2007)
12. B. Klartag, Power-law estimates for the central limit theorem for convex sets. J. Funct. Anal. **245**, 284–310 (2007)
13. B. Klartag, V.D. Milman, Geometry of log-concave functions and measures. Geom. Dedicata **112**, 169–182 (2005)
14. G.V. Livshyts, Maxima surface area of a convex set in R^n with respect to exponential rotation invariant measures. J. Math. Anal. Appl. **404**, 231–238 (2013)
15. Q. Maurmann, F. Morgan, Isoperimetric comparison theorems for manifolds with density. Calc. Var. **36**, 1–5 (2009)
16. V.D. Milman, A. Pajor, Isotropic position and interia ellipsoids and zonoids of the unit ball of a normed n−dmensional space, in *Geometric Aspects of Functional Analysis*. Lecture Notes in Mathematics vol. 1376 (1989), pp. 64–104
17. F.L. Nazarov, On the maximal perimeter of a convex set in $\mathbb{R}^n$ with respect to Gaussian measure. Geom. Aspects Funct. Anal. **1807**, 169–187 (2003)
18. V.N. Sudakov, B.S. Tsirel'son, Extremal properties of half-spaces for spherically invariant measures. Problems in the theory of probability distributions, II. Zap. Nauch. Leningrad Otdel. Mat. Inst. Steklov **41**, 14–24 (1974) (in Russian)

On the Equivalence of Modes of Convergence for Log-Concave Measures

Elizabeth S. Meckes and Mark W. Meckes

Abstract An important theme in recent work in asymptotic geometric analysis is that many classical implications between different types of geometric or functional inequalities can be reversed in the presence of convexity assumptions. In this note, we explore the extent to which different notions of distance between probability measures are comparable for log-concave distributions. Our results imply that weak convergence of isotropic log-concave distributions is equivalent to convergence in total variation, and is further equivalent to convergence in relative entropy when the limit measure is Gaussian.

1 Introduction and Statements of Results

An important theme in recent work in asymptotic geometric analysis is that many classical implications between different types of geometric or functional inequalities can be reversed in the presence of convexity. A particularly striking recent example is the work of E. Milman [11–13], showing for example that, on a Riemannian manifold equipped with a probability measure satisfying a convexity assumption, the existence of a Cheeger inequality, a Poincaré inequality, and exponential concentration of Lipschitz functions are all equivalent. Important earlier examples of this theme are C. Borell's 1974 proof of reverse Hölder inequalities for log-concave measures [4], and K. Ball's 1991 proof of a reverse isoperimetric inequality for convex bodies [1].

In this note, we explore the extent to which different notions of distance between probability measures are comparable in the presence of a convexity assumption. Specifically, we consider **log-concave** probability measures; that is, Borel probability measures μ on $\mathbb{R}^n$ such that for all nonempty compact sets $A, B \subseteq \mathbb{R}^n$ and every $\lambda \in (0, 1)$,

$$\mu(\lambda A + (1 - \lambda)B) \geq \mu(A)^\lambda \mu(B)^{1-\lambda}.$$

E.S. Meckes • M.W. Meckes (✉)
Department of Mathematics, Case Western Reserve University, 10900 Euclid Ave., Cleveland, OH 44106, USA
e-mail: elizabeth.meckes@case.edu; mark.meckes@case.edu

© Springer International Publishing Switzerland 2014
B. Klartag, E. Milman (eds.), *Geometric Aspects of Functional Analysis*,
Lecture Notes in Mathematics 2116, DOI 10.1007/978-3-319-09477-9__24

We moreover consider only those log-concave probability measures μ on $\mathbb{R}^n$ which are **isotropic**, meaning that if $X \sim \mu$ then

$$\mathbb{E}X = 0 \qquad \text{and} \qquad \mathbb{E}XX^T = I_n.$$

The following distances between probability measures μ and ν on $\mathbb{R}^n$ appear below.

1. The **total variation distance** is defined by

$$d_{TV}(\mu, \nu) := 2 \sup_{A \subseteq \mathbb{R}^n} |\mu(A) - \nu(A)|,$$

where the supremum is over Borel measurable sets.
2. The **bounded Lipschitz distance** is defined by

$$d_{BL}(\mu, \nu) := \sup_{\|g\|_{BL} \leq 1} \left| \int g \, d\mu - \int g \, d\nu \right|,$$

where the bounded-Lipschitz norm $\|g\|_{BL}$ of $g : \mathbb{R}^n \to \mathbb{R}$ is defined by

$$\|g\|_{BL} := \max \left\{ \|g\|_\infty, \sup_{x \neq y} \frac{|g(x) - g(y)|}{\|x - y\|} \right\}$$

and $\|\cdot\|$ denotes the standard Euclidean norm on $\mathbb{R}^n$. The bounded-Lipschitz distance is a metric for the weak topology on probability measures (see, e.g., [6, Theorem 11.3.3]).
3. The L_p **Wasserstein distance** for $p \geq 1$ is defined by

$$W_p(\mu, \nu) := \inf_\pi \left[\int \|x - y\|^p \, d\pi(x, y) \right]^{\frac{1}{p}},$$

where the infimum is over couplings π of μ and ν; that is, probability measures π on $\mathbb{R}^{2n}$ such that $\pi(A \times \mathbb{R}^n) = \mu(A)$ and $\pi(\mathbb{R}^n \times B) = \nu(B)$. The L_p Wasserstein distance is a metric for the topology of weak convergence plus convergence of moments of order p or less. (See [15, Sect. 6] for a proof of this fact, and a lengthy discussion of the many fine mathematicians after whom this distance could reasonably be named.)
4. If μ is absolutely continuous with respect to ν, the **relative entropy**, or **Kullback–Leibler divergence** is defined by

$$H(\mu \mid \nu) := \int \left(\frac{d\mu}{d\nu} \right) \log \left(\frac{d\mu}{d\nu} \right) d\nu = \int \log \left(\frac{d\mu}{d\nu} \right) d\mu.$$

It is a classical fact that for any probability measures μ and ν on $\mathbb{R}^n$,

$$d_{BL}(\mu, \nu) \leq d_{TV}(\mu, \nu). \tag{1}$$

This follows from a dual formulation of total variation distance: the Riesz representation theorem implies that

$$d_{TV}(\mu, \nu) = \sup \left\{ \left| \int g \, d\mu - \int g \, d\nu \right| : g \in C(\mathbb{R}^n), \ \|g\|_\infty \leq 1 \right\}. \tag{2}$$

In the case that μ and ν are log-concave, there is the following complementary inequality.

Proposition 1. *Let μ and ν be log-concave isotropic probability measures on $\mathbb{R}^n$. Then*

$$d_{TV}(\mu, \nu) \leq C \sqrt{n d_{BL}(\mu, \nu)}.$$

In this result and below, C, c, etc. denote positive constants which are independent of n, μ, and ν, and whose values may change from one appearance to the next.

In the special case in which $n = 1$ and $\nu = \gamma_1$, Brehm, Hinow, Vogt and Voigt proved a similar comparison between total variation distance and Kolmogorov distance d_K.

Proposition 2 ([5, Theorem 3.3]). *Let μ be a log-concave measure on $\mathbb{R}$. Then*

$$d_{TV}(\mu, \gamma_1) \leq C \sqrt{\max\{1, \log(1/d_K(\mu, \gamma_1))\} \, d_K(\mu, \gamma_1)}.$$

Together with (1), Proposition 1 implies the following.

Corollary 3. *On the family of isotropic log-concave probability measures on $\mathbb{R}^n$, the topologies of weak convergence and of total variation coincide.*

Corollary 3 will probably be unsurprising to experts, but we have not seen it stated in the literature.

Proposition 1 and Corollary 3 are false without the assumption of isotropicity. For example, a sequence of nondegenerate Gaussian measures $\{\mu_k\}_{k \in \mathbb{N}}$ on $\mathbb{R}^n$ may weakly approach a Gaussian measure μ supported on a lower-dimensional subspace, but $d_{TV}(\mu_k, \mu) = 2$ for every k. It may be possible to extend Corollary 3 to a class of log-concave probability measures with, say, a nontrivial uniform lower bound on the smallest eigenvalue of the covariance matrix, but we will not pursue this here.

The Kantorovitch duality theorem (see [15, Theorem 5.10]) gives a dual formulation of the L_1 Wasserstein distance similar to the formulation of total variation distance in (2):

$$W_1(\mu, \nu) = \sup_g \left| \int g \, d\mu - \int g \, d\nu \right|,$$

where the supremum is over 1-Lipschitz functions $g : \mathbb{R}^n \to \mathbb{R}$. An immediate consequence is that for any probability measures μ and ν,

$$d_{BL}(\mu, \nu) \leq W_1(\mu, \nu).$$

The following complementary inequality holds in the log-concave case.

Proposition 4. *Let μ and ν be log-concave isotropic probability measures on $\mathbb{R}^n$. Then*

$$W_1(\mu, \nu) \leq C \max \left\{ \sqrt{n}, \log \left(\frac{\sqrt{n}}{d_{BL}(\mu, \nu)} \right) \right\} d_{BL}(\mu, \nu). \tag{3}$$

The following graph of $f(x) = \max \left\{ 1, \log \left(\frac{1}{x} \right) \right\} x$ may be helpful in visualizing the bounds in Proposition 4 and the results below.

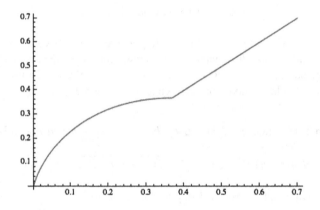

In particular, when d_{BL} is moderate, we simply have $W_1 \leq C \sqrt{n} d_{BL}$. When d_{BL} is small, the right hand side of (3) is not quite linear in d_{BL}, but is $o\left(n^{\varepsilon/2} d_{BL}^{1-\varepsilon}\right)$ for each $\varepsilon > 0$.

From Hölder's inequality, it is immediate that if $p \leq q$, then $W_p(\mu, \nu) \leq W_q(\mu, \nu)$. In the log-concave case, we have the following.

Proposition 5. *Let μ and ν be isotropic log-concave probability measures on $\mathbb{R}^n$ and let $1 \leq p < q$. Then*

$$W_q(\mu, \nu)^q \leq C \left(\max \left\{ \sqrt{n}, \log \left(\frac{(c \max\{q, \sqrt{n}\})^q}{W_p(\mu, \nu)^p} \right) \right\} \right)^{q-p} W_p(\mu, \nu)^p.$$

Because the bounded-Lipschitz distance metrizes the weak topology, and convergence in L_p Wasserstein distance implies convergence of moments of order smaller than p, Propositions 4 and 5 imply the following.

Corollary 6. *Let* μ, $\{\mu_k\}_{k\in\mathbb{N}}$ *be isotropic log-concave probability measures on* $\mathbb{R}^n$ *such that* $\mu_k \to \mu$ *weakly. Then all moments of the* μ_k *converge to the corresponding moments of* μ.

The following, known as the Csiszár–Kullback–Pinsker inequality, holds for any probability measures μ and ν:

$$d_{TV}(\mu,\nu) \leq \sqrt{2H(\mu \mid \nu)}. \tag{4}$$

(See [3] for a proof, generalizations, and original references.) Unlike the other notions of distance considered above, $H(\cdot \mid \cdot)$ is not a metric, and $H(\mu \mid \nu)$ can only be finite if μ is absolutely continuous with respect to ν. Nevertheless, it is frequently used to quantify convergence; (4) shows that convergence in relative entropy is stronger than convergence in total variation. Convergence in relative entropy is particularly useful in quantifying convergence to the Gaussian distribution, and it is in that setting that (4) can be essentially reversed under an assumption of log-concavity.

Proposition 7. *Let* μ *be an isotropic log-concave probability measure on* $\mathbb{R}^n$, *and let* γ_n *denote the standard Gaussian distribution on* $\mathbb{R}^n$. *Then*

$$H(\mu \mid \gamma_n) \leq C \max\left\{\log^2\left(\frac{n}{d_{TV}(\mu,\gamma_n)}\right), n\log(n+1)\right\} d_{TV}(\mu,\gamma_n).$$

The proof of Proposition 7 uses a rough bound on the isotropic constant $L_f = \|f\|_\infty^{1/n}$ of the density f of μ. Better estimates are available but only result in a change in the absolute constants in our bound. In the case that the isotropic constant is bounded independent of n (e.g. if μ is the uniform measure on an unconditional convex body, or if the hyperplane conjecture is proved), then the bound above can be improved slightly to

$$H(\mu \mid \gamma_n) \leq C \max\left\{\log^2\left(\frac{n}{d_{TV}(\mu,\gamma_n)}\right), n\right\} d_{TV}(\mu,\gamma_n).$$

Corollary 8. *Let* $\{\mu_k\}_{k\in\mathbb{N}}$ *be isotropic log-concave probability measures on* $\mathbb{R}^n$. *The following are equivalent:*

1. $\mu_k \to \gamma_n$ *weakly.*
2. $\mu_k \to \gamma_n$ *in total variation.*
3. $H(\mu_k \mid \gamma_n) \to 0$.

It is worth noting that Proposition 7 implies that B. Klartag's central limit theorem for convex bodies (proved in [8, 9] in total variation) also holds in the *a priori* stronger sense of entropy, with a polynomial rate of convergence.

2   Proofs of the Results

The proof of Proposition 1 uses the following deconvolution result of R. Eldan and B. Klartag.

Lemma 9 ([7, Proposition 10]). *Suppose that f is the density of an isotropic log-concave probability measure on $\mathbb{R}^n$, and for $t > 0$ define*

$$\varphi_t(x) = \frac{1}{(2\pi t^2)^{n/2}} e^{-\|x\|^2/2t^2}.$$

Then

$$\|f - f * \varphi_t\|_1 \le cnt.$$

Proof (Proof of Proposition 1). Let $g \in C(\mathbb{R}^n)$ with $\|g\|_\infty \le 1$. For $t > 0$, let $g_t = g * \varphi_t$, where φ_t is as in Lemma 9. It follows from Young's inequality that $\|g_t\|_\infty \le 1$ and that g_t is $1/t$-Lipschitz. We have

$$\left| \int g \, d\mu - \int g \, dv \right| \le \left| \int (g - g_t) \, d\mu \right| + \left| \int g_t \, d\mu - \int g_t \, dv \right| + \left| \int (g_t - g) \, dv \right|.$$

It is a classical fact due to C. Borell [4] that a log-concave probability measures which is not supported on a proper affine subspace of $\mathbb{R}^n$ has a density. If f is the density of μ, then by Lemma 9,

$$\left| \int (g - g_t) \, d\mu \right| = \left| \int g(f - f * \varphi_t) \right| \le \|f - f * \varphi_t\|_1 \le cnt,$$

and

$$\left| \int (g - g_t) \, dv \right| \le cnt$$

similarly. Furthermore,

$$\left| \int g_t \, d\mu - \int g_t \, dv \right| \le d_{BL}(\mu, v) \|g_t\|_{BL} \le d_{BL}(\mu, v) \max\{1, 1/t\}.$$

Combining the above estimates and taking the supremum over g yields

$$d_{TV}(\mu, v) \le d_{BL}(\mu, v) \max\{1, 1/t\} + cnt$$

for every $t > 0$. The proposition follows by picking $t = \sqrt{d_{BL}(\mu, v)/2n} \le 1$.

The remaining propositions all depend in part on the following deep concentration result due to G. Paouris.

Proposition 10 ([14]). *Let X be an isotropic log-concave random vector in $\mathbb{R}^n$. Then*

$$\mathbb{P}\left[\|X\| \geq R\right] \leq e^{-cR}$$

for every $R \geq C\sqrt{n}$, and

$$\left(\mathbb{E}\,\|X\|^p\right)^{1/p} \leq C\max\{\sqrt{n}, p\}$$

for every $p \geq 1$.

The following simple optimization lemma will also be used in the remaining proofs.

Lemma 11. *Given $A, B, M, k > 0$,*

$$\inf_{t \geq M}\left(At^k + Be^{-t}\right) \leq A\left(1 + (\max\{M, \log(B/A)\})^k\right).$$

Proof. Set $t = \max\{M, \log(B/A)\}$.

Proof (Proof of Proposition 4). Let $g : \mathbb{R}^n \to \mathbb{R}$ be 1-Lipschitz and without loss of generality assume that $g(0) = 0$, so that $|g(x)| \leq \|x\|$. For $R > 0$ define

$$g_R(x) = \begin{cases} -R & \text{if } g(x) < -R, \\ g(x) & \text{if } -R \leq g(x) \leq R, \\ R & \text{if } g(x) > R, \end{cases}$$

and observe that $\|g_R\|_{BL} \leq \max\{1, R\}$. Let $X \sim \mu$ and $Y \sim \nu$. Then

$$|\mathbb{E}g(X) - \mathbb{E}g(Y)| \leq \mathbb{E}\,|g_R(X) - g_R(Y)| + \mathbb{E}\,|g(X) - g_R(X)| + \mathbb{E}\,|g(Y) - g_R(Y)|$$

$$\leq \max\{1, R\}d_{BL}(\mu, \nu) + \mathbb{E}\,\|X\|\,\mathbb{1}_{\|X\| \geq R} + \mathbb{E}\,\|Y\|\,\mathbb{1}_{\|Y\| \geq R}.$$

By the Cauchy–Schwarz inequality and Proposition 10,

$$\mathbb{E}\,\|X\|\,\mathbb{1}_{\|X\| \geq R} \leq \sqrt{n\mathbb{P}\left[\|X\| \geq R\right]} \leq \sqrt{n}e^{-cR}$$

for $R \geq C\sqrt{n}$, and the last term is bounded similarly. Combining the above estimates and taking the supremum over g yields

$$W_1(\mu, \nu) \leq \max\{1, R\}d_{BL}(\mu, \nu) + 2\sqrt{n}e^{-cR}$$

for every $R \geq C\sqrt{n}$. The proposition follows using Lemma 11.

Proof (Proof of Proposition 5). Let (X, Y) be a coupling of μ and ν on $\mathbb{R}^n \times \mathbb{R}^n$. Then for each $R > 0$,

$$\mathbb{E} \, \|X - Y\|^q \leq R^{q-p} \mathbb{E} \left[\|X - Y\|^p \, \mathbb{1}_{\|X-Y\| \leq R} \right] + \sqrt{\mathbb{P} \left[\|X - Y\| \geq R \right] \mathbb{E} \, \|X - Y\|^{2q}}.$$

By Proposition 10,

$$\mathbb{P} \left[\|X - Y\| \geq R \right] \leq \mathbb{P} \left[\|X\| \geq R/2 \right] + \mathbb{P} \left[\|Y\| \geq R/2 \right] \leq e^{-cR}$$

when $R \geq C \sqrt{n}$, and

$$\left(\mathbb{E} \, \|X - Y\|^{2q} \right)^{1/2q} \leq \left(\mathbb{E} \, \|X\|^{2q} \right)^{1/2q} + \left(\mathbb{E} \, \|Y\|^{2q} \right)^{1/2q} \leq C \max\{q, \sqrt{n}\},$$

so that

$$\mathbb{E} \, \|X - Y\|^q \leq R^{q-p} \mathbb{E} \, \|X - Y\|^p + \left(C \max\{q, \sqrt{n}\} \right)^q e^{-cR}$$

for every $R \geq C \sqrt{n}$. Taking the infimum over couplings and then applying Lemma 11 completes the proof.

The proof of Proposition 7 uses the following variance bound which follows from a more general concentration inequality due to Bobkov and Madiman.

Lemma 12 (See [2, Theorem 1.1]). *Suppose that μ is an isotropic log-concave probability measure on $\mathbb{R}^n$ with density f, and let $Y \sim \mu$. Then*

$$\mathrm{Var}\big(\log f(Y)\big) \leq Cn.$$

Proof (Proof of Proposition 7). Let f be the density of μ, and let $\varphi(x) = (2\pi)^{-n/2} e^{-\|x\|^2/2}$ be the density of γ_n. Let $Z \sim \gamma_n$, $Y \sim \mu$, $X = \frac{f(Z)}{\varphi(Z)}$, and $W = \frac{f(Y)}{\varphi(Y)}$. Then

$$H(\mu \mid \gamma_n) = \mathbb{E} X \log X.$$

In general, if μ and ν have densities f_μ and f_ν, it is an easy exercise to show that $d_{TV}(\mu, \nu) = \int |f_\mu - f_\nu|$; from this, it follows that

$$d_{TV}(\mu, \gamma_n) = \mathbb{E} \, |X - 1| = \frac{1}{2} \mathbb{E}(X - 1) \mathbb{1}_{X \geq 1}.$$

Let $h(x) = x \log x$. Since h is convex and $h(1) = 0$, we have that $h(x) \leq a(x - 1)$ for $1 \leq x \leq R$ as long as a is such that $h(R) \leq a(R - 1)$. Let $R \geq 2$, so that $\frac{R}{R-1} \leq 2$. Then

$$h(R) = R \log R \leq 2(R - 1) \log R = a(R - 1)$$

for $a = 2 \log R$. Thus

$$
\begin{aligned}
\mathbb{E} X \log X &\leq \mathbb{E}(X \log X) \mathbb{1}_{X \geq 1} \\
&\leq a\mathbb{E}(X - 1)\mathbb{1}_{X \geq 1} + \mathbb{E}(X \log X)\mathbb{1}_{X \geq R} \\
&= (\log R)d_{TV}(\mu, \gamma_n) + \mathbb{E}(X \log X)\mathbb{1}_{X \geq R}.
\end{aligned}
$$

The Cauchy–Schwarz inequality implies that

$$
\mathbb{E}(X \log X)\mathbb{1}_{X \geq R} = \mathbb{E}(\log W)\mathbb{1}_{W \geq R} \leq \sqrt{\mathbb{E}(\log W)^2}\sqrt{\mathbb{P}[W \geq R]}.
$$

By the L^2 triangle inequality, we have

$$
\begin{aligned}
\sqrt{\mathbb{E}(\log W)^2} &= \sqrt{\mathbb{E}\,|\log f(Y) - \log \varphi(Y)|^2} \\
&\leq \sqrt{\mathbb{E}\,|\log f(Y)|^2} + \sqrt{\mathbb{E}\,|\log \varphi(Y)|^2},
\end{aligned}
$$

and by Proposition 10,

$$
\mathbb{E}\,|\log \varphi(Y)|^2 = \mathbb{E}\left(\frac{n}{2}\log 2\pi + \frac{\|Y\|^2}{2}\right)^2 \leq Cn^2.
$$

By Lemma 12,

$$
\mathbb{E}\,|\log f(Y)|^2 \leq (\mathbb{E}\log f(Y))^2 + Cn.
$$

Recall that the **entropy** of μ is

$$
-\int f(y) \log f(y)\, dy = -\mathbb{E}\log f(Y) \geq 0,
$$

and that γ_n is the maximum-entropy distribution with identity covariance, so that

$$
(\mathbb{E}\log f(Y))^2 \leq (\mathbb{E}\log \varphi(Z))^2 = \left(n \log \sqrt{2\pi e}\right)^2.
$$

Thus

$$
\sqrt{\mathbb{E}(\log W)^2} \leq Cn.
$$

By [10, Theorem 5.14(e)], $\|f\|_\infty \leq 2^{8n}n^{n/2}$, and so

$$
\begin{aligned}
\mathbb{P}[W \geq R] = \mathbb{P}\left[\frac{f(Y)}{\varphi(Y)} \geq R\right] &\leq \mathbb{P}\left[e^{\|Y\|^2/2} \geq (2^{17}\pi n)^{-n/2}R\right] \\
&= \mathbb{P}\left[\|Y\| \geq \sqrt{2\log\left((2^{17}\pi n)^{-n/2}R\right)}\right]
\end{aligned}
$$

for each $R \geq (2^{17}\pi n)^{n/2}$. Proposition 10 now implies that

$$\mathbb{P}[W \geq R] \leq e^{-c\sqrt{\log R - \frac{n}{2}\log(2^{17}\pi n)}} \leq e^{-c'\sqrt{\log R}}$$

for $\log R \geq Cn \log(n+1)$.
Substituting $S = c\sqrt{\log R}$, all together this shows that

$$H(\mu \mid \gamma_n) \leq C \left(S^2 d_{TV}(\mu, \gamma_n) + ne^{-S}\right)$$

for every $S \geq c\sqrt{n \log(n+1)}$. The result follows using Lemma 11.

Acknowledgements This research was partially supported by grants from the Simons Foundation (#267058 to E.M. and #264103 to M.M.) and the U.S. National Science Foundation (DMS-1308725 to E.M.). This work was carried out while the authors were visiting the Institut de Mathématiques de Toulouse at the Université Paul Sabatier; the authors thank them for their generous hospitality.

References

1. K. Ball, Volume ratios and a reverse isoperimetric inequality. J. Lond. Math. Soc. (2) **44**(2), 351–359 (1991)
2. S. Bobkov, M. Madiman, Concentration of the information in data with log-concave distributions. Ann. Probab. **39**(4), 1528–1543 (2011)
3. F. Bolley, C. Villani, Weighted Csiszár-Kullback-Pinsker inequalities and applications to transportation inequalities. Ann. Fac. Sci. Toulouse Math. (6) **14**(3), 331–352 (2005)
4. C. Borell, Convex measures on locally convex spaces. Ark. Mat. **12**, 239–252 (1974)
5. U. Brehm, P. Hinow, H. Vogt, J. Voigt, Moment inequalities and central limit properties of isotropic convex bodies. Math. Z. **240**(1), 37–51 (2002)
6. R.M. Dudley, *Real Analysis and Probability*. Cambridge Studies in Advanced Mathematics, vol. 74 (Cambridge University Press, Cambridge, 2002). Revised reprint of the 1989 original
7. R. Eldan, B. Klartag, Pointwise estimates for marginals of convex bodies. J. Funct. Anal. **254**(8), 2275–2293 (2008)
8. B. Klartag, A central limit theorem for convex sets. Invent. Math. **168**(1), 91–131 (2007)
9. B. Klartag, Power-law estimates for the central limit theorem for convex sets. J. Funct. Anal. **245**(1), 284–310 (2007)
10. L. Lovász, S. Vempala, The geometry of logconcave functions and sampling algorithms. Random Struct. Algorithms **30**(3), 307–358 (2007)
11. E. Milman, On the role of convexity in functional and isoperimetric inequalities. Proc. Lond. Math. Soc. (3) **99**(1), 32–66 (2009)
12. E. Milman, On the role of convexity in isoperimetry, spectral gap and concentration. Invent. Math. **177**(1), 1–43 (2009)
13. E. Milman, Isoperimetric and concentration inequalities: equivalence under curvature lower bound. Duke Math. J. **154**(2), 207–239 (2010)
14. G. Paouris, Concentration of mass on convex bodies. Geom. Funct. Anal. **16**(5), 1021–1049 (2006)
15. C. Villani, *Optimal Transport: Old and New*. Grundlehren der Mathematischen Wissenschaften (Fundamental Principles of Mathematical Sciences), vol. 338 (Springer, Berlin, 2009)

A Remark on the Diameter of Random Sections of Convex Bodies

Shahar Mendelson

Abstract We obtain a new upper estimate on the Euclidean diameter of the intersection of the kernel of a random matrix with iid rows with a given convex body. The proof is based on a small-ball argument rather than on concentration and thus the estimate holds for relatively general matrix ensembles.

1 Introduction

In this note we revisit the following problem.

Let μ be an isotropic measure[1] on $\mathbb{R}^n$, and by 'isotropic' we mean a symmetric measure with respect to zero that satisfies

$$\int_{\mathbb{R}^n} \langle x, t \rangle^2 d\mu(t) = \|x\|_{\ell_2^n}^2 \text{ for every } x \in \mathbb{R}^n.$$

Given a random vector X distributed according to μ and for $X_1, \ldots, X_k$ that are independent copies of X, let Γ be the $k \times n$ random matrix $k^{-1/2} \sum_{i=1}^k \langle X_i, \cdot \rangle e_i$, whose rows are $X_1, \ldots, X_k$.

Question 1.1. *If $T \subset \mathbb{R}^n$ is a convex body (that is, a convex, centrally-symmetric set with a non-empty interior), what is the typical diameter of $T \cap \ker(\Gamma)$?*

The origin of this problem was the study of the geometry of convex bodies, and in particular, Milman's low-M^* estimate [11] and subsequent estimates on the Gelfand widths of convex bodies, due to Pajor and Tomczak-Jaegermann [12, 13].

The focus of the original question had been the existence of a section of T of codimension k and of a small Euclidean diameter, and was established by estimating

---

[1] We will abuse notation and not distinguish between the measure μ and the random vector X that has μ as its law.

S. Mendelson (✉)
Department of Mathematics, Technion IIT, Haifa, Israel
e-mail: shahar@tx.technion.ac.il

© Springer International Publishing Switzerland 2014   395
B. Klartag, E. Milman (eds.), *Geometric Aspects of Functional Analysis*,
Lecture Notes in Mathematics 2116, DOI 10.1007/978-3-319-09477-9_25

$\mathbb{E}$diam$(T \cap E)$ from above, relative to the uniform measure on the Grassmann manifold $G_{n-k,n}$.

In recent years, more emphasis has been put on other choices of measures on the Grassmann manifold, for example, using the distribution generated by kernels of matrices selected from some random ensemble—like $\Gamma = k^{-1/2} \sum_{i=1}^{k} \langle X_i, \cdot \rangle e_i$ defined above.

The standard way of estimating $\mathbb{E}$diam$(T \cap \ker(\Gamma))$ for such matrix ensembles is based on the quadratic empirical processes indexed by linear forms associated with T.

It is straightforward to show (see, for example, the discussion in [8]) that given $r > 0$, if

$$\sup_{x \in T \cap rS^{n-1}} \left| \frac{1}{k} \sum_{i=1}^{k} \langle X_i, x \rangle^2 - \mathbb{E}\langle X, x \rangle^2 \right| \leq \frac{r^2}{2}, \tag{1}$$

one has

$$\frac{1}{2} \|x\|_{\ell_2^n}^2 \leq \|\Gamma x\|_{\ell_2^k}^2 \leq \frac{3}{2} \|x\|_{\ell_2^n}^2$$

for every $x \in T$ of ℓ_2^n norm larger than r. Hence, on the event given by (1), diam$(T \cap \ker(\Gamma)) \leq r$.

Setting $r_0(k, \delta)$ to be the smallest for which

$$Pr\left(\sup_{x \in T \cap rS^{n-1}} \left| \frac{1}{k} \sum_{i=1}^{k} \langle X_i, x \rangle^2 - \mathbb{E}\langle X, x \rangle^2 \right| \leq \frac{r^2}{2} \right) \geq 1 - \delta,$$

it follows that with probability at least $1 - \delta$,

$$\text{diam}(T \cap \ker(\Gamma)) \leq r_0,$$

and a similar argument may be used to control $\mathbb{E}$diam$(T \cap \ker(\Gamma))$.

Unfortunately, estimating the quadratic empirical process is a difficult task. In fact, one has a satisfactory estimate that holds for every convex body $T \subset \mathbb{R}^n$ only for measures that are subgaussian or unconditional log-concave.

Definition 1.2. Given a real random variable Z, set $\|Z\|_{\psi_2} = \inf\{c > 0 : \mathbb{E}\exp(|Z|^2/c^2) \leq 2\}$.

The measure μ is L-subgaussian if $\|\langle x, \cdot \rangle\|_{\psi_2(\mu)} \leq L\|\langle x, \cdot \rangle\|_{L_2(\mu)}$ for every $x \in \mathbb{R}^n$.

Theorem 1.3 ([6]). *There exist absolute constants c_1, c_2 and c_3 for which the following holds. Let μ be an isotropic, L-subgaussian measure (and, in particular, for every $x \in \mathbb{R}^n$, $\|\langle X, x \rangle\|_{\psi_2(\mu)} \leq L\|x\|_{\ell_2^n}$).*

Let $T \subset \mathbb{R}^n$ and set $d_T = \sup_{t \in T} \|t\|_{\ell_2^n}$. For $u \geq c_1$, with probability at least

$$1 - 2 \exp(-c_2 u^2 (\mathbb{E}\|G\|_{T^\circ}/d_T)^2),$$

$$\sup_{x \in T} \left| \sum_{i=1}^k (\langle X_i, x \rangle^2 - \mathbb{E}\langle X, x \rangle^2) \right| \leq c_3 L^2 u^2 \left(u\sqrt{k} d_T \mathbb{E}\|G\|_{T^\circ} + (\mathbb{E}\|G\|_{T^\circ})^2 \right),$$

where $G = (g_1, \ldots, g_n)$ is the standard gaussian vector in $\mathbb{R}^n$ and $\mathbb{E}\|G\|_{T^\circ} = \mathbb{E}\sup_{t \in T} |\langle G, t \rangle|$.

A version of Theorem 1.3 has been established in [8] when $T \subset S^{n-1}$ and with a weaker probability estimate.

Theorem 1.3 follows from a general bound on the quadratic empirical process that is based on a global complexity parameter of the indexing set [9], and that will not be defined here. Thanks to Talagrand's Majorizing Measures Theorem (see the book [17] for a detailed survey on this topic), this complexity parameter is upper bounded by $\sim \mathbb{E}\|G\|_{T^\circ}$ in the subgaussian case, thus leading to Theorem 1.3. However, in other cases, controlling the complexity parameter is nontrivial.

One other case in which the global complexity may be upper bounded using a mean-width of T, is when X is isotropic, unconditional and log-concave. Using the Bobkov-Nazarov Theorem [1], X is dominated by $Y = (y_1, \ldots, y_n)$, a vector with independent, standard, exponential coordinates. One may show [9] that with high probability,

$$\sup_{x \in T} \left| \sum_{i=1}^k (\langle X_i, x \rangle^2 - \mathbb{E}\langle X, x \rangle^2) \right| \lesssim \sqrt{k} d_T \mathbb{E}\|Y\|_{T^\circ} + (\mathbb{E}\|Y\|_{T^\circ})^2. \tag{2}$$

The proof of (2) is based on two additional observations. First, that when X is isotropic, unconditional and log-concave, the global complexity parameter of T may be bounded using a mixture of Talagrand's γ_α functionals, and second, that this mixture is equivalent to $\mathbb{E}\|Y\|_{T^\circ}$ [16].

Additional bounds on the quadratic process are known for more general measures, but only for very specific choices of sets T. The most important example is when T is the Euclidean ball, and the quadratic empirical process may be used to obtain a Bai-Yin type estimate on the largest and smallest singular values of Γ [9, 10].

At this point, it should be noted that (1) is a much stronger statement than what is actually needed to bound the diameter of $T \cap \ker(\Gamma)$. Clearly, any sort of a positive *lower* bound on

$$\inf_{x \in T \cap rS^{n-1}} \|\Gamma x\|_{\ell_2^k} \tag{3}$$

would suffice—rather than the two-sided, 'almost isometric' bound that follows from bounds on the quadratic process.

Here, we will show that (3) holds for rather general matrix ensembles.

Theorem 1.4. *Let X be an isotropic vector on $\mathbb{R}^n$ and assume that linear forms satisfy the following small-ball condition: that there is some $\lambda > 0$ for which*

$$Pr(|\langle x, X \rangle| \geq \lambda \|x\|_{\ell_2^n}) \geq 99/100 \quad \text{for every } x \in \mathbb{R}^n.$$

Then, there exist a constant c that depends only on λ, for which, with probability at least $3/4$,

$$\text{diam}(T \cap \ker(\Gamma)) \leq \frac{c}{\sqrt{k}} \cdot \max \left\{ \mathbb{E}\|G\|_{T^\circ}, \mathbb{E}\|k^{-1/2} \sum_{i=1}^{k} X_i\|_{T^\circ} \right\}.$$

Theorem 1.4 can be improved and extended in various ways.

First of all, the 'correct' upper estimate on the diameter should be based on a fixed point condition defined using the norms $\| \ \|_{(T \cap rB_2^n)^\circ}$ rather than the norm $\| \ \|_{T^\circ}$. Also, the constant probability estimate of $3/4$ may be improved significantly to $1 - 2\exp(-ck)$ with a slightly more involved proof (see [4] and [7] for a similar argument). We will formulate, without proof, a more general version of Theorem 1.4 at the end of the note.

Examples.

1. If X is an isotropic L-subgaussian vector, it is standard to verify that $k^{-1/2} \sum_{i=1}^{k} X_i$ is isotropic and cL-subgaussian for a suitable absolute constant c. Therefore,

$$\mathbb{E}\|k^{-1/2} \sum_{i=1}^{k} X_i\|_{T^\circ} \leq c_1 L \mathbb{E}\|G\|_{T^\circ},$$

and by Theorem 1.4, with probability at least $3/4$,

$$\text{diam}(T \cap \ker(\Gamma)) \leq c_1(\lambda, L) \frac{\mathbb{E}\|G\|_{T^\circ}}{\sqrt{k}}.$$

This coincides with the estimate from [8] (up to the 'localization' of using $\| \ \|_{(T \cap rB_2^n)^\circ}$ instead of $\| \ \|_{T^\circ}$ mentioned above) and with the classical result of [12] when X is the standard gaussian vector.

2. If X is an isotropic, unconditional, log-concave measure then so is $Z = k^{-1/2} \sum_{i=1}^{k} X_i$. By the Bobkov-Nazarov Theorem [1], both Z and G are strongly dominated by Y, the random vector with independent, standard exponential coordinates. Therefore, by Theorem 1.4, with probability at least $3/4$,

$$\text{diam}(T \cap \ker(\Gamma)) \le c_2(\lambda) \frac{\mathbb{E}\|Y\|_{T^\circ}}{\sqrt{k}}.$$

3. Theorem 1.4 leads to a 'heavy tails' result in some cases. Since X is symmetric, $\sum_{i=1}^{k} X_i$ has the same distribution as $\sum_{i=1}^{k} \varepsilon_i X_i$, where $(\varepsilon_i)_{i=1}^{k}$ are independent, symmetric $\{-1, 1\}$-valued random variables that are independent of $(X_i)_{i=1}^{k}$. If T° has a Rademacher type 2 constant $R_2(T^\circ)$, then

$$\mathbb{E}\|k^{-1/2} \sum_{i=1}^{k} X_i\|_{T^\circ} = \mathbb{E}\|k^{-1/2} \sum_{i=1}^{k} \varepsilon_i X_i\|_{T^\circ} \le R_2(T^\circ)(\mathbb{E}\|X\|_{T^\circ}^2)^{1/2},$$

and with probability at least $3/4$,

$$\text{diam}(T \cap \ker(\Gamma)) \le \frac{c_3(\lambda)}{\sqrt{k}} \cdot \max\left\{\mathbb{E}\|G\|_{T^\circ}, R_2(T^\circ)(\mathbb{E}\|X\|_{T^\circ}^2)^{1/2}\right\}.$$

For example, if $T = B_1^n$ and $X \in \beta B_\infty^n$ almost surely, then $\|X\|_{T^\circ} = \|X\|_{\ell_\infty^n} \le \beta$, $R_2(\ell_\infty^n) \le \sqrt{\log n}$ and $\mathbb{E}\|G\|_{\ell_\infty^n} \lesssim \sqrt{\log n}$. Therefore,

$$\text{diam}(B_1^n \cap \ker(\Gamma)) \le c_3(\lambda)\beta \sqrt{\frac{\log n}{k}}.$$

2   Proof of Theorem 1.4

Lemma 2.1. *Let ζ be a random variable that satisfies*

$$Pr(|\zeta| \ge \lambda \|\zeta\|_{L_2}) \ge 1 - \varepsilon \tag{4}$$

for constants $0 < \varepsilon < 1/12$ and $\lambda > 0$.

If $\zeta_1, \ldots, \zeta_k$ are independent copies of ζ, then with probability at least $1 - 2^{-6\varepsilon k}$ there is a subset $J \subset \{1, \ldots, k\}$ of cardinality at least $(1 - 6\varepsilon)k$, and for every $j \in J$,

$$|\zeta_j| \ge \lambda \|\zeta\|_{L_2}.$$

Proof. It suffices to show that no more than $6\varepsilon k$ of the $|\zeta_i|$'s are smaller than $\lambda \|\zeta\|_{L_2}$. By a binomial estimate, if $6\varepsilon k \le k/2$,

$$Pr\left(\exists J \subset \{1, \ldots, k\}, |J| = 6\varepsilon k, |\zeta_j| \le \lambda \|\zeta\|_{L_2} \text{ if } j \in J\right)$$

$$\le \binom{k}{6\varepsilon k} Pr^{6\varepsilon k} (|\zeta| \le \lambda \|\zeta\|_{L_2}) \le \left(\frac{e}{6\varepsilon}\right)^{6\varepsilon k} \cdot \varepsilon^{6\varepsilon k} \le 2^{-6\varepsilon k}.$$

■

Let $\{\zeta^i : 1 \le i \le N\}$ be a collection of random variables, and for every i let $Z_i \in \mathbb{R}^k$ be a random vector with independent coordinates, distributed according to the random variable ζ^i. Denote by $Z_i(j)$ the j-th coordinate of Z_i.

Corollary 2.2. *If each ζ^i satisfies the small-ball condition* (4) *and $N \le 2^{3\varepsilon k}$, then with probability at least $1 - 2^{-3\varepsilon k}$, for every $1 \le i \le N$ there is a subset $J_i \subset \{1,\ldots,k\}$, of cardinality at least $(1 - 6\varepsilon)k$, and*

$$|Z_i(j)| \ge \lambda \|\zeta^i\|_{L_2} \text{ for every } j \in J_i.$$

Proof of Theorem 1.4. Let $\varepsilon = 1/600$ and observe that by the small ball assumption and since $\|\langle X, x\rangle\|_{L_2} = \|x\|_{\ell_2^n}$ then,

$$Pr(|\langle X, x\rangle| \ge \lambda \|x\|_{\ell_2^n}) \ge (1 - \varepsilon)$$

for every $x \in \mathbb{R}^n$.

Fix $r > 0$ to be named later and set $T_r = T \cap rS^{n-1}$. Let

$$\rho = c \frac{\mathbb{E}\|G\|_{T_r^\circ}}{\sqrt{\varepsilon k}}$$

for a suitable absolute constant c and set $V_r \subset T_r$ to be a maximal ρ-separated subset of T_r with respect to the ℓ_2^n norm. Sudakov's inequality (see, e.g. [5, 14]) shows that for the right choice of c, $|V_r| \le 2^{3\varepsilon k}$.

Let

$$\zeta^i = \langle X, v_i\rangle, \quad v_i \in V_r, \quad 1 \le i \le 2^{3\varepsilon k},$$

and set $Z_i = (Z_i(j))_{j=1}^k = (\langle X_j, v_i\rangle)_{j=1}^k$ where $X_1,\ldots,X_k$ are independent copies of X.

Applying Corollary 2.2 to the set $\{Z_i : 1 \le i \le 2^{3\varepsilon k}\}$, it follows that with probability at least $1 - 2^{-3\varepsilon k}$, for every $v \in V_r$ there is a subset $J_v \subset \{1,\ldots,k\}$, $|J_v| \ge (1 - 6\varepsilon)k = 99k/100$, and for every $j \in J_v$,

$$|\langle X_j, v\rangle| \ge \lambda \|v\|_{\ell_2^n} = \lambda r, \tag{5}$$

and the last equality holds because $V_r \subset rS^{n-1}$.

For every $x \in T_r$, let $\pi(x)$ be the nearest point to x in V_r with respect to the ℓ_2^n norm. Therefore,

$$\mathbb{E}|\langle X, x - \pi(x)\rangle| \le \|\langle x - \pi(x), \cdot\rangle\|_{L_2(\mu)} = \|x - \pi(x)\|_{\ell_2^n} \le \rho.$$

By the Giné-Zinn symmetrization inequality [3], the contraction inequality for Bernoulli processes (see, e.g., [5]), and since $x - \pi(x) \in 2T \cap \rho B_2^n$ for every $x \in T_r$,

$$\mathbb{E}\sup_{x\in T_r}\frac{1}{k}\sum_{i=1}^{k}|\langle X_i, x-\pi(x)\rangle|$$

$$\leq \rho + \mathbb{E}\sup_{x\in T_r}\left|\frac{1}{k}\sum_{i=1}^{k}|\langle X_i, x-\pi(x)\rangle| - \mathbb{E}|\langle X_i, x-\pi(x)\rangle|\right|$$

$$\leq \rho + \frac{2}{k}\mathbb{E}\sup_{x\in T_r}\left|\sum_{i=1}^{k}\varepsilon_i|\langle X_i, x-\pi(x)\rangle|\right| \leq \rho + \frac{2}{k}\mathbb{E}\sup_{x\in T_r}\left|\sum_{i=1}^{k}\varepsilon_i\langle X_i, x-\pi(x)\rangle\right|$$

$$\leq \rho + \frac{2}{k}\mathbb{E}\left\|\sum_{i=1}^{k}\varepsilon_i X_i\right\|_{(2T\cap\rho B_2^n)^\circ} \qquad \leq \rho + \frac{4}{k}\mathbb{E}\left\|\sum_{i=1}^{k}X_i\right\|_{(T\cap\rho B_2^n)^\circ}.$$

Hence, by the choice of ρ and the trivial inclusion $T \cap \rho B_2^n \subset T$,

$$\mathbb{E}\sup_{x\in T_r}\frac{1}{k}\sum_{i=1}^{k}|\langle X_i, x-\pi(x)\rangle| \leq c\frac{\mathbb{E}\|G\|_{T^\circ}}{\sqrt{\varepsilon k}} + \frac{4}{k}\mathbb{E}\left\|\sum_{i=1}^{k}X_i\right\|_{T^\circ}. \tag{6}$$

Set $A = \mathbb{E}\sup_{x\in T_r} k^{-1}\sum_{i=1}^{k}|\langle X_i, x-\pi(x)\rangle|$ and let

$$W_r = \left\{(\langle X_i, x-\pi(x)\rangle)_{i=1}^{k} : x \in T_r\right\}.$$

Note that for $0 < \delta < 1$, with probability at least $1 - \delta$,

$$W_r \subset k(A/\delta)B_1^k.$$

On that event, if $(w_i)_{i=1}^{k} \in W_r$ and $(w_i^*)_{i=1}^{k}$ is a non-increasing rearrangement of $(|w_i|)_{i=1}^{k}$,

$$w_{k/100}^* \leq \frac{\|w\|_{\ell_1^k}}{k/100} \leq \frac{100}{\delta}A.$$

Thus, for every $x \in T_r$ there is a subset $J_x' \subset \{1,\ldots,k\}$ of cardinality at least $99k/100$, and for every $j \in J_x'$,

$$|\langle X_i, x-\pi(x)\rangle| \leq \frac{100}{\delta}A. \tag{7}$$

Fix $X_1,\ldots,X_k$ in the intersection of the two events defined in (5) and (7). For every $x \in T_r$ set $I_x = J_x' \cap J_{\pi(x)}$. Observe that $|I_x| \geq 98k/100$ and that for every $i \in I_x$,

$$|\langle X_i, x \rangle| \geq |\langle X_i, \pi(x) \rangle| - |\langle X_i, x - \pi(x) \rangle| \geq \lambda r - \frac{100}{\delta} A$$

$$\geq \lambda r - \frac{c_1}{\delta \sqrt{k}} \cdot \left(\mathbb{E}\|G\|_{T^\circ} + \mathbb{E} \left\| \frac{1}{\sqrt{k}} \sum_{i=1}^{k} X_i \right\|_{T^\circ} \right).$$

Therefore, if

$$r \geq \frac{c_2(\lambda, \delta)}{\sqrt{k}} \cdot \max \left\{ \mathbb{E}\|G\|_{T^\circ}, \mathbb{E} \left\| \frac{1}{\sqrt{k}} \sum_{i=1}^{k} X_i \right\|_{T^\circ} \right\},$$

then with probability at least $1 - \delta - 2^{-3\varepsilon k} = 1 - \delta - 2^{-k/200}$, for each $x \in T_r$, $|\langle X_i, x \rangle| \geq (\lambda/2)\|x\|_{\ell_2^n}$ on at least $98k/100$ coordinates; Thus,

$$\inf_{x \in T_r} \|\Gamma x\|_{\ell_2^k} \gtrsim \lambda \|x\|_{\ell_2^n}. \tag{8}$$

Finally, using the convexity of T and since the condition in (8) is positive-homogeneous, (8) holds for any $x \in T$ with $\|x\|_{\ell_2^n} \geq r$, as claimed. ∎

3 Concluding Comments

The proof of Theorem 1.4 has two components. The first is based on a small-ball estimate for linear functionals and does not require additional information on their tails. Thus, this part holds even for heavy-tailed ensembles.

The more restrictive condition is on the random vector $k^{-1/2} \sum_{i=1}^{k} X_i$. Still, it is far easier to handle the norm $\| \sum_{i=1}^{k} X_i \|_{T^\circ}$ than the supremum of the quadratic empirical process indexed by T.

The estimate in Theorem 1.4 can be improved using what is, by now, a standard argument. First, observe that all the inequalities leading to (6) hold in probability and not just in expectation (see, for example, [2, 15]). Keeping the 'localization' level r, one can define two fixed points:

$$\rho_k(\delta, Q_1) = \inf \left\{ \rho : Pr \left(\|k^{-1/2} \sum_{i=1}^{k} X_i \|_{(T \cap \rho B_2^n)^\circ} \geq Q_1 \rho \sqrt{k} \right) \leq \delta \right\},$$

and

$$r_k(Q_2) = \inf \{ r : \mathbb{E}\|G\|_{(T \cap rS^{n-1})^\circ} \leq Q_2 r \sqrt{k} \}.$$

It is straightforward to verify that there are constants Q_1 and Q_2 that depend only on λ, for which, with probability at least $1 - \delta - 2^{-k/200}$, if

$$\|x\|_{\ell_2^n} \gtrsim \max\{\rho_k(\delta, Q_1), r_k(Q_2)\},$$

then

$$\|\Gamma x\|_{\ell_2^k} \gtrsim \lambda \|x\|_{\ell_2^n}.$$

Thus, on the same event,

$$\operatorname{diam}(T \cap \ker(\Gamma)) \lesssim \max\{\rho_k(\delta, Q_1), r_k(Q_2)\}.$$

Finally, it is possible to use a slightly more involved, empirical processes based method, that leads to an exponential probability estimate of $1 - 2\exp(-ck)$ in Theorem 1.4. Result of a similar flavour may by found in [4] and in [7].

Since the goal in this note was to present the idea of using a simple small-ball argument, rather than pursuing an optimal result, we have opted to present this proof.

Acknowledgements The work was partially supported by the Mathematical Sciences Institute – The Australian National University and by the Israel Science Foundation grant 900/10.

References

1. S.G. Bobkov, F.L. Nazarov, On convex bodies and log-concave probability measures with unconditional basis, in *Geometric Aspects of Functional Analysis*. Lecture Notes in Mathematics, vol. 1807, (2003), pp. 53–69
2. R.M. Dudley, *Uniform Central Limit Theorems*. Cambridge Studies in Advanced Mathematics, vol. 63 (Cambridge University Press, Cambridge, 1999)
3. E. Giné, J. Zinn, Some limit theorems for empirical processes. Ann. Probab. **12**(4), 929–989 (1984)
4. V. Koltchinskii, S. Mendelson, Bounding the smallest singular value of a random matrix without concentration. Preprint
5. M. Ledoux, M. Talagrand, *Probability in Banach spaces. Isoperimetry and processes*. Ergebnisse der Mathematik und ihrer Grenzgebiete (3), vol. 23 (Springer, Berlin, 1991)
6. S. Mendelson, On the geometry of subgaussian coordinate projections. Preprint
7. S. Mendelson, Learning without concentration. Preprint
8. S. Mendelson, A. Pajor, N. Tomczak-Jaegermann, Reconstruction and subgaussian operators. Geom. Funct. Anal. **17**, 1248–1282 (2007)
9. S. Mendelson, G. Paouris, On generic chaining and the smallest singular values of random matrices with heavy tails. J. Funct. Anal. **262**(9), 3775–3811 (2012)
10. S. Mendelson, G. Paouris, On the singular values of random matrices. J. Eur. Math. Soc. **16**(4), 823–834 (2014)
11. V. Milman, Random subspaces of proportional dimension of finite dimensional normed spaces: approach through the isoperimetric inequality. Lect. Notes Math. **1166**, 106–115 (1985)
12. A. Pajor, N. Tomczak-Jaegermann, Subspaces of small codimension of finite-dimensional Banach spaces, Proc. Am. Math. Soc. **97**(4), 637–642 (1986)

13. A. Pajor, N. Tomczak-Jaegermann, Nombres de Gelfand et sections euclidiennes de grande dimension. (French) (Gelfand numbers and high-dimensional Euclidean sections). Séminaire d'Analyse Fonctionelle 1984/1985, Publ. Math. Univ. Paris VII, 26, Univ. Paris VII, Paris (1986), pp. 37–47
14. G. Pisier, *The Volume of Convex Bodies and Banach Space Geometry* (Cambridge University Press, Cambridge, 1989)
15. A.W. Van der Vaart, J.A. Wellner, *Weak Convergence and Empirical Processes* (Springer, Berlin, 1996)
16. M. Talagrand, The supremum of some canonical processes. Am. J. Math. **116**, 283–325 (1994)
17. M. Talagrand, *Upper and Lower Bounds for Stochastic Processes: Modern Methods and Classical Problems* (Springer, Berlin, 2014)

A Note on Certain Convolution Operators

Piotr Nayar and Tomasz Tkocz

Abstract In this note we consider a certain class of convolution operators acting on the L_p spaces of the one dimensional torus. We prove that the identity minus such an operator is nicely invertible on the subspace of functions with mean zero.

1 Introduction

Let $\mathbb{T} = \mathbb{R}/\mathbb{Z}$ be the one dimensional torus viewed as a compact group with the addition modulo 1, $x \oplus y = (x + y) \bmod 1$, $x, y \in \mathbb{R}$, equipped with the Haar measure, i.e. the unique invariant probability measure (the Lebesgue measure). To begin with, fix $1 \leq p \leq \infty$ and consider the averaging operator U_t acting on $L_p(\mathbb{T})$ (with the usual norm $\|f\| = \left(\int_{\mathbb{T}} |f|^p\right)^{1/p}$ for $p < \infty$, and $\|f\| = \operatorname{ess\,sup}_{\mathbb{T}} |f|$ for $p = \infty$),

$$(U_t f)(x) = \frac{1}{2t} \int_{-t}^{+t} f(x \oplus s) \, \mathrm{d}s, \quad t \in (0, 1). \tag{1}$$

If t is small, is the operator $I - U_t$ invertible, or, in other words, how much does $U_t f$ differ from f? Of course, averaging a constant function does not change it, but excluding such a trivial case, we get a quantitative answer.

Theorem 1. *Let $t \in (0, 1)$. There exists a universal constant c such that for every $1 \leq p \leq \infty$ and every $f \in L_p(\mathbb{T})$ with $\int_{\mathbb{T}} f = 0$ we have*

P. Nayar (✉)
Institute of Mathematics, University of Warsaw, Banacha 2, 02-097 Warszawa, Poland

Institute for Mathematics and its Applications, University of Minnesota, 207 Church Street SE, 432 Lind Hall, Minneapolis, MN 55455, USA
e-mail: nayar@mimuw.edu.pl

T. Tkocz
Institute of Mathematics, University of Warsaw, Banacha 2, 02-097 Warszawa, Poland

Mathematics Institute, University of Warwick, Coventry CV4 7AL, UK
e-mail: tkocz@mimuw.edu.pl

© Springer International Publishing Switzerland 2014   405
B. Klartag, E. Milman (eds.), *Geometric Aspects of Functional Analysis*,
Lecture Notes in Mathematics 2116, DOI 10.1007/978-3-319-09477-9_26

$$\| f - U_t f \| \ge ct^2 \| f \|,\tag{2}$$

where $\|\cdot\|$ denotes the L_p norm.

Note that if p was equal to 2, then, with the aid of the Fourier analysis, the above estimate would be trivial. However, $\|\cdot\|$ is set to be the L_p norm for some $1 \le p \le \infty$, the constant does not depend on p, therefore the situation is more subtle.

When $p = 1$, if we further estimate the left hand side of (2) using the Sobolev inequality, see [4], we obtain the following corollary.

Corollary 1. *Let us consider $t \in (0, 1)$ and assume that f belongs to the Sobolev space $W^{1,1}(\mathbb{T})$ with $\int_\mathbb{T} f = 0$. Then we have*

$$\int_\mathbb{T} \left| f'(x) - \frac{f(x \oplus t) - f(x \oplus -t)}{2t} \right| dx \ge ct^2 \int_\mathbb{T} |f(x)| \, dx, \tag{3}$$

where $c > 0$ is a universal constant.

Remark 1. Setting $t = 1/2$, inequality (3) becomes the usual Sobolev inequality, so (3) can be viewed as a certain generalization of the Sobolev inequality.

Remark 2. Set $f(x) = \cos(2\pi x)$. Then $\| f - U_t f \| = \| f \| \left(1 - \frac{1}{2\pi t} \sin(2\pi t) \right) \approx \frac{2}{3} \pi^2 t^2 \| f \|$, for small t. Therefore, the inequality in Theorem 1 is sharp in a sense.

In this note we give a proof of a generalization of Theorem 1. We say that a $\mathbb{T}$-valued random variable Z is *c-good* with some positive constant c if $\mathbb{P}(Z \in A) \ge c|A|$ for all measurable $A \subset \mathbb{T}$. Equivalently, by Lebesgue's decomposition theorem it means that the absolutely continuous part of Z (with respect to the Lebesgue measure) has a density bounded below by a positive constant. We say that a real random variable Y is *ℓ-decent* if $Y_1 + \ldots + Y_\ell$ has a nontrivial absolutely continuous part, where $Y_1, Y_2, \ldots$ are i.i.d. copies of Y. Our main result reads

Theorem 2. *Given $t \in (0, 1)$ and an ℓ-decent real random variable Y, consider the operator A_t given by*

$$(A_t f)(x) = \mathbb{E} f(x \oplus tY). \tag{4}$$

Then there exists a positive constant c which depends only on the distribution of the random variable Y such that for every $1 \le p \le \infty$ and every $f \in L_p(\mathbb{T})$ with $\int_\mathbb{T} f = 0$ we have

$$\| f - A_t f \| \ge ct^2 \| f \|,$$

where $\|\cdot\|$ denotes the L_p norm.

Remark 3. One cannot hope to prove a statement similar to Theorem 2 for purely atomic measures. Indeed, just consider the case $p = 1$ and let Y be distributed according to the law $\mu_Y = \sum_{i=1}^{\infty} p_i \delta_{x_i}$. Then for every $\varepsilon > 0$ and every $t \in (0, 1)$ there exists $f \in L_1(\mathbb{T})$ such that $\|f - A_t(f)\| < \varepsilon$ and $\|f\| = 1$. To see this take N such that $\sum_{i=N+1}^{\infty} p_i < \varepsilon/4$ and let $f_n(x) = \frac{\pi}{2} \sin(2\pi nx)$. Then $\|f_n\| = 1$. Let $n_0 \geq 8\pi/\varepsilon$. Consider the sequence $\big((\pi n t x_1 \mod 2\pi, \dots, \pi n t x_N \mod 2\pi)\big)_n$ for $n = 0, 1, 2, \dots, n_0^N$ and observe that by the pigeonhole principle there exist $0 \leq n_1 < n_2 \leq n_0^N$ such that for all $1 \leq i \leq N$ we have $\mathrm{dist}(\pi t x_i (n_1 - n_2), 2\pi\mathbb{Z}) \leq \frac{2\pi}{n_0}$. Taking $n = n_2 - n_1$ we obtain

$$\|f_n - A_t(f_n)\| \leq \frac{\pi}{2} \sum_{i=1}^{N} p_i \|\sin(2\pi nx) - \sin(2\pi n(x + tx_i))\| + \frac{\varepsilon}{2}$$

$$= \pi \sum_{i=1}^{N} p_i |\sin(\pi n t x_i)| \cdot \|\cos(2\pi nx \oplus \pi n t x_i)\| + \frac{\varepsilon}{2}$$

$$\leq 2 \sum_{i=1}^{N} p_i |\sin(\pi n t x_i)| + \frac{\varepsilon}{2} \leq \frac{4\pi}{n_0} \sum_{i=1}^{N} p_i + \frac{\varepsilon}{2} \leq \varepsilon.\square$$

Our result gives the bound for the norm of an operator of the form $(I - A_t)^{-1}$. The main difficulty is that this operator is not globally invertible. Of course, boundedness of a resolvent operator $R_\lambda(A) = (A - zI)^{-1}$ has been thoroughly studied (see e.g. [1, 3] which feature Hilbert space settings for Hilbert-Schmidt and Schatten-von Neumann operators). Let us also mention that the first part of the book [8] is a set of related articles concerning mainly the problem of finding the inverse formula for certain Toeplitz-type operators. The paper [6] contains the famous Gohberg-Semencul formula for the inverse of a non-Hermitian Toeplitz matrix. In [5] the authors generalized the results of [8] to the case of Toeplitz matrices whose entries are taken from some noncommutative algebra with a unit. The operators of the form $I - K$ (acting e.g. on $L_1([0, 1])$), where K is a certain operator with a kernel $k(t-s)$, are continuous versions of the operators given by Toeplitz matrices. Paper [6] deals also with this kind of operators, namely

$$(I - K)(f)(x) = f(x) - \int_0^1 k(t - s) f(s) \, ds,$$

where $k \in L_1([-1, 1])$. In the case of $I - K$ being invertible, the authors give a formula for the inverse operator $(I - K)^{-1}$ in terms of solutions of certain four integral equations. See also Article 3 in [8] for generalizations of these formulas.

2 Proof of Theorem 2

We begin with two lemmas.

Lemma 1. *Suppose Y is an ℓ-decent random variable. Let $Y_1, Y_2, \ldots$ be independent copies of Y. Then there exist a positive integer $N = N(Y)$ and numbers $c = c(Y) > 0$, $C_0 = C_0(Y) \geq 1$ such that for all $C \geq C_0$ and $n \geq N$ the random variable*

$$X_n^{(C)} = \left(C \cdot \frac{Y_1 + \ldots + Y_n}{\sqrt{n}} \right) \mod 1 \tag{5}$$

is c-good.

Proof. We prove the lemma in a few steps considering more and more general assumptions about Y.

Step I. Suppose that the characteristic function of Y belongs to $L_p(\mathbb{R})$ for some $p \geq 1$. In this case, by a certain version of the Local Central Limit Theorem, e.g. Theorem 19.1 in [2, p. 189], we know that the density q_n of $(Y_1 + \ldots + Y_n - n\mathbb{E}Y)/\sqrt{n}$ exists for sufficiently large n, and satisfies

$$\sup_{x \in \mathbb{R}} \left| q_n(x) - \frac{1}{\sqrt{2\pi}\sigma} e^{-x^2/2\sigma^2} \right| \xrightarrow[n \to \infty]{} 0, \tag{6}$$

where $\sigma^2 = \mathrm{Var}(Y)$. Observe that the density $g_n^{(C)}$ of $X_n^{(C)}$ equals

$$g_n^{(C)}(x) = \sum_{k \in \mathbb{Z}} \frac{1}{C} q_n\left(\frac{1}{C}(x + k) - \sqrt{n}\mathbb{E}Y \right), \qquad x \in [0, 1].$$

Using (6), for $\delta = \frac{e^{-2/\sigma^2}}{\sqrt{2\pi}\sigma}$ we can find $N = N(Y)$ such that

$$q_n(x) > \frac{1}{\sqrt{2\pi}\sigma} e^{-x^2/2\sigma^2} - \delta/8, \qquad x \in \mathbb{R}, \ n \geq N.$$

Therefore, to be close to the maximum of the Gaussian density we sum over only those k's for which $x + k \in (-2C, 2C) + C\sqrt{n}\mathbb{E}Y$ for all $x \in [0, 1]$. Since there are at least C and at most $4C$ such k's, we get that

$$g_n^{(C)}(x) > \frac{1}{C} \frac{1}{\sqrt{2\pi}\sigma} e^{-2/\sigma^2} \cdot C - \frac{1}{C} \frac{\delta}{8} \cdot 4C = \frac{1}{2\sqrt{2\pi}\sigma} e^{-2/\sigma^2}.$$

In particular, it implies that $X_n^{(C)}$ is c-good with $c = \frac{1}{2\sqrt{2\pi}\sigma} e^{-2/\sigma^2}$. Thus, in this case, it suffices to set $C_0 = 1$.

Step II. Suppose that the law of Y is of the form $q\mu + (1-q)\nu$ for some $q \in (0, 1]$ and some Borel probability measures μ, ν on $\mathbb{R}$ such that the characteristic function of μ belongs to $L_p(\mathbb{R})$ for some $p \geq 1$. Notice that

$$\mu_{Y_1+\ldots+Y_N} = \mu_Y^{\star N} = (q\mu + (1-q)\nu)^{\star N} = \sum_{k=0}^{N} \binom{N}{k} q^k (1-q)^{N-k} \mu^{\star k} \star \nu^{\star(N-k)}$$

$$\geq \sum_{k=N_0}^{N} \binom{N}{k} q^k (1-q)^{N-k} \mu^{\star k} \star \nu^{\star(N-k)} = c_{N,N_0} \left(\mu^{\star N_0} \star \rho_{N,N_0} \right),$$

where

$$\rho_{N,N_0} = \frac{1}{c_{N,N_0}} \sum_{k=N_0}^{N} \binom{N}{k} q^k (1-q)^{N-k} \mu^{\star(k-N_0)} \star \nu^{\star(N-k)}$$

is a probability measure, and

$$c_{N,N_0} = \sum_{k=N_0}^{N} \binom{N}{k} q^k (1-q)^{N-k}$$

is a normalisation constant. Choosing $N_0 = \lfloor qN - C_1\sqrt{q(1-q)N} \rfloor$ we can guarantee that $c_{N,N_0} \geq 1/2$ eventually, say for $N \geq \tilde{N}$. Denoting by $\bar{Y}, Z$ the random variables with the law μ, ρ_{N,N_0} respectively and by $\bar{Y}_i$ i.i.d. copies of $\bar{Y}$, we get

$$\mathbb{P}\left(X_N^{(C)} \in A \right) \geq c_{N,N_0} \mathbb{P}\left(\left(C\frac{\bar{Y}_1 + \ldots + \bar{Y}_{N_0}}{\sqrt{N}} + C\frac{Z_{N,N_0}}{\sqrt{N}} \right) \mod 1 \in A \right).$$

By Step I, the first bit $C(\bar{Y}_1 + \ldots + \bar{Y}_{N_0})/\sqrt{N}$ is c-good for some $c > 0$ and $C \geq C_0^{(II)} = \sup_{N \geq \tilde{N}} \sqrt{N/N_0}$. Moreover, note that if U is a c-good $\mathbb{T}$-valued r.v., then so is $U \oplus V$ for every $\mathbb{T}$-valued r.v. V which is independent of U. As a result, $X_N^{(C)}$ is $c/2$-good.

Step III. Now we consider the general case, i.e. Y is ℓ-decent for some $\ell \geq 1$. For $n \geq \ell$ we can write

$$C \cdot \frac{Y_1 + \ldots + Y_n}{\sqrt{n}} = C\sqrt{\frac{\lfloor n/\ell \rfloor}{n}} \cdot \frac{\tilde{Y}_1 + \ldots + \tilde{Y}_{\lfloor n/\ell \rfloor}}{\sqrt{\lfloor n/\ell \rfloor}} + C\frac{\tilde{R}}{\sqrt{n}}$$

with $\tilde{Y}_j = Y_{(j-1)\ell+1} + \ldots + Y_{j\ell}$ for $j = 1, \ldots, \lfloor n/\ell \rfloor$, and $\tilde{R} = Y_{\lfloor n/\ell \rfloor \ell+1} + \ldots + Y_n$. Since the absolutely continuous part of the law μ of $\tilde{Y}_j$ is nontrivial, then μ is of the form $qv_1 + (1-q)v_2$ with $q \in (0, 1]$ and the characteristic function of v_1 belonging to some L_p. Indeed, μ has a bit which is a uniform distribution on some measurable set whose characteristic function is in L_2. Therefore, applying Step II for $\tilde{Y}_j$'s we get that $X_n^{(C)}$ is c-good when $C \sqrt{\frac{\lfloor n/\ell \rfloor}{n}} \geq C_0^{(II)}$. So we can set $C_0 = C_0^{(II)} \sqrt{2\ell}$.

Lemma 2. *Suppose Z is a $\mathbb{T}$-valued c-good random variable and B_Z is the operator defined by $(B_Z f)(x) = \mathbb{E} f(x \oplus Z)$. Then for every $1 \leq p \leq \infty$ and every $f \in L_p(\mathbb{T})$ with $\int_{\mathbb{T}} f = 0$ we have $\|B_Z f\| \leq (1-c)\|f\|$, where $\|\cdot\|$ is the L_p norm.*

Proof. Fix $1 \leq p < \infty$. Let μ be the law of Z. Define the measure $v(A) = (\mu(A) - c|A|)/(1-c)$ for measurable $A \subset \mathbb{T}$. Since μ is c-good, v is a Borel probability measure on $\mathbb{T}$. Take $f \in L_p(\mathbb{T})$ with mean zero. Then by Jensen's inequality we have

$$\|B_Z f\|^p = \int_0^1 \left| \int_0^1 f(x \oplus s) \, d\mu(s) \right|^p dx$$

$$= (1-c)^p \int_0^1 \left| \int_0^1 f(x \oplus s) \, dv(s) \right|^p dx$$

$$\leq (1-c)^p \int_0^1 \int_0^1 |f(x \oplus s)|^p \, dv(s) \, dx$$

$$= (1-c)^p \|f\|^p \int_0^1 dv(s) = (1-c)^p \|f\|^p.$$

Since c does not depend on p we get the same inequality for $p = \infty$ by passing to the limit.

Now we are ready to give the proof of Theorem 2.

Proof. Fix $1 \leq p \leq \infty$. Let $Y_1, Y_2, \ldots$ be independent copies of Y. Observe that

$$(A_t^n f)(x) = \mathbb{E} f(x \oplus (tY_1 \oplus \ldots \oplus tY_n))$$

$$= \mathbb{E} f\left(x \oplus \left(t\sqrt{n}\left(\frac{Y_1 + \ldots + Y_n}{\sqrt{n}}\right) \bmod 1\right)\right).$$

Take $n(t) = C_0^2 \lceil 1/t^2 \rceil N$, where C_0 and N are the numbers given by Lemma 1. Therefore, with $X_{n(t)}^{(C)}$ defined by (5), we can write

$$(A_t^{n(t)} f)(x) = \mathbb{E} f\left(x \oplus X_{n(t)}^{(C)}\right),$$

where $C = t\sqrt{n(t)} = tC_0 \sqrt{\lceil 1/t^2 \rceil N} \geq C_0 \sqrt{N} \geq C_0$. Thus $X_{n(t)}^{(C)}$ is $c(Y)$-good with some constant $c(Y) \in (0, 1)$. From Lemma 2 we have

$$\left\| A_t^{n(t)} f \right\| \leq (1 - c(Y)) \, \|f\|$$

for all f satisfying $\int_{\mathbb{T}} f = 0$.

The operator A_t is a contraction, namely $\|A_t f\| \leq \|f\|$ for all $f \in L_p(\mathbb{T})$. Using this observation and the triangle inequality we obtain

$$\|f - A_t f\| \geq \frac{1}{n} \left(\|f - A_t f\| + \|A_t f - A_t^2 f\| + \ldots + \|A_t^{n-1} f - A_t^n f\| \right)$$

$$\geq \frac{1}{n} \|f - A_t^n f\| \, .$$

Taking $n = n(t)$ we arrive at

$$\frac{1}{n(t)} \left\| f - A_t^{n(t)} f \right\| \geq \frac{1}{t^{-2} + 1} \cdot \frac{1}{C_0^2 \cdot N} \left(\|f\| - \left\| A_t^{n(t)} f \right\| \right) \geq \frac{c(Y)}{2 C_0^2 \cdot N} t^2 \|f\| \, .$$

It suffices to take $c = c(Y)/(2 C_0^2 \cdot N)$.

Remark 4. Consider an ℓ-decent random variable Y. As it was noticed in the proof of Lemma 1 (Step III), the law $Y_1 + \ldots + Y_\ell$ has a bit whose characteristic function is in L_2. Conversely, if the law of $S_m = Y_1 + \ldots + Y_m$ has the form $q\mu + (1-q)\nu$ with $q \in (0, 1]$ and the characteristic function of μ belonging to L_p for some $p \geq 1$, then the characteristic function of the bit $\mu^{\star \lceil p/2 \rceil}$ of the sum of $\lceil p/2 \rceil$ i.i.d. copies of S_m is in L_2. In particular, that bit has a density function in $L_1 \cap L_2$. Thus Y is $(m \lceil p/2 \rceil)$-decent.

Remark 5. The idea to study the operators A_t (see (4)) stemmed from the following question posed by Gideon Schechtman (personal communication): *given $\varepsilon > 0$, is it true that there exists a natural number $k = k(\varepsilon)$ such that for any bounded linear operator $T : L_1[0, 1] \to L_1[0, 1]$ with $\|T\|_{L_1 \to L_1} \leq 1$ which has the property*

$$\forall f \in L_1[0, 1] \ (\, |\mathrm{supp} f| \leq 1/2 \implies \|Tf\|_1 \geq \varepsilon \|f\|_1)$$

there exist $\delta > 0$ and functions $g_1, \ldots, g_k \in L_\infty[0, 1]$ such that

$$\|Tf\|_1 \geq \delta \|f\|_1 \quad \text{for any } f \in L_1[0, 1] \text{ satisfying } \int_0^1 fg_j = 0, \, j = 1, \ldots, k?$$

This, in an equivalent form, was asked by Bill Johnson in relation with a question on Mathoverflow (http://mathoverflow.net/questions/101253). Our hope was that an operator $T = I - A_t$, for some Y, would provide a negative answer to Schechtman's question. However, Theorem 2 says that if Y is an ℓ-decent random variable, then T is nicely invertible on the subspace of functions $f \in L_p$ such that $\int f \cdot 1 = 0$. Some time after this paper had been written, the question was answered in the negative (see [7]).

Acknowledgements This work was initiated while the authors were visiting the Weizmann Institute of Science in Rehovot, Israel. We thank Prof. Gideon Schechtman for supervision and making our stay possible.

We are grateful to Prof. Krzysztof Oleszkiewicz for his remarks which led to the present general statement of Theorem 2. We thank Prof. Keith Ball for helping us to simplify the proof of Lemma 2. We also appreciate all the valuable comments Prof. Stanisław Kwapień gave us.

Research of the First named author partially supported by NCN Grant no. 2011/01/N/ST1/01839. Research of the second named author partially supported by NCN Grant no. 2011/01/N/ST1/05960.

References

1. O. Bandtlow, Estimates for norms of resolvents and an application to the perturbation of spectra. Math. Nachr. **267**, 3–11 (2004)
2. R.N. Bhattacharya, R. Ranga Rao, *Normal Approximation and Asymptotic Expansions*. Classics in Applied Mathematics, vol. 64 (Society for Industrial and Applied Mathematics, Philadelphia, 2010)
3. M.I. Gil', Estimates of the norm of the resolvent of a completely continuous operator (Russian). Mat. Zametki **26**(5), 713–717, 814 (1979)
4. D. Gilbarg, N.S. Trudinger, *Elliptic Partial Differential Equations of Second Order*. Reprint of the 1998 edition. Classics in Mathematics. (Springer, Berlin, 2001)
5. I. Gohberg, G. Heinig, Inversion of finite Toeplitz matrices consisting of elements of a noncommutative algebra, in *Convolution Equations and Singular Integral Operators: Selected Papers of Israel Gohberg and Georg Heinig, Israel Gohberg and Nahum Krupnik*. Operator Theory: Advances and Applications, vol. 206 (Birkhäuser Verlag, Basel, 2010)
6. I. Gohberg, L. Semencul, The inversion of finite Toeplitz matrices and their continual analogues (in Russian). Matem. Issled. **7**(2(24)), 201–223 (1972)
7. W. Johnson, A. Nasseri, G. Schechtman, T. Tkocz, Injective Tauberian operators on L_1 and operators with dense range on ℓ_∞ (preprint)
8. L. Lerer, L.M. Spitkovsky, V. Olshevsky, *Convolution Equations and Singular Integral Operators: Selected Papers of Israel Gohberg and Georg Heinig, Israel Gohberg and Nahum Krupnik*. Operator Theory: Advances and Applications, vol. 206 (Birkhäuser Verlag, Basel, 2010)

On Isotropicity with Respect to a Measure

Liran Rotem

Abstract A body C is said to be isotropic with respect to a measure μ if the function

$$\theta \to \int_C \langle x, \theta \rangle^2 \, d\mu(x)$$

is constant on the unit sphere. In this note, we extend a result of Bobkov, and prove that every body can be put in isotropic position with respect to any rotation invariant measure.

When the body C is convex, and the measure μ is log-concave, we relate the isotropic position with respect to μ to the famous M-position, and give bounds on the isotropic constant.

1 Introduction

Let μ be a finite Borel measure on $\mathbb{R}^n$ with finite second moments. For simplicity, we will always assume our measures are *even*, i.e. measures which satisfy $\mu(A) = \mu(-A)$ for every Borel set A. We will say that such a measure is isotropic if the function

$$\theta \mapsto \int \langle x, \theta \rangle^2 \, d\mu(x)$$

is constant on the unit sphere $S^{n-1} = \{x : |x| = 1\}$.

L. Rotem (✉)
School of Mathematical Sciences, Sackler Faculty of Exact Sciences, Tel Aviv University, Ramat Aviv, Tel Aviv 69978, Israel
e-mail: liranro1@post.tau.ac.il

© Springer International Publishing Switzerland 2014
B. Klartag, E. Milman (eds.), *Geometric Aspects of Functional Analysis*,
Lecture Notes in Mathematics 2116, DOI 10.1007/978-3-319-09477-9_27

In particular, let C be an origin-symmetric and compact set in $\mathbb{R}^n$ with non-empty interior. From now on, such sets will simply be called *bodies*. Let λ_C be the restriction of the Lebesgue measure λ to the set C:

$$\lambda_C(A) = \lambda(A \cap C).$$

We say that C is isotropic if the measure λ_C is isotropic, i.e. if the integrals

$$\int_C \langle x, \theta \rangle^2 \, dx$$

are independent of $\theta \in S^{n-1}$.

For a discussion of isotropic bodies and measures see, for example, [9] and [1]. Notice that at the moment we do not assume our measures and bodies satisfy any convexity properties, nor do we assume any normalization condition. This will change in Sect. 3.

In this note we will study the following notion:

Definition 1. Let μ be an even locally finite Borel measure on $\mathbb{R}^n$. Let C be a body with $\mu(C) > 0$. We say that C is isotropic with respect to μ, or that the pair (C, μ) is isotropic, if

$$\int_C \langle x, \theta \rangle^2 \, d\mu(x)$$

is independent of $\theta \in S^{n-1}$.

From a formal point of view, this is not a new definition. Isotropicity of the pair (C, μ) is nothing more than isotropicity of the measure μ_C, where μ_C is the restriction of μ to C. In particular, C is isotropic with respect to the Lebesgue measure if and only if it is isotropic. However, this new notation is better suited for our needs, as we want to separate the roles of μ and C.

Let us demonstrate this point by discussing the notion of an *isotropic position*. It is well known that for any measure μ one can find a linear map $T \in \mathrm{SL}(n)$ such that the push-forward $T_\# \mu$ is isotropic. The proof may be written in several ways (again, see for example the proofs in [9] and [1]), but in any case this is little more than an exercise in linear algebra. Since $T_\# (\mu_C) = (T_\# \mu)_{T(C)}$ we see that for every pair (C, μ) one can find a map $T \in \mathrm{SL}(n)$ such that $(TC, T_\# \mu)$ is isotropic.

However, we are interested in a different problem. For us, the measure μ is a fixed "universal" measure, which we are unwilling to change. Given a body C, we want to put it in an isotropic position with respect to this given μ. In other words, we want to find a map $T \in \mathrm{SL}(n)$ such that (TC, μ) is isotropic. This is already a non-linear problem, and it is far less obvious that such a T actually exists.

Of course, there is one choice of μ for which the problem is trivial. For the Lebesgue measure λ we know that $T_\# \lambda = \lambda$ for any $T \in \mathrm{SL}(n)$. Hence the two problems coincide, and there is nothing new to prove.

There is one non-trivial case where the problem was previously solved. Let γ be the Gaussian measure on $\mathbb{R}^n$, defined by

$$\gamma(A) = (2\pi)^{-\frac{n}{2}} \int_A e^{-|x|^2/2} dx.$$

In [3], Bobkov proved the following result:

Theorem 1. *Let C be a body in $\mathbb{R}^n$. If $\gamma(C) \geq \gamma(TC)$ for all $T \in SL(n)$, then C is isotropic with respect to γ.*

If C is assumed to be convex, then the converse is also true.

A simple compactness argument shows that the map $T \mapsto \gamma(TC)$ attains a maximum on $SL(n)$ for some map T_0. Theorem 1 implies that $T_0 C$ is isotropic with respect to γ.

The main goal of this note is to discuss isotropicity with respect any rotation invariant measure. In the next section we will extend Bobkov's argument, and prove that C can be put in isotropic position with respect to any rotation invariant measure μ. Then in Sect. 3 we will restrict our attention to the case where C is convex and μ is log-concave (all relevant definitions will be given there). We will relate the isotropicity of (C, μ) to the M-position, and give an upper bound on the isotropic constant of C with respect to μ.

2 Existence of Isotropic Position

In this section we will assume that μ is rotation invariant:

Definition 2. We say that μ is rotation invariant if there exists a bounded function $f : [0, \infty) \to [0, \infty)$ such that

$$\frac{d\mu}{dx} = f(|x|).$$

We will always assume that f has a finite first moment, i.e. $\int_0^\infty tf(t)dt < \infty$.

Our goal is to prove that if μ is rotation invariant, then for every C one can find a map $T \in SL(n)$ such that (TC, μ) is isotropic. To do so we will need the following definitions:

Definition 3. Let $\mu = f(|x|) dx$ be a rotation invariant measure on $\mathbb{R}^n$. Then:

1. The associated measure $\hat{\mu}$ is the measure on $\mathbb{R}^n$ with density $g(|x|)$, where $g : [0, \infty) \to [0, \infty)$ is defined by

$$g(t) = \int_t^\infty sf(s)ds.$$

2. Given a body C we define the associated functional $J_{\mu,C} : \mathrm{SL}(n) \to \mathbb{R}$ by

$$J_{\mu,C}(T) = \hat{\mu}(TC) = \int_C g(|Tx|)\, dx.$$

When the measure μ is obvious from the context, we will write J_C instead of $J_{\mu,C}$.

As one example of the definitions, notice that for the Gaussian measure γ we have $\hat{\gamma} = \gamma$. Hence the functional $J_{\gamma,C}$ is exactly the one being maximized in Bobkov's Theorem 1.

In the general case, we have the following Proposition:

Proposition 1. *Fix a rotation invariant measure μ, a body C, and $T \in \mathrm{SL}(n)$. Then (TC, μ) is isotropic if and only if T is a critical point of $J_{\mu,C}$.*

Proof. We will first show that the identity matrix I is a critical point for J_C if and only if (C, μ) is isotropic. Indeed, I is a critical point if and only if

$$\frac{d}{dt}\bigg|_{t=0} J_C\left(e^{tA}\right) = 0$$

for all maps $A \in \mathfrak{sl}(n)$, i.e. all linear maps A with $\mathrm{Tr}\, A = 0$.

An explicit calculation of the derivative gives

$$\frac{d}{dt}\bigg|_{t=0} J_C\left(e^{tA}\right) = \frac{d}{dt}\bigg|_{t=0} \left(\int_C g\left(|e^{tA}x|\right) dx\right) = \int_C \left(\frac{d}{dt}\bigg|_{t=0} g\left(|e^{tA}x|\right)\right) dx$$

$$= \int_C g'(|x|) \cdot \left\langle \frac{x}{|x|}, Ax\right\rangle dx = -\int_C |x| f(|x|) \left\langle \frac{x}{|x|}, Ax\right\rangle dx$$

$$= -\int_C \langle x, Ax\rangle\, d\mu(x).$$

Hence we see that I is a critical value for J_C if and only if

$$\int_C \langle x, Ax\rangle\, d\mu(x) = 0$$

for all maps A with $\mathrm{Tr}\, A = 0$. This condition is known and easily seen to be equivalent to isotropicity of (C, μ).

So far we have proved the result only for $T = I$. For general case notice that $J_{TC}(S) = J_C(ST)$ for every $S, T \in \mathrm{SL}(n)$. Hence T is a critical point for J_C if and only if I is a critical point for J_{TC}, which holds if and only if (TC, μ) is isotropic. $\quad\square$

From here it is easy to deduce the main result:

Theorem 2. *Let μ be a rotation invariant measure on $\mathbb{R}^n$, and let C be a body in $\mathbb{R}^n$. Then there exists a map $T \in \mathrm{SL}(n)$ such that TC is isotropic with respect to μ.*

Proof. Write $s_n(T)$ for the smallest singular value of a map $T \in SL(n)$. It is not hard to see that $J_{\mu,C}(T) \to 0$ as $s_n(T) \to 0$. Since

$$\{T \in SL(n) : s_n(T) \geq \varepsilon\}$$

is compact, it follows that $J_{\mu,C}$ attains a global maximum at some point T. In particular, T is a critical point for $J_{\mu,C}$, so (TC, μ) is isotropic.

Remark 1. When $\mu = \gamma$ and the body C is convex, the functional $J_{\mu,C}$ has a unique positive definite critical point. In other words, there exists a positive definite matrix $S \in SL(n)$ such that the set of critical points of $J_{\mu,C}$ is exactly $\{US : U \in O(n)\}$. Moreover, every such critical point is a global maximum. The proof of these facts, which appears in [3], is based on the so-called (B) conjecture, proved by Cordero-Erausquin et al. [6]. This fact explains the second half of Theorem 1. It also implies that the isotropic position with respect to γ is unique, up to rotations and reflections.

For general measures μ, we have no analog of the (B) conjecture, and so $J_{\mu,C}$ may have critical points which are not the global maximum. Hence we define:

Definition 4. We say that C is in principle isotropic position with respect to μ if $J_{\mu,C}$ is maximized at the identity matrix I.

Proposition 1 shows that if C is in principle isotropic position with respect to μ, then it is also isotropic with respect to μ in the sense of Definition 1. If the (B) conjecture happens to hold for the measure μ, then these two notions coincide. However, we currently know the (B) conjecture for very few measures: the original result concerns the Gaussian measure, and Livne Bar-On has recently proved the conjecture when μ is a uniform measure in the plane [2].

Let us conclude this section with one application of Theorem 2 for isotropicity of bodies:

Proposition 2. *Let B be a Euclidean ball of some radius $r > 0$. Then for every body C one can find a map $T \in SL(n)$ such that $TC \cap B$ is isotropic.*

Proof. Let $\mu = \lambda_B$ be the uniform measure on B. μ is rotation invariant, so we can apply Theorem 2 and find a map $T \in SL(n)$ such that TC is isotropic with respect to μ. This just means that $\mu_{TC} = \lambda_{TC \cap B}$ is an isotropic measure, or that $TC \cap B$ is an isotropic body.

Following Proposition 2, Prof. Bobkov asked about an interesting variant concerning Minkowski addition. Recall that the Minkowski sum of sets $A, B \subseteq \mathbb{R}^n$ is defined by

$$A + B = \{a + b : a \in A, b \in B\}.$$

Bobkov then posed the following question:

Problem 1. Let B be a Euclidean ball of some radius $r > 0$. Given a convex body C, is it always possible to find $T \in SL(n)$ such that $TC + B$ is isotropic?

Unfortunately, we do not know the answer to this question.

3   Properties of Isotropic Pairs

Let μ be an (even) isotropic measure with density f. The isotropic constant of μ is defined as

$$L_\mu = \frac{f(0)^{\frac{1}{n}}}{\mu(\mathbb{R}^n)^{\frac{1}{n}+\frac{1}{2}}} \left(\int \langle x, \theta \rangle^2 \, d\mu(x) \right)^{\frac{1}{2}}.$$

We define the isotropic constant of the pair (C, μ) to be the isotropic constant of μ_C, so

$$L(C, \mu) = \frac{f(0)^{\frac{1}{n}}}{\mu(C)^{\frac{1}{n}+\frac{1}{2}}} \left(\int_C \langle x, \theta \rangle^2 \, d\mu(x) \right)^{\frac{1}{2}}.$$

A major open question, known as the slicing problem, asks if $L_K = L(K, \lambda)$ is bounded from above by a universal constant for every dimension n and every isotropic convex body K in $\mathbb{R}^n$ (see [9] for a much more information). It turns out that for certain rotation invariant measures μ it is possible to give an upper bound on $L(K, \mu)$ whenever K is a convex body in principle isotropic position with respect to μ. In the Gaussian case, this was done by Bobkov in [3]. We will demonstrate how his methods can be extended, starting with the case where μ is a uniform measure on the Euclidean ball.

In order to prove our result, we will need the notion of M-position. Let B be the Euclidean ball of volume (Lebesgue measure) 1. A convex body K of volume 1 is said to be in M-position with constant $C > 0$ if

$$|K \cap B| \geq C^{-n}.$$

There are many other equivalent ways to state this definition, but this definition will be the most convenient for us. A remarkable theorem of Milman shows that for every convex body K of volume 1 there exists a map $T \in SL(n)$ such that TK is in M-position, with some universal constant C (see [8]).

After giving all the definitions, we are ready to prove the following:

Theorem 3. *Let K be a convex body of volume 1 such that $K \cap B$ is in principle isotropic position (i.e. K is in principle isotropic position with respect to μ, when μ is the uniform measure on B). Then*

1. K is in M-position. In fact

$$|K \cap B| \geq \left(\frac{1}{2}\right)^{n+1} \sup_{T \in SL(n)} |TK \cap B| \geq C^{-n}$$

for some universal $C > 0$.

2. The isotropic constant of $K \cap B$ is bounded by an absolute constant.

Proof. We should understand how $J_{K,\mu}$ looks in this special case. If we denote the radius of B by r, then $d\mu = f(|x|)dx$, where $f = \mathbf{1}_{[0,r]}$. Hence

$$g(t) = \int_t^\infty s \cdot \mathbf{1}_{[0,r]}(s)ds = \begin{cases} \frac{r^2-t^2}{2} & t \leq r \\ 0 & t > r, \end{cases}$$

and

$$J_K(T) = \int_{TK} g(|x|)\, dx = \int_{TK \cap B} \frac{r^2 - |x|^2}{2} dx.$$

Notice that $J_K(T)$ is almost the same as the volume $|TK \cap B|$ (properly normalized) for every $T \in SL(n)$. Indeed, on the one hand we have

$$J_K(T) = \int_{TK \cap B} \frac{r^2 - |x|^2}{2} dx \leq \int_{TK \cap B} \frac{r^2}{2} dx = \frac{r^2}{2} |TK \cap B|,$$

and on the other hand we have

$$J_K(T) = \int_{TK \cap B} \frac{r^2 - |x|^2}{2} dx \geq \int_{\frac{TK \cap B}{2}} \frac{r^2 - |x|^2}{2} dx$$

$$\geq \int_{\frac{TK \cap B}{2}} \frac{r^2 - \left(\frac{r}{2}\right)^2}{2} dx = \frac{3}{8} r^2 \left| \frac{TK \cap B}{2} \right| \geq \left(\frac{1}{2}\right)^{n+1} \frac{r^2}{2} |TK \cap B|.$$

Since K is in principle isotropic position we know that for every $T \in SL(n)$ we have $J_K(T) \leq J_K(I)$, and then

$$|TK \cap B| \leq 2^{n+1} \cdot \frac{2}{r^2} J_K(T) \leq 2^{n+1} \frac{2}{r^2} J_K(I) \leq 2^{n+1} |K \cap B|,$$

so the first inequality of (1) is proven. In particular, by Milman's theorem we get that $|K \cap B| \geq C^{-n}$ for some universal constant $C > 0$.

Finally, in order to prove (2), Notice that it follows from the definition that

$$L_{K \cap B} = \frac{1}{|K \cap B|^{\frac{1}{n}+\frac{1}{2}}} \left(\int_{K \cap B} \frac{|x|^2}{n} dx \right)^{\frac{1}{2}}.$$

Since B has radius $\leq C\sqrt{n}$ we get that

$$L_{K\cap B} \leq \frac{1}{|K\cap B|^{\frac{1}{n}+\frac{1}{2}}} \cdot \left(\int_{K\cap B} \frac{C^2 n}{n} dx\right)^{\frac{1}{2}} = \frac{C}{|K\cap B|^{\frac{1}{n}}} \leq C',$$

and the theorem is proven.

What happens for general rotation invariant measures μ? In order to prove a similar estimate, we will need to assume that the measure μ is log-concave:

Definition 5. A Borel measure μ on $\mathbb{R}^n$ is log-concave if for every Borel sets A and B and every $0 < \lambda < 1$ we have

$$\mu(\lambda A + (1-\lambda)B) \geq \mu(A)^\lambda \mu(B)^{1-\lambda}.$$

Borel [4, 5] gave a simple and useful characterization of log-concave measures: assume μ is not supported on any affine hyperplane. Then μ is log-concave if and only if μ has a density f, which is log-concave. Log-concavity of f just means that $(-\log f)$ is a convex function.

For log-concave measures we have the following bound on the isotropic constant of (K, μ):

Proposition 3. *Let μ be a log-concave, rotation invariant measure on $\mathbb{R}^n$, and let $K \subseteq \mathbb{R}^n$ be a convex body. Then*

$$L(K, \mu) \leq C \cdot \mu(\mathbb{R}^n)^{\frac{1}{n}} \cdot \mu(K)^{-\frac{1}{n}}$$

for some universal constant $C > 0$.

Proof. Write $d\mu(x) = f(|x|)dx$. Since both sides of the inequality are invariant to a scaling of f, we may assume without loss of generality that $f(0) = 1$.

Recall the following construction of Ball [1]: If μ is a log-concave measure with density f, and $p \geq 1$, we define

$$K_p(\mu) = \left\{ x \in \mathbb{R}^n : \int_0^\infty f(rx)r^{p-1}dr \geq \frac{f(0)}{p} \right\}.$$

Ball proved that $K_p(\mu)$ is a convex body, but we won't need this fact in our proof. We will need that fact that if $p = n+1$ and $f(0) = 1$, then $L_{K_{n+1}(\mu)} \asymp L_\mu$ and $|K_{n+1}(\mu)| \asymp \mu(\mathbb{R}^n)$. This is proven, for example, in Lemma 2.7 of [7]. Here the notation $A \asymp B$ means that $\frac{A}{B}$ is bounded from above and from below by universal constants.

Since our μ is rotation invariant, $K_{n+1}(\mu)$ is just a Euclidean ball. If we denote its radius by R, then

$$\mu(\mathbb{R}^n)^{\frac{1}{n}} \asymp |K_{n+1}(\mu)|^{\frac{1}{n}} \asymp \frac{R}{\sqrt{n}}.$$

Now we turn our attention to the body $\widetilde{K} = K_{n+1}(\mu_K)$. It is obvious that $\widetilde{K} \subseteq K_{n+1}(\mu)$. Hence we get

$$L(K,\mu) \asymp L_{\widetilde{K}} = \frac{1}{|\widetilde{K}|^{\frac{1}{n}+\frac{1}{2}}} \left(\int_{\widetilde{K}} \frac{|x|^2}{n} dx \right)^{\frac{1}{2}} \leq \frac{1}{|\widetilde{K}|^{\frac{1}{n}+\frac{1}{2}}} \left(\int_{\widetilde{K}} \frac{R^2}{n} dx \right)^{\frac{1}{2}}$$

$$= \frac{R}{\sqrt{n}} \cdot |\widetilde{K}|^{-\frac{1}{n}} \asymp \mu(\mathbb{R}^n)^{\frac{1}{n}} \cdot \mu_K(\mathbb{R}^n)^{-\frac{1}{n}} = \mu(\mathbb{R}^n)^{\frac{1}{n}} \cdot \mu(K)^{-\frac{1}{n}},$$

and the proof is complete.

Therefore in order to bound $L(K,\mu)$ from above we need to bound $\mu(K)$ from below. Notice that we have three distinct functionals on $\mathrm{SL}(n)$:

$$T \mapsto \mu(TK)$$
$$T \mapsto J_{K,\mu}(T) = \hat{\mu}(TK)$$
$$T \mapsto |TK \cap B|.$$

The estimates of Theorem 3 only depend on these functionals being close to each other. More concretely, we have the following:

Theorem 4. *Let μ be a log-concave rotation invariant measure on $\mathbb{R}^n$, and let $K \subseteq \mathbb{R}^n$ be a convex body of volume 1, which is in principle isotropic position with respect to μ. Assume that*

$$\sup_{T \in \mathrm{SL}(n)} \frac{J_{K,\mu}(T)}{|TK \cap B|} \leq a^n \cdot \inf_{T \in \mathrm{SL}(n)} \frac{J_{K,\mu}(T)}{|TK \cap B|},$$

and that

$$\mu(K) \geq b^{-n} \cdot |K \cap B|.$$

Then:

1. K is in M-position with constant $C \cdot a$ for some universal $C > 0$.
2. $L(K,\mu) \leq C\mu(\mathbb{R}^n)^{\frac{1}{n}} \cdot ab$ for some universal constant $C > 0$.

Proof. There is very little to prove here. For (1), define for simplicity

$$m = \inf_{T \in \mathrm{SL}(n)} \frac{J_{K,\mu}(T)}{|TK \cap B|}.$$

Then for every $T \in \mathrm{SL}(n)$ we have

$$|TK \cap B| \leq \frac{1}{m} \cdot J_{K,\mu}(T) \leq \frac{1}{m} \cdot J_{K,\mu}(I) \leq \frac{a^n m}{m} \cdot |K \cap B| = a^n |K \cap B|.$$

By this estimate and Milman's theorem, it follows that K is in M-position with constant $C \cdot a$.

Now for (2) we use Proposition 3 together with part (1) and immediately obtain

$$L(K, \mu) \leq C\mu \left(\mathbb{R}^n\right)^{\frac{1}{n}} \mu(K)^{-\frac{1}{n}} \leq C\mu \left(\mathbb{R}^n\right)^{\frac{1}{n}} \cdot b \, |K \cap B|^{-\frac{1}{n}}$$

$$\leq C\mu \left(\mathbb{R}^n\right)^{\frac{1}{n}} \cdot b \cdot (Ca) = C'\mu \left(\mathbb{R}^n\right)^{\frac{1}{n}} \cdot ab$$

Of course, in general there is no reason for a and b to be small. However, for any specific μ, one may try and compute explicit values for these constants. Both Theorem 3 and Bobkov's theorem in the Gaussian case follow from this general scheme.

Acknowledgements I would like to thank Prof. Sergey Bobkov for fruitful and interesting discussions, and my advisor, Prof. Vitali Milman, for his help and support. I would also like to thank the referee for his useful remarks and corrections. The author is supported by the Adams Fellowship Program of the Israel Academy of Sciences and Humanities. The author is also supported by ISF grant 826/13 and BSF grant 2012111.

References

1. K. Ball, Logarithmically concave functions and sections of convex sets in $\mathbb{R}^n$. Stud. Math. **88**(1), 69–84 (1988)
2. A.L. Bar-On, The (B) conjecture for uniform measures in the plane, in: *Geometric Aspects of Functional Analysis: Israel Seminar (GAFA) 2011–2013*, ed. by B. Klartag, E. Milman (Springer, Heidelberg, 2014, this volume)
3. S. Bobkov, On Milman's ellipsoids and M-position of convex bodies, in *Concentration, Functional Inequalities and Isoperimetry*, ed. by C. Houdré, M. Ledoux, E. Milman, M. Milman. Contemporary Mathematics, vol. 545 (American Mathematical Society, Providence, 2011), pp. 23–33
4. C. Borell, Convex measures on locally convex spaces. Ark. Mat. **12**(1–2), 239–252 (1974)
5. C. Borell, Convex set functions in d-space. Period. Math. Hung. **6**(2), 111–136 (1975)
6. D. Cordero-Erausquin, M. Fradelizi, B. Maurey, The (B) conjecture for the Gaussian measure of dilates of symmetric convex sets and related problems. J. Funct. Anal. **214**(2), 410–427 (2004)
7. B. Klartag, On convex perturbations with a bounded isotropic constant. Geom. Funct. Anal. **16**(6), 1274–1290 (2006)
8. V. Milman, An inverse form of the Brunn-Minkowski inequality, with applications to the local theory of normed spaces. C. R. Acad. Sci. I Math. **302**(1), 25–28 (1986)
9. V. Milman, A. Pajor, Isotropic position and inertia ellipsoids and zonoids of the unit ball of a normed n-dimensional space, in *Geometric Aspects of Functional Analysis, Israel Seminar 1987–1988*, ed. by J. Lindenstrauss, V. Milman. Lecture Notes in Mathematics, vol. 1376 (Springer, Berlin, 1989), pp 64–104

A Formula for Mixed Volumes

Rolf Schneider

Abstract An identity for mixed volumes, discovered by Gusev and Esterov, which involves together with some convex bodies also the convex hull of their union, is given a new proof, using only the classical approach to mixed volumes.

In the introduction to his paper [2], Esterov writes (with slightly different notation) the following: "Counting Euler characteristics of the discriminant of the quadratic equation in terms of Newton polytopes in two different ways, Gusev [3] found an unexpected relation for mixed volumes of two polytopes P_1 and P_2 in $\mathbb{R}^n$ and the convex hull P of their union. For instance, assuming $n = 2$ and denoting the mixed area of polygons A, B by $V(A, B)$, this relation specializes to

$$V(P, P) - V(P, P_1) - V(P, P_2) + V(P_1, P_2) = 0.$$

We call it unexpected because it is not a priori invariant under parallel translations of P_1."

Esterov [2] proved a multidimensional generalization of this equality. To formulate it, we denote the support function of a convex body $K \subset \mathbb{R}^n$ by h_K and write the mixed volume $V(K_1, \ldots, K_n)$ of the convex bodies $K_1, \ldots, K_n$ in the form $V(h_{K_1}, \ldots, h_{K_n})$, that is, we consider it equivalently as a functional of support functions. By linearity in each argument, the mixed volume can then be extended to differences of support functions. This extension appears already in the work of Aleksandrov [1, Sect. 6]; see also [4, Sect. 5.2]. Esterov's extension of Gusev's formula (in a slightly more general form, which is needed for the inductive proof below) then reads as follows. Here $\mathcal{K}^n$ denotes the set of convex bodies (nonempty, compact, convex subsets) of $\mathbb{R}^n$.

Theorem. *Let* $2 \leq k \leq n$, *let* $A_1, \ldots, A_k, B_1, \ldots, B_{n-k}, C_1, \ldots, C_{n-k} \in \mathcal{K}^n$, *write*

$$A := \mathrm{conv}(A_1 \cup \cdots \cup A_k),$$

R. Schneider (✉)
Mathematisches Institut, Albert-Ludwigs-Universität, 79104 Freiburg i. Br., Germany
e-mail: rolf.schneider@math.uni-freiburg.de

© Springer International Publishing Switzerland 2014   423
B. Klartag, E. Milman (eds.), *Geometric Aspects of Functional Analysis*,
Lecture Notes in Mathematics 2116, DOI 10.1007/978-3-319-09477-9__28

and abbreviate the $(n-k)$-tuple $(h_{B_1} - h_{C_1}, \ldots, h_{B_{n-k}} - h_{C_{n-k}})$ by $\mathcal{H}_{n-k}$. Then

$$V(h_A - h_{A_1}, \ldots, h_A - h_{A_k}, \mathcal{H}_{n-k}) = 0. \tag{1}$$

Note that

$$h_A = \max\{h_{A_1}, \ldots, h_{A_k}\},$$

hence, at each point of the unit sphere, at least one the functions $h_A - h_{A_1}, \ldots, h_A - h_{A_k}$ appearing in (1) vanishes. This observation may serve as an intuitive hint to the validity of (1), but it is, of course, not a proof.

Using multilinearity, Eq. (1) can be written as an identity involving mixed volumes of the bodies $A_1, \ldots, A_k$ and the convex hull of their union, and of the bodies B_i, C_i. By the symmetry of the mixed volume, the special role played by the first k arguments in (1) can be taken over by any k arguments.

In addition to Esterov's dictum of being 'unexpected', this identity may well be called a surprise, since it deals with the well-established topic of mixed volumes, but within this classical theory it had not been noticed before. In the following, we give a proof of the identity (1) that, other than the proofs of Gusev and Esterov, stays entirely within the classical theory.

This proof makes use of the representation of the mixed volume of $K_1, \ldots, K_n \in \mathcal{K}^n$ by

$$V(K_1, \ldots, K_n) = \frac{1}{n} \int_{\mathbb{S}^{n-1}} h_{K_1} \, dS(K_2, \ldots, K_n, \cdot), \tag{2}$$

where $\mathbb{S}^{n-1}$ is the unit sphere of $\mathbb{R}^n$ and $S(K_2, \ldots, K_n, \cdot)$ denotes the mixed area measure of the bodies $K_2, \ldots, K_n$ (see, e.g. [4, Sect. 5.1]). By linearity in each argument, also the mixed area measures and formula (2) extend to differences of support functions (also this extension appears already in [1]).

The following lemma expresses in a precise way how the mixed area measures are 'determined locally'.

Lemma. *Let $K, L, K_2, \ldots, K_{n-1} \in \mathcal{K}^n$, let $\omega \subset \mathbb{S}^{n-1}$ be an open set. If*

$$h_K(u) = h_L(u) \quad \text{for all } u \in \omega,$$

then

$$S(K, K_2, \ldots, K_{n-1}, \omega') = S(L, K_2, \ldots, K_{n-1}, \omega')$$

for all Borel sets $\omega' \subset \omega$.

Proof. Let $u \in \omega$. Since the functions h_K and h_L coincide in a neighbourhood of u, their directional derivatives at u also coincide. Therefore, Theorem 1.7.2 and the second displayed formula on p. 88 of [4] show that $\tau(K, \omega) = \tau(L, \omega)$, where

$\tau(K, \omega)$ denotes the reverse spherical image of K at ω. This implies the equality $S_{n-1}(K, \cdot) \llcorner \omega = S_{n-1}(L, \cdot) \llcorner \omega$ for the restrictions of the surface area measures S_{n-1} to the set ω; see [4, p. 215]. In this equation, K and L may be replaced (with the same proof) by $K + \lambda_2 K_2 + \cdots + \lambda_{n-1} K_{n-1}$ and $L + \lambda_2 K_2 + \cdots + \lambda_{n-1} K_{n-1}$, respectively, with any fixed $\lambda_2, \ldots, \lambda_{n-1} \geq 0$. Now the assertion of the lemma follows from [4], formula (5.21). $\qquad \square$

In the lemma, we can again use multilinearity to replace $K_2, \ldots, K_{n-1}$ by differences of support functions. Then the lemma immediately yields the case $k = 2$ of (1). For this, we put

$$\omega_1 := \{u \in \mathbb{S}^{n-1} : h_{A_1}(u) \geq h_{A_2}(u)\},$$

$$\omega_2 := \{u \in \mathbb{S}^{n-1} : h_{A_1}(u) < h_{A_2}(u)\}.$$

Then $h_A(u) = h_{A_1}(u)$ for $u \in \omega_1$ and $h_A(u) = h_{A_2}(u)$ for $u \in \omega_2$. Since ω_2 is open, the lemma gives

$$S(h_A - h_{A_2}, \mathcal{H}_{n-2}, \cdot) \llcorner \omega_2 = 0.$$

It follows that

$$nV(h_A - h_{A_1}, h_A - h_{A_2}, \mathcal{H}_{n-2})$$

$$= \int_{\mathbb{S}^{n-1}} (h_A - h_{A_1}) \, dS(h_A - h_{A_2}, \mathcal{H}_{n-2}, \cdot)$$

$$= \int_{\omega_1} (h_A - h_{A_1}) \, dS(h_A - h_{A_2}, \mathcal{H}_{n-2}, \cdot) + \int_{\omega_2} (h_A - h_{A_1}) \, dS(h_A - h_{A_2}, \mathcal{H}_{n-2}, \cdot)$$

$$= 0.$$

This is formula (1) for $k = 2$.

The general case of (1) is now proved by induction on k. Let $3 \leq k \leq n$ and suppose the assertion has been proved for the convex hull of less than k convex bodies. We abbreviate

$$h^{(m)} := h_{\mathrm{conv}(A_1 \cup \cdots \cup A_m)}.$$

Then the identity

$$V(h^{(k)} - h_1, \ldots, h^{(k)} - h_k, \mathcal{H}_{n-k})$$

$$= \sum_{j=0}^{k-2} V(h^{(k-1)} - h_1, \ldots, h^{(k-1)} - h_j, h^{(k)} - h^{(k-1)}, h^{(k)} - h_{j+2}, \ldots, h^{(k)} - h_k, \mathcal{H}_{n-k})$$

$$+ V(h^{(k-1)} - h_1, \ldots, h^{(k-1)} - h_{k-1}, h^{(k)} - h_k, \mathcal{H}_{n-k}) \tag{3}$$

holds. For the proof, we note that the summand with $j = k - 2$, which is

$$V(h^{(k-1)} - h_1, \ldots, h^{(k-1)} - h_{k-2}, h^{(k)} - h^{(k-1)}, h^{(k)} - h_k, \mathcal{H}_{n-k}),$$

and the last term (3) add up to

$$V(h^{(k-1)} - h_1, \ldots, h^{(k-1)} - h_{k-2}, h^{(k)} - h_{k-1}, h^{(k)} - h_k, \mathcal{H}_{n-k}).$$

Continuing in this way, we obtain the identity.

Each term in the sum vanishes, by the induction hypothesis for two convex bodies, using the differences $h^{(k)} - h^{(k-1)}$ and $h^{(k)} - h_k$ and observing that

$$\mathrm{conv}(\mathrm{conv}(A_1 \cup \cdots \cup A_{k-1}) \cup A_k) = \mathrm{conv}(A_1 \cup \cdots \cup A_k).$$

The last term (3) vanishes by the induction hypothesis for $k - 1$ convex bodies.

Acknowledgements I thank Vitali Milman for drawing my attention to the identity (1) of Gusev and Esterov.

References

1. A.D. Aleksandrov, To the theory of mixed volumes of convex bodies, Part I: Extension of certain concepts of the theory of convex bodies (in Russian). Mat. Sbornik N. S. **2** (1937), 947–972 (English transl. in *Selected Works. Part I: Selected Scientific Papers*, ed. by Y.G. Reshetnyak, S.S. Kutateladze. Classics of Soviet Mathematics, vol. 4. Gordon and Breach, Amsterdam, 1996, pp. 31–59)
2. A. Esterov, Tropical varieties with polynomial weights and corner loci of piecewise polynomials. Mosc. Math. J. **12**, 55–76 (2012). arXiv:1012.5800v3
3. G. Gusev, Euler characteristic of the bifurcation set for a polynomial of degree 2 or 3. arXiv:1011.1390v2. Translated from: Monodromy zeta-functions and Newton diagrams (in Russian); Ph.D. thesis, MSU, Moscow (2008)
4. R. Schneider, *Convex Bodies: The Brunn–Minkowski Theory*, 2nd edn. (Cambridge University Press, Cambridge, 2014)

On Convergence of Blaschke and Minkowski Symmetrization Through Stability Results

Alexander Segal

Abstract We show how existing results of stability for Brunn-Minkowski and related inequalities imply results regarding rate of convergences of Minkowski and Blaschke symmetrization processes to the Euclidean ball. To be more precise, the results imply that the amount of symmetrizations needed to approach the Euclidean ball within some distance ϵ, a polynomial number of symmetrizations (in the dimension and $\frac{1}{\epsilon}$) suffice.

1 Introduction

1.1 Blaschke and Minkowski Symmetrization

Given a convex set K and $u \in S^{n-1}$ (where S^{n-1} is the boundary of the Euclidean unit ball D_n) one may consider two possible symmetrizations of K with respect to u. The first one is well known and called the Minkowski symmetral of K, defined as follows:

$$\tau_u(K) = \frac{K + R_u(K)}{2},$$

where R_u is reflection with respect to the hyperplane defined by $u^\perp$.

Another possibility of symmetrization process, which to our knowledge, hasn't been considered so far, is using the Blaschke sum. Before we introduce the Blaschke sum, let us discuss the so called Minkowski existence problem. Consider a measure μ given on S^{n-1}. What can one say about the existence and uniqueness of a convex set K such that μ is its surface area measure? Recall that given a convex set K, the surface area measure of K is defined to be

$$S_K(A) = \sigma(\{x \in \partial K : N(x) \in A\}),$$

where $A \subset S^{n-1}$ is a Borel set and $N(x)$ is unit normal at point x.

A. Segal (✉)
Tel Aviv University, Tel Aviv, Israel
e-mail: segalale@gmail.com

© Springer International Publishing Switzerland 2014   427
B. Klartag, E. Milman (eds.), *Geometric Aspects of Functional Analysis*,
Lecture Notes in Mathematics 2116, DOI 10.1007/978-3-319-09477-9_29

This question was answered first by Minkowski for some special cases. Minkowski's proof contained the essential ideas for the general case which was later shown by Alexandrov [1, 2] and independently by Fenchel and Jessen [7] (see [19, pp. 390–393]). Namely, he showed the following:

Theorem 1.1. *Let μ be a finite Borel measure on S^{n-1} such that*

1. $\int u d\mu(u) = 0$.
2. μ is not contained in any great sub-sphere.

Then, there exists a convex body K, unique up to a translation, such that μ is its surface area measure.

Using Theorem 1.1 one may define, up to translation, what is called the Blaschke sum of two convex bodies K, L. The Blaschke sum $K \sharp L$ is defined as the convex body whose surface area measure is the sum of the surface area measures of K and L. Additionally, one may define multiplication by constant $\lambda \cdot K$ as the convex body corresponding to the measure $\lambda S_K(\cdot)$.

Thus, given a set K and $u \in S^{n-1}$ the Blaschke symmetral of K will be

$$B_u(K) = \frac{1}{2} \cdot (K \sharp R_u(K)), \tag{1}$$

where $\lambda \cdot K$ is defined as the set whose surface area measure is λS_K. Intuitively, applying successive Minkowski or Blaschke symmetrization to a convex body K, one may expect a sequence that converges to the Euclidean ball. Indeed, it is well known that for each convex body K, there exists a sequence of directions $\{u_i\} \subset S^{n-1}$ such that $\tau_{u_k}(\ldots \tau_{u_1}(K))$ converges to some Euclidean ball in the Hausdorff metric. This fact gives rise to the question of the rate of convergence to Euclidean ball. That is, given $\epsilon > 0$ how many Minkowski symmetrizations are required in order to transform a convex set K into a new convex set K' which is within Hausdorff distance of at most ϵ from the corresponding Euclidean ball. Before we discuss results regarding Minkowski symmetrization, we will need to introduce several important notions often used in convex analysis. Our first notion is the supporting functional of a convex body K, defined as follows:

$$h_K(u) = \sup\{\langle x, u \rangle : x \in K\},$$

where $u \in S^{n-1}$ and $\langle \cdot, \cdot \rangle$ is the standard scalar product. Since $h_K(u) + h_K(-u)$ describes the width of K in direction u, one may define mean-width of K by averaging:

$$\omega(K) = \int_{S^{n-1}} (h_K(u) + h_K(-u)) d\sigma(u) = 2 \int_{S^{n-1}} h_K(u) d\sigma(u),$$

where σ is the normalized Haar measure on S^{n-1}. However, in this note it would be more comfortable to consider the mean radius of K rather than the mean diameter.

We will denote this by $M^*(K)$, which obviously differs from $\omega(K)$ by a factor of 2. This parameter is broadly used in asymptotic geometric analysis (see [11] for detailed discussion).

Obviously, Minkowski symmetrization is a process that preserves the mean width of a body. Additionally, by the Brunn-Minkowski inequality, Minkowski symmetrization increases the volume. Thus, if one may choose a sequence of Minkowski symmetrizations that converges to a ball, the Urysohn's inequality would follow:

$$M^*(K) \geq \left(\frac{|K|}{\kappa_n}\right)^{1/n},$$

where $|\cdot|$ denotes the standard Lebesgue measure on $\mathbb{R}^n$ and $\kappa_n := |D_n|$. The first result containing quantitative estimates for Minkowski symmetrization was given by Bourgain, Lindenstrauss and Milman in [5]. The authors showed the following:

Theorem 1.2 (Bourgain et al. [5]). *Let $0 < \epsilon < 1$ and $n > n_0(\epsilon)$. Given a convex body $K \subset \mathbb{R}^n$ there exist $cn \log n + c(\epsilon)n$ Minkowski symmetrizations that transform K into a body K' such that*

$$(1 - \epsilon)M^*(K)D_n \subset K' \subset (1 + \epsilon)M^*(K)D_n,$$

where $c(\epsilon), n_0(\epsilon)$ are of order $exp(c\epsilon^{-2} \log \frac{1}{\epsilon})$ and $c > 0$ is a universal constant.

Theorem 1.2 was improved by Klartag in [14] whose result reads the following:

Theorem 1.3 (Klartag). *Let $0 < \epsilon < \frac{1}{2}$. Given a convex body $K \subset \mathbb{R}^n$ there exist $cn \log \frac{1}{\epsilon}$ Minkowski symmetrizations that transform K into a body K' such that*

$$(1 - \epsilon)M^*(K)D_n \subset K' \subset (1 + \epsilon)M^*(K)D_n,$$

where $c > 0$ is a universal constant.

While the optimal rate of convergence of Minkowski symmetrizations to Euclidean ball in Hausdorff metric is known, the proof involves heavy analysis. Additionally, no similar results exist for Blaschke symmetrization. Moreover, there is no proof that one may converge to the Euclidean ball applying successive Blaschke symmetrizations. We will deal with these questions by using some stability results we mention now.

1.2 Stability

The well known Brunn-Minkowski inequality states that

$$|K + L|^{1/n} \geq |K|^{1/n} + |L|^{1/n}. \tag{2}$$

It is well known that equality holds only when K and L are homothetic. However, given that the right-hand side of inequality (2) is "close" to the left-hand side, one may ask if K and L must be close in some sense (up to homothety), as well. Dealing with stability of this inequality dates back to Diskant [6], where he provides a quantitative answer in term of the Hausdorff metric. Diskant showed the following:

Theorem 1.4. *Let K, L be convex sets with equal volume. Denote $\Phi(K, L, t) = |(1-t)K + tL|^{1/n} - (1-t)|K|^{1/n} - t|L|^{1/n}$. If $\Phi(K, L, t) < \epsilon$ for some $\epsilon > 0$ and for all $0 < t < 1$, then $d_H(K, L) < C\epsilon^{1/n}$ for some universal constant $C > 0$.*

However, two convex sets can look very "similar" from the volume distribution point of view and have a large Hausdorff distance (and vice vera). A more reasonable metric to measure distance of convex sets is the symmetric difference, also known as Nikodym metric. The first quantitative result of Brunn-Minkowski inequality in terms of symmetric difference appeared in [8, 9]. Before we state the result, let use introduce several notations. The first one is a way to measure similarity of two convex sets, known as the Fraenkel asymmetry measure:

$$A(K, L) := \inf_{x_0 \in \mathbb{R}^n} \frac{|K \Delta(x_0 + rL)|}{|K|},$$

where $K \Delta L = (K \setminus L) \cup (L \setminus K)$ and $r^n|L| = |K|$. The second notion is the Brunn-Minkowski deficit which is self explanatory:

$$\beta(K, L) = \frac{|K + L|^{1/n}}{|K|^{1/n} + |L|^{1/n}} - 1.$$

Using this notation, the authors of [8, 9] showed that

$$A(K, L) \leq C(n)\sqrt{\beta(K, L)}, \tag{3}$$

or equivalently,

$$|K + L|^{\frac{1}{n}} \geq (|K|^{\frac{1}{n}} + |L|^{\frac{1}{n}})\left(1 + \frac{A(K, L)^2}{C(n)^2}\right).$$

In [9] it was shown that the constant $C(n)$ is of order n^7 and that the power of β is optimal. In [16] the constant $C(n)$ was improved to be of order $n^{3.5}$ in the general case, and n^3 when the sets are known to be centrally symmetric. It is easy to see that inequality (2) implies the classical Brunn-Minkowski inequality, and that it provides a stronger result in many cases.

The Lebesgue measure retains a similar concavity property with respect to Blaschke sum. This is known as the Knesser-Suss inequality (see [19, p. 394]) which states that given convex sets K, L we have

$$|K \sharp L|^{1-\frac{1}{n}} \geq |K|^{1-\frac{1}{n}} + |L|^{1-\frac{1}{n}}. \tag{4}$$

Moreover, Bucur, Fragala and Lamboley showed in [4] that the stability inequality (3) implies the same stability for (4) with the same constant. That is, given two convex sets with the same volume,

$$|K\sharp L|^{1-\frac{1}{n}} \geq \left(|K|^{1-\frac{1}{n}} + |L|^{1-\frac{1}{n}}\right)\left(1 + \frac{A(K,L)^2}{C(n)^2}\right). \tag{5}$$

In this paper we will use both stability results to show convergence rate of Blaschke and Minkowski symmetrization. Namely, we show the following:

Theorem 1.5. *Let K be a convex set such that $M^*(K) = 1$ and let $\epsilon > 0$. Then, there exist $\frac{Cn^7}{\epsilon^2}$ Minkowski symmetrizations that transform K into a convex set K' such that $A(K', D_n) < \epsilon$.*

Theorem 1.6. *Let K be a convex set with surface area $S(K) = n\kappa_n$ and let $\epsilon > 0$. Then, there exist $\frac{Cn^7}{\epsilon^2}\log n$ Blaschke symmetrizations that transform K into a convex set K' such that $A(K', D_n) < \epsilon$.*

Remark 1.7. The assumption that K and L are of equal volume in (3) and (5) is actually not needed, and the inequalities hold in the general case up to some scaling factor. However, we will make use only of the case where K and L are of equal volume.

2   Some Tools

2.1   Properties of Blaschke Symmetrization

Lemma 2.1. *Let K be a convex body and denote by $S_K(\cdot)$ its surface area measure. Then, for any $g \in SO(n)$ we have*

1. $S_{gK}(\omega) = S_K(g^{-1}\omega)$.
2. $g(K\sharp L) = gK\sharp gL$.
3. $g(\lambda \cdot K) = \lambda \cdot gK$.

Proof. The first part is clear from the definition of surface area measure. The second part follows immediately:

$$S_{g(K\sharp L)}(\omega) = S_{K\sharp L}(g^{-1}\omega) = S_K(g^{-1}\omega) + S_L(g^{-1}\omega) = S_{gK}(\omega) + S_{gL}(\omega)$$
$$= S_{gK\sharp gL}(\omega).$$

The third claim follows similarly.

Corollary 2.2. *Let K be a convex body and let $u \in S^{n-1}$. Then, $B_u(K)$ (as defined in (1)) is symmetric with respect to $u^\perp$.*

Proof. This follows immediately from Lemma 2.1. Indeed, let R_u be the reflection with respect to $u^\perp$. Then, $R_u(B_u(K)) = \frac{1}{2} \cdot (R_u(K) \sharp R_u(R_u(K))) = B_u(K)$.

Additionally, notice that since reflections with respect to orthogonal directions commute, apply Blaschke symmetrization to $B_u(K)$ with respect to any direction orthogonal to u does not affect the symmetry of $B_u(K)$.

For more details we refer the reader to [20].

2.2 Symmetric Surface Area Measure

We will use a result by Schneider for symmetric surface area measures (see [18]), which was shown in a more general form by Averkov, Makai and Martini (see [3], Theorem 3.1):

Theorem 2.3. *Given a direction $u \in S^{n-1}$ denote by $S_u^+ = S^{n-1} \cap \{x \in \mathbb{R}^n : \langle x, u \rangle \geq 0\}$ and $S_u^- = S^{n-1} \cap \{x \in \mathbb{R}^n : \langle x, u \rangle \leq 0\}$. Let K be a convex set and S_K its surface area measure. If*

$$S_K(S_u^+) = S_K(S_u^-)$$

holds for every $u \in S^{n-1}$, then the set K is centrally symmetric.

2.3 Stability of the Euclidean Ball

Here we show some "stability" property of a Euclidean ball with respect to reflections. It is easy to see that the Euclidean ball is the only convex set which is symmetric with respect to all possible reflections. Thus, it would make sense to claim that a convex set which is close in some sense to all of its reflections should be close to the Euclidean ball:

Lemma 2.4 ([13]). *Let K be a centrally symmetric convex set in $\mathbb{R}^n$ and let $\epsilon > 0$. If for any reflection R_u we have that $A(K, R_u(K)) < \epsilon$ then $A(K, D_n) < 4\epsilon$.*

Proof. Using spherical coordinates we may write

$$A(K, R_u(K)) = \frac{|K \triangle R_u(K)|}{|K|} = \frac{|D_n|}{|K|} \int_{S^{n-1}} |f(x)^n - f(R_u(x))^n| d\sigma(x) < \epsilon, \tag{6}$$

where $f : S^{n-1} \to \mathbb{R}^+$ is the radial function of K and σ is the normalized Haar measure on the unit sphere. Since we know that (6) holds for every direction $u \in S^{n-1}$, averaging over all the directions yields:

$$\frac{|D_n|}{|K|} \int_{S^{n-1}} \int_{S^{n-1}} |f(x)^n - f(y)^n| d\sigma(x) d\sigma(y) < \epsilon.$$

Denote by r_0 the radius of the ball that satisfies $|K| = |r_0 D_n|$. This implies that $|K \setminus r_0 D_n| = |r_0 D_n \setminus K|$.

Denote the sets $A, B \subset S^{n-1}$ such that $A = \{x \in S^{n-1} : f(x)^n \geq r_0^n\}$ and $B = \{x \in S^{n-1} : f(x)^n \leq r_0^n\}$. Obviously,

$$A(K, D_n) = \frac{|D_n|}{|K|} \int_{S^{n-1}} |f(x)^n - r_0^n| d\sigma(x) = 2\frac{|D_n|}{|K|} \int_A |f(x)^n - r_0^n| d\sigma(x)$$

$$= 2\frac{|D_n|}{|K|} \int_B |f(x)^n - r_0^n| d\sigma(x).$$

Then, the following inequality holds:

$$\epsilon > \frac{|D_n|}{|K|} \int_{S^{n-1}} \int_{S^{n-1}} |f(x)^n - f(y)^n| d\sigma(x) d\sigma(y)$$

$$\geq \frac{|D_n|}{|K|} \int_B \int_A |f(x)^n - f(y)^n| d\sigma(x) d\sigma(y)$$

$$\geq \frac{|D_n|}{|K|} \int_B \int_A |f(x)^n - r_0^n| d\sigma(x) d\sigma(y) = \frac{1}{2}\sigma(B)A(K, D_n).$$

In the same way

$$\epsilon > \frac{1}{2}\sigma(A)A(K, D_n).$$

Combining the above we get that

$$A(K, D_n) < 4\epsilon.$$

2.4   Projection Body and Blaschke Sum

Given a convex centrally symmetric body K, its projection body ΠK is defined by

$$h_{\Pi K}(u) = Vol_{n-1}(Proj_{u^\perp} K) = \frac{1}{2} \int_{S^{n-1}} |\langle u, v \rangle| dS_K(v).$$

It is not hard to check that for any $A \in GL(n)$ we have

$$\Pi(AK) = |det(A)||A^{-T} \Pi K,$$

and $h_{\Pi K}(A^{-t}u) = h_{A^{-t}(\Pi K)}(u)$. Thus, if τ is a reflection operator we have that $\Pi(\tau K) = \tau \Pi K$.

Additionally, the map $K \to \Pi K$ is injective on the class of centrally symmetric convex bodies. That is,

Theorem 2.5 (Alexandrov (see [10], p 142)). *Let K, L be centrally symmetric convex bodies such that $\Pi K = \Pi L$. Then $K = L$.*

The projection body is closely related to Blaschke sum. More precisely, given two convex centrally symmetric sets K, L, the following relation holds (see [10, p. 183])

$$\Pi(K \sharp L) = \Pi K + \Pi L.$$

Using this fact we conclude that applying Blaschke symmetrization to a convex body is equivalent to applying Minkowski symmetrization to its projection body:

$$\tau_u(\Pi K) = \frac{1}{2}(\Pi K + R_u(\Pi K)) = \frac{1}{2}(\Pi K + (\Pi(R_u K))) \qquad (7)$$

$$= \frac{1}{2}\Pi(K \sharp R_u K) = \Pi(\frac{1}{2} \cdot (K \sharp R_u K)) = \Pi(B_u(K)).$$

The projection body is closely related to Shephard's problem: If the area of the shadow on the hyperplane of one convex body is always less than that of another, can the same be said about their volumes? Formally, given two convex bodies K, T such that $\Pi K \subseteq \Pi T$, does this imply that $|K| \leq |T|$? Although the answer to this question is negative for dimension $n \geq 3$ (see [15, 17]), it is known that if T is a zonoid the result is affirmative. Thus, since ellipsoids are zonoids, one may use John's theorem to show that the answer to Shephard's problem is positive up to a multiplicative term of $\sqrt{n}$ (see [10, p. 163]):

Theorem 2.6. *Let K, T be centrally symmetric convex bodies such that $\Pi K \subseteq \Pi T$. Then $|K| \leq \sqrt{n}|T|$.*

We will use Theorem 2.6 to get lower bounds for the volume of a convex body after Blaschke symmetrization. Another useful note regarding the projection body is that the following holds:

$$M^*(\Pi K) = \frac{\kappa_{n-1}}{n\kappa_n} S(K), \qquad (8)$$

where $S(K)$ is the surface area of K. Indeed, by Cauchy surface area formula (see [10, p. 406]):

$$\frac{\kappa_{n-1}}{n\kappa_n} S(K) = \int_{S^{n-1}} |Proj_{u^\perp}| d\sigma(u) = \int_{S^{n-1}} h_{\Pi K}(u) d\sigma(u) = M^*(\Pi K).$$

3 Minkowski Symmetrization

Using the tools described above we can estimate the amount of Minkowski symmetrizations required to transform any convex set into a set close to the unit ball (with a small asymmetry index) and prove Theorem 1.5. However, we will first show a preliminary rate of convergence which also depends on the body we want to symmetrize:

Theorem 3.1. Let K be a convex set such that $M^*(K) = 1$ and let $\epsilon > 0$. Then, there exist $\left\lceil \frac{Cn^5}{\epsilon^2} \log \frac{|D_n|}{|K|} \right\rceil$ Minkowski symmetrizations that transform K into a convex set K' such that $A(K', D_n) < \epsilon$.

Proof. By Urysohn's inequality we know that

$$1 \geq \frac{|K|}{|D_n|}.$$

Without loss of generality, we may assume that K is centrally symmetric. Indeed, otherwise, we may choose an orthonormal basis in $\mathbb{R}^n$ and apply n-Minkowski symmetrizations with respect to the basis. The resulting set is centrally symmetric. Consider the following process. Assume we have some fixed $\epsilon > 0$. If there exists $u \in S^{n-1}$ such that $A(K, R_u(K)) > \epsilon$ then define $K_1 = \tau_u(K)$. By the stability result we have

$$|K_1|^{1/n} \geq |K|^{1/n} \left(1 + \frac{\epsilon^2}{Cn^6}\right).$$

Notice that K is centrally symmetric, hence by Remark 4.2 in [16], we may use (3) with $C(n) = Cn^3$. If it is possible to repeat the above procedure m times we get a convex set K_m that satisfies:

$$|D_n|^{1/n} \geq |K_m|^{1/n} \geq |K|^{1/n} \left(1 + \frac{\epsilon^2}{Cn^6}\right)^m.$$

From this inequality we see that after at most $\frac{\log(|D_n|/|K|)}{n \log\left(1 + \frac{\epsilon^2}{Cn^6}\right)}$, we get a set K_m that satisfies $A(K_m, R_u(K_m)) < \epsilon$ for each reflection R_u.

To sum it up, we get that after at most $\frac{Cn^5}{\epsilon^2} \log \frac{|D_n|}{|K|}$ Minkowski symmetrizations we obtain a set L that is close to all of it reflections:

$$A(L, R_u(L)) < \epsilon.$$

Applying Lemma 2.4 we get that after at most

$$\left\lceil \frac{Cn^5}{\epsilon^2} \log \frac{|D_n|}{|K|} \right\rceil$$

Minkowski symmetrizations, $A(L, D_n) < 4\epsilon$.

3.1 Removing Dependence on Volume of K

In order to remove the dependence of the volume of the initial set we symmetrize, we would like to show some lower bound for this volume after a fixed number of symmetrizations. To achieve this we use the following lemma.

Lemma 3.2. *Let K be a centrally symmetric convex and denote its mean width by m_0. Then, there exist n Minkowski symmetrizations that transform K into a new set K' such that $\frac{m_0}{2^{n-2}} Q_n \subset K'$, where Q_n is the centered unit cube..*

Proof. Since $M^*(K) = m_0$ there exists a direction u such that $K \cap \mathbb{R}u$ is of length $2m_0$. Choose $V = \{v_1, \ldots v_{n-1}\} \in S^{n-1}$ and $u_i := R_{v_i} u$ such that the set $U = \{u_1, \ldots u_{n-1}, u\}$ is orthonormal, and apply Minkowski symmetrizations to K with respect to the directions in V. Obviously, by Monotonicity of Minkowski symmetrization, the resulting set K' contains the cube obtained by symmetrizating the interval $K \cap \mathbb{R}u$ with respect to the same directions. It is easy to see that the resulting set is the cube $\frac{m_0}{2^{n-2}} Q_n$.

Applying Lemma 3.2 to the set K in Theorem 1.5 provides us with the estimate

$$\log \frac{|D_n|}{|K|} \leq Cn^2.$$

In total, we get that for every convex set $\frac{Cn^7}{\epsilon^2}$ Minkowski symmetrizations suffice to approach the Euclidean ball.

4 Blaschke Symmetrization

Once again, to prove Theorem 1.6 we will first show a preliminary result:

Theorem 4.2. *Let K be a convex set with surface area $S(K) = n\kappa_n$ and let $\epsilon > 0$. Then, there exist $\left\lceil \frac{Cn^6}{\epsilon^2} \log \frac{|D_n|}{|K|} \right\rceil$ Blaschke symmetrizations that transform K into a convex set K' such that $A(K', D_n) < \epsilon$.*

Proof. As before, we may assume that K is centrally symmetric, since by applying n symmetrization with respect to some orthogonal basis we get a surface area measure that satisfies the conditions of Theorem 2.3. By the isoperimetric inequality we know that $\kappa_n \geq |K|$.

Fixing some $\epsilon > 0$ and applying the same symmetrization process as in Theorem 3.1, with respect to Blaschke symmetrization, we conclude that after m steps we obtain a set K_m such that

$$\kappa_n^{1-1/n} \geq |K|^{1-1/n} \left(1 + \frac{\epsilon^2}{Cn^6}\right)^m.$$

Hence, after at most $m = \frac{\log(\kappa_n/|K|)}{\log\left(1 + \frac{\epsilon^2}{Cn^6}\right)}$ we obtain a set K_m that satisfies

$$A(K_m, R_u(K_m)) < \epsilon$$

for each reflection R_u.

Thus, after no more than $\frac{Cn^6}{\epsilon^2} \log \frac{|D_n|}{|K|}$ Blaschke symmetrizations we obtain a set L such that

$$A(L, R_u(L)) < \epsilon,$$

for every reflection R_u. As before, applying Lemma 2.4 we get that after at most

$$\left\lceil \frac{Cn^6}{\epsilon^2} \log \frac{|D_n|}{|K|} \right\rceil$$

Blaschke symmetrizations, $A(L, D_n) < 4\epsilon$.

Note that in both proofs the number of required symmetrizations is 0 if $|K| = |D_n|$. This is indeed correct, as we know that the equality case for both, the isoperimetric and Urysohn's inequalities is if and only if $K = D_n$.

4.1 Removing the Dependence on Volume of K

Since it is not clear how to create a large cube inside K using Blaschke symmetrization, we will use Minkowski symmetrization to create a large cube in ΠK and show that this is enough for our purpose. Since $S(K) = n\kappa_n$, by (8) we have that $M^*(\Pi K) = \kappa_{n-1}$. By Lemma 3.2 we know that there exist $n - 1$ Minkowski symmetrizations $\tau_{v_1}, \ldots \tau_{v_{n-1}}$ applied to ΠK transform it into a new body P' such that $\frac{\kappa_{n-1}}{2^{n-2}} Q_n \subset P'$, where Q_n is the centered unit cube. Apply the corresponding Blaschke symmetrizations $B_{v_1}, \ldots B_{v_{n-1}}$ to K. By (7) we obtain a new body K' such that $\Pi(K') = P'$. Additionally, since $\Pi Q_n = 2Q_n$ (see [10, p. 145]), it is not hard to see that $\Pi\left(\left(\frac{\kappa_{n-1}}{2^{n-1}}\right)^{\frac{1}{n-1}} Q_n\right) = \frac{\kappa_{n-1}}{2^{n-2}} Q_n \subseteq \Pi(K')$. Thus, by Theorem 2.6 we get that

$$|K'| \geq \frac{1}{\sqrt{n}} \left(\frac{\kappa_{n-1}}{2^{n-1}} \right)^{\frac{n}{n-1}},$$

which in turn gives us the bound

$$\log \frac{\kappa_n}{|K'|} \leq Cn \log n.$$

This completes the proof of Theorem 1.6.

Remark 4.2. Notice that in proof of Lemma 2.4 any value of r_0 such that $|K \setminus r_0 D_n| > 0$ and $|r_0 D_n \setminus K| > 0$ will provide the same result. Also, since K is not contained in the unit ball, we may choose $r_0 = 1$ and get that

$$\frac{|K \Delta D_n|}{|K|} < 4\epsilon.$$

Thus we get a similar convergence result in the Nikodym metric as well.

5   Appendix: Relation to Hausdorff Metric

First let us check that we can get a result similar to 1.5 by applying the following theorem of Groemer (see [12]) to Theorem 1.3:

Theorem 5.1 (Groemer). *Assume that K and L be convex bodies and let $D = \max\{diam(K), diam(L)\}$. Then, for $n \geq 2$ we have*

$$|K \Delta L| \leq c_n d_H(K, L),$$

where $c_n = \frac{\kappa_n}{2^{1/n}-1} \left(\frac{D}{2} \right)^{n-1}$.

Now, assume we have a convex set K such that $M^*(K) = 1$. By Theorem 1.3 we may transform K into K' such that $d_H(K', D_n) < \epsilon$ using at most $Cn \log \frac{1}{\epsilon}$ Minkowski symmetrizations. Thus, in this case we have that $diam(K') < 2 + 2\epsilon$ and $(1 - \epsilon)^n \kappa_n \leq |K'| \leq \kappa_n$. Denote by r_0 the volume radius of K'. Then using triangle inequality we have

$$A(K', D_n) = \frac{|K' \Delta r_0 D_n|}{|K'|} \leq \frac{|K' \Delta D_n| + (1 - r_0)^n \kappa_n}{(1 - \epsilon)^n \kappa_n} \leq \frac{c_n \epsilon + \epsilon^n \kappa_n}{(1 - \epsilon)^n \kappa_n}$$

$$= \frac{\frac{2}{2^{1/n}-1} (1 + \epsilon)^{n-1} \epsilon + \epsilon^n}{(1 - \epsilon)^n}.$$

Thus, for ϵ small enough, say less than $\frac{1}{n^2}$, we get the required estimate.

Although no similar theorem exists for Blaschke symmetrizations, one may show convergence in the Hausdorff metric using Theorem 1.6.

Lemma 5.2 ([13]). *Let K be a centrally symmetric convex set such that $|K| = |D_n|$ and $A(K, D_n) < \epsilon$, for some $\epsilon > 0$. Then $d_H(K, D_n) < C\epsilon^{1/n}$ where $C > 0$ is a universal constant.*

Proof. Let R be the outer radius of K and $u \in S^{n-1}$ such that $Ru \in \partial K$. Denote by K_t the homothety centered at Ru, where $t = \frac{R-1}{2R}$. That is, $K_t = t(K - Ru) + Ru$. By convexity, $tK \subseteq K$ and by the choice of u, $K_t \cap int(D_n)$ is empty. Thus $K_t \subset K \triangle D_n$ which implies that $|K_t| < \epsilon|D_n|$. Since $|K_t| = t^n|K|$ we get,

$$\left(1 - \frac{1}{R}\right)^n < 2^n\epsilon.$$

Thus, $R < \frac{1}{1 - 2\epsilon^{1/n}}$. Similarly, we get a lower bound for the inradius. Indeed, let r be the inradius of K and $v \in S^{n-1}$ such that $rv \in \partial K$. Denote by D_t the homothety centered at v of D_n where $t = \frac{1-r}{2}$. Again, $D_t \subset D_n$ and by the choice of v, $D_t \cap int(K)$ is empty. Thus, $D_t \subset K \triangle D_n$ which implies that $|D_t| < \epsilon|D_n|$ and

$$(1 - r)^n < 2^n\epsilon.$$

Equivalently, $r > 1 - 2\epsilon^{1/n}$ and for ϵ small enough we get that

$$d_H(K, D_n) < C\epsilon^{1/n},$$

where C is a universal constant.

Thus, Theorem 1.6 and Lemma 5.2 imply the following proposition:

Proposition 5.3. *Let K be a convex body such that $S(K) = n\kappa_n$. Then, there exist $C\frac{n^8}{\epsilon^2}\log n$ Blaschke symmetrizations that transform K into a new body K' such that $d_H(K', D_n) < C'\epsilon^{1/n}$, where $C, C' > 0$ are universal constants.*

Acknowledgements The author would like to thank professor Sergei Ivanov for providing the main ideas for the proofs of Lemmas 2.4 and 5.2 and professor Vitali Milman for useful advice. The author was partially supported by the ISF grant no. 387/09.

References

1. A.D. Alexandrov, Extension of two theorems of Minkowski on convex to arbitrary convex bodies. Mat. Sb. (N.S.) **3**, 27–46 (1938) (Russian)
2. A.D. Alexandrov, On the surface area function of a convex body. Mat. Sb. (N.S.) **6**, 167–174 (1939) (Russian)
3. G. Averkov, E. Makai, H. Martini, Characterizations of Central Symmetry for Convex Bodies in Minkowski Spaces. Stud. Sci. Math. Hung. **46**(4), 493–514 (2009)

4. D. Bucur, I. Fragal, J. Lamboley, Optimal convex shapes for concave functionals. Control Optim. Calc. Var. **18**(3), 693–711 (2012)
5. J. Bourgain, J. Lindenstrauss, V.D. Milman, Minkowski sums and symmetrizations. *Geometric Aspects of Functional Analysis - Israel Seminar (1986–87)*, ed. by J. Lindenstrauss, V.D. Milman. Lecture Notes in Mathematics, vol. 1317 (Springer, Berlin, 1988), pp. 44–66
6. V.I. Diskant, Stability of the solution of a Minkowski equation (Russian). Sibirsk. Mat. Z. **14**, 669–673, 696 (1973)
7. W. Fenchel, B. Jessen, Mengenfunktionen und konvexe K'orper. Danske Vid. Selskab. Mat.-Fys. Medd. **16**, 1–31 (1938)
8. A. Figalli, F. Maggi, F. Pratelli, A refined Brunn-Minkowski inequality for convex sets. Ann. Inst. H. Poincaré Anal. Non Linaire **26**(6), 2511–2519 (2009)
9. A. Figalli, F. Maggi, F. Pratelli, A mass transportation approach to quantitative isoperimetric inequalities. Invent. Math. **182**(1), 167–211 (2010)
10. R. Gardner, *Geometric Tomography*, 2nd edn. (Cambridge University Press, Cambridge, 2006)
11. A.A. Giannopoulos, V.D. Milman, Euclidean structure in finite dimensional, in *Normed Spaces*, ed. by W.B. Johnson, J. Lindenstrauss. Handbook of the Geometry of Banach Spaces, vol. 1 (Elsevier, Amsterdam, 2001), pp. 707–779
12. H. Groemer, On the symmetric difference metric for convex bodies. Contrib. Algebra Geom. **41**(1), 107–114 (2000)
13. S. Ivanov, private communication (2013)
14. B. Klartag, Rate of convergence of geometric symmetrization. Geom. and Funct. Anal. **14**(6) 1322–1338 (2004)
15. C.M. Petty, Projection bodies, in *Proc. Coll. Convexity (Copenhagen 1965)*, Kobenhavns Univ. Mat. Inst. (1967), pp. 234–241
16. A. Segal, Remark on stability of Brunn-Minkowski and isoperimetric inequalities for convex bodies, in *Geometric Aspects of Functional Analysis*. Lecture Notes in Mathematics, vol. 2050 (Springer, Heidelberg, 2012), pp. 381–391
17. R. Schneider, Zur einem Problem von Shephard uber die Projektionen konvexer Korper. Math. Z. **101**, 71–82 (1967)
18. R. Schneider, Ueber eine Integralgleichung in der Theorie der konvexen Koerper. Math. Nachr. **44**, 55–75 (1970)
19. R. Schneider, *Convex Bodies: The Brunn Minkowski Theorey* (Cambridge University Press, Cambridge, 1993)
20. H. Zouaki, Convex set symmetry measurement using Blaschke addition. Pattern Recognit. **36**(3), 753–763 (2003)

Positive Temperature Versions of Two Theorems on First-Passage Percolation

Sasha Sodin

Abstract The estimates on the fluctuations of first-passsage percolation due to Talagrand (a tail bound) and Benjamini–Kalai–Schramm (a sublinear variance bound) are transcribed into the positive-temperature setting of random Schröedinger operators.

1 Introduction

Let $H = -\frac{1}{2d}\Delta + V$ be a random Schrödinger operator on $\mathbb{Z}^d$ with non-negative potential $V \geq 0$:

$$(H\psi)(x) = (1 + V(x))\psi(x) - \frac{1}{2d}\sum_{y \sim x}\psi(y)\,, \quad \psi \in \ell^2(\mathbb{Z}^d)\,.$$

Assume that the entries of V are independent, identically distributed, and satisfy

$$\mathbb{P}\{V(x) > 0\} > 0. \tag{1}$$

The inverse $G = H^{-1}$ of H defines a random metric

$$\rho(x, y) = \log\frac{\sqrt{G(x, x)G(y, y)}}{G(x, y)} \tag{2}$$

on $\mathbb{Z}^d$ (see Lemma 3 below for the verification of the triangle inequality). We are interested in the behaviour of $\rho(x, y)$ for large $\|x - y\|$ (here and forth $\|\cdot\|$ stands for the ℓ_1 norm); to simplify the notation, set $\rho(x) = \rho(0, x)$.

Zerner proved [25, Theorem A], using Kingman's subadditive ergodic theorem [13], that if V satisfies (1) and

$$\mathbb{E}\log^d(1 + V(x)) < \infty\,. \tag{3}$$

S. Sodin (✉)
Department of Mathematics, Princeton University, Princeton, NJ 08544, USA
e-mail: asodin@princeton.edu

then

$$\rho(x) = \|x\|_V (1 + o(1)), \quad \|x\| \to \infty, \tag{4}$$

where $\| \cdot \|_V$ is a deterministic norm on $\mathbb{R}^d$ determined by the distribution of V. As to the fluctuations of $\rho(x)$, Zerner showed [25, Theorem C] that (1), (3), and

$$\text{if } d = 2, \text{ then } \mathbb{P}\{V(x) = 0\} = 0$$

imply the bound

$$\text{Var}\,\rho(x) \leq C_V \|x\|. \tag{5}$$

In dimension $d = 1$, the bound (5) is sharp; moreover, ρ obeys a central limit theorem

$$\frac{\rho(x) - \mathbb{E}\rho(x)}{\sigma_V |x|^{1/2}} \xrightarrow[|x|\to\infty]{D} N(0, 1),$$

which follows from the results of Furstenberg and Kesten [10]. In higher dimension, the fluctuations of ρ are expected to be smaller: the fluctuation exponent

$$\chi = \limsup_{\|x\|\to\infty} \frac{\frac{1}{2}\log \text{Var}\,\rho(x)}{\log \|x\|}$$

is expected to be equal to $1/3$ in dimension $d = 2$, and to be even smaller for $d \geq 3$; see Krug and Spohn [14].

These conjectures are closely related to the corresponding conjectures for first-passage percolation. In fact, ρ is a positive-temperature counterpart of the (site) first-passage percolation metric corresponding to $\omega = \log(1 + V)$; we refer to Zerner [25, Sect. 3] for a more elaborate discussion of this connection.

The rigorous understanding of random metric fluctuations in dimension $d \geq 2$ is for now confined to a handful of (two-dimensional) integrable models, where $\chi = 1/3$ (see Corwin [8] for a review), and to several weak-disorder models in dimension $d \geq 4$, for which $\chi = 0$ (see Imbrie and Spencer [11], and Bolthausen [6]).

It is a major problem to find the value of the exponent χ beyond these two classes of models. We refer to the works of Chatterjee [7] and Auffinger–Damron [2, 3] for some recent results (in arbitrary dimension) establishing a connection between the fluctuation exponent χ and the wandering exponent ξ describing transversal fluctuations of the geodesics.

Here we carry out a much more modest task: verifying that the upper bounds on the fluctuations in (bond) first-passage percolation due to Talagrand [23] and Benjamini, Kalai, and Schramm [5] are also valid for the random metric (2). Zerner's bound (5) is a positive-temperature counterpart of Kesten's estimate [12]. Kesten showed that the (bond) first-passage percolation ρ_{FPP} satisfies

$$\operatorname{Var} \rho_{\text{FPP}}(x) \leq C \, \|x\| \; ; \tag{6}$$

furthermore, if the underlying random variables have exponential tails, then so does $(\rho_{\text{FPP}}(x) - \mathbb{E}\rho_{\text{FPP}}(x))/\sqrt{\|x\|}$. Talagrand improved the tail bound to

$$\mathbb{P}\{|\rho_{\text{FPP}}(x) - \mathbb{E}\rho_{\text{FPP}}(x)| \geq t\} \leq C \exp\left\{-\frac{t^2}{C\,\|x\|}\right\} , \quad 0 \leq t \leq \|x\| .$$

Benjamini, Kalai, and Schramm [5] proved, in dimension $d \geq 2$, the sublinear bound

$$\operatorname{Var} \rho_{\text{FPP}}(x) \leq C \, \|x\| / \log(\|x\| + 2) , \tag{7}$$

for the special case of Bernoulli-distributed potential. Benaïm and Rossignol [4] extended this bound to a wider class of distributions ("nearly gamma" in the terminology of [4]), and complemented it with an exponential tail estimate. Damron, Hanson, and Sosoe [9] proved (7) for arbitrary potential with $2 + \log$ moments. Extensions of the Benjamini–Kalai–Schramm bound to other models have been found by van der Berg and Kiss [24], Matic and Nolen [17], and Alexander and Zygouras [1].

Theorem 1 below is a positive temperature analogue of Talagrand's bound (in order to use a more elementary concentration inequality from [21, 23] instead of a more involved one from [23], we establish a slightly stronger conclusion under a slightly stronger assumption). Theorem 2 is a positive temperature analogue of the Benjamini–Kalai–Schramm bound.

The strategy of the proof is very close to the original arguments; the modification mainly enters in a couple of deterministic estimates. Compared to the closely related work of Piza [19] on directed polymers, we economise on the use of the random walk representation (16), with the hope that the savings will eventually suffice to address an extension discussed in Sect. 4.

Set $\mu(x) = \mathbb{E}\rho(x)$.

Theorem 1. *Suppose the entries of V are independent, identically distributed, bounded from below by $\epsilon > 0$, and from above by $0 < M < \infty$. Then*

$$\mathbb{P}\{\rho(x) \leq \mu(x) - t\} \leq C \exp\left\{-\frac{t^2}{C(\epsilon, M)(\mu(x) + 1)}\right\} , \tag{8}$$

and

$$\mathbb{P}\{\rho(x) \geq \mu(x) + t\} \leq C \exp\left\{-\frac{t^2}{C(\epsilon, M)(\mu(x) + t + 1)}\right\} , \tag{9}$$

for every $t \geq 0$.

Remark 1. The assumption $\epsilon \leq V \leq M$ yields the deterministic estimate

$$C_\epsilon^{-1} \|x\| \leq \rho(x) \leq C_M \|x\| , \tag{10}$$

which, in conjunction with (8) and (9), implies the inequality

$$\mathbb{P}\{|\rho(x) - \mu(x)| \geq t\} \leq C \exp\left\{-\frac{t^2}{C(\epsilon, M)\|x\|}\right\} .$$

Theorem 2. *Assume that the distribution of the potential is given by*

$$\mathbb{P}\{V(x) = a\} = \mathbb{P}\{V(x) = b\} = 1/2$$

for some $0 < a < b$, *and that* $d \geq 2$. *Then*

$$\operatorname{Var} \rho(x) \leq C_{a,b} \frac{\|x\|}{\log(\|x\| + 2)} . \tag{11}$$

We conclude the introduction with a brief comment on lower bounds. In dimension $d = 2$, Newman and Piza [18] proved the logarithmic lower bound

$$\operatorname{Var} \rho_{\mathrm{FPP}}(x) \geq \frac{1}{C} \log(\|x\| + 1) . \tag{12}$$

A version for directed polymers (the positive temperature counterpart of directed first passage percolation) was proved by Piza [19]; the argument there is equally applicable to the undirected polymers which are the subject of the current note. We are not aware of any non-trivial lower bounds in dimension $d \geq 3$.

2   Proof of Theorem 1

The proof of Theorem 1 is based on Talagrand's concentration inequality [21, 23]. We state this inequality as

Lemma 1 (Talagrand). *Assume that* $\{V(x) \mid x \in \mathscr{X}\}$ *are independent random variables, the distribution of every one of which is supported in* $[0, M]$. *Then, for every convex (or concave) L-Lipschitz function* $f : \mathbb{R}^{\mathscr{X}} \to \mathbb{R}$.

$$\mathbb{P}\{f \geq \mathbb{E}f + t\} \leq C \exp\left\{-\frac{t^2}{CM^2L^2}\right\} ,$$

where $C > 0$ *is a constant.*

Denote $g(x) = G(0, x)$. To apply Lemma 1, we first compute the gradient of $\log g$, and then estimate its norm.

Lemma 2. *For any* $x, y \in \mathbb{Z}^d$,

$$\frac{\partial}{\partial V(y)} \log g(x) = -\frac{G(0, y)G(y, x)}{G(0, x)} .$$

Proof. Let $P_y = \delta_y \delta_y^*$ be the projector on the y-th coordinate. Set $H_h = H + hP_y$, $G_h = H_h^{-1}$. By the resolvent identity

$$G_h = G - hGP_y G_h ,$$

hence

$$\frac{d}{dh}\Big|_{h=0} G_h = -GP_y G$$

and

$$\frac{d}{dh}\Big|_{h=0} G_h(0, x) = -G(0, y)G(y, x) .$$

$\square$

Our next goal is to prove

Proposition 1. *Suppose* $V \geq \epsilon > 0$. *Then*

$$\sum_y \left[\frac{G(0, y)G(y, x)}{G(0, x)} \right]^2 \leq A_\epsilon (\rho(x) + 1) , \tag{13}$$

where A_ϵ *depends only on* ϵ.

The proof consists of two ingredients. The first one, equivalent to the triangle inequality for ρ, yields an upper bound on every term in the left-hand side of (13).

Lemma 3. *For any* $x, y \in \mathbb{Z}^d$,

$$\frac{G(0, y)G(y, x)}{G(0, x)} \leq G(y, y) \leq C_\epsilon .$$

Proof. Let H_y be the operator obtained by erasing the edges that connect y to its neighbours, and let $G_y = H_y^{-1}$. By the resolvent identity,

$$G(0, x) = G_y(0, x) + \frac{1}{2d} \sum_{y' \sim y} G_y(0, y')G(y, x) .$$

In particular,

$$G(0, y) = \frac{1}{2d} \sum_{y' \sim y} G_y(0, y')G(y, y) \,.$$

Therefore

$$G(0, x) = G_y(0, x) + \frac{G(0, y)G(y, x)}{G(y, y)} \,.$$

□

The second ingredient is

Lemma 4. *For any $x \in \mathbb{Z}^d$,*

$$\sum_y \frac{G(0, y)G(y, x)}{G(0, x)} \leq C_\epsilon(\rho(x) + 1) \,.$$

The proof of Lemma 4 requires two more lemmata. Denote

$$g_2(x) = G^2(0, x) = \sum_y G(0, y)G(y, x) \,, \quad u(x) = \frac{g_2(x)}{g(x)} \,.$$

Lemma 5. *For any $x \in \mathbb{Z}^d$,*

$$\sum_{y \sim x} \frac{g(y)}{2d(1 + V(x))g(x)} = 1 - \frac{\delta(x)}{(1 + V(0))g(0)} \tag{14}$$

and

$$u(x) = \sum_{y \sim x} u(y) \frac{g(y)}{2d(1 + V(x))g(x)} + \frac{1}{1 + V(x)} \,. \tag{15}$$

Proof. The first formula follows from the relation $Hg = \delta$, and the second one— from the relation $Hg_2 = g$. □

Set $\widetilde{\rho}(x) = \log \frac{G(0,0)}{G(0,x)}$.

Lemma 6. *For any $x \in \mathbb{Z}^d$,*

$$\widetilde{\rho}(x) \geq \sum_{y \sim x} \widetilde{\rho}(y) \frac{g(y)}{2d(1 + V(x))g(x)}$$

$$+ \log(1 + V(x)) + \log\left(1 - \frac{1}{(1 + V(0))g(0)}\right)\delta(x) \,.$$

Proof. For $x \neq 0$, (14) and the concavity of logarithm yield

$$\sum_{y \sim x} \frac{g(y)}{2d(1 + V(x))g(x)} \log \frac{2d(1 + V(x))g(x)}{g(y)} \leq \log(2d) .$$

Using (14) once again, we obtain

$$-\widetilde{\rho}(x) + \sum_{y \sim x} \widetilde{\rho}(y) \frac{g(y)}{2d(1 + V(x))g(x)} + \log(1 + V(x)) \leq 0 .$$

The argument is similar for $x = 0$.   □

Proof (Proof of Lemma 4). Let $A \geq \log^{-1}(1 + \epsilon)$. Then from Lemmata 5 and 6 the function $u_A = u - A\widetilde{\rho}$ satisfies

$$u_A(x) \leq \sum_{y \sim x} u_A(y) \frac{g(y)}{2d(1 + V(x))g(x)} - A \log \left(1 - \frac{1}{(1 + V(0))g(0)} \right) \delta(x) .$$

By a finite-volume approximation argument (which is applicable due to the deterministic bound (10)),

$$\max u_A(x) = u_A(0) \leq -\frac{A}{1 - \frac{1}{(1+V(0))g(0)}} \log \left(1 - \frac{1}{(1 + V(0))g(0)} \right) \leq A'_\epsilon ,$$

whence

$$u(x) \leq A'_\epsilon + A\widetilde{\rho}(x) \leq C_\epsilon(1 + \rho(x)) .$$

□

Proof (Proof of Proposition 1). By Lemma 3 ,

$$L = \sum_y \left[\frac{G(0, y)G(y, x)}{G(0, x)} \right]^2$$

$$\leq \max_y G(y, y) \sum_y \frac{G(0, y)G(y, x)}{G(0, x)} = \max_y G(y, y) u(x) .$$

The inequality $V \geq \epsilon$ implies $G(y, y) \leq A''_\epsilon$, and Lemma 4 implies

$$u(x) \leq C_\epsilon(\rho(x) + 1) .$$

□

Next, we need

Lemma 7. *For any $x \in \mathbb{Z}^d$, $\log g(x)$, $\log \frac{G(0,x)}{G(0,0)}$, and $\log \frac{G(0,x)}{G(x,x)}$ are convex functions of the potential. Consequently,*

$$\rho(x) = -\frac{1}{2}\left[\log \frac{G(0,x)}{G(0,0)} + \log \frac{G(0,x)}{G(x,x)}\right]$$

is a concave function of the potential.

Proof. The first statement follows from the random walk expansion:

$$g(x) = \sum \frac{1}{1 + V(x_0)} \frac{1}{2d} \frac{1}{1 + V(x_1)} \frac{1}{2d} \cdots \frac{1}{2d} \frac{1}{1 + V(x_k)}, \tag{16}$$

where the sum is over all paths $w : x_0 = 0, x_1, \cdots, x_{k-1}, x_k = x$. Indeed, for every w

$$T_w = \log \frac{1}{1 + V(x_0)} \frac{1}{2d} \frac{1}{1 + V(x_1)} \frac{1}{2d} \cdots \frac{1}{2d} \frac{1}{1 + V(x_k)}$$

is a convex function of V, hence also $\log g(x) = \log \sum_w e^{T_w}$ is convex.

To prove the second statement, observe that

$$G(0,x) = \frac{1}{2d} G(0,0) \sum_{y \sim 0} G_0(y,x),$$

where G_0 is obtained by deleting the edges adjacent to 0. Therefore

$$\log \frac{G(0,x)}{G(0,0)} = -\log(2d) + \log \sum_{y \sim 0} G_0(y,x);$$

for every y, $\log G_0(y,x)$ is a convex function of V, hence so is $\log \frac{G(0,x)}{G(0,0)}$. □

Proof (Proof of Theorem 1). Denote $\rho_0(x) = \min(\rho(x), \mu(x))$. Then by Lemma 2 and Proposition 1

$$\|\nabla_V \rho_0(x)\|_2^2 \leq A_\epsilon(\mu(x) + 1),$$

A_ϵ depends only on ϵ. By Lemma 7, ρ_0 is concave, therefore by Lemma 1

$$\mathbb{P}\{\rho(x) \leq \mu(x) - t\} \leq \exp\left\{-\frac{t^2}{CM^2 A_\epsilon(\mu(x) + 1)}\right\}.$$

Similarly, set $\rho_t(x) = \min(\rho(x), \mu(x) + t)$. Then

$$\|\nabla_V \rho_t(x)\|_2^2 \le A_\epsilon(\mu(x) + t + 1) ,$$

therefore by Lemma 1

$$\mathbb{P}\{\rho(x) \ge \mu(x) + t\} = \mathbb{P}\{\rho_t(x) \ge \mu(x) + t\}$$

$$\underset{\Delta}{\le} \exp\left\{-\frac{t^2}{CM^2 A_\epsilon(\mu(x) + t + 1)}\right\} .$$

□

3   Proof of Theorem 2

The proof follows the strategy of Banjamini, Kalai, and Schramm [5]. Without loss of generality we may assume that $\|x\| \ge 2$; set $m = \lfloor \|x\|^{1/4} \rfloor + 1$.

To implement the Benjamini–Kalai–Schramm averaging argument, set

$$F = -\frac{1}{\#B} \sum_{z \in B} \log G(z, x + z) ,$$

where

$$B = B(0, m) = \{z \in \mathbb{Z}^d \mid \|z\| \le m\}$$

is the ball of radius m about the origin (cf. Alexander and Zygouras [1]). According to Lemma 3,

$$G(0, x) \ge \frac{G(z, x + z)G(0, z)G(x, x + z)}{G(z, z)G(x + z, x + z)} ,$$

therefore $\rho(x) \le F + C_{a,b}m$; similarly, $\rho(x) \ge F - C_{a,b}m$. It is therefore sufficient to show that

$$\mathrm{Var}\, F \le C_{a,b} \frac{\|x\|}{\log \|x\|} .$$

We use another inequality due to Talagrand [22] (see Ledoux [16] for a semigroup derivation). Let $\mathcal{X}$ be a (finite or countable) set. Let $\sigma_x^+ : \{a, b\}^{\mathcal{X}} \to \{a, b\}^{\mathcal{X}}$ be the map setting the x-th coordinate to b, and $\sigma_x^- : \{a, b\}^{\mathcal{X}} \to \{a, b\}^{\mathcal{X}}$ –the map setting the x-th coordinate to a. Denote

$$\partial_x f = f \circ \sigma_x^+ - f \circ \sigma_x^- .$$

Lemma 8 (Talagrand). *For any function f on $\{a,b\}^{\mathcal{X}}$,*

$$\operatorname{Var} f \le C_{a,b} \sum_{x \in \mathcal{X}} \frac{\mathbb{E}|\partial_x f|^2}{1 + \log \frac{\mathbb{E}|\partial_x f|^2}{(\mathbb{E}|\partial_x f|)^2}} . \tag{17}$$

Let us estimate the right-hand side for $f = F$, $\mathcal{X} = \mathbb{Z}^d$. Denote

$$\sigma_x^t = t\sigma_x^+ + (1-t)\sigma_x^- ;$$

then

$$\partial_x F = \int_0^1 \frac{\partial F}{\partial V(x)} \circ \sigma_x^t \, dt .$$

According to Lemma 2,

$$\frac{\partial F}{\partial V(y)} = \frac{1}{\#B} \sum_{z \in B} \frac{G(z,y)G(y, x+z)}{G(z, x+z)} .$$

Therefore

$$\begin{aligned}
\mathbb{E}\frac{\partial F}{\partial V(y)} \circ \sigma_y^t &= \mathbb{E}\frac{1}{\#B} \sum_{z \in B} \frac{G(z,y)G(y, x+z)}{G(z, x+z)} \circ \sigma_y^t \\
&= \mathbb{E}\frac{1}{\#B} \sum_{z \in B} \frac{G(0, y-z)G(y-z, x)}{G(0, x)} \circ \sigma_{y-z}^t \\
&= \mathbb{E}\frac{1}{\#B} \sum_{v \in y+B} \frac{G(0, v)G(v, x)}{G(0, x)} \circ \sigma_v^t .
\end{aligned}$$

Lemma 9. *For any $Q \subset \mathbb{Z}^d$ and any $x', x \in \mathbb{Z}^d$,*

$$\sum_{v \in Q} \frac{G(x', v)G(v, x)}{G(x', x)} \le C_a(\operatorname{diam}_\rho Q + 1) \le C_{a,b}(\operatorname{diam} Q + 1) . \tag{18}$$

Let us first conclude the proof of Theorem 2 and then prove the lemma. Set $\delta = m^{-\frac{1}{2}}$, and let

$$A = \left\{ y \in \mathbb{Z}^d \;\middle|\; \mathbb{E}\left(\partial_y F\right)^2 \le \delta \, \mathbb{E}\partial_y F \right\} .$$

Then the contribution of coordinates in A to the right-hand side of (17) is at most $C\delta\|x\|$ by Lemma 4. For y in the complement of A, Lemma 9 yields

$$\mathbb{E}\partial_y F \le \frac{Cm}{\#B},$$

hence

$$\mathbb{E}\left(\partial_y F\right)^2 \ge \delta\,\mathbb{E}\partial_y F \ge \frac{\delta\#B}{Cm}\left(\mathbb{E}\partial_y F\right)^2,$$

and

$$\log \frac{\mathbb{E}\left(\partial_y F\right)^2}{\left(\mathbb{E}\partial_y F\right)^2} \ge \log \frac{\delta}{Cm} \ge \log(\|x\|/C')$$

by the inequality $\#B \ge Cm^2$ (which holds with d-independent C). The contribution of the complement of A to (17) is therefore at most $C'\frac{\|x\|}{\log\|x\|}$. Thus finally

$$\mathrm{Var}\,F \le \frac{C''\|x\|}{\log\|x\|}.$$

$\square$

Proof (Proof of Lemma 9). For $Q \subset \mathbb{Z}^d$ and $x', x \in \mathbb{Z}^d$, set

$$u_Q(x', x) = \frac{(G\mathbb{1}_Q G)(x', x)}{G(x', x)} = \frac{\sum_{q\in Q} G(x', q)G(q, x)}{G(x', x)}.$$

Similarly to Lemma 5,

$$u_Q(x', x) = \sum_{y\sim x} u_Q(x', y)\frac{G(x', y)}{2d(1 + V(x))G(x', x)} + \frac{\mathbb{1}_Q(x)}{1 + V(x)}.$$

By a finite-volume approximation argument, it is sufficient to prove the estimate (18) in a finite box. Then $\max_x u_Q(x', x)$ is attained for some $x_{\max} \in Q$. By symmetry, $\max_{x', x} u_Q(x', x)$ is attained when both x' and x are in Q. On the other hand, for $x', x \in Q$

$$u_Q(x', x) \le u_{\mathbb{Z}^d}(x', x) \le C(1 + \log\frac{1}{G(x', x)}) \le C'(1 + \mathrm{diam}_\rho Q)$$

by Lemma 4.

$\square$

Remark 2. To extend Theorem 2 to the generality of the work of Benaïm and Rossignol [4], one may use the modified Poincaré inequality of [4] instead of Talagrand's inequality (17); this argument also yields a tail bound as in [4].

One may also hope that the even more general methods of Damron, Hanson, and Sosoe [9] could be adapted to the setting of the current paper.

4 A Remark

Let $H = -\frac{1}{2d}\Delta + V$ be a random Schrödinger operator on $\mathbb{Z}^d$. For $z \in \mathbb{C} \setminus \mathbb{R}$, set $G_z = (H - z)^{-1}$. The analysis of Somoza, Ortuño, and Prior [20] (see further Le Doussal [15]) suggests that, in dimension $d = 2$,

$$\mathrm{Var}\log|G_z(x, y)| \asymp (\|x - y\| + 1)^{2/3},$$

and that a similar estimate is valid for the boundary values $G_{\lambda+i0}$ (which exist for almost every $\lambda \in \mathbb{R}$) even when λ is in the spectrum of H.

Having this circle of questions in mind, it would be interesting to study the fluctuations of $\log|G_z(x, y)|$ for $z \in \mathbb{C} \setminus \mathbb{R}$. In dimension $d = 1$, the results of Furstenberg and Kesten [10] imply that

$$\mathrm{Var}\log|G_z(x, y)| \asymp |x - y| + 1.$$

We are not aware of any rigorous bounds in dimension $d \geq 2$. In particular, we do not know a proof of the estimate

$$\mathrm{Var}\log|G_z(x, y)| = o(\|x - y\|^2), \quad \|x - y\| \to \infty, \tag{19}$$

even when z is such that the random walk representation (16) is convergent.

Acknowledgements I am grateful to Thomas Spencer for helpful conversations, and to Itai Benjamini, Michael Damron, Alexander Elgart, and Gil Kalai for their comments on a preliminary version of this note.

The author acknowledges partial support by NSF under grant PHY 1305472.

References

1. K.S. Alexander, N. Zygouras, Subgaussian concentration and rates of convergence in directed polymers. Electron. J. Probab. **18**(5), 28 pp. (2013)
2. A. Auffinger, M. Damron, A simplified proof of the relation between scaling exponents in first-passage percolation. Ann. Probab. **42**(3), 1197–1211 (2014)
3. A. Auffinger, M. Damron, The scaling relation $\chi = 2\xi - 1$ for directed polymers in a random environment. ALEA, Lat. Am. J. Probab. Math. Stat. **10**(2), 857–880 (2013)
4. M. Benaïm, R. Rossignol, Exponential concentration for first passage percolation through modified Poincaré inequalities. Ann. Inst. Henri Poincaré Probab. Stat. **44**(3), 544–573 (2008)
5. I. Benjamini, G. Kalai, O. Schramm, First passage percolation has sublinear distance variance. Ann. Probab. **31**(4), 1970–1978 (2003)

6. E. Bolthausen, A note on the diffusion of directed polymers in a random environment. Commun. Math. Phys. **123**(4), 529–534 (1989)
7. S. Chatterjee, The universal relation between scaling exponents in first-passage percolation. Ann. Math. **177**(2), 663–697 (2013)
8. I. Corwin, The Kardar–Parisi–Zhang equation and universality class. Random Matrices Theory Appl. **1**(1), 1130001, 76 pp. (2012)
9. M. Damron, J. Hanson, Ph. Sosoe, Sublinear variance in first-passage percolation for general distributions. arXiv:1306.1197
10. H. Furstenberg, H. Kesten, Products of random matrices. Ann. Math. Stat. **31**, 457–469 (1960)
11. J.Z. Imbrie, T. Spencer, Diffusion of directed polymers in a random environment. J. Stat. Phys. **52**(3–4), 609–626 (1988)
12. H. Kesten, On the speed of convergence in first-passage percolation. Ann. Appl. Probab. **3**(2), 296–338 (1993)
13. J.F.C. Kingman, Subadditive ergodic theory. Ann. Probab. **1**, 883–909 (1973)
14. J. Krug, H. Spohn, Kinetic roughening of growing surfaces, in *Solids Far from Equilibrium: Growth, Morphology, and Defects*, ed. by C. Godèrche (Cambridge University Press, Cambridge, 1991), pp. 412–525
15. P. Le Doussal, Universal statistics for directed polymers and the KPZ equation from the replica Bethe Ansatz (2012) http://www.newton.ac.uk/seminar/20120920101010501.html
16. M. Ledoux, Deviation inequalities on largest eigenvalues, in *Geometric Aspects of Functional Analysis*. Lecture Notes in Mathematics, vol. 1910 (Springer, Berlin, 2007), pp. 167–219
17. I. Matic, J. Nolen, A sublinear variance bound for solutions of a random Hamilton-Jacobi equation. J. Stat. Phys. **149**(2), 342–361 (2012)
18. C.M. Newman, M.S.T. Piza, Divergence of shape fluctuations in two dimensions. Ann. Probab. **23**(3), 977–1005 (1995)
19. M.S.T. Piza, Directed polymers in a random environment: some results on fluctuations. J. Stat. Phys. **89**(3–4), 581–603 (1997)
20. A.M. Somoza, M. Ortuño, J. Prior, Universal distribution functions in two-dimensional localized systems. Phys. Rev. Lett. **99**(11), 116602 (2007)
21. M. Talagrand, An isoperimetric theorem on the cube and the Kintchine–Kahane inequalities. Proc. Am. Math. Soc. **104**(3), 905–909 (1988)
22. M. Talagrand, On Russo's approximate zero-one law. Ann. Probab. **22**(3), 1576–1587 (1994)
23. M. Talagrand, Concentration of measure and isoperimetric inequalities in product spaces. Inst. Hautes Études Sci. Publ. Math. **81**, 73–205 (1995)
24. J. van den Berg, D. Kiss, Sublinearity of the travel-time variance for dependent first-passage percolation. Ann. Probab. **40**(2), 743–764 (2012)
25. M.P.W. Zerner, Directional decay of the Green's function for a random nonnegative potential on $\mathbb{Z}^d$. Ann. Appl. Probab. **8**(1), 246–280 (1998)

The Randomized Dvoretzky's Theorem in l_∞^n and the χ-Distribution

Konstantin E. Tikhomirov

Abstract Let $\varepsilon \in (0, 1/2)$. We prove that if for some $n > 1$ and $k > 1$, a majority of k-dimensional sections of the ball in l_∞^n is $(1 + \varepsilon)$-spherical then necessarily $k \le C\varepsilon \ln n / \ln \frac{1}{\varepsilon}$, where C is a universal constant. The bound for k is optimal up to the choice of C.

1 Introduction

The classical theorem of A. Dvoretzky in the version improved and strengthened by V. Milman, states: *there is a function $c(\varepsilon) > 0$ such that for all $\varepsilon \in (0, 1/2)$, $n > 1$ and $1 \le k \le c(\varepsilon) \ln n$, any n-dimensional normed space admits a k-dimensional subspace which is $(1 + \varepsilon)$-Euclidean.*

See [2] and [6], respectively, for the original theorems. A broad perspective of the subject and its developments can be found in the books [7] and [8], as well as in a recent survey [11] and references therein.

The bound $k \le c(\varepsilon) \ln n$ is in general optimal with respect to n, but the form of the function $c(\varepsilon)$ is not clear up to this day. The original formula for $c(\varepsilon)$ from [6] was subsequently improved to $c(\varepsilon) = c\varepsilon^2$ in [4] and then to $c(\varepsilon) = c\varepsilon/(\ln \frac{1}{\varepsilon})^2$ in [9]; this is the best known general result so far (note that in the class of n-dimensional spaces with a 1-symmetric basis, $c(\varepsilon) = c / \ln \frac{1}{\varepsilon}$ [12]).

The problem of optimal dependence on ε in Dvoretzky's theorem can be "randomized" as follows: given an n-dimensional normed space X, determine all k such that a random k-dimensional subspace of X is $(1 + \varepsilon)$-Euclidean with a high probability. Of course, the solution depends on the definition of "randomness". For example, in [5] the question was considered for $X = l_\infty^n$ and a certain probabilistic model which gives $(1 + \varepsilon)$-Euclidean subspaces with a large probability for all $k \le c \ln n / \ln \frac{1}{\varepsilon}$. However, the distribution of the random subspaces in [5] is not invariant under rotations. The (unique) rotation invariant distribution of subspaces of l_∞^n was studied in [10].

K.E. Tikhomirov (✉)
632 CAB, University of Alberta, Edmonton, AB, Canada T6G2G1
e-mail: ktikhomi@ualberta.ca

© Springer International Publishing Switzerland 2014   455
B. Klartag, E. Milman (eds.), *Geometric Aspects of Functional Analysis*,
Lecture Notes in Mathematics 2116, DOI 10.1007/978-3-319-09477-9_31

It was proved in [10] that the standard Gaussian vector $g = (g_1, g_2, \ldots, g_n)$ in $\mathbb{R}^n$ satisfies

$$\mathbb{P}\{\|g\|_\infty < (1-\varepsilon)M_\infty\} \leq 2\exp(-n^{c\varepsilon}),$$

$$\mathbb{P}\{\|g\|_\infty > (1+\varepsilon)M_\infty\} \leq 2n^{-c\varepsilon},$$

where M_∞ is the median of the norm of g in l_∞^n and $c > 0$ is a universal constant. A usual "ε-net" argument then implies that a random k-dimensional subspace $E \subset l_\infty^n$, uniformly distributed on the Grassmannian $G_{n,k}$, is $(1 + \varepsilon)$-spherical with probability at least $1 - 2n^{-\tilde{c}\varepsilon}$, provided that $k \leq \tilde{c}\varepsilon \ln n / \ln \frac{1}{\varepsilon}$. Of course, a natural question is whether $\ln \frac{1}{\varepsilon}$ in the upper bound for k can be removed. In [10] it was claimed that it is indeed possible.

The main purpose of this note is to show that in fact $\ln \frac{1}{\varepsilon}$ is necessary and the bound $k \leq \tilde{c}\varepsilon \ln n / \ln \frac{1}{\varepsilon}$ is optimal (up to the choice of the constant). In other words, if $k \geq \max(2, C\varepsilon \ln n / \ln \frac{1}{\varepsilon})$ then with a substantial probability the random k-dimensional subspace $E \subset l_\infty^n$ uniformly distributed on $G_{n,k}$ is *not* $(1 + \varepsilon)$-spherical. To achieve our goal, we shall link the geometry of "typical" subspaces of l_∞^n to certain properties of the χ-distribution.

2 Preliminaries

Let us start with some notation. For any $n \geq 1$ and any vector $x \in \mathbb{R}^n$, $\|x\|_2$ and $\|x\|_\infty$ denote the canonical Euclidean norm and l_∞^n-norm of x, respectively. The probability space $(\mathbb{P}, \Sigma, \Omega)$ is fixed. Everywhere in the text, $g_1, g_2, \ldots$ are independent standard Gaussian variables and $g = (g_1, g_2, \ldots, g_n)$ is the standard Gaussian vector in $\mathbb{R}^n$. By M_2 (M_∞) we will denote the median of $\|g\|_2$ (respectively, the median of $\|g\|_\infty$). Finally, with some abuse of terminology, we will call a subspace $E \subset l_\infty^n$ $(1 + \varepsilon)$-*spherical* if

$$\max_{\substack{x \in E \\ \|x\|_2 = 1}} \|x\|_\infty / \min_{\substack{x \in E \\ \|x\|_2 = 1}} \|x\|_\infty \leq 1 + \varepsilon.$$

From well known estimates for the Gaussian distribution (see, for example, [1, p. 264] or [3, Lemma VII.1.2]) it follows that for $\alpha \to 0$,

$$\frac{1}{\alpha}\mathbb{P}\left\{|g_1| \geq \sqrt{2\ln(1/\alpha)} - \frac{\ln\ln(1/\alpha)}{4\sqrt{\ln(1/\alpha)}}\right\} \longrightarrow 0; \qquad (1)$$

$$\frac{1}{\alpha}\mathbb{P}\left\{|g_1| \geq \sqrt{\ln(1/\alpha)}\right\} \longrightarrow \infty. \qquad (2)$$

Fix some $k > 1$. The variable $\xi^{(k)} = \sqrt{\sum_{i=1}^{k} g_i^2}$ has the χ-distribution with k degrees of freedom; the distribution density f_k of $\xi^{(k)}$ is given by

$$f_k(t) = \begin{cases} \frac{t^{k-1}e^{-t^2/2}}{2^{k/2-1}\Gamma(k/2)}, & t \geq 0; \\ 0, & \text{otherwise.} \end{cases}$$

where Γ is the Gamma function. It is obvious that $\mathbb{P}\{\xi^{(k)} \geq \tau\} \geq \mathbb{P}\{|g_1| \geq \tau\}$ for all $\tau > 0$, so in view of (2) for all sufficiently small $\alpha > 0$,

$$\mathbb{P}\{\xi^{(k)} \geq \sqrt{\ln(1/\alpha)}\} \geq \alpha. \tag{3}$$

We shall use the formula for f_k to improve the last estimate. Suppose that $\tau = \tau(\alpha)$ satisfies $\mathbb{P}\{\xi^{(k)} \geq \tau\} = \alpha$, i.e.

$$\alpha = \int_\tau^\infty f_k(t)\, dt.$$

By (3), for small α we have $\tau \geq \sqrt{\ln(1/\alpha)}$ and

$$\alpha = \int_\tau^\infty \frac{t^{k-1}e^{-t^2/2}}{2^{k/2-1}\Gamma(k/2)}\, dt = \int_{\tau^2}^\infty \frac{t^{k/2-1}e^{-t/2}}{2^{k/2}\Gamma(k/2)}\, dt \geq \int_{\tau^2}^\infty \frac{(\ln(1/\alpha))^\ell e^{-t/2}}{2^{k/2}\Gamma(k/2)}\, dt$$

$$= \frac{(\ln(1/\alpha))^\ell e^{-\tau^2/2}}{2^\ell \Gamma(k/2)},$$

where $\ell = \frac{k}{2} - 1$. Hence,

$$\tau \geq \sqrt{2\ln(1/\alpha) + 2\ln\frac{(\ln(1/\alpha))^\ell}{2^\ell \Gamma(k/2)}} \geq \sqrt{2\ln(1/\alpha) + 2\ell \ln\frac{\ln(1/\alpha)}{2\ell+1}}. \tag{4}$$

Clearly, for $\ell \ll \ln(1/\alpha)$ we have $\ln(1/\alpha) \geq \ell \ln\frac{\ln(1/\alpha)}{2\ell+1}$, and (4) implies

$$\tau \geq \sqrt{2\ln(1/\alpha)} + \frac{\ell \ln\frac{\ln(1/\alpha)}{2\ell+1}}{2\sqrt{\ln(1/\alpha)}}.$$

Thus, we have shown the following:

Lemma 1. *There are absolute constants $\alpha_0 > 0$ and $c_0 > 0$ such that whenever $\alpha \in (0, \alpha_0)$ and $2 \leq k \leq c_0 \ln(1/\alpha)$ then*

$$\mathbb{P}\left\{\xi^{(k)} \geq \sqrt{2\ln(1/\alpha)} + \frac{(k/2 - 1)\ln\frac{\ln(1/\alpha)}{k-1}}{2\sqrt{\ln(1/\alpha)}}\right\} \geq \alpha. \tag{5}$$

3 Random Subspaces of l_∞^n

For natural numbers n and k, let $\Gamma_{nk} : \mathbb{R}^k \to \mathbb{R}^n$ be the standard Gaussian operator given by

$$\Gamma_{nk}(z) = \Gamma_{nk}(z_1, z_2, \ldots, z_k) = \sum_{i=1}^{n}\left(\sum_{j=1}^{k} g_{ij}z_j\right) e_i$$

for $z = (z_1, z_2, \ldots, z_k) \in \mathbb{R}^k$, where $e_1, e_2, \ldots, e_n$ is the canonical basis in $\mathbb{R}^n$ and $\{g_{ij}\}$ are independent standard Gaussian variables.

The following proposition, together with the results of [10] on the distribution of $\|g\|_\infty$, is the main tool in proving the central result of the note. It shows that with a substantial probability, the number

$$\max_{z \in S^{k-1}} \|\Gamma_{nk}z\|_\infty / \min_{z \in S^{k-1}} \|\Gamma_{nk}z\|_\infty$$

is noticeably farther from 1 than $\|g\|_\infty / M_\infty$:

Proposition 1. *There are universal constants $c > 0$ and $n_0 \in \mathbb{N}$ such that for all $n \geq n_0$ and all $k > 1$,*

$$\max_{z \in S^{k-1}} \|\Gamma_{nk}z\|_\infty / \min_{z \in S^{k-1}} \|\Gamma_{nk}z\|_\infty > 1 + \frac{ck \ln \frac{c \ln n}{k}}{\ln n} \tag{6}$$

with probability greater than $1/2$.

Proof. By (1), there exists $\alpha_1 > 0$ such that for all $n \geq \alpha_1^{-1}$ and $k \geq 1$,

$$\mathbb{P}\left\{\|\Gamma_{nk}(1, 0, \ldots, 0)\|_\infty \leq \sqrt{2\ln n} - \frac{\ln \ln n}{4\sqrt{\ln n}}\right\}$$

$$= \mathbb{P}\left\{|g_1| \leq \sqrt{2\ln n} - \frac{\ln \ln n}{4\sqrt{\ln n}}\right\}^n > \frac{1}{2} + \frac{1}{e}. \tag{7}$$

Define $n_0 = \lceil \max(\alpha_0^{-1}, \alpha_1^{-1}) \rceil$ and $c = \min(c_0, 1/24)$, where α_0 and c_0 are taken from Lemma 1. Now, fix any $n \geq n_0$ and $k > 1$. Note that for $k \geq c_0 \ln n$ the statement is trivial so we will assume that $k < c_0 \ln n$. For any point of the probability space $\omega \in \Omega$,

$$\max_{z \in S^{k-1}} \|\Gamma_{nk}(\omega)z\|_\infty = \max_i \max_{(z_1,\ldots,z_k) \in S^{k-1}} |z_1 g_{i1}(\omega) + \cdots + z_k g_{ik}(\omega)|$$

$$= \max_i \sqrt{g_{i1}(\omega)^2 + \cdots + g_{ik}(\omega)^2}$$

$$= \max_i \xi_i(\omega),$$

where $\xi_1, \xi_2, \ldots, \xi_n$ are independent random variables having the χ-distribution with k degrees of freedom. Letting $\alpha = 1/n$ in (5), we get:

$$\mathbb{P}\left\{\max_i \xi_i \geq \sqrt{2 \ln n} + \frac{(k-2) \ln \frac{\ln n}{k-1}}{4\sqrt{\ln n}}\right\}$$

$$= 1 - \mathbb{P}\left\{\xi_1 \leq \sqrt{2 \ln n} + \frac{(k-2) \ln \frac{\ln n}{k-1}}{4\sqrt{\ln n}}\right\}^n$$

$$\geq 1 - \left(1 - \frac{1}{n}\right)^n$$

$$\geq 1 - \frac{1}{e}.$$

Combining the last estimate with (7), we get the result. $\qquad\square$

Further, we will need the following version of the "θ-net" argument: *Let $k > 1$ and $\mathcal{N}$ be a θ-net (with respect to the Euclidean norm $\|\cdot\|_2$ in $\mathbb{R}^k$) on S^{k-1} for some $\theta < 1/2$. Next, let X be a normed space and $T : \mathbb{R}^k \to X$ be a linear operator such that for some $M > 0$ and $\delta \in [0, 1)$*

$$(1-\delta)M \leq \|Ty\|_X \leq (1+\delta)M \quad \text{for all } y \in \mathcal{N}.$$

Then for any $z \in S^{k-1}$

$$(1 - 2\delta - 2\theta)M \leq \|Tz\|_X \leq (1 + 2\delta + 2\theta)M. \tag{8}$$

For convenience of a less experienced reader we give a short proof. For every $z \in S^{k-1}$, there is $y \in \mathcal{N}$ such that $\|y - z\|_2 \leq \theta$. Then

$$\|Tz\|_X \leq \|Ty\|_X + \|T(y - z)\|_X \leq (1 + \delta)M + \theta\|T\|,$$

where $\|T\|$ denotes the operator norm from ℓ_2^k to X. Taking the maximum over z, we get $\|T\| \leq (1 + \delta)(1 - \theta)^{-1}M$. In particular this implies the right hand side inequality in (8). For the left hand side we start with

$$\|Tz\|_X \ge \|Ty\|_X - \|T(y-z)\|_X \ge (1-\delta)M - \theta\|T\|,$$

and then use the estimate for $\|T\|$ obtained above.

The next statement expresses the well known fact that, for $k \ll n$, with a large probability $\|\Gamma_{nk}z\|_2$ is almost a constant on the sphere S^{k-1}. Let $C_2 > 0$ be such that for all $n > 1$ the dimension of any $3/2$-spherical subspace of l_∞^n is bounded from above by $C_2 \ln n$.

Lemma 2. *There is a universal constant $n_1 \in \mathbb{N}$ such that for all $n \ge n_1$ and $k \le C_2 \ln n$*

$$\mathbb{P}\left\{ \max_{z \in S^{k-1}} \|\Gamma_{nk}z\|_2 \Big/ \min_{z \in S^{k-1}} \|\Gamma_{nk}z\|_2 \le 1 + \frac{1}{n^{1/4}} \right\} \ge \frac{3}{4}. \tag{9}$$

Proof. By a concentration inequality for Gaussian vectors (see, for example, [8, Theorem 4.7] or [7, Theorem V.1]) and since $M_2 \approx \sqrt{n}$, we have for some $C_1 > 0$, all $n \ge 1$ and the standard Gaussian vector g in $\mathbb{R}^n$

$$\mathbb{P}\{|\|g\|_2 - M_2| > \theta M_2\} \le 2\exp(-C_1\theta^2 n) \quad \text{for any } \theta > 0. \tag{10}$$

Then we choose the constant $n_1 \in \mathbb{N}$ so that for all $n \ge n_1$

$$1 - 2(48n^{1/4})^{C_2 \ln n} \exp(-C_1\sqrt{n}/256) \ge 3/4.$$

Fix any $n \ge n_1$ and $1 \le k \le C_2 \ln n$; let $\theta = \frac{1}{16n^{1/4}}$ and $\mathcal{N}$ be a θ-net on S^{k-1} of cardinality at most $(3/\theta)^k$. For any point $z \in S^{k-1}$, $\Gamma_{nk}z$ is the standard Gaussian vector in $\mathbb{R}^n$, so in particular

$$\mathbb{P}\{|\|\Gamma_{nk}y\|_2 - M_2| \le \theta M_2 \text{ for all } y \in \mathcal{N}\}$$
$$\ge 1 - (3/\theta)^k \mathbb{P}\{|\|g\|_2 - M_2| > \theta M_2\}$$
$$\ge 1 - 2(3/\theta)^k \exp(-C_1\theta^2 n).$$

The "θ-net" argument implies that

$$\mathbb{P}\left\{ \max_{z \in S^{k-1}} \|\Gamma_{nk}z\|_2 \Big/ \min_{z \in S^{k-1}} \|\Gamma_{nk}z\|_2 \le 1 + 16\theta \right\}$$
$$\ge 1 - 2(3/\theta)^k \exp(-C_1\theta^2 n)$$
$$\ge 1 - 2(48n^{1/4})^{C_2 \ln n} \exp(-C_1\sqrt{n}/256)$$
$$\ge \frac{3}{4},$$

and the result follows. $\square$

To emphasize the geometric character of our main result we shall present it in terms of the Grassmann manifolds. Note that the probabilistic formulations used until now—which are more convenient for calculations—still remain in the proof of the theorem.

For natural numbers $1 \leq k \leq n$, the Grassmann manifold of all k-dimensional subspaces of $\mathbb{R}^n$ is denoted by $G_{n,k}$; by $\mu_{n,k}$ we denote the normalized rotation invariant Haar measure on $G_{n,k}$. In view of invariance of the distribution of the Gaussian vector under rotations, we have for any Borel subset $A \subset G_{n,k}$

$$\mathbb{P}\{\mathrm{Im}\ \Gamma_{nk} \in A\} = \mu_{n,k}(A). \tag{11}$$

Theorem 1. *Let $\varepsilon \in (0, 1/2)$ and $n > 1$. Then*

(1) There is an absolute constant $\tilde{c} > 0$ such that whenever $k \leq \tilde{c}\varepsilon \ln n / \ln \frac{1}{\varepsilon}$, then

$$\mu_{n,k}\{E \in G_{n,k} : E \text{ is } (1+\varepsilon)\text{-spherical subspace of } l_\infty^n\} \geq 1 - 2n^{-\tilde{c}\varepsilon}; \tag{12}$$

(2) Conversely, if for some $k > 1$

$$\mu_{n,k}\{E \in G_{n,k} : E \text{ is } (1+\varepsilon)\text{-spherical subspace of } l_\infty^n\} \geq \frac{3}{4} \tag{13}$$

then necessarily $k \leq C\varepsilon \ln n / \ln \frac{1}{\varepsilon}$, where $C > 0$ is an absolute constant.

Proof. The first part of the theorem is essentially proved in [10]. Indeed, by [10, Proposition 1], for some constant $c_1 > 0$

$$\mathbb{P}\{|\|g\|_\infty - M_\infty| > \varepsilon M_\infty\} \leq 2n^{-c_1\varepsilon}. \tag{14}$$

When n is small or $\varepsilon < \frac{1}{\ln n}$, (12) is obvious (for a well-chosen constant $\tilde{c}$), so we can assume that

$$\frac{c_1(64 \ln n)^2}{C_1 n} \leq 1, \quad \varepsilon \geq \frac{1}{\ln n}, \tag{15}$$

where C_1 is taken from (10). Pick a natural number $k \leq \frac{c_1 \varepsilon \ln n}{18 \ln(1/\varepsilon)}$. As before, Γ_{nk} is the Gaussian operator. Note that event $\{\mathrm{Im}\ \Gamma_{nk}$ is $(1 + \varepsilon) -$ spherical$\}$ contains the event

$$\left\{ \max_{z \in S^{k-1}} \|\Gamma_{nk}z\|_2 / \min_{z \in S^{k-1}} \|\Gamma_{nk}z\|_2 \leq 1 + \frac{\varepsilon}{4} \text{ and} \right.$$

$$\left. \max_{z \in S^{k-1}} \|\Gamma_{nk}z\|_\infty / \min_{z \in S^{k-1}} \|\Gamma_{nk}z\|_\infty \leq 1 + \frac{\varepsilon}{4} \right\}.$$

Then in view of (11), (10), (14) and the "θ-net" argument for $\theta = \varepsilon/64$,

$$\mu_{n,k}\{E \in G_{n,k} : E \text{ is } (1+\varepsilon)\text{-spherical}\}$$

$$\geq \mathbb{P}\Big\{ \max_{z \in S^{k-1}} \|\Gamma_{nk}z\|_2 / \min_{z \in S^{k-1}} \|\Gamma_{nk}z\|_2 \leq 1 + \frac{\varepsilon}{4} \text{ and}$$

$$\max_{z \in S^{k-1}} \|\Gamma_{nk}z\|_\infty / \min_{z \in S^{k-1}} \|\Gamma_{nk}z\|_\infty \leq 1 + \frac{\varepsilon}{4}\Big\}$$

$$\geq 1 - (3/\theta)^k \mathbb{P}\{|\|g\|_2 - M_2| > \theta M_2\}$$

$$- (3/\theta)^k \mathbb{P}\{|\|g\|_\infty - M_\infty| > \theta M_\infty\}$$

$$\geq 1 - 2(3/\theta)^k \exp(-C_1\theta^2 n) - 2(3/\theta)^k n^{-c_1\varepsilon}.$$

By (15), $C_1\theta^2 n \geq c_1\varepsilon \ln n$, hence

$$\mathbb{P}\{\operatorname{Im}\Gamma_{nk} \text{ is } (1+\varepsilon) - \text{spherical}\} \geq 1 - 4n^{-c_1\varepsilon} n^{\frac{c_1}{18}\varepsilon \ln \frac{3}{\theta}/\ln\frac{1}{\varepsilon}} \geq 1 - 4n^{-c_1\varepsilon/2}.$$

The statement follows by properly defining $\tilde{c}$.

Now, we turn to the second part of the theorem. Suppose that $k > 1$ satisfies (13). This implies, in particular, that l_∞^n contains $(1 + \varepsilon)$-Euclidean subspaces of dimension k, so $k \leq C_3 \ln n / \ln\frac{1}{\varepsilon}$ for an absolute constant C_3 (see, for example, [11, Claim 3.3]). Let n_0, n_1, C_2 and c be as they were defined in Proposition 1 and Lemma 2. The cases when n or $1/\varepsilon$ is small, can be treated in a trivial way, so further we assume

$$n \geq \max(n_0, n_1), \quad \frac{3\ln n}{cn^{1/4}} \leq 1, \quad k \leq \frac{c}{e}\ln n, \quad \varepsilon < \frac{c^2}{3}. \tag{16}$$

Obviously, $k \leq C_2 \ln n$. Then (9) and (13) give

$$\mathbb{P}\Big\{ \max_{z \in S^{k-1}} \|\Gamma_{nk}z\|_\infty / \min_{z \in S^{k-1}} \|\Gamma_{nk}z\|_\infty \leq (1+\varepsilon)(1+1/n^{1/4})\Big\}$$

$$\geq \mathbb{P}\Big\{ \max_{z \in S^{k-1}} \|\Gamma_{nk}z\|_2 / \min_{z \in S^{k-1}} \|\Gamma_{nk}z\|_2 \leq 1 + \frac{1}{n^{1/4}} \text{ and}$$

$$\max_{\substack{x \in \operatorname{Im}\Gamma_{nk} \\ \|x\|_2=1}} \|x\|_\infty / \min_{\substack{x \in \operatorname{Im}\Gamma_{nk} \\ \|x\|_2=1}} \|x\|_\infty \leq 1 + \varepsilon\Big\}$$

$$\geq \frac{1}{2}.$$

Hence, by Proposition 1,

$$1 + \frac{ck \ln\frac{c\ln n}{k}}{\ln n} \leq (1+\varepsilon)(1+1/n^{1/4}). \tag{17}$$

If $\varepsilon \le 1/n^{1/4}$ then, in view of (16) and (17), $k \le k \ln \frac{c \ln n}{k} \le \frac{3 \ln n}{cn^{1/4}} \le 1$, leading to contradiction. Hence, $\varepsilon > 1/n^{1/4}$, and (17) yields

$$k \le \frac{3\varepsilon \ln n}{c \ln \frac{c \ln n}{k}}. \tag{18}$$

In particular (18) implies $\frac{k}{\ln n} \le \frac{3}{c}\varepsilon$, so $\ln \frac{c \ln n}{k} \ge \ln \frac{c^2}{3\varepsilon}$. Substituting it back to (18) we get

$$k \le \frac{3\varepsilon \ln n}{c \ln \frac{c^2}{3\varepsilon}}.$$

$\square$

Remark 1. The probability $3/4$ in the second part of the Theorem can be replaced with any (fixed) positive number; this only affects the constant.

Acknowledgements I would like to thank Prof. N. Tomczak-Jaegermann for introducing me to this topic and for valuable suggestions on the text. Also, I thank Prof. G. Schechtman for a fruitful conversation.

References

1. H.A. David, *Order Statistics*, 2nd edn. (Wiley, New York, 1981)
2. A. Dvoretzky, Some results on convex bodies and Banach spaces, in *Proceedings of the International Symposium on Linear Spaces*, Jerusalem (1961), pp. 123–160
3. W. Feller, *An Introduction to Probability Theory and Its Applications*, vol. 1, 3rd edn. (Wiley, New York, 1968)
4. Y. Gordon, Some inequalities for Gaussian processes and applications. Israel J. Math. **50**(4), 265–289 (1985)
5. Y. Gordon, A.E. Litvak, A. Pajor, N. Tomczak-Jaegermann, Random ε-nets and embeddings in l_∞^N. Studia Math. **178**(1), 91–98 (2007)
6. V.D. Milman, A new proof of the theorem of A. Dvoretzky on sections of convex bodies. Funct. Anal. Appl. **5**(4), 288–295 (1971)
7. V.D. Milman, G. Schechtman, *Asymptotic Theory of Finite-Dimensional Normed Spaces*. Lecture Notes in Mathematics, vol. 1200 (Springer, Berlin, 1986)
8. G. Pisier, *The Volumes of Convex Bodies and Banach Space Geometry* (Cambridge University Press, Cambridge, 1989)
9. G. Schechtman, Two observations regarding embedding subsets of Euclidean spaces in normed spaces. Adv. Math. **200**, 125–135 (2006)
10. G. Schechtman, The random version of Dvoretzky's theorem in l_∞^n, in *Geometric Aspects of Functional Analysis*. Lecture Notes in Mathematics, vol. 1910 (Springer, Berlin, 2007), pp. 265–270
11. G. Schechtman, Euclidean sections of convex bodies, in *Asymptotic Geometric Analysis*, ed. by M. Ludwig et al. Fields Institute Communications, vol. 68 (Springer, New York, 2013), 271–288
12. K.E. Tikhomirov, Almost Euclidean sections in symmetric spaces and concentration of order statistics. J. Funct. Anal. **265**, 2074–2088 (2013)

LECTURE NOTES IN MATHEMATICS

Edited by J.-M. Morel, B. Teissier; P.K. Maini

Editorial Policy (for Multi-Author Publications: Summer Schools / Intensive Courses)

1. Lecture Notes aim to report new developments in all areas of mathematics and their applications - quickly, informally and at a high level. Mathematical texts analysing new developments in modelling and numerical simulation are welcome. Manuscripts should be reasonably selfcontained and rounded off. Thus they may, and often will, present not only results of the author but also related work by other people. They should provide sufficient motivation, examples and applications. There should also be an introduction making the text comprehensible to a wider audience. This clearly distinguishes Lecture Notes from journal articles or technical reports which normally are very concise. Articles intended for a journal but too long to be accepted by most journals, usually do not have this "lecture notes" character.

2. In general SUMMER SCHOOLS and other similar INTENSIVE COURSES are held to present mathematical topics that are close to the frontiers of recent research to an audience at the beginning or intermediate graduate level, who may want to continue with this area of work, for a thesis or later. This makes demands on the didactic aspects of the presentation. Because the subjects of such schools are advanced, there often exists no textbook, and so ideally, the publication resulting from such a school could be a first approximation to such a textbook. Usually several authors are involved in the writing, so it is not always simple to obtain a unified approach to the presentation.

   For prospective publication in LNM, the resulting manuscript should not be just a collection of course notes, each of which has been developed by an individual author with little or no coordination with the others, and with little or no common concept. The subject matter should dictate the structure of the book, and the authorship of each part or chapter should take secondary importance. Of course the choice of authors is crucial to the quality of the material at the school and in the book, and the intention here is not to belittle their impact, but simply to say that the book should be planned to be written by these authors jointly, and not just assembled as a result of what these authors happen to submit.

   This represents considerable preparatory work (as it is imperative to ensure that the authors know these criteria before they invest work on a manuscript), and also considerable editing work afterwards, to get the book into final shape. Still it is the form that holds the most promise of a successful book that will be used by its intended audience, rather than yet another volume of proceedings for the library shelf.

3. Manuscripts should be submitted either online at www.editorialmanager.com/lnm/ to Springer's mathematics editorial, or to one of the series editors. Volume editors are expected to arrange for the refereeing, to the usual scientific standards, of the individual contributions. If the resulting reports can be forwarded to us (series editors or Springer) this is very helpful. If no reports are forwarded or if other questions remain unclear in respect of homogeneity etc, the series editors may wish to consult external referees for an overall evaluation of the volume. A final decision to publish can be made only on the basis of the complete manuscript; however a preliminary decision can be based on a pre-final or incomplete manuscript. The strict minimum amount of material that will be considered should include a detailed outline describing the planned contents of each chapter.

   Volume editors and authors should be aware that incomplete or insufficiently close to final manuscripts almost always result in longer evaluation times. They should also be aware that parallel submission of their manuscript to another publisher while under consideration for LNM will in general lead to immediate rejection.

4. Manuscripts should in general be submitted in English. Final manuscripts should contain at least 100 pages of mathematical text and should always include

   – a general table of contents;
   – an informative introduction, with adequate motivation and perhaps some historical remarks: it should be accessible to a reader not intimately familiar with the topic treated;
   – a global subject index: as a rule this is genuinely helpful for the reader.

   Lecture Notes volumes are, as a rule, printed digitally from the authors' files. We strongly recommend that all contributions in a volume be written in the same LaTeX version, preferably LaTeX2e. To ensure best results, authors are asked to use the LaTeX2e style files available from Springer's web-server at
   ftp://ftp.springer.de/pub/tex/latex/svmonot1/ (for monographs) and
   ftp://ftp.springer.de/pub/tex/latex/svmultt1/ (for summer schools/tutorials).
   Additional technical instructions, if necessary, are available on request from:
   lnm@springer.com.

5. Careful preparation of the manuscripts will help keep production time short besides ensuring satisfactory appearance of the finished book in print and online. After acceptance of the manuscript authors will be asked to prepare the final LaTeX source files and also the corresponding dvi-, pdf- or zipped ps-file. The LaTeX source files are essential for producing the full-text online version of the book. For the existing online volumes of LNM see:
   http://www.springerlink.com/openurl.asp?genre=journal&issn=0075-8434.
   The actual production of a Lecture Notes volume takes approximately 12 weeks.

6. Volume editors receive a total of 50 free copies of their volume to be shared with the authors, but no royalties. They and the authors are entitled to a discount of 33.3 % on the price of Springer books purchased for their personal use, if ordering directly from Springer.

7. Commitment to publish is made by letter of intent rather than by signing a formal contract. Springer-Verlag secures the copyright for each volume. Authors are free to reuse material contained in their LNM volumes in later publications: a brief written (or e-mail) request for formal permission is sufficient.

Addresses:
Professor J.-M. Morel, CMLA,
École Normale Supérieure de Cachan,
61 Avenue du Président Wilson, 94235 Cachan Cedex, France
E-mail: morel@cmla.ens-cachan.fr

Professor B. Teissier, Institut Mathématique de Jussieu,
UMR 7586 du CNRS, Équipe "Géométrie et Dynamique",
175 rue du Chevaleret,
75013 Paris, France
E-mail: teissier@math.jussieu.fr

For the "Mathematical Biosciences Subseries" of LNM:

Professor P. K. Maini, Center for Mathematical Biology,
Mathematical Institute, 24-29 St Giles,
Oxford OX1 3LP, UK
E-mail: maini@maths.ox.ac.uk

Springer, Mathematics Editorial I,
Tiergartenstr. 17,
69121 Heidelberg, Germany,
Tel.: +49 (6221) 4876-8259
Fax: +49 (6221) 4876-8259
E-mail: lnm@springer.com